Graduate Texts in Mathematics 20

Graduate Texts in Mathematics

continued after index

Dale Husemoller

Fibre Bundles

Third Edition

Springer Science+Business Media, LLC

Dale Husemoller
Department of Mathematics
Haverford College
Haverford, PA 19041
USA

With four figures

Mathematics Subject Classification (1991): 14F05, 14F15, 18F15, 18F25, 55RXX

Library of Congress Cataloging-in-Publication Data
Husemoller, Dale.
    Fibre bundles / Dale Husemoller. — 3rd ed.
        p.   cm. — (Graduate texts in mathematics ; 20)
    Includes bibliographical references and index.
    ISBN 978-1-4757-2263-5   ISBN 978-1-4757-2261-1 (eBook)
    DOI 10.1007/978-1-4757-2261-1
    1. Fiber bundles (Mathematics) I. Title. II. Series.
    QA612.6.H87   1993
    514′.224—dc20   93-4694

Printed on acid-free paper.

Production managed by Henry Krell; manufacturing supervised by Jacqui Ashri.
Typeset by Asco Trade Typesetting Ltd., Hong Kong.

9 8 7 6 5 4 3 2 1

ISBN 978-1-4757-2263-5

To the memory of my mother and my father

Preface to the Third Edition

In this edition, we have added two new chapters, Chapter 7 on the gauge group of a principal bundle and Chapter 19 on the definition of Chern classes by differential forms. These subjects have taken on special importance when we consider new applications of the fibre bundle theory especially to mathematical physics. For these two chapters, the author profited from discussions with Professor M. S. Narasimhan.

The idea of using the term bundle for what is just a map, but is eventually a fibre bundle projection, is due to Grothendieck.

The bibliography has been enlarged and updated. For example, in the Seifert reference [1932] we find one of the first explicit references to the concept of fibrings.

The first edition of the *Fibre Bundles* was translated into Russian under the title "Расслоенные Пространства" in 1970 by В. А. Йсковских with general editor М. М. Постникова. The remarks and additions of the editor have been very useful in this edition of the book. The author is very grateful to A. Voronov, who helped with translations of the additions from the Russian text.

Part of this revision was made while the author was a guest of the Max Planck Institut from 1988 to 89, the ETH during the summers of 1990 and 1991, the University of Heidelberg during the summer of 1992, and the Tata Institute for Fundamental Research during January 1990, 1991, and 1992. It is a pleasure to acknowledge all these institutions as well as the Haverford College Faculty Research Fund.

1993   Dale Husemoller

Preface to the Second Edition

In this edition we have added a section to Chapter 15 on the Adams conjecture and a second appendix on the suspension theorems. For the second appendix the author profitted from discussion with Professors Moore, Stasheff, and Toda.

I wish to express my gratitude to the following people who supplied me with lists of corrections to the first edition: P. T. Chusch, Rudolf Fritsch, David C. Johnson, George Lusztig, Claude Schocket, and Robert Sturg.

Part of the revision was made while the author was a guest of the I.H.E.S in January, May, and June 1974.

1974   Dale Husemoller

Preface to the First Edition

The notion of a fibre bundle first arose out of questions posed in the 1930s on the topology and geometry of manifolds. By the year 1950, the definition of fibre bundle had been clearly formulated, the homotopy classification of fibre bundles achieved, and the theory of characteristic classes of fibre bundles developed by several mathematicians: Chern, Pontrjagin, Stiefel, and Whitney. Steenrod's book, which appeared in 1950, gave a coherent treatment of the subject up to that time.

About 1955, Milnor gave a construction of a universal fibre bundle for any topological group. This construction is also included in Part I along with an elementary proof that the bundle is universal.

During the five years from 1950 to 1955, Hirzebruch clarified the notion of characteristic class and used it to prove a general Riemann-Roch theorem for algebraic varieties. This was published in his Ergebnisse Monograph. A systematic development of characteristic classes and their applications to manifolds is given in Part III and is based on the approach of Hirzebruch as modified by Grothendieck.

In the early 1960s, following lines of thought in the work of A. Grothendieck, Atiyah and Hirzebruch developed K-theory, which is a generalized cohomology theory defined by using stability classes of vector bundles. The Bott periodicity theorem was interpreted as a theorem in K-theory, and J. F. Adams was able to solve the vector field problem for spheres, using K-theory. In Part II, an introduction to K-theory is presented, the nonexistence of elements of Hopf invariant 1 proved (after a proof of Atiyah), and the proof of the vector field problem sketched.

I wish to express gratitude to S. Eilenberg, who gave me so much encouragement during recent years, and to J. C. Moore, who read parts of the manuscript and made many useful comments. Conversations with J. F. Adams, R. Bott, A. Dold, and F. Hirzebruch helped to sharpen many parts of the manuscript. During the writing of this book, I was particularly influenced by the Princeton notes of J. Milnor and the lectures of F. Hirzebruch at the 1963 Summer Institute of the American Mathematical Society.

1966   Dale Husemoller

Contents

CHAPTER 1
Preliminaries on Homotopy Theory

In this introductory chapter, we consider those aspects of homotopy theory that will be used in later sections of the book. This is done in outline form. References to the literature are included.

Two books on homotopy theory, those by Hu [1]† and Hilton [1], contain much of the background material for this book. In particular, chapters 1 to 5 of Hu [1] form a good introduction to the homotopy needed in fibre bundle theory.

1. Category Theory and Homotopy Theory

A homotopy $f_t: X \to Y$ is a continuous one-parameter family of maps, and two maps f and g are homotopically equivalent provided there is a homotopy f_t with $f = f_0$ and $g = f_1$. Since this is an equivalence relation, one can speak of a homotopy class of maps between two spaces.

As with the language of set theory, we use the language of category theory throughout this book. For a good introduction to category theory, see MacLane [2].

We shall speak of the category **sp** of (topological) spaces, (continuous) maps, and composition of maps. The category **H** of spaces, homotopy classes of maps, and composition of homotopy classes is a quotient category. Similarly, we speak of maps and of homotopy classes of maps that preserve base points. The associated categories of pointed spaces (i.e., spaces with base points) are denoted $\mathbf{sp}_0$ and $\mathbf{H}_0$, respectively.

The following concept arises frequently in fibre bundle theory.

---

† Bracketed numbers refer to bibliographic entries at end of book.

1.1 Definition. Let X be a set, and let Φ be a family of spaces M whose underlying sets are subsets of X. The Φ-topology on X is defined by requiring a set U in X to be open if and only if $U \cap M$ is open in M for each $M \in \Phi$. If X is a space and if Φ is a family of subspaces of X, the topology on X is said to be Φ-defined provided the Φ-topology on the set X is the given topology on X.

For example, if X is a Hausdorff space and if Φ is a family of compact subspaces, X is called a k-space if the topology of X is Φ-defined. If $M_1 \subset M_2 \subset \cdots \subset X$ is a sequence of spaces in a set X, the inductive topology on X is the Φ-topology, where $\Phi = \{M_1, M_2, \ldots\}$.

The following are examples of unions of spaces which are given the inductive topology.

$$\mathbf{R}^1 \subset \mathbf{R}^2 \subset \cdots \subset \mathbf{R}^n \subset \cdots \subset \mathbf{R}^\infty = \bigcup_{1 \leq n} \mathbf{R}^n$$

$$\mathbf{C}^1 \subset \mathbf{C}^2 \subset \cdots \subset \mathbf{C}^n \subset \cdots \subset \mathbf{C}^\infty = \bigcup_{1 \leq n} \mathbf{C}^n$$

$$S^1 \subset S^2 \subset \cdots \subset S^n \subset \cdots \subset S^\infty = \bigcup_{1 \leq n} S^n$$

$$RP^1 \subset RP^2 \subset \cdots \subset RP^n \subset \cdots \subset RP^\infty = \bigcup_{1 \leq n} RP^n$$

$$CP^1 \subset CP^2 \subset \cdots \subset CP^n \subset \cdots \subset CP^\infty = \bigcup_{1 \leq n} CP^n$$

Above, RP^n denotes the real projective space of lines in $\mathbf{R}^{n+1}$, and CP^n denotes the complex projective space of complex lines in $\mathbf{C}^{n+1}$. We can view RP^n as the quotient of S^n with x and $-x$ identified, and we can view CP^n as the quotient of $S^{2n+1} \subset \mathbf{C}^{n+1}$, where the circle $ze^{i\theta}$ for $0 \leq \theta \leq 2\pi$ is identified to a point.

It is easily proved that each locally compact space is a k-space. The spaces S^∞, RP^∞, and CP^∞ are k-spaces that are not locally compact.

2. Complexes

The question of whether or not a map defined on a subspace prolongs to a larger subspace frequently arises in fibre bundle theory. If the spaces involved are CW-complexes and the subspaces are subcomplexes, a satisfactory solution of the problem is possible.

A good introduction to this theory is the original paper of J. H. C. Whitehead [1, secs. 4 and 5]. Occasionally, we use relative cell complexes (X, A), where A is a closed subset of X and $X - A$ is a disjoint union of open cells with attaching maps. The reader can easily generalize the results of Whitehead [1] to relative cell complexes. In particular, one can speak of relative CW-complexes. If X^n is the n-skeleton of a CW-complex, then (X, X^n) is a relative CW-complex.

The prolongation theorems for maps defined on CW-complexes follow from the next proposition.

2.1 Proposition. *Let (X, A) be a relative CW-complex having one cell C with an attaching map $u_C: I^n \to X = A \cup C$, and let $f: A \to Y$ be a map. Then f extends to a map $g: X \to Y$ if and only if $fu_C: \partial I^n \to Y$ is null homotopic.*

A space Y is said to be connected in dimension n provided every map $S^{n-1} \to Y$ is null homotopic or, in other words, prolongs to a map $B^n \to Y$. From (2.1) we easily get the following result.

2.2 Theorem. *Let (X, A) be a relative CW-complex, and let Y be a space that is connected in each dimension for which X has cells. Then each map $A \to Y$ prolongs to a map $X \to Y$.*

As a corollary of (2.2), a space is contractible, i.e., homotopically equivalent to a point, if and only if it is connected in each dimension.

The above methods yield the result that the homotopy extension property holds for CW-complexes; see Hilton [1, p. 97].

The following theorems are useful in considering vector bundles over CW-complexes. Since they do not seem to be in the literature, we give details of the proofs.

If C is a cell in a CW-complex X and if $u_C: B^n \to X$ is the attaching map, then $u_C(0)$ is called the center of C.

2.3 Theorem. *Let (X, A) be a finite-dimensional CW-complex. Then there exists an open subset V of X with $A \subset V \subset X$ such that A is a strong deformation retract of V with a homotopy h_t. This can be done so that V contains the center of no cell C of X, and if U_A is an open subset of A, there is an open subset U_X of X with $U_X \cap A = U_A$ and $h_t(U_X) \subset U_X$ for $t \in I$.*

Proof. We prove this theorem by induction on the dimension of X. For $\dim X = -1$, the result is clear. For $X^n = X$, let V' be an open subset of X^{n-1} with $A \subset V' \subset X^{n-1}$ and a contracting homotopy $h'_t: V' \to V'$. Let U' be the open subset of V' with $U' \cap A = U_A$ and $h_t(U') \subset U'$ for $t \in I$. This is given by the inductive hypothesis.

For each n-cell C, let $u_C: B^n \to X$ be the attaching map of C, and let V'_C denote the open subset $u_C^{-1}(V')$ of ∂B^n and U'_C denote $u_C^{-1}(U')$. Let M_C denote the closed subset of all ty for $t \in [0, 1]$ and $y \in \partial B^n - V'_C$. There is an open subset V of X with $V \cap X_{n-1} = V'$ and $u_C^{-1}(V) = B^n - M_C$, that is, $y \in u_C^{-1}(V)$ if and only if $y \neq 0$ and $y/\|y\| \in V'_C$, and there is an open subset U_X of V with $U_X \cap X^{n-1} = U'$ and $y \in u_C^{-1}(U_X)$ if and only if $y \neq 0$ and $y/\|y\| \in U'_C$.

We define a contracting homotopy $h_t: V \to V$ by the following requirements: $h_t(u_C(y)) = u_C(2ty/\|y\| + (1 - 2t)y)$ for $y \in B^n$, $t \in [0, \frac{1}{2}]$, $h_t(x) = x$ for $x \in V'$, $t \in [0, \frac{1}{2}]$, $h_t(x) = h'_{2t-1}(h_{1/2}(x))$ for $t \in [\frac{1}{2}, 1]$, where h'_t is defined in the first paragraph. Then A is a strong deformation retract of V, and $h_t(U_X) \subset U_X$

by the character of the radial construction. Finally, we have $u_C(0) \notin V$ for each cell C of X. This proves the theorem.

2.4 Remark. With the notation of Theorem (2.3), if U_A is contractible, U_X is contractible.

2.5 Theorem. *Let X be a finite CW-complex with m cells. Then X can be covered by m contractible open sets.*

Proof. We use induction on m. For $m = 1$, X is a point, and the statement is clearly true. Let C be a cell of maximal dimension. Then X equals a subcomplex A of $m - 1$ cells with C attached by a map u_C. There are $V_1', \ldots, V_{m-1}'$ contractible open sets in A which prolong by (2.3) and (2.4) to contractible open sets $V_1, \ldots, V_{m-1}$ of X which cover A. If V_m denotes $C = u_C(\text{int } B^n)$, then $V_1, \ldots, V_m$ forms an open contractible covering of X.

2.6 Theorem. *Let X be a CW-complex of dimension n. Then X can be covered by $n + 1$ open sets $V_0, \ldots, V_n$ such that each path component of V_i is contractible.*

Proof. For $n = 0$ the statement of the theorem clearly holds, and we use induction on n. Let $V_0', \ldots, V_n'$ be an open covering of the $(n - 1)$-skeleton of X, where each component of V_i' is a contractible set. Let V be an open neighborhood of X^{n-1} in X with a contracting homotopy leaving X^{n-1} elementwise fixed $h_t : V \to V$ onto X^{n-1}. Using (2.3), we associate with each component of V_i' an open contractible set in V. The union of these disjoint sets is defined to be V_i. Let V_n be the union of the open n cells of X. The path components of V_n are the open n cells. Then the open covering $V_0, \ldots, V_n$ has the desired properties.

3. The Spaces Map (X, Y) and $\text{Map}_0 (X, Y)$

For two spaces X and Y, the set Map (X, Y) of all maps $X \to Y$ has several natural topologies. For our purposes the compact-open topology is the most useful. If $\langle K, V \rangle$ denotes the subset of all $f \in$ Map (X, Y) with $f(K) \subset V$ for $K \subset X$ and $V \subset Y$, the compact-open topology is generated by all sets $\langle K, V \rangle$ such that K is a compact subset of X and V is an open subset of Y.

The subset $\text{Map}_0 (X, Y)$ of base point preserving maps is given the subspace topology.

The spaces Map (X, Y) are useful for homotopy theory because of the natural map

$$\theta : \text{Map} (Z \times X, Y) \to \text{Map} (Z, \text{Map} (X, Y))$$

which assigns to $f(z, x)$ the map $Z \to \text{Map}(X, Y)$, where the image of $z \in Z$ is the map $x \mapsto f(z, x)$. This map

$$\text{Map}(Z \times X, Y) \to \text{Map}(Z, \text{Map}(X, Y))$$

is a homeomorphism onto its image set for Hausdorff spaces. Moreover, we have the following proposition by an easy proof.

3.1 Proposition. *For two spaces X and Y, the function*

$$\theta \colon \text{Map}(Z \times X, Y) \to \text{Map}(Z, \text{Map}(X, Y))$$

is bijective if and only if the substitution function $\sigma \colon \text{Map}(X, Y) \times X \to Y$, where $\sigma(f, x) = f(x)$, is continuous.

The substitution function $\sigma \colon \text{Map}(X, Y) \times X \to Y$ is continuous for X locally compact. By applying (3.1) to the case $Z = I$, the closed unit interval, we see that a homotopy from X to Y, that is, a map $X \times I \to Y$, can be viewed as a path in $\text{Map}(X, Y)$.

A map similar to θ can be defined for base point preserving maps defined on compact spaces X and Z, using the reduced product $Z \wedge X = (Z \times X)/ (Z \vee X)$. Here $Z \vee X$ denotes the disjoint union of Z and X with base points identified. The space $Z \vee X$ is also called the wedge product. The map corresponding to θ is defined:

$$\text{Map}_0 (Z \wedge X, Y) \to \text{Map}_0 (Z, \text{Map}_0 (X, Y))$$

It is a homeomorphism for Z and X compact spaces or for Z and X two CW-complexes.

Let 0 be the base point of $I = [0, 1]$, and view S^1 as $[0, 1]/\{0, 1\}$. The following functors $\textbf{sp}_0 \to \textbf{sp}_0$ are very useful in homotopy theory.

3.2 Definition. The cone over X, denoted $C(X)$, is $X \wedge I$; the suspension of X, denoted $S(X)$, is $X \wedge S^1$; the path space of X, denoted $P(X)$, is $\text{Map}_0 (I, X)$; and the loop space of X, denoted $\Omega(X)$, is $\text{Map}_0 (S^1, X)$.

A point of $C(X)$ or $S(X)$ is a class $\langle x, t \rangle$ determined by a pair $(x, t) \in X \times I$, where $\langle x_0, t \rangle = \langle x, 0 \rangle =$ base point of $C(X)$ or $S(X)$ and, in addition, $\langle x, 1 \rangle =$ base point of $S(X)$. If $f \colon X \to Y$ is a map, $C(f)(\langle x, t \rangle) = \langle f(x), t \rangle$ defines a map $C(f) \colon C(X) \to C(Y)$, and $S(f)(\langle x, t \rangle) = \langle f(x), t \rangle$ defines a map $S(f) \colon S(X) \to S(Y)$; with these definitions, $C \colon \textbf{sp}_0 \to \textbf{sp}_0$ and $S \colon \textbf{sp}_0 \to \textbf{sp}_0$ are functors. Also, we consider the map $\omega \colon X \to C(X)$, where $\omega(x) = \langle x, 1 \rangle$. Then $S(X)$ equals $C(X)/\omega(X)$. Since S^1 is $[0, 1]$ with its two end points pinched to a point, one can easily check that the equal sets $S(X)$ and $C(X)/\omega(X)$ have the same topologies.

Path space $P(X)$ can be viewed as the subspace of paths $u \colon I \to X$ such that $u(0) = x_0$, and $\Omega(X)$ as the subspace of paths $u \colon I \to X$ such that $u(0) = u(1) = x_0$. If $f \colon X \to Y$ is a map, then $P(f)u = fu$ defines a map $P(f) \colon P(X) \to P(Y)$, and $\Omega(f)u = fu$ defines a map $\Omega(f) \colon \Omega(X) \to \Omega(Y)$. With these definitions, $P \colon \textbf{sp}_0 \to \textbf{sp}_0$ and $\Omega \colon \textbf{sp}_0 \to \textbf{sp}_0$ are functors. Also, we consider the map $\pi: P(X) \to X$, where $\pi(u) = u(1)$. Then $\Omega(X)$ equals $\pi^{-1}(x_0)$ as a subspace.

3.3 Proposition. *The functions $\omega: 1_{\mathbf{s}\mathbf{p}_0} \to C$ and $\pi: P \to 1_{\mathbf{s}\mathbf{p}_0}$ are morphisms of functors.*

Proof. If $f: X \to Y$ is a map, then $\omega(f(x)) = \langle f(x), 1 \rangle = C(f)\omega(x)$ for each $x \in X$, and $f\pi(u) = fu(1) = \pi(P(f)u)$ for each $u \in P(X)$.

3.4 Proposition. *The following statements are equivalent for a base point preserving map $f: X \to Y$.*

(1) *The map f is homotopic to the constant.*
(2) *There exists a map $g: C(X) \to Y$ with $g\omega = f$.*
(3) *There exists a map $h: X \to P(Y)$ with $\pi h = f$.*

Proof. Condition (1) says that there is a map $f^*: X \times I \to Y$ with $f^*(x, 0) = y_0$, $f^*(x, 1) = f(x)$, and $f^*(x_0, t) = y_0$. The existence of f^* is equivalent to the existence of $g: C(X) \to Y$, where $g\langle x, 1 \rangle = f(x)$. The existence of f^* is equivalent to the existence of $h: X \to P(Y)$, where $h(x)(1) = f(x)$.

3.5 Proposition. *The spaces $C(X)$ and $P(X)$ are contractible.*

Proof. Let $h_s: C(X) \to C(X)$ be the homotopy defined by $h_s(\langle x, t \rangle) = \langle x, st \rangle$. Then h_1 is the identity, and h_0 is constant. Similarly, let $k_s: P(X) \to P(X)$ be the homotopy defined by $k_s(u)(t) = u(st)$.

As an easy application of Proposition (3.1), we have the next theorem.

3.6 Theorem. *There exists a natural bijection $\alpha: [S(X), Y]_0 \to [X, \Omega(Y)]_0$, where $\alpha[f\langle x, t \rangle] = [(\theta f)(x)(t)]$.*

4. Homotopy Groups of Spaces

Let $[X, Y]_0$ denote base point preserving homotopy classes of maps $X \to Y$. A multiplication on a pointed space Y is a map $\phi: Y \times Y \to Y$. The map θ defined a function $\phi_X: [X, Y]_0 \times [X, Y]_0 \to [X, Y]_0$ for each space X, by composition. If $([X, Y]_0, \phi_X)$ is a group for each X, then (Y, ϕ) is called a homotopy associative H-space. The loop space ΩY is an example of a homotopy associative H-space, where $\phi: \Omega Y \times \Omega Y \to \Omega Y$ is given by the following relation:

$$\phi(u, v)(t) = \begin{cases} u(2t) & \text{for } 0 \leq t \leq \frac{1}{2} \\ v(2t - 1) & \text{for } \frac{1}{2} \leq t \leq 1 \end{cases}$$

A comultiplication on a pointed space X is a map $\psi: X \to X \vee X$. The map ψ defines a function $\psi^Y: [X, Y]_0 \times [X, Y]_0 \to [X, Y]_0$ for each space Y, by composition. If $([X, Y]_0, \psi^Y)$ is a group for each Y, then (X, ψ) is called a homotopy associative coH-space. The suspension SX is an example of a homotopy associative coH-space, where $\psi: SX \to SX \vee SX$ is given by the following relation:

$$\psi(\langle x, t \rangle) = \begin{cases} (\langle x, 2t \rangle, *) & \text{for } 0 \leq t \leq \frac{1}{2} \\ (*, \langle x, 2t - 1 \rangle) & \text{for } \frac{1}{2} \leq t \leq 1 \end{cases}$$

The following result is very useful and easily proved.

4.1 Proposition. *Let (X, ψ) be a homotopy associative coH-space and (Y, ϕ) a homotopy associative H-space. Then the group structures on $[X, Y]_0$ derived for X are equal, and this structure is commutative.*

The sphere S^n equals $S(S^{n-1})$, and there is a natural homotopy associative coH-space structure on S^n.

4.2 Definition. The nth homotopy group $(n \geq 1)$ of a space X, denoted $\pi_n(X)$, is $[S^n, X]_0$, with the group structure derived from the coH-space structure of S^n.

The $X \mapsto \pi_n(X)$ is a functor for pointed spaces and homotopy classes of maps preserving base points to groups.

The following proposition is useful in computing the homotopy groups of S^m, RP^m, and CP^m.

4.3 Proposition. *Let X be a union of subspaces X_q such that $X_q \subset X_{q+1}$. We assume that each compact subset K of X is a subset of some X_q. If for each n there exists an integer $q(n)$ such that the inclusion $X_q \to X_k$ induces an isomorphism $\pi_n(X_q) \to \pi_n(X_k)$ for $q(n) \leq q \leq k$, the inclusion $X_q \to X$ induces an isomorphism $\pi_n(X_q) \to \pi_n(X)$ for $q \geq q(n)$.*

A reference for this section is Hu [1, chap. 4].

5. Fibre Maps

A map $p: E \to B$ has the homotopy lifting property for a space W, provided for each map $g: W \to E$ and each homotopy $f_t: W \to B$ with $pg = f_0$ there exists a homotopy $g_t: W \to E$ with $g_0 = g$ and $pg_t = f_t$ for all $t \in I$.

5.1 Definition. A map is a fibre map provided it has the homotopy lifting property for CW-complexes.

To check whether or not a map is a fibre map, one need only verify that it has the homotopy lifting property for cells.

The next theorem is useful in finding examples of fibre maps.

5.2 Theorem. *Let* $p: E \to B$ *be a map, and let* $\{U_i\}$ *for* $i \in I$ *be an open covering of* B *such that* $p: p^{-1}(U_i) \to U_i$ *is a fibre map for all* $i \in I$. *Then* $p: E \to B$ *is a fibre map.*

The next theorem is the most useful elementary property of fibre maps.

5.3 Theorem. *Let* $p: E \to B$ *be a fibre map, and let* $x_0 \in p^{-1}(b_0) = F$, *the fibre of* p *over* $b_0 \in B$. *Then there is a natural group morphism* $\partial: \pi_n(B, b_0) \to \pi_{n-1}(F, x_0)$ *such that the following sequence of groups is exact:*

$$\to \pi_n(E, x_0) \xrightarrow{p*} \pi_n(B, b_0) \xrightarrow{\partial} \pi_{n-1}(F, x_0) \to \pi_{n-1}(E, x_0) \to$$

The reader is invited to apply this theorem to the following examples of fibre maps.

(1) The exponential $p: \mathbf{R} \to S^1$ given by $p(t) = \exp 2\pi i t$ with fibre $\mathbf{Z}$.
(2) The map $p: S^n \to RP^n$ which assigns to x the real line through x. The fibre is Z_2.
(3) The Hopf map $p: S^{2n+1} \to CP^n$ which assigns to x the complex line through x. The fibre is S^1.

PART I

THE GENERAL THEORY
OF FIBRE BUNDLES

CHAPTER 2

Generalities on Bundles

A bundle is just a map viewed as an object in a particular category. It is the basic underlying structure for the more complicated notions of vector bundle and fibre bundle. In this chapter we study the category of bundles in a manner that leads us to the additional structures on a bundle described in the next two chapters. Examples are given to illustrate the concept of a bundle and the various enrichments of this concept.

1. Definition of Bundles and Cross Sections

1.1 Definition. A *bundle* is a triple (E, p, B), where $p: E \to B$ is a map. The space B is called the *base space*, the space E is called the *total space*, and the map p is called the *projection* of the bundle. For each $b \in B$, the space $p^{-1}(b)$ is called the *fibre* of the bundle over $b \in B$.

Intuitively, one thinks of a bundle as a union of fibres $p^{-1}(b)$ for $b \in B$ parametrized by B and "glued together" by the topology of the space E. Usually a Greek letter ($\xi, \eta, \zeta, \lambda$, etc.) is used to denote a bundle; then $E(\xi)$ denotes the total space of ξ, and $B(\xi)$ denotes the base space of ξ.

1.2 Example. The product bundle over B with fibre F is $(B \times F, p, B)$, where p is the projection on the first factor.

In the next section we consider further examples of bundles.

1.3 Definition. A bundle (E', p', B') is a subbundle of (E, p, B) provided E' is a subspace of E, B' is a subspace of B, and $p' = p|E': E' \to B'$.

Many of the examples in the next section arise as subbundles of product bundles. Before taking up examples of bundles, we consider the general notion of cross section. Cross sections of certain bundles can be identified with familiar geometric objects.

1.4 Definition. A cross section of a bundle (E, p, B) is a map $s: B \to E$ such that $ps = 1_B$. In other words, a cross section is a map $s: B \to E$ such that $s(b) \in p^{-1}(b)$, the fibre over b, for each $b \in B$.

Let (E', p', B) be a subbundle of (E, p, B), and let s be a cross section of (E, p, B). Then s is a cross section of $(E', p'B)$ if and only if $s(b) \in E'$ for each $b \in B$.

1.5 Proposition. *Every cross section s of a product bundle $(B \times F, p, B)$ has the form $s(b) = (b, f(b))$, where $f: B \to F$ is a map uniquely defined by s.*

Proof. Every map $s: B \to B \times F$ has the form $s(b) = (s'(b), f(b))$, where $s': B \to B$ and $f: B \to F$ are maps uniquely defined by s. Since $ps(b) = s'(b)$, s is a cross section if and only if $s(b) = (b, f(b))$ for each $b \in B$.

The proposition says that the function that assigns to each cross section s of the product bundle $(B \times F, p, B)$ the map $pr_2 s: B \to F$ is a bijection from the set of all cross sections of $(B \times F, p, B)$ to the set of maps $B \to F$.

If (E, p, B) is a subbundle of the product bundle $(B \times F, p, B)$, the cross sections s of (E, p, B) have the form $s(b) = (b, f(b))$, where $f: B \to F$ is a map such that $(b, f(b)) \in E$ for each $b \in B$.

2. Examples of Bundles and Cross Sections

Let $(x|y)$ denote the euclidean inner product on $\mathbf{R}^n$, and let

$$\|x\| = \sqrt{(x|x)}$$

be the euclidean norm.

2.1 Example. The tangent bundle over S^n, denoted $\tau(S^n) = (T, p, S^n)$, and the normal bundle over S^n, denoted $v(S^n) = (N, q, S^n)$, are two subbundles of the product bundle $(S^n \times \mathbf{R}^{n+1}, p, S^n)$ whose total spaces are defined by the relation $(b, x) \in T$ if and only if the inner product $(b|x) = 0$ and by $(b, x) \in N$ if and only if $x = kb$ for some $k \in R$.

An element $(b, x) \in T$ is called a tangent vector to S^n at b, and an element $(b, x) \in N$ is called a normal vector to S^n at b. The fibres $p^{-1}(b) \subset T$ and $q^{-1}(b)$ are vector spaces of dimensions n and 1, respectively. A cross section of $\tau(S^n)$ is called a (tangent) vector field on S^n, and a cross section of $v(S^n)$ is called a normal vector field on S^n.

2.2 Example. The bundle of (orthonormal) k-frames $\tau_k(S^n)$ over S^n for $k \leq n$, denoted (E, p, S^n), is a subbundle of the product bundle $(S^n \times (S^n)^k, p, S^n)$

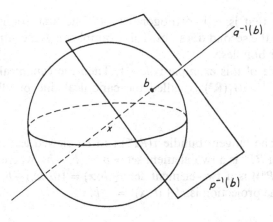

Figure 1

whose total space E is the subspace of $(b, v_1, \ldots, v_k) \in S^n \times (S^n)^k$ such that $(b|v_i) = 0$ and $(v_i|v_j) = \delta_{i,j}$ for $1 \leq i, j \leq k$.

An element $(b, v_1, \ldots, v_k) \in E$ is an orthonormal system of k tangent vectors to S^n at $b \in S^n$. A cross section of $\tau_k(S^n)$ is called a field of k-frames. For the bundle $\tau_k(S^n)$ the existence of a cross section is a difficult problem, and it is considered in a later chapter. Byy projecting on the first k-factors, the existence of a cross section of $\tau_l(S^n)$ implies the existence of a cross section of $\tau_k(S^n)$ for $k \leq l \leq n$.

2.3 Definition. The Stiefel variety of (orthonormal) k-frames in $\mathbf{R}^n$, denoted $V_k(\mathbf{R}^n)$, is the subspace of $(v_1, \ldots, v_k) \in (S^{n-1})^k$ such that $(v_i|v_j) = \delta_{i,j}$.

Since $V_k(\mathbf{R}^n)$ is a closed subset of a compact space, it is a compact space. With each k-frame $(v_1, \ldots, v_k)$ there is associated the k-dimensional subspace $\langle v_1, \ldots, v_k \rangle$ with basis $v_1, \ldots, v_k$. Each k-dimensional subspace of $\mathbf{R}^n$ is of the form $\langle v_1, \ldots, v_k \rangle$.

2.4 Definition. The Grassmann variety of k-dimensional subspaces of $\mathbf{R}^n$, denoted $G_k(\mathbf{R}^n)$, is the set of k-dimensional subspaces of $\mathbf{R}^n$ with the quotient topology defined by the function $(v_1, \ldots, v_k) \rightarrow \langle v_1, \ldots, v_k \rangle$ of $V_k(\mathbf{R}^n)$ onto $G_k(\mathbf{R}^n)$.

Then $G_k(\mathbf{R}^n)$ is a compact space. Note that $V_1(\mathbf{R}^n) = S^{n-1}$ and $G_1(\mathbf{R}^n) = RP^{n-1}$. By a natural inclusion we have $G_k(\mathbf{R}^n) \subset G_k(\mathbf{R}^{n+1})$, and we form $G_k(\mathbf{R}^\infty) = \bigcup_{k \leq n} G_k(\mathbf{R}^n)$ and give it the inductive topology.

2.5 Example. The canonical k-dimensional vector bundle γ_k^n on $G_k(\mathbf{R}^n)$ is the subbundle of the product bundle $(G_k(\mathbf{R}^n) \times \mathbf{R}^n, p, G_k(\mathbf{R}^n))$ with the total space consisting of the subspace of pairs $(V, x) \in G_k(\mathbf{R}^n) \times \mathbf{R}^n$ with $x \in V$. Similarly, the orthogonal complement vector bundle $*\gamma_k^n$ is the subbundle of $(G_k(\mathbf{R}^n) \times \mathbf{R}^n, p, G_k(\mathbf{R}^n))$ with the total space consisting of the subspace of pairs (V, x)

with $(V|x) = 0$; that is, x is orthogonal to V. The first example holds for $n = \infty$ whereas the second does not. This example plays a central role in the theory of vector bundles.

A special case of this example is $k = 1$. Then the canonical vector bundle γ_1^n on $RP^{n-1} = G_1(\mathbf{R}^n)$ is called the canonical line bundle (it is one-dimensional).

2.6 Example. The tangent bundle $\tau(RP^n)$ can be viewed as the quotient of $\tau(S^n)$. A point of RP^n is a two-element set $\pm b = \{b, -b\}$, where $b \in S^n$, and a point of $E(\tau(RP^n))$ is a two-element set $\pm(b, x) = \{(b, x), (-b, -x)\}$, where $(b, x) \in \tau(S^n)$. The projection is $p(\pm(b, x)) = \pm b$.

3. Morphisms of Bundles

A bundle morphism is, roughly speaking, a fibre preserving map. In the next definition we make this idea precise.

3.1 Definition. Let (E, p, B) and (E', p', B') be two bundles. A bundle morphism $(u, f): (E, p, B) \to (E', p', B')$ is a pair of maps $u: E \to E'$ and $f: B \to B'$ such that $p'u = fp$.

The relation $p'u = fp$ is, in effect, the requirement that the following diagram be commutative.

$$
\begin{array}{ccc}
E & \xrightarrow{\;u\;} & E' \\
{\scriptstyle p}\downarrow & & \downarrow{\scriptstyle p'} \\
B & \xrightarrow{\;f\;} & B'
\end{array}
$$

The bundle morphism condition $p'u = fp$ can also be expressed by the relation $u(p^{-1}(b)) \subset (p')^{-1}(f(b))$ for each $b \in B$; that is, the fibre over $b \in B$ is carried into the fibre over $f(b)$ by u. It should be observed that the map f is uniquely determined by u when p is surjective.

3.2 Definition. Let (E, p, B) and (E', p', B) be two bundles over B. A bundle morphism over B (or B-morphism) $u: (E, p, B) \to (E', p', B)$ is a map $u: E \to E'$ such that $p = p'u$.

The relation $p = p'u$ is, in effect, the requirement that the following diagram be commutative.

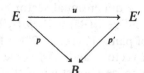

The bundle morphism condition $p'u = p$ can also be expressed by the relation $u(p^{-1}(b)) \subset (p')^{-1}(b)$ for each $b \in B$; that is, u is fibre preserving. The bundle morphisms u over B are just the bundle morphisms $(u, 1_B)$.

3.3 Examples. If (E', p', B') is a subbundle of (E, p, B) and if $f: B' \to B$ and $u: E' \to E$ are inclusion maps, then $(u, f): (E', p', B') \to (E, p, B)$ is a bundle morphism. The cross sections of (E, p, B) are precisely the B-morphisms $s: (B, 1, B) \to (E, p, B)$. Consequently, every general property of morphisms applies to sections.

The pair $(1_E, 1_B): (E, p, B) \to (E, p, B)$ is a bundle morphism that is a B-morphism. If $(u, f): (E, p, B) \to (E', p', B')$ and $(u', f')(E', p', B') \to (E'', p'', B'')$ are bundle morphisms, we have the following commutative diagram:

$$
\begin{array}{ccccc}
E & \xrightarrow{\ u\ } & E' & \xrightarrow{\ u'\ } & E'' \\
\downarrow{\scriptstyle p} & & \downarrow{\scriptstyle p'} & & \downarrow{\scriptstyle p''} \\
B & \xrightarrow{\ f\ } & B' & \xrightarrow{\ f'\ } & B''
\end{array}
$$

Consequently, the compositions define a bundle morphism $(u'u, f'f)$: $(E, p, B) \to (E'', p'', B'')$ which is defined to be the composition $(u', f')(u, f)$ of (u, f) and (u', f').

3.4 Definition. The category of bundles, denoted **Bun**, has as its objects all bundles (E, p, B) and as morphisms from (E, p, B) to (E', p', B') the set of all bundle morphisms. Composition is composition of bundle morphisms as defined above. For each space B, the subcategory of bundles over B, denoted **Bun**$_B$, has as its objects bundles with base space B and B-morphisms as its morphisms.

From general properties in a category, a bundle morphism (u, f): $(E, p, B) \to (E', p', B')$ is an isomorphism if and only if there exists a morphism $(u', f'): (E', p', B') \to (E, p, B)$ with $f'f = 1_B$, $ff' = 1_{B'}$, $u'u = 1_E$, and $uu' = 1_{E'}$. The notion of two bundles being isomorphic has a well-defined meaning.

3.5 Definition. A space F is the fibre of a bundle (E, p, B) provided every fibre $p^{-1}(b)$ for $b \in B$ is homeomorphic to F. A bundle (E, p, B) is trivial with fibre F provided (E, p, B) is B-isomorphic to the product bundle $(B \times F, p, B)$.

4. Products and Fibre Products

4.1 Definition. The product of two bundles (E, p, B) and (E', p, B') is the bundle $(E \times E', p \times p', B \times B')$.

As with spaces, the reader can easily describe the operation of the product as a functor **Bun** $\times$ **Bun** $\to$ **Bun**. Moreover, the concept clearly extends to an arbitrary family of bundles. This is the product in the category **Bun** in the sense of category theory.

4.2 Definition. The fibre product $\xi_1 \oplus \xi_2$ of two bundles

$$\xi_1 = (E_1, p_1, B) \qquad \text{and} \qquad \xi_2 = (E_2, p_2, B)$$

over B is $(E_1 \oplus E_2, q, B)$, where $E_1 \oplus E_2$ is the subspace of all $(x, x') \in E_1 \times E_2$ with $p_1(x) = p_2(x')$ and $q(x, x') = p_1(x) = p_2(x')$.

The fibre product is sometimes called the Whitney sum. The fibre $q^{-1}(b)$ of $(E_1 \oplus E_2, q, B)$ over $b \in B$ is $p_1^{-1}(b) \times p_2^{-1}(b) \subset E_1 \times E_2$. This is the reason for the term fibre product.

We define $\oplus: \mathbf{Bun}_B \times \mathbf{Bun}_B \to \mathbf{Bun}_B$ as a functor. Let $u_1: (E_1, p_1, B) \to (E_1', p_1', B)$ and $u_2: (E_2, p_2, B) \to (E_2', p_2', B)$ be two B-morphisms. Then we define the B-morphism $u_1 \oplus u_2: (E_1 \oplus E_2, q, B) \to (E_1' \oplus E_2', q', B)$ by the relation $(u_1 \oplus u_2)(x_1, x_2) = (u_1(x_1), u_2(x_2))$. Since $p_1' u_1(x_1) = p_1(x_1) = p_2(x_2) = p_2' u_2(x_2)$, $u_1 \oplus u_2$ is a well-defined morphism; clearly, the relation $1_{E_1} \oplus 1_{E_2} = 1_{E_1 \oplus E_2}$ holds. If $v_1: (E_1', p_1', B) \to (E_1'', p_1'', B)$ and $v_2: (E_2', p_2', B) \to (E_2'', p_2'', B)$ are also B-morphisms, then we have $(v_1 \oplus v_2)(u_1 \oplus u_2) = (v_1 u_1) \oplus (v_2 u_2)$. Consequently, $\oplus$ is a functor. The fibre product is the product in the category $\mathbf{Bun}_B$ in the sense of category theory.

The map $u: B \times F_1 \times F_2 \to (B \times F_1) \oplus (B \times F_2)$ defined by the relation $u(b, y_1, y_2) = (b, y_1, y_2) = (b, y_1, b, y_2)$ is a homeomorphism and defines a B-isomorphism of product bundles $u: (B \times F_1 \times F_2, q, B) \to (B \times F_1, p_1, B) \oplus (B \times F_2, p_2, B)$. Using this isomorphism and the functorial properties of $\oplus$, we have the next proposition.

4.3 Proposition. *If (E_1, p_1, B) is a trivial bundle with fibre F_1 and if (E_2, p_2, B) is a trivial bundle with fibre F_2, then $(E_1, p_1, B) \oplus (E_2, p_2, B)$ is a trivial bundle with fibre $F_1 \times F_2$.*

In the next proposition we compute the cross sections of a fibre product.

4.4 Proposition. *The cross sections s of a fibre product $(E_1 \oplus E_2, q, B)$ are of the form $s(b) = (s_1(b), s_2(b))$, where s_1 is a cross section of (E_1, p, B) and s_2 is a cross section of (E_2, p_2, B) uniquely defined by s.*

Proof. Each cross section s is a map $s: B \to E_1 \oplus E_2 \subset E_1 \times E_2$; therefore, s is of the form $s(b) = (s_1(b), s_2(b))$, where $s_1: B \to E_1$ and $s_2: B \to E_2$. For s to be a cross section, $b = qs(b) = p_1 s_1(b) = p_2 s_2(b)$ for each $b \in B$; that is, s_1 and s_2 are cross sections.

Finally, we consider three calculations of fibre products. Let θ^k denote the product bundle $(B \times \mathbf{R}^k, p, B)$.

4.5 Example. There is an isomorphism $u: \gamma_{kn} \oplus (*\gamma_{kn}) \to \theta^n$ defined by $u((V, x), (V, x')) = (V, x + x')$ for $V \in G_k(\mathbf{R}^n)$, $(V, x) \in E(\gamma_{kn})$, and $(V, x') \in E(*\gamma_{kn})$; see (2.5). Since every $y \in \mathbf{R}^n$ can be written $y = x + x'$, where $x \in V$, x' orthogonal to V, and since this decomposition is continuous in V, the map u is a $G_k(\mathbf{R}^n)$-isomorphism.

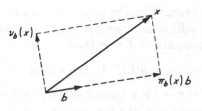

Figure 2

4.6 For the next two examples, we use the following notations. For each $b \in \mathbf{R}^n$, $b \neq 0$, there are two linear functions $v_b: \mathbf{R}^n \to \mathbf{R}^n$, the normal map, and $\pi_b: \mathbf{R}^n \to \mathbf{R}$, the projection on b, such that $x = v_b(x) + \pi_b(x)x$, $\pi_b(x) = (b|x)/(b|b)$, $v_b(x) = x - \pi_b(x)b$, and $(b|v_b(x)) = 0$ for each $x \in \mathbf{R}^n$.

4.7 Example. An isomorphism $u: \tau(S^n) \oplus v(S^n) \to \theta^{n+1}$ is defined by the relation $u((b,x),(b,x')) = (b, x + x')$ for $(b,x) \in E(\tau(S^n))$ and $(b,x') \in E(v(S^n))$. The inverse of u is the B-morphism v defined by the relation $v(b,x) = ((b, v_b(x)), (b, \pi_b(x)b))$.

Let λ denote the canonical line bundle on RP^n.

4.8 Example. Over RP^n, there is an isomorphism $u: (n+1)\lambda \to \tau(RP^n) \oplus \theta^1$ with inverse v defined by the following relations:

$$u(\pm b, (a_0 b, \ldots, a_n b)) = (\pm(b, v_b(a_0, \ldots, a_n)), (\pm b, \pi_b(a_0, \ldots, a_n)))$$

$$v(\pm(b,x),(\pm b, k)) = (\pm b, p_0(x + kb)b, \ldots, p_n(x + kb)b)$$

These maps are well defined and are inverses of each other from the relations $a = v_b(a) + \pi_b(a)b$ and $-a = -v_b(a) + \pi_b(a)(-b)$, where $-v_b(a) = v_b(-a) = v_{-b}(-a)$ and $\pi_b(a) = \pi_{-b}(-a) = -\pi_{-b}(a)$. Here we use the notations of (2.6) and (4.6), and $k\xi = \xi \oplus \cdots (k) \cdots \oplus \xi$.

5. Restrictions of Bundles and Induced Bundles

5.1 Definition. Let $\xi = (E, p, B)$ be a bundle, and let A be a subset of B. Then the restriction of ξ to A, denoted $\xi|A$, is the bundle (E', p', A), where $E' = p^{-1}(A)$ and $p|E'$.

5.2 Examples. In a natural way we can consider $G_k(\mathbf{R}^n) \subset G_k(\mathbf{R}^{n+m})$. Then for the canonical k-dimensional vector bundle over the grassmannians, $\gamma_k^{n+m}|G_k(\mathbf{R}^n) = \gamma_{kn}$.

If ξ is the product bundle over B with fibre F and if A is a subset of B, then $\xi|A$ is the product bundle over A with fibre F.

Restriction of bundles satisfies the following transitivity property. If $A_1 \subset A \subset B$ and if ξ is a bundle over B, we have $\xi|A_1 = (\xi|A)|A_1$, and $\xi|B = \xi$. If $u: \xi \to \eta$ is a B-morphism and if $A \subset B$, then

$$u_A = u|E(\xi|A): \xi|A \to \eta|A$$

is an A-morphism. If $v: \eta \to \xi$ is a second B-morphism, we have $(vu)_A = v_A u_A$ and $(1_\xi)_A = 1_{\xi|A}$. Consequently, the functions $\xi \mapsto \xi|A$ and $u \mapsto u_A$ define a functor $\mathbf{Bun}_B \to \mathbf{Bun}_A$.

In the next definition we generalize the process of restriction.

5.3 Definition. Let $\xi = (E, p, B)$ be a bundle, and let $f: B_1 \to B$ be a map. The induced bundle of ξ under f, denoted $f^*(\xi)$, has as base space B_1, as total space E_1 which is the subspace of all pairs $(b_1, x) \in B_1 \times E$ with $f(b_1) = p(x)$, and as projection p_1 the map $(b_1, x) \mapsto b_1$.

5.4 Example. Let ξ be a bundle over B, and let A be a subspace of B with inclusion map $j: A \to B$. Then $\xi|A$ and $j^*(\xi)$ are A-isomorphic. In effect, we define $u: \xi|A \to j^*(\xi)$ by $u(x) = (p(x), x)$, and this is clearly an A-isomorphism.

If $f^*(\xi)$ is the induced bundle of ξ under $f: B_1 \to B$, then $f_\xi: E(f^*(\xi)) \to E(\xi)$, defined by $f_\xi(b_1, x) = x$, together with f define a morphism $(f_\xi, f): f^*(\xi) \to \xi$, which is referred to as the canonical morphism of an induced bundle.

5.5 Proposition. If $(f_\xi, f): f^*(\xi) \to \xi$ is the canonical morphism from the bundle of ξ under a map $f: B_1 \to B$, then for each $b_1 \in B_1$ the restriction $f_\xi: p_1^{-1}(b_1) \to p^{-1}(f(b_1))$ is a homeomorphism. Moreover, if $(v, f): \eta \to \xi$ is any bundle morphism, there exists a B_1-morphism $w: \eta \to f^*(\xi)$ such that $f_\xi w = v$. The morphism w is unique with respect to this property.

Proof. The fibre $p_1^{-1}(b_1) \subset b_1 \times E$ is the subspace of $(b_1, x) \in b_1 \times E$ with $p(x) = f(b_1)$. Consequently, $f_\xi: b_1 \times p^{-1}(f(b_1)) \to p^{-1}(f(b_1))$ defined by $f_\xi(b_1, x) = x$ is clearly a homeomorphism.

For the second statement, let $w(y) = (p_\eta(y), v(y))$. Since (v, f) is a morphism, we have $f(p_\eta(y)) = p(v(y))$, and, consequently, $w: E(\eta) \to E(f^*(\xi))$ is a B_1-morphism. Clearly, we have $f_\xi w = v$. For uniqueness, the relation $p_1(w(y)) = p_\eta(y)$, which holds for any B_1-morphism w, and the relation $f_\xi w = v$ imply that $w(y) = (p_\eta(y), v(y))$ for each $y \in E(\eta)$. This proves the proposition.

If $u: \xi \to \eta$ is a B-morphism and if $f: B_1 \to B$ is a map, there is a B_1-morphism $f^*(u): f^*(\xi) \to f^*(\eta)$ defined by the relation $f^*(u)(b_1, x) = (b_1, u(x))$. Clearly, we have $f^*(1_\xi) = 1_{f^*(\xi)}$, and if $v: \eta \to \zeta$ is a second B-morphism, then $f^*(vu)(b_1, x) = (b_1, vu(x)) = f^*(v)(b_1, u(x)) = f^*(v)f^*(u)(b_1, x)$. Therefore, we have the next proposition.

5.6 Proposition. *For each map* $f: B_1 \to B$, *the family of functions* f^*: $\mathbf{Bun}_B \to$ $\mathbf{Bun}_{B_1}$ *defines a functor. Moreover, for a B-morphism* $u: \xi \to \eta$ *the following diagram is commutative.*

$$
\begin{array}{ccc}
E(f^*(\eta)) & \xrightarrow{\ f_\eta\ } & E(\eta) \\
\end{array}
$$

$$
\begin{array}{ccc}
E(f^*(\xi)) & \xrightarrow{\ f_\xi\ } & E(\xi) \\
& & \\
B_1 & \xrightarrow{\ f\ } & B
\end{array}
$$

Proof. We must check the last statement. Let $(b_1, x) \in E(f^*(\xi))$, and compute $u(f_\xi(b_1, x)) = u(x) = f_\eta(b_1, u(x)) = f_\eta(f^*(u))(b_1, x))$. We have $uf_\xi = f_\eta f^*(u)$.

Finally, we have the following transitivity relation.

5.7 Proposition. *Let* $g: B_2 \to B_1$ *and* $f: B_1 \to B$ *be two maps, and let* ξ *be a bundle over B. Then* $1^*(\xi)$ *and* ξ *are B-isomorphic, and* $g^*(f^*(\xi))$ *and* $(fg)^*(\xi)$ *are* B_2-isomorphic.

Proof. Define $u: \xi \to 1^*(\xi)$ by the relation $u(x) = (p(x), x)$, and u is clearly an isomorphism. Next, let $v: (fg)^*(\xi) \to g^*(f^*(\xi))$ be defined by $v(b_2, x) = (b_2, (g(b_2), x))$. Then v is clearly an isomorphism.

5.8 Corollary. *Let* $f: (B_1, A_1) \to (B, A)$ *be a map of pairs, let* $g = f|A_1: A_1 \to$ A, *and let* ξ *be a bundle over B. Then* $g^*(\xi|A)$ *and* $f^*(\xi)|A_1$ *are* A_1-isomorphic.

Proof. Let $j: A \to B$ and $j_1: A_1 \to B_1$ be the respective inclusion maps. Then $fj_1 = jg$, and, in view of (5.4), (5.6), and (5.7), there is the following sequence of A_1-isomorphisms:

$$
f^*(\xi)|A_1 \cong j_1^* f^*(\xi) \cong (fj_1)^*(\xi) \cong (jg)^*(\xi) \cong g^*(j^*(\xi)) \cong g^*(\xi|A)
$$

The next result is useful in discussing fibre bundles.

5.9 Proposition. *Let* $\xi = (E, p, B)$ *be a bundle, let* $f: B_1 \to B$ *be a map, and let* $f^*(\xi) = (E_1, p_1, B_1)$ *be the induced bundle of* ξ *under f. If p is an open map,* p_1 *is an open map.*

Proof. Let W be an open neighborhood of $(b_1, x) \in E_1$, where $E_1 \subset B_1 \times E$. We must find a neighborhood V of $b_1 = p_1(b_1, x)$ with $p_1(W) \supset V$. From the definition of the topology of E_1 there exist open neighborhoods V_1 of $b_1 \in B$ and U of $x \in E$ with $(V_1 \times U) \cap E_1 \subset W$. Let $V = V_1 \cap f^{-1}(p(U))$. Then for each $b_1 \in V$ there exists $x \in U$ with $p(x) = f(b_1)$, that is, $(b_1, x) \in W$ and $b_1 = p_1(b_1, x) \in V$. Therefore, we have $p_1(W) \supset V$.

The following relation between cross sections and induced bundles is useful in Sec. 7.

5.10 Proposition. *Let $\xi = (E, p, B)$ be a bundle, let $f: B_1 \to B$ be a map, and let $(f_\xi, f): f^*(\xi) \to \xi$ be the canonical morphism of the induced bundle. If s is a cross section of ξ, then $\sigma: B_1 \to E(f^*(\xi))$ defined by $\sigma(b_1) = (b_1, sf(b_1))$ is a cross section with $f_\xi \sigma = sf$. If f is an identification map and if σ is a cross section of $f^*(\xi)$ such that $f_\xi \sigma$ is constant on all sets $f^{-1}(b)$ for $b \in B$, there is a cross section s of ξ such that $sf = f_\xi \sigma$.*

Proof. We have $p_1 \sigma(b_1) = p_1(b_1, sf(b_1)) = b_1$ and $f(b_1) = psf(b_1)$; consequently, σ is a cross section of $f^*(\xi)$. The relation $f_\xi \sigma(b_1) = f_\xi(b_1, sf(b_1)) = sf(b_1)$ also follows.

For the second statement, we have a factorization of $f_\xi \sigma$ by f, giving a map $s: B \to E$ with $sf = f_\xi \sigma$. Moreover, $psf = pf_\xi \sigma = fp_1 \sigma = f$ and $ps = 1_B$ since f is surjective. Then s is the desired cross section.

6. Local Properties of Bundles

6.1 Definition. Two bundles ξ and η over B are locally isomorphic provided for each $b \in B$ there exists an open neighborhood U of b such that $\xi | U$ and $\eta | U$ are U-isomorphic.

Clearly, two isomorphic bundles are locally isomorphic.

6.2 Definition. A bundle ξ over B is locally trivial with fibre F provided ξ is locally isomorphic with the product bundle $(B \times F, p, B)$.

The next proposition makes the idea of a local property meaningful.

6.3 Proposition. *The relation of being locally isomorphic is an equivalence relation on the class of all bundles over B.*

Proof. The transitivity of the relation is the nontrivial part. Let U and V be two open neighborhoods of $b \in B$ such that $\xi | U$ and $\eta | U$ are U-isomorphic and $\eta | V$ and $\zeta | V$ are V-isomorphic. By (5.7), the bundles $\xi | (U \cap V)$, $\eta | (U \cap V)$, and $\zeta | (U \cap V)$ are $(U \cap V)$-isomorphic.

6.4 Corollary. *If ξ is locally isomorphic to a locally trivial bundle, ξ is locally trivial.*

A local property of bundles is a property of bundles that is unchanged between locally isomorphic bundles. The property that the projection is a fibre map is a local property by 1 (5.2).

6.5 Proposition. *Let ξ and η be two bundles over B, and let $f: B_1 \to B$ be a map. If ξ and η are locally isomorphic, then $f^*(\xi)$ and $f^*(\eta)$ are locally isomorphic over B_1.*

Proof. By Corollary (5.8), we have $f^*(\xi|U) \cong f^*(\xi)|f^{-1}(U)$ for each open set $U \subset B$. If $\xi|U$ and $\eta|U$ are U-isomorphic, $f^*(\xi)|f^{-1}(U)$ and $f^*(\eta)|f^{-1}(U)$ are $f^{-1}(U)$-isomorphic.

6.6 Corollary. *Let ξ and η be two locally isomorphic bundles over B, and let $A \subset B$. Then $\xi|A$ and $\eta|A$ are locally isomorphic.*

6.7 Corollary. *Let ξ be a locally trivial bundle over B with fibre F, let $f: B_1 \to B$ be a map, and let A be a subset of B. Then $f^*(\xi)$ and $\xi|A$ are locally trivial with fibre F.*

7. Prolongation of Cross Sections

In this section we generalize the prolongation theorems for maps (see Chap. 1, Sec. 2) to cross sections of locally trivial bundles. This prolongation theorem is the fundamental step in the classification theory of fibre bundles over CW-complexes. Although we prove the homotopy classifcation theorem for fibre bundles over an arbitrary space, the results of this section are used to give more precise information about homotopy properties of fibre bundles over CW-complexes.

7.1 Theorem. *Let $\xi = (E, p, B)$ be a locally trivial bundle with fibre F, where (B, A) is a relative CW-complex. Then all cross sections s of $\xi|A$ prolong to a cross section s^* of ξ under either of the following hypotheses:*

(H1) *The space F is $(m - 1)$-connected for each $m \leq \dim B$.*
(H2) *There is a relative CW-complex (Y, X) such that $B = Y \times I$ and $A = (X \times I) \cap (Y \times 0)$, where $I = [0, 1]$.*

Proof. First, we prove the theorem under hypothesis (H1). We assume the theorem is true for all B with $\dim B < n$. This is the case for $n = 0$ because $B = A$. We let B be of dimension n. By the inductive hypothesis we have a cross section s' of $\xi|B_{n-1}$ with $s'|A = s$. We let C be an n-cell of B with attaching map $u_C: I^n \to B$. The bundle $u_C^*(\xi)$ over I^n is locally trivial, and since I^n is compact, we can dissect I^n into equal cubes K of length $1/k$ such that $u_C^*(\xi)|K$ is trivial. By (5.10) the cross section s' defines a cross section σ' of $u_C^*(\xi)|\partial I^n$. Applying the inductive hypothesis to σ', we can assume that σ' is defined on the $(n - 1)$-skeleton of I^n decomposed into cubes K of length $1/k$. The cross section σ' now defined on ∂K is given by a map $\partial K \to F$ [see [1.5]], which by the connectivity hypothesis on F prolongs to K. This prolonged map yields a prolongation σ of σ' over each cell K and, therefore, a cross section σ of $u_C^*(\xi)$. Using the natural morphism $u_C^*(\xi) \to \xi$ over u_C and (5.10), we have a cross section s_C of $\xi|\overline{C}$ such that $s_C|(\overline{C} \cap B_{n-1}) = s'|(\overline{C} \cap B_{n-1})$. We define a cross section s^* of ξ by the requirements that $s^*|B_{n-1} = s'$ and $s^*|\overline{C} = s_C$. By the weak topology property, s^* is continuous.

Finally, if dim $B = \infty$, F is n-connected for each n, and we construct inductively cross sections s_n of $\xi|B_n$ such that $s_n|B_{n-1} = s_{n-1}$ and $s_{-1} = s$. We define a cross section s^* of ξ by the requirement that $s^*|B_n = s_n$.

Second, we prove the theorem under hypothesis (H2). We assume the theorem is true for all Y with dim $Y < n$. This is the case for $n = 0$ because $x = Y$ and $A = B$. We let Y be of dimension n. By the inductive hypothesis we have a cross section s' of $\xi|[(Y \times 0) \cup (Y_{n-1} \times I)]$ with $s'|[(Y \times 0) \cup (X \times I)] = s$. We let C be an n-cell of Y with attaching map $u_C\colon I^n \to Y$. The bundle $(u_C \times 1_I)^*(\xi)$ over $I^n \times I$ is locally trivial, and since $I^n \times I$ is compact, we can dissect $I^n \times I$ into equal cubes $K \times [(i-1)/k, i/k]$ of length $1/k$ for $1 \leq i \leq k$ such that $(u_C \times 1_I)^*(\xi)$ is trivial over each of these cubes. By (5.10) the cross section s' defines a cross section σ' of $(u_C \times 1_I)^*(\xi)|[(I^n \times 0) \cup (\partial I^n \times I)]$. Applying the inductive hypothesis to σ' with respect to $I^n \times [0, 1/k]$, we can assume that σ' is defined on each $\partial K \times [0, 1/k]$ making up $(I^n)_{n-1} \times [0, 1/k]$. The cross section σ' on $(\partial K \times [0, 1/k]) \cup (K \times 0)$ is given by a map $(\partial K \times [0, 1/k]) \cup (K \times 0) \to F$ [see (1.5)], which prolongs to $K \times [0, 1/k]$. This prolonged map yields a prolongation σ of σ' over each cell $K \times [0, 1/k]$ and, therefore, a cross section of $(u_C \times 1_I)^*(\xi)$ over $I^n \times [0, 1/k]$. Continuing this process k times, we have a cross section σ of $(u_C \times 1_I)^*(\xi)$. Using the natural morphism $(u_C \times 1_I)^*(\xi) \to \xi$ over $u_C \times 1_I$ and (5.10), we have a cross section s_C of $\xi|(\overline{C} \times I)$ such that $s_C|(\overline{C} \times 0) = s|(\overline{C} \times 0)$ and $s_C|(Y_{n-1} \times I) = s'|(Y_{n-1} \times I)$. We define a cross section s^* of ξ by the requirements that $s^*|(Y_{n-1} \times I) = s'$ and $s^*|(\overline{C} \times I) = s_C$. By the weak topology property, s^* is continuous.

Finally, if dim $Y = \infty$, we construct inductively cross sections s_n of $\xi|(Y_n \times I)$ such that $s_n|(Y_{n-1} \times I) = s|(Y_n \times 0)$, and $s_{-1} = s|(X \times I)$. We define a cross section s^* of ξ by the requirement that $s^*|(Y_n \times I) = s_n$. This proves the theorem.

Note: There is a parallel in the two proofs under the two hypotheses of (7.1). The proofs differ only in the character of the prolongation over "small" cells.

Exercises

1. Prove that $\tau(S^{n+q})|S^n$ is isomorphic to $\tau(S^n) \oplus \theta^q$, where θ^q is the trivial bundle with fibre $\mathbf{R}^q$ and $S^n \subset S^{n+q}$ is the standard inclusion.

2. Prove that $\gamma_k^{n+q}|G_k(\mathbf{R}^n) \cong \gamma_{k^n}$, where $G_k(\mathbf{R}^n) \subset G_k(\mathbf{R}^{n+q})$ in a natural way. Let $G_k(\mathbf{R}^n) \subset G_{k+q}(\mathbf{R}^{n+q})$ by the map $V \to V \oplus W$, where W is the q-dimensional subspace with basis $e_{n+1}, \ldots, e_{n+q}$ in $\mathbf{R}^{n+q}$. Prove that $\gamma_{k+q}^{n+q}|G_k(\mathbf{R}^n) \cong \gamma_{k^n} \oplus \theta^q$.

3. Using the fact that S^{2n-1} is the set of unit vectors in $\mathbf{C}^n$, prove that S^{2n-1} has one unit vector field on it. Using the fact that S^{4n-1} is the set of unit vectors in $\mathbf{H}^n$, prove that S^{4n-1} has three unit vector fields on it which are orthonormal at each point. *Hint*: Do cases S^1 and S^3 first.

4. Prove that if S^n has a vector field which is everywhere nonzero the identity and the antipodal map $x \mapsto -x$ of $S^n \to S^n$ are homotopic.

5. Let $\xi = (E, p, B)$ be a bundle with fibre F over $B = B_1 \cup B_2$, where $B_1 = A \times [a, c]$ and $B_2 = A \times [c, b]$. Prove that if $\xi | B_1$ and $\xi | B_2$ are trivial ξ is trivial.

CHAPTER 3

Vector Bundles

A vector bundle is a bundle with an additional vector space structure on each fibre. The concept arose from the study of tangent vector fields to smooth geometric objects, e.g., spheres, projective spaces, and, more generally, manifolds. The vector bundle structure is so rich that the set of isomorphism classes of k-dimensional vector bundles over a paracompact space B is in a natural bijective correspondence with the set of homotopy classes of mappings of B into the Grassmann manifold of k-dimensional subspaces in infinite-dimensional space.

1. Definition and Examples of Vector Bundles

Let F denote the field of real numbers **R**, complex numbers **C**, or quaternions **H**.

1.1 Definition. A k-dimensional vector bundle ξ over F is a bundle (E, p, B) together with the structure of a k-dimensional vector space over F on each fibre $p^{-1}(b)$ such that the following local triviality condition is satisfied. Each point of B has an open neighborhood U and a U-isomorphism $h: U \times F^k \to p^{-1}(U)$ such that the restriction $b \times F^k \to p^{-1}(b)$ is a vector space isomorphism for each $b \in U$.

An F-vector bundle is called a real vector bundle if $F = \mathbf{R}$, a complex vector bundle if $F = \mathbf{C}$, and a quaternionic vector bundle if $F = \mathbf{H}$. The U-isomorphism $h: U \times F^k \to p^{-1}(U)$ is called a local coordinate chart of ξ.

Examples 2(2.1), 2(2.5), and 2(2.6) admit the structure of a vector bundle in a natural way.

1.2 Example. The k-dimensional product bundle over a space B is the bundle $(B \times F^k, p, B)$ with the vector space structure of F^k defining the vector space structure on $b \times F^k = p^{-1}(b)$ for $b \in B$. The local triviality condition is realized by letting $U = B$ and $h = 1$.

1.3 Example. The tangent bundle $\tau(S^n)$ has a natural real vector space structure on each fibre since it is a subspace of $\mathbf{R}^{n+1}$. In the quotient bundle $\tau(RP^n)$ there is a vector space structure on each fibre. As for local triviality, let U_i be the open subset of $x \in S^n$ with $x_i \neq 0$, $0 \leq i \leq n$, and let $u_i: \mathbf{R}^n \to \mathbf{R}^{n+1}$ be the linear injection $u_i(x_1, \ldots, x_n) = (x_1, \ldots, x_i, 0, x_{i+1}, \ldots, x_n)$. Then $h_i: U_i \times \mathbf{R}^n \to p^{-1}(U_i) \subset E(\tau(S^n))$, where $h_i(b, x) = (b, v_b(u_i(x)))$ [see 2(4.6)] has the desired properties. Also, this construction proves that $\tau(RP^n)$ is locally trivial.

1.4 Example. The function $\pi: G_k(F^m) \times F^m \to F^m$, where $\pi(V, x)$ is the orthogonal projection of x into V, is a map. For $H \subset \{1, 2, \ldots, m\}$, a subset of k elements, we have a linear map $u_H: F^k \to F^m$ by placing 0 in each coordinate not in H. With these maps, we prove that $\gamma_k^m = (E, p, G_k(F^m))$ is locally trivial. Since E is the subspace of $G_k(F^m) \times F^m$ consisting of pairs (V, x) with $x \in V$, the fibre over V is $\{V\} \times V$, and the vector space structure is determined by the subspace V. Let U_H be the open subspace of $G_k(F^m)$ consisting of $V \in G_k(F^m)$ such that $\pi(V, -): u_H(F^k) \to V$ is a bijection. Then $h_H: U_H \times F^k \to p^{-1}(U_H)$ is defined by the relation $h_H(V, x) = (V, \pi(V, x))$, and h_H is an isomorphism that is linear on each fibre. For more details of the above argument, see Chap. 7. For the present, the above is an exercise.

From the local triviality of a vector bundle we have the following continuity properties.

1.5 Proposition. *Let* $\xi = (E, p, B)$ *be a* k-*dimensional vector bundle. Then* p *is an open map. The fibre preserving functions* $a: E \oplus E \to E$ *and* $s: F \times E \to E$ *defined by the algebraic operations* $a(x, x') = x + x'$ *and* $s(k, x) = kx$, $k \in F$, *are continuous.*

Proof. For each local coordinate $h: U \times F^k \to p^{-1}(U)$, the above statements hold for the above functions restricted to $p^{-1}(U)$ or $p^{-1}(U) \oplus p^{-1}(U)$ for s or a, respectively. Since the family of $p^{-1}(U)$ is an open covering of E, the above statements are true for ξ.

Using the ideas connected with this proposition, we are able to put an algebraic structure on the set of cross sections of a vector bundle.

1.6 Proposition. *Let* s *and* s' *be two cross sections of a vector bundle* $\xi = (E, p, B)$, *and let* $\phi: B \to F$ *be a map. Then the function* $s + s'$ *defined by* $(s + s')(b) = s(b) + s'(b)$ *is a cross section of* ξ, *the function* ϕs *defined by* $(\phi s)(b) = \phi(b)s(b)$ *is a cross section of* ξ, *and the map* $b \mapsto 0 \in p^{-1}(b)$ *is a cross section (the zero cross section).*

Proof. Let $h: U \times F^k \to p^{-1}(U)$ be a local coordinate of ξ over U, and let $h^{-1}s(b) = (b, f(b))$ and $h^{-1}s'(b) = (b, f'(b))$ for $b \in B$, where $f: U \to F^k$ and $f': U \to F^k$ are maps. Then $h^{-1}(s + s')(b) = (b, f(b) + f'(b))$, $h^{-1}(\phi s)(b) = (b, \phi(b)f(b))$, and $h^{-1}(0)(b) = (b, 0)$ for $b \in U$. Consequently, $s + s'$, ϕs, and 0 are continuous maps and, therefore, cross sections.

Proposition (1.6) says that the set of cross sections of ξ form a module over the ring $C_F(B(\xi))$ of continuous F-valued functions on $B(\xi)$.

2. Morphisms of Vector Bundles

A vector bundle morphism is, roughly speaking, a fibre preserving map that is linear on each fibre. In the next definition we make this idea precise.

2.1 Definition. Let $\xi = (E, p, B)$ and $\xi' = (E', p', B')$ be two vector bundles. A morphism of vector bundles $(u, f): \xi \to \xi'$ is a morphism of the underlying bundles; that is, $u: E \to E'$ and $f: B \to B'$ are maps such that $p'u = fp$, and the restriction $u: p^{-1}(b) \to p^{-1}(f(b))$ is linear for each $b \in B$.

2.2 Definition. Let $\xi = (E, p, B)$ and $\xi' = (E', p', B)$ be two vector bundles over a space B. A B-morphism of vector bundles $u: \xi \to \xi'$ is defined by a morphism of the form $(u, 1_B): \xi \to \xi'$.

If $u: \xi \to \xi'$ is a B-morphism, then $p'u = p$, and the restriction $u: p^{-1}(b) \to (p')^{-1}(b)$ is linear for each $b \in B$.

2.3 Example. Let ξ be the product bundle $(B \times F^k, p, B)$, and let η be the product bundle $(B \times F^m, p, B)$. The B-morphisms have the form $u(b, x) = (b, f(b, x))$, where $f: B \times F^k \to F^m$ is a map such that $f(b, x)$ is linear in x. Let $\mathbf{L}(F^k, F^m)$ denote the vector space of all linear transformations $F^k \to F^m$. By matrix representation, $\mathbf{L}(F^k, F^m)$ is isomorphic to F^{km}. Then $f: B \times F^k \to F^m$ is continuous if and only if $b \mapsto f(b, -)$ as a function $B \to \mathbf{L}(F^k, F^m)$ is continuous; i.e., each matrix element is continuous.

As with bundles [see 2(3.3)], identities are B-morphisms of vector bundles, and the composition of vector bundle morphisms is a vector morphism. Let $\xi = (E, p, B)$ and $\xi' = (E', p', B')$ be two vector bundles, and let $f: B \to B'$ be a map. Then $u: E \to E'$ is defined to be $u(x) = 0$ in $(p')^{-1}(f(p(b)))$, for each $b \in B$ combines with f to define a morphism of vector bundles $(u, f): \xi \to \xi'$.

2.4 Definition. The category of vector bundles, denoted **VB**, has as its objects vector bundles. Its morphisms are defined in (2.1). Composition is composition of morphisms of vector bundles.

For each space B, let $\mathbf{VB}_B$ denote the subcategory of vector bundles over B and B-morphisms. For each integer $k \geq 0$, let $\mathbf{VB}^k$ denote the full subcategory of k-dimensional vector bundles. Finally, the subcategory $\mathbf{VB}_B^k$ of k-dimensional vector bundles over B is the intersection $\mathbf{VB}_B \cap \mathbf{VB}^k$.

An isomorphism of vector bundles over B is a morphism $u\colon \xi \to \xi'$ such that there exists a morphism $v\colon \xi' \to \xi$ with $vu = 1_\xi$ and $uv = 1_{\xi'}$. In the next theorem we derive a criterion for a B-morphism to be an isomorphism.

2.5 Theorem. *Let* $u\colon \xi \to \xi'$ *be a B-morphism between two vector bundles. Then u is an isomorphism if and only if $u\colon p^{-1}(b) \to (p')^{-1}(b)$ is a vector space isomorphism for each $b \in B$.*

Proof. The direct implication is immediate because the inverse of $u\colon p^{-1}(b) \to (p')^{-1}(b)$ is the restriction to $(p')^{-1}(b)$ of the inverse of u. Conversely, let $v\colon \xi' \to \xi$ be the function defined by the requirement that $v|(p')^{-1}(b)$ be the inverse of the restricdted linear transformation $u\colon p^{-1}(b) \to (p')^{-1}(b)$. The function v will be the desired inverse of u provided v is continuous. Let U be an open subset of B, let $h\colon U \times F^k \to p^{-1}(U)$ be a local coordinate of ξ, and let $h'\colon U \times F^k \to (p')^{-1}(U)$ be a local coordinate of ξ'. It suffices to prove $v\colon (p')^{-1}(U) \to p^{-1}(U)$ is continuous for every such U. By (2.3), $(h')^{-1}uh$ has the form $(b, x) \mapsto (b, f_b(x))$, where $b \mapsto f_b$ is a map $U \to \mathbf{L}(F^k, F^k)$. Then $h^{-1}vh'$ has the form $(b, x) \mapsto (b, f_b^{-1}(x))$, where $b \mapsto f_b^{-1}$ is a map $U \to \mathbf{L}(F^k, F^k)$. Therefore, the restriction $v\colon (p')^{-1}(U) \to p^{-1}(U)$ is continuous. This proves the theorem.

Finally, we observe that the fibre product $\xi_1 \oplus \xi_2$ of two vector bundles ξ_1 and ξ_2 over a space B is a vector bundle over B. The vector space structure on $q^{-1}(b) = p_1^{-1}(b) \times p_2^{-1}(b)$ is that of the direct sum of two vector spaces. If $h_1\colon U \times F^n \to p_1^{-1}(U)$ is a local chart of ξ_1 and if $h_2\colon U \times F^m \to p_2^{-1}(U)$ is a local chart of ξ_2, then $h_1 \oplus h_2\colon U \times F^{n+m} \to q^{-1}(U)$ is a local chart of $\xi_1 \oplus \xi_2$.

2.6 Definition. The Whitney sum of two vector bundles ξ_1 and ξ_2 over B, denoted $\xi_1 \oplus \xi_2$, is the fibre product of the underlying bundles ξ_1 and ξ_2 with the above vector bundle structure.

3. Induced Vector Bundles

In this section, we demonstrate that the results of Chap. 2, Sec. 5, apply to the category of vector bundles.

3.1 Proposition. *Let ξ be a k-dimensional vector bundle over B, and let $f\colon B_1 \to B$ be a map. Then $f^*(\xi)$ admits the structure of a vector bundle, and $(f_\xi, f)\colon f^*(\xi) \to \xi$ is a vector bundle morphism. Moreover, this structure is unique, and $f_\xi\colon p_1^{-1}(b_1) \to p^{-1}(b)$ is a vector space isomorphism.*

Proof. The fibre $p_1^{-1}(b_1)$ of $f^*(\xi) = (E_1, p_1, B_1)$ over $b_i \in B_1$ is $b_1 \times p^{-1}(f(b_1)) \subset E_1 \subset B_1 \times E$. For (b_1, x), $(b_1, x') \in p_1^{-1}(b_1)$, we require $(b_1, x) + (b_1, x') = (b_1, x + x')$ and $k(b_1, x) = (b_1, kx)$, where $k \in F$. Since $f_\xi(b_1, x) = x$, the restriction $f_\xi: p_1^{-1}(b_1) \to p^{-1}(b)$ is a linear isomorphism, and this requirement uniquely defines the vector space structure of $p_1^{-1}(b_1)$.

Finally, we exhibit the local triviality of $f^*(\xi)$. If $h: U \times F^k \to p^{-1}(U)$ is avector bundle isomorphism over U, then $h': f^{-1}(U) \times F^k \to p_1^{-1}(f^{-1}(U))$, where $h'(b_1, x) = (b_1, h(f(b_1), x))$, is a vector bundle isomorphism over $f^{-1}(U)$.

In connection with the factorization in 2(5.5), we observe that if $(u, f): \eta \to \xi$ is a vector bundle morphism then u factors as a composition $f_\xi v$, where $\eta \overset{v}{\to} f^*(\xi) \overset{f_\xi}{\to} \xi$, $v(y) = (p_\eta(y), u(y))$, and $f_\xi(b_1, x) = x$. Moreover, v is a vector bundle morphism over $B(\eta)$. In view of Theorem (2.5), the $B(\eta)$-morphism v is an isomorphism if and only if v is an isomorphism on each fibre, which, in turn, is equivalent to u being a fibrewise isomorphism; that is, $u: p_\eta^{-1}(b) \to p_\xi^{-1}(f(b))$ is an isomorphism for each $b \in B(\eta)$.

We formulate this result in the following statement.

3.2 Theorem. *Let ξ and η be two vector bundles. For a map $f: B(\eta) \to B(\xi)$, the vector bundles η and $f^*(\xi)$ are $B(\eta)$-isomorphic if and only if there exists a morphism $(u, f): \eta \to \xi$ such that u is an isomorphism on each fibre of η.*

If $u: \xi \to \eta$ is a B-morphism of vector bundles and if $f: B_1 \to B$ is a map, then $f^*(u): f^*(\xi) \to f^*(\eta)$ is a B_1-morphism of vector bundles. This is seen immediately from the formula $f^*(u)(b_1, x) = (b_1, u(x))$; that is, the linearity of u over $f(b_1)$ implies the linearity of $f^*(u)$ over b_1. Therefore, $f^*: \mathbf{VB}_B \to \mathbf{VB}_{B_1}$ is a functor. Let $g: B_2 \to B_1$ and $f: B_1 \to B$ be two maps, and let ξ be a vector bundle over B. Then, as vector bundles, $1^*(\xi)$ and ξ are B-isomorphic, and $g^*(f^*(\xi))$ and $(fg)^*(\xi)$ are B_2-isomorphic.

The above results apply to the restriction of a vector bundle ξ to a subspace $A \subset B(\xi)$.

4. Homotopy Properties of Vector Bundles

The first two lemmas concerning vector bundles are the analogues of Exercise 5 in Chap. 2 and the first step in the proof of 2(7.1) under the second hypothesis.

4.1 Lemma. *Let $\xi = (E, p, B)$ be a vector bundle of dimension k over $B = B_1 \cup B_2$, where $B_1 = A \times [a, c]$ and $B_2 = A \times [c, b]$, $a < c < b$. If $\xi|B_1 = (E_1, p_1, B_1)$ and $\xi|B_2 = (E_2, p_2, B_2)$ are trivial, ξ is trivial.*

Proof. Let $u_i: B_i \times F^k \to E_i$ be a B_i-isomorphism for $i = 1, 2$, and let $v_i = u_i|((B_1 \cap B_2) \times F^k)$, $i = 1, 2$. Then $h = v_2^{-1} v_1$ is an $A \times \{c\}$-isomorphism of trivial bundles, and therefore h has the form $h(x, y) = (x, g(x)y)$, where $(x, y) \in$ $(B_1 \cap B_2) \times F^k$ and $g: A \to GL(k, F)$ is a map. We prolong h to a B_2-isomorphism $w: B_2 \times F^k \to B_2 \times F^k$ by the formula $w(x, t, y) = (x, t, g(x)y)$ for each $x \in A$, $y \in F^k$, and $t \in [c, b]$. Then the bundle isomorphisms $u_1: B \times F^k \to E_1$ and $u_2 w: B_2 \times F^k \to E_2$ are equal on $(B_1 \cap B_2) \times F^k$, which is a closed set. Therefore, there exists an isomorphism $u: B \times F^k \to E$ with $u|B_1 \times F^k = u_1$ and $u|B_2 \times F^k = u_2 w$.

4.2 Lemma. *Let ξ be a vector bundle over $B \times I$. Then there exists an open covering $\{U_i\}$, $i \in I$, of B such that $\xi|(U_i \times I)$ is trivial.*

Proof. For each $b \in B$ and $t \in I$ there is an open neighborhood $U(t)$ of b in B and $V(t)$ of t in $[0, 1]$ such that $\xi|(U(t) \times V(t))$ is trivial. Therefore, by the compactness of $[0, 1]$, there exist a finite sequence of numbers $0 = t_0 < t_1 < \cdots < t_n = 1$ and open neighborhoods $U(i)$ of b in B such that $\xi|(U(i) \times [t_{i-1}, t_i])$ is trivial for $1 \leq i \leq n$. Let $U = \bigcap_{1 \leq i \leq n} U(i)$. Then the bundle $\xi|(U \times [0, 1])$ is trivial by an application of Lemma (4.1) $n - 1$ times. Therefore, there is an open covering $\{U_i\}$, $i \in I$, of B such that $\xi|(U_i \times I)$ is trivial.

The next theorem is the first important step in the development of the homotopy properties of vector bundles.

4.3 Theorem. *Let $r: B \times I \to B \times I$ be defined by $r(b, t) = (b, 1)$ for $(b, t) \subset B \times I$, and let $\xi^k = (E, p, B \times I)$ be a vector bundle over $B \times I$, where B is a paracompact space. There is a map $u: E \to E$ such that $(u, r): \xi \to \xi$ is a morphism of vector bundles and u is an isomorphism on each fibre.*

Proof. Let $\{U_i\}$, $i \in I$, be a locally finite open covering of B such that $\xi|(U_i \times I)$ is trivial. This covering exists by (4.2) and the paracompactness of B. Let $\{\eta_i\}$, $i \in I$, be an envelope of unity subordinate to the open covering $\{U_i\}$, $i \in I$, that is, the support of η_i is a subset of U_i and $1 = \max_{i \in I} \eta_i(b)$ for each $b \in B$. Let $h_i: U_i \times I \times F^k \to p^{-1}(U_i \times I)$ be a $(U_i \times I)$-isomorphism of vector bundles.

We define a morphism $(u_i, r_i): \xi \to \xi$ by the relations $r_i(b, t) = (b, \max(\eta_i(b), t))$, u_i is the identity outside $p^{-1}(U_i \times I)$, and $u_i(h_i(b, t, x)) = h_i(b, \max(\eta_i(b), t), x)$ for each $(b, t, x) \in U_i \times I \times F^k$. We well order the set I. For each $b \in B$, there is an open neighborhood $U(b)$ of b such that $U_i \cap U(b)$ is nonempty for $i \in I(b)$, where $I(b)$ is a finite subset of I. On $U(b) \times I$, we define $r = r_{i(n)} \cdots r_{i(1)}$, and on $p^{-1}(U(b) \times I)$, we define $u = u_{i(n)} \cdots u_{i(1)}$, where $I(b) = \{i(1), \ldots, i(n)\}$ and $i(1) < i(2) < \cdots < i(n)$. Since r_i on $U(b) \times I$ and u_i on $p^{-1}(U(b) \times I)$ are identities for $i \notin I(b)$, the maps r and u are infinite compositions of maps where all but a finite number of terms are identities near a point. Since each u_i is an isomorphism on each fibre, the composition u is an isomorphism on each fibre.

4.4 Corollary. *With the notations of Theorem* (4.3), $\xi \cong r^*(\xi|(B \times 1))$ *over* $B \times I$.

Proof. This result is a direct application of Theorem (3.2) to Theorem (4.3).

Let $\xi = (E, p, B)$ be a vector bundle, and let Y be a space. We use the notation $\xi \times Y$ for the vector bundle $(E \times Y, p \times 1_Y, B \times Y)$. The fibre over $(b, y) \in B \times Y$ is $p^{-1}(b) \times y$, which has a natural vector space structure that it derives from $p^{-1}(b)$. If $h: U \times F^k \to p^{-1}(U)$ is a U-isomorphism, the $h \times 1_Y$: $U \times Y \times F^k \to p^{-1}(U) \times Y = (p \times 1_Y)^{-1}(U \times Y)$ is a $(U \times Y)$-isomorphism. Consequently, $\xi \times Y$ is a vector bundle, and this leads to the following version of (4.3).

4.5 Corollary. *With the notations of Theorem* (4.3),

$$\xi \cong (\xi|(B \times 1)) \times I$$

are vector bundles over $B \times I$.

Proof. For this, it suffices to observe that $r^*(\xi|(B \times 1)) = (\xi|B \times 1) \times I$. In both cases the total space of the bundles is the subspace of $(b, t, x) \in B \times I \times E(\xi|(B \times 1))$ such that $(b, 1) = p(x)$, and the projection is the map $(b, t, x) \mapsto (b, t)$.

4.6 Corollary. *With the notations of Theorem* (4.3), *there exists, after restriction, an isomorphism* $(u, r): \xi|(B \times 0) \to \xi|(B \times 1)$.

Proof. This is a direct application of Theorem (2.5) to the situation described in (4.3) where $r = 1$ on $B \times 0 = B \times 1 = B$.

Finally, we have the following important application of (4.6) in the framework of homotopy theory.

4.7 Theorem. *Let* $f, g: B \to B'$ *be two homotopic maps, where* B *is a paracompact space, and let* ξ *be a vector bundle over* B'. *Then* $f^*(\xi)$ *and* $g^*(\xi)$ *are* B-*isomorphic.*

Proof. Let $h: B \times I \to B'$ be a map with $h(x, 0) = f(x)$ and $h(x, 1) = g(x)$. Then $f^*(\xi) \cong h^*(\xi)|(B \times 0)$ over B, and $g^*(\xi) \cong h^*(\xi)|(B \times 1)$ over B. By (4.6), $h^*(\xi)|(B \times 0)$ and $h^*(\xi)|(B \times 1)$ are B-isomorphic, and, therefore, $f^*(\xi)$ and $g^*(\xi)$ are B-isomorphic.

4.8 Corollary. *Every vector bundle over a contractible paracompact space* B *is trivial.*

Proof. Let $f: B \to B$ be the identity, and let $g: B \to B$ be a constant map. For each vector bundle ξ over B, $f^*(\xi)$ is B-isomorphic to ξ, and $g^*(\xi)$ is B-

isomorphic to the product bundle $(B \times F^k, p, B)$. Since f and g are homotopic, ξ is isomorphic to the product bundle $(B \times F^k, p, B)$, by (4.7).

Theorem (4.7) is the first of the three main theorems on the homotopy classification of vector bundles.

5. Construction of Gauss Maps

5.1 Definition. A *Gauss map* of a vector bundle ξ^k in F^m $(k \le m \le +\infty)$ is a map $g: E(\xi^k) \to F^m$ such that g is a linear monomorphism when restricted to any fibre of ξ.

Recall that $E(\gamma_k^m)$ is the subspace of $(V, x) \in G_k(F^m) \times F^m$ with $x \in V$. Then the projection $q: E(\gamma_k^m) \to F^m$, given by the relation $q(V, x) = x$, is a Gauss map. In the next proposition, we see that every Gauss map can be constructed from this map and vector bundle morphisms.

5.2 Proposition. *If* $(u, f): \xi^k \to \gamma_k^m$ *is a vector bundle morphism that is an isomorphism when restricted to any fibre of* ξ^k, *then* $qu: E(\xi^k) \to F^m$ *is a Gauss map. Conversely, if* $g: E(\xi^k) \to F^m$ *is a Gauss map, there exists a vector bundle morphism* $(u, f): \xi^k \to \gamma_k^m$ *such that* $qu = g$.

Proof. The first statement is clear. For the second, let $f(b) = g(p^{-1}(b)) \in G_k(F^m)$, and let $u(x) = (f(p(x)), g(x)) \in E(\gamma_k^m)$ for $x \in E(\xi^k)$. We see that f is continuous by looking at a local coordinate of ξ, and from this u is also continuous.

5.3 Corollary. *There exists a Gauss map* $g: E(\xi) \to F^m$ $(k \le m \le +\infty)$ *if and only if* ξ *is* $B(\xi)$-*isomorphic with* $f^*(\gamma_k^m)$ *for some map* $f: B(\xi) \to G_k(F^m)$.

Proof. This follows from Proposition (5.2) and Theorem (3.2).

In Theorem (5.5), we construct a Gauss map for each vector bundle over a paracompact space. First, we need a preliminary result concerning the open sets over which a vector bundle is trivial.

5.4 Proposition. *Let* ξ *be a vector bundle over a paracompact space* B *such that* $\xi | U_i$, $i \in I$, *is trivial, where* $\{U_i\}$, $i \in I$, *is an open covering. Then there exists a countable open covering* $\{W_j\}$, $1 \le j$, *of* B *such that* $\xi | W_j$ *is trivial. Moreover, if each* $b \in B$ *is a member of at most* n *sets* U_i, *there exists a finite open covering* $\{W_j\}$, $1 \le j \le n$, *of* B *such that* $\xi | W_j$ *is trivial.*

Proof. By paracompactness, let $\{\eta_i\}$, $i \in I$, be a partition of unity with $V_i = \eta_i^{-1}(0, 1] \subset U_i$. For each $b \in B$, let $S(b)$ be the finite set of $i \in I$ with $\eta_i(b) > 0$. For each finite subset $S \subset I$, let $W(S)$ be the open subset of all $b \in B$ such that $\eta_i(b) > \eta_j(b)$ for each $i \in S$ and $j \notin S$.

If S and S' are two distinct subsets of I each with m elements, then $W(S) \cap W(S')$ is empty. In effect, there exist $i \in S$ with $i \notin S'$ and $j \in S'$ with $j \notin S$. For $b \in W(S)$ we have $\eta_i(b) > \eta_j(b)$, and for $b \in W(S')$ we have $\eta_j(b) > \eta_i(b)$. Therefore, $W(S) \cap W(S')$ is empty.

Let W_m be the union of all $W(S(b))$ such that $S(b)$ has m elements. Since $i \in S(b)$ yields the relation $W(S(b)) \subset V_i$, the bundle $\xi|W(S(b))$ is trivial, and since W_m is a disjoint union, $\xi|W_m$ is trivial. Finally, under the last hypothesis, W_j is empty for $n < j$.

5.5 Theorem. *For each vector bundle ξ^k over a paracompact space B there is a Gauss map $g: E(\xi) \to F^\infty$. Moreover, if B has an open covering of sets $\{U_i\}$, $1 \leq i \leq n$, such that $\xi|U_i$ is trivial, ξ has a Gauss map $g: E(\xi) \to F^{kn}$.*

Proof. Let $\{U_i\}$ be the countable or finite open covering of B such that $\xi|U_i$ is trivial, let $h_i: U_i \times F^k \to \xi|U_i$ be U_i-isomorphisms, and let $\{\eta_i\}$ be a partition of unity with closure of $\eta_i^{-1}((0,1]) \subset U_i$. We define $g: E(\xi) \to \sum_i F^k$ as

$g = \sum_i g_i$, where $g_i|E(\xi|U_i)$ is $(\eta_i p)(p_2 h_i^{-1})$ and $p_2: U \times F^k \to F^k$ is the projection on the second factor. Outside $E(\xi|U_i)$, the map g_i is zero.

Since each $g_i: E(\xi) \to F^k$ is a monomorphicm on the fibres of $E(\xi)$ over b with $\eta_i(b) > 0$, and since the images of g_i are in complementary subspaces, the map g is a Gauss map. In general, $\sum_i F^k$ is F^∞, but if there are only n sets U_i, then $\sum_i F^k$ is F^{kn}.

Theorem (5.5) with Corollary (5.6) is the second main homotopy classification theorem for vector bundles.

5.6 Corollary. *Every vector bundle ξ^k over a paracompact space B is B-isomorphic to $f^*(\gamma_k)$ for some $f: B \to G_k(F^\infty)$.*

The following concept was suggested by Theorem (5.5).

5.7 Definition. A vector bundle ξ is of finite type over B provided there exists a finite open covering $U_1, \ldots, U_n$ of B such that $\xi|U_i$ is trivial, $1 \leq i \leq n$.

In the next theorem we derive other formulations of the notion of finite type. By 1(2.6) and (4.8) every vector bundle over a finite-dimensional CW-complex is of the finite type.

5.8 Proposition. *For a vector bundle ξ over a space B, the following are equivalent.*

(1) *The bundle ξ is of the finite type.*
(2) *There exists a map $f: B \to G_k(F^m)$ for some m such that $f^*(\gamma_k^m)$ and ξ are B-isomorphic.*
(3) *There exists a vector bundle η over B such that $\xi \oplus \eta$ is trivial.*

Proof. By the construction in (5.5), statement (1) implies (2). Since $\gamma_k^m \oplus *\gamma_k^m$ is trivial over $G_k(F^m)$, then $f^*(\gamma_k^m) \oplus f^*(*\gamma_k^m)$ and θ^m are B-isomorphic. Let η be $f^*(*\gamma_k^m)$. Since $f^*(\gamma_k^m \oplus *\gamma_k^m)$ is trivial, the bundle $\xi \oplus \eta$ is trivial. Finally, the composition $E(\xi) \to E(\xi \oplus \eta) \to B \times F^m \to F^m$ is a Gauss map.

6. Homotopies of Gauss Maps

Let F^{ev} denote the subspace of $x \in F^\infty$ with $x_{2i+1} = 0$, and F^{odd} with $x_{2i} = 0$ for $i \geq 0$. For these subspaces, $F^\infty = F^{ev} \oplus F^{odd}$. Two homotopies $g^e: F^n \times I \to F^{2n}$ and $g^o: F^n \times I \to F^{2n}$ are defined by the following formulas:

$$g_t^e(x_0, x_1, x_2, \ldots) = (1 - t)(x_0, x_1, x_2, \ldots) + t(x_0, 0, x_1 0, x_2, \ldots)$$

$$g_t^o(x_0, x_1, x_2, \ldots) = (1 - t)(x_0, x_1, x_2, \ldots) + t(0, x_0, 0, x_1 0, x_2, \ldots)$$

The properties of these homotopies are contained in the following proposition. In the above formulas and in the next proposition, we have $1 \leq n \leq +\infty$.

6.1 Proposition. *With the above notations, these homotopies have the following properties:*

(1) *The maps g_0^e and g_0^o each equal the inclusion $F^n \to F^{2n}$.*
(2) *For $t = 1$, $g_1^e(F^n) = F^{2n} \cap F^{ev}$ and $g_1^o(F^n) = F^{2n} \cap F^{odd}$.*
(3) *There are vector bundle morphisms $(u^e, f^e): \gamma_{k^n} \to \gamma_k^{2n}$ and $(u^o, f^o): \gamma_{k^n} \to \gamma_k^{2n}$ such that $qu^e = g_1^e$, $qy^o = g_1^o$.*
(4) *f^e and f^o are homotopic to the inclusion $G_k(F^n) \to G_k(F^{2n})$.*

Proof. Statements (1) and (2) follow immediately from the formulas for g_t^e and g_t^o. For (3), we use (5.2). Finally, the homotopies g_t^e and g_t^o define homotopies of f^e and f^o with 1.

The next theorem describes to what extent Gauss maps are unique in terms of homotopy properties of their associated bundle morphisms. We use the above notations.

6.2 Theorem. *Let $f, f_1: B \to G_k(F^n)$ be two maps such that $f^*(\gamma_{k^n})$ and $f_1^*(\gamma_{k^n})$ are B-isomorphic and let $j: G_k(F^n) \to G_k(F^{2n})$ be the natural inclusion. Then the maps $j\, f$ and $j\, f_1$ are homotopic for $1 \leq n \leq +\infty$.*

Proof. By hypothesis, there is a vector bundle ξ over B and two morphisms $(u, f): \xi \to \gamma_{k^n}$ and $(u_1, f_1): \xi \to \gamma_{k^n}$ which are isomorphisms when restricted to the fibres of ξ. Let $g = qu: E(\xi) \to F^n$ and $g_1 = qu_1: E(\xi) \to F^n$ be the associated Gauss maps. Composing with the above maps, we have morphisms $(u^e u, f^e f): \xi \to \gamma_k^{2n}$ with a Gauss map $g_1^e g: E(\xi) \to F^{ev} \cap F^{2n}$ and $(u^o u, f^o f)$:

$\xi \to \gamma_k^{2n}$ with a Gauss map $g_1^o g_1 \colon E(\xi) \to F^{\mathrm{odd}} \cap F^{2n}$. We define a Gauss map $h \colon E(\xi) \times I \to F^{2n}$ by the relation $h_t(x) = (1 - t)g_1^e g(x) + t g_1^o g_1(x)$. For a fibre $p^{-1}(b) \subset E(\xi)$, the linear maps $g_1^e g \colon p^{-1}(b) \to F^{\mathrm{ev}}$ and $g_1^o g_1 \colon p^{-1}(b) \to F^{\mathrm{odd}}$ are monomorphisms, and since $F^{\mathrm{ev}} \cap F^{\mathrm{odd}} = 0$, the map $h_t \colon p^{-1}(b) \to F^{2n}$ is a linear monomorphism. Therefore, there is a Gauss map $h \colon E(\xi) \times I \to F^{2n}$ which determines a bundle morphism $(w, k) \colon \xi \to \gamma_k^{2n}$. The map $k \colon B \times I \to G_k(F^{2n})$ is a homotopy from $f^e f$ to $f^o f_1$. Since f and $f^e f$ are homotopic and $f^o f_1$ and f_1 are homotopic, f and f_1 are homotopic. This proves the theorem.

Theorem (6.2) is the third of the three main homotopy classification theorems.

7. Functorial Description of the Homotopy Classification of Vector Bundles

Let **P** denote the category paracompact spaces and homotopy classes of maps. Let **ens** denote, as usual, the category of sets and functions.

Let $\mathrm{Vect}_k(B)$ denote the set of B-isomorphism classes of k-dimensional vector bundles over B. For a k-dimensional bundle ξ, we denote by $\{\xi\}$ the class in $\mathrm{Vect}_k(B)$ determined by ξ. If $[f] \colon B_1 \to B$ is a homotopy class of maps between paracompact spaces, we define a function $\mathrm{Vect}_k([f]) \colon \mathrm{Vect}_k(B) \to \mathrm{Vect}_k(B_1)$ by the relation $\mathrm{Vect}_k([f])(\{\xi\}) = \{f^*(\xi)\}$. By the remarks at the end of Sec. 3 and Theorem (4.7), $\mathrm{Vect}_k([f])$ is a well-defined function.

7.1 Proposition. *The family of functions* $\mathrm{Vect}_k \colon \mathbf{P} \to \mathbf{ens}$ *is a cofunctor.*

Proof. Since $1^*(\xi)$ and ξ are B-isomorphic, the function $\mathrm{Vect}_k([1])$ is the identity. If $[f] \colon B_1 \to B$ and $[g] \colon B_2 \to B_1$ are two homotopy classes of maps, $g^*(f^*(\xi))$ and $(fg)^*(\xi)$ are B_2-isomorphic. Consequently, $\mathrm{Vect}_k([f][g]) = \mathrm{Vect}_k([g]) \mathrm{Vect}_k([f])$, and Vect_k satisfies the axioms for being a cofunctor.

For each B, we define a function $\phi_B \colon [B, G_k(F^\infty)] \to \mathrm{Vect}_k(B)$ by the relation $\phi_B([f]) = \{f^*(\gamma_k)\}$. Again by Theorem (4.7), ϕ_B is a well-defined function. The next theorem, together with the definition of Vect_k and ϕ_B, brings together all aspects of the homotopy classification theory of vector bundles.

7.2 Theorem. *The family* ϕ *of functions* ϕ_B *defines an isomorphism of cofunctors* $\phi \colon [-, G_k(F^\infty)] \to \mathrm{Vect}_k$.

Proof. First, we prove that ϕ is a morphism of cofunctors. For this, let $[f] \colon B_1 \to B$ be a homotopy class of maps. Then the following diagram is commutative.

$$[B, G_k(F^\infty)] \xrightarrow{\phi_B} \mathrm{Vect}_k(B)$$

$$[[f], G_k(F^\infty)] \downarrow \qquad\qquad \downarrow \mathrm{Vect}_k([f])$$

$$[B_1, G_k(F^\infty)] \xrightarrow{\phi_{B_1}} \mathrm{Vect}_k(B_1)$$

In effect, if $[g] \in [B, G_k(F^\infty)]$, we have

$$\mathrm{Vect}_k([f])\phi_B([g]) = \mathrm{Vect}_k([f])\{g^*(\gamma_k)\} = \{f^*g^*(\gamma_k)\}$$

and $\phi_{B_1}[[f], G_k(F^\infty)][g] = \phi_{B_1}([g][f]) = \{(gf)^*(\gamma_k)\}$.

Finally, ϕ is an isomorphism because each ϕ_B is a bijection. The function ϕ_B is surjective by (5.5) and (5.6), and it is injective by (6.2). This proves the theorem.

7.3 The isomorphism $\phi: [-, G_k(F^\infty)] \to \mathrm{Vect}_k$ is called a corepresentation of the cofunctor Vect_k. The preceding four sections have been dedicated to proving that the cofunctor Vect_k is corepresentable. In this way the problem of classifying vector bundles, i.e., of computing $\mathrm{Vect}_k(B)$, has been reduced to the calculation of sets of homotpy classes of maps, i.e., the sets $[B, G_k(F^\infty)]$.

8. Kernel, Image, and Cokernel of Morphisms with Constant Rank

Let $u: \xi \to \eta$ be a morphism of vector bundles over B. We define three bundles $\ker u$, which is a subbundle of ξ; $\mathrm{im}\, u$, which is a subbundle of η; and $\mathrm{coker}\, u$, which is a quotient bundle of η. The total space of $\ker u$ is the subspace of $x \in E(\xi)$ such that $u(x) = 0$ in η over $p_\xi(x)$. The total space of $\mathrm{im}\, u$ is the subspace of $u(x)$, $x \in E(\xi)$. The total space of $\mathrm{coker}\, u$ is the quotient space of $E(\eta)$ by the following relation: y, $y' \in E(\eta)$ are related provided $p_\eta(y') = p_\eta(y)$ and $y - y' = u(x)$ for some $x \in E(\xi)$. The projection of η factors through $E(\mathrm{coker}\, u)$ to define the projection of $\mathrm{coker}\, u$.

In general, $\ker u$ and $\mathrm{coker}\, u$ are *not* vector bundles because they do not satisfy the property of local triviality. The following example illustrates this: Let $u: [0, 1] \times \mathbf{R} \to [0, 1] \times \mathbf{R}$ be the $[0, 1]$-morphism defined by $u(t, x) = (t, tx)$. Then $\ker(u)_b = 0$ for $b \neq 0$, and $\ker(u)_0 = \mathbf{R}$ for $b = 0$ and $\mathrm{im}(u)_0 = 0$, and $\mathrm{coker}(u)_b = 0$ for $b \neq 0$, and $\mathrm{coker}(u)_0 = \mathbf{R}$. In the next definition we describe those vector bundle morphisms u, where $\ker u$, $\mathrm{im}\, u$, and $\mathrm{coker}\, u$ are vector bundles.

Recall that the rank of a linear transformation $f: V \to W$ is $\dim V - \dim\{\ker f\}$ which equals $\dim\{\mathrm{im}\, f\}$.

8.1 Definition. Let $u: \xi \to \eta$ be a B-morphism. Then u is of constant rank k provided $u_b: p^{-1}(b) \to p^{-1}(b)$ is of rank k for each $b \in B$.

8.2 Theorem. *Let* $u: \xi^n \to \eta^m$ *be a B-morphism of vector bundles of constant rank* k. *Then* $\ker u$, $\operatorname{im} u$, *and* $\operatorname{coker} u$ *are vector bundles over* B.

Proof. Since the statement refers to a local question, u, ξ^n, and η^m can be restricted to a coordinate neighborhood. Consequently, there is a B-morphism $u: B \times F^n \to B \times F^m$, and it has the form $u(b, x) = (b, u_b(x))$, where $b \mapsto u_b$ is a map $B \to \mathbf{L}(F^n, F^m)$. For each $b \in B$, the rank of u_b is k.

At $a \in B$, $u_a: F^n = V_1 \oplus V_2 \to F^m = W_1 \oplus W_2$, where $V_2 = \ker u_a$, $W_1 = \operatorname{im} u_a$, $\dim V_1 = \dim W_1 = k$, $\dim V_2 = n - k$, and $\dim W_2 = m - k$. For each $b \in B$, we define

$$V = F^n \oplus W_2 = V_1 \oplus V_2 \oplus W_2 \overset{w_b}{\to} W_1 \oplus W_2 \oplus V_2 = F^m \oplus V_2 = W$$

by the requirement that $w_b | V_1$ be $(u_b | V_1) \oplus 0$, $w_b | V_2$ be $(u_b | V_2) \oplus 1_{V_2}$, and $w_b | W_2$ be $0 \oplus 1_{W_2} \oplus 0$. Since $u_a | V_1: V_1 \to W_1$ is an isomorphism, the linear transformation w_a is an isomorphism. Since the isomorphisms form an open subset of $\mathbf{L}(V, W)$ and since $b \mapsto w_b$ is continuous, the function w_b is a linear isomorphism for each $b \in U$, where U is some open neighborhood of a. Let $v_b: W \to V$ be the inverse of w_b for each $b \in U$. Then $b \mapsto v_b$ is continuous.

First, we prove the triviality of $\ker u | U$. We observe that $(x_1, x_2) \in \ker u_b$ if and only if $w_b(x_1, x_2, 0) = x_2$, $x_2 \in V_2$; that is, $(x_1, x_2) = v_b(x_2)$, and we have $\ker u_b = v_b(V_2)$. Therefore, the map $(b, x_2) \mapsto (b, v_b(x_2))$ is a U-isomorphism $U \times V_2 \to E(\ker u | U)$ with inverse $(b, x) \mapsto (b, w_b(x))$.

Next, we prove the triviality of $\operatorname{im} u | U$. For this, we observe that $u_b(x_1) = 0$ if and only if $w_b(x_1) = 0$ for $x_1 \in V_1$. Therefore, $u_b | V_1: V_1 \to \operatorname{im} u_b$ is an isomorphism for each $b \in U$ since w_b is a monomorphism. Therefore, $(b, x) \mapsto (b, u_b(x))$ is an isomorphism $U \times V_1 \to E(\operatorname{im} u | U)$ with inverse $(b, y) \mapsto (b, v_b(y))$ since $u_b | V_1 = w_b | V_1$ for each $b \in U$.

Finally, we prove the triviality of $\operatorname{coker} u | U$. For this, we observe that $\operatorname{im} u_b \cap W_2 = 0$. For $u_b(x_1, x_2) \in \operatorname{im} u_b \cap W_2$, we have $w_b(x_1, x_2, y) = 0$, and since w_b is injective, we have $x_1 = x_2 = y = 0$. Then the quotient map $(b, y) \to (b, y \bmod (\operatorname{im} u))$ is a monomorphism $U \times W_2 \to E(\operatorname{coker} u | U)$, and, for reasons of dimension, it is a U-isomorphism. Its inverse U-morphism is the factorization of the projection $U \times (W_1 \oplus W_2) \to U \times W_2$ through $\operatorname{coker} u$. This proves the theorem.

8.3 Corollary. *Let* $u: \xi^n \to \eta^m$ *be a B-morphism that is injective, or, equivalently, it is a monomorphism on each fibre of* ξ. *Then* $\operatorname{im} u$ *and* $\operatorname{coker} u$ *are vector bundles.*

Proof. The B-morphism has constant rank n.

8.4 Corollary. *Let* $u: \xi^n \to \eta^m$ *be a B-morphism that is surjective, or, equivalently, it is an epimorphism on each fibre of* ξ. *Then* $\ker u$ *is a vector bundle.*

Proof. The B-morphism has constant rank m.

8.5 Remark. The usual terminology of exact sequences carries over to vector bundles and morphisms of constant rank. A sequence $\xi \to \eta \to \zeta$ is exact provided $\operatorname{im} u = \ker v$. A sequence

$$\xi_0 \overset{u_1}{\to} \xi_1 \overset{u_2}{\to} \xi_2 \to \cdots \to \xi_{n-1} \overset{u_n}{\to} \xi_n$$

is exact provided $\operatorname{im} u_i = \ker u_{i+1}$ for each i, $1 \leq i \leq n - 1$. If $0 \to \xi \overset{u}{\to} \eta \overset{v}{\to} \zeta \to 0$ is an exact sequence, ξ and $\ker v$ are isomorphic, and ζ and $\operatorname{coker} v$ are isomorphic. This is called a short exact sequence. The bundle 0 is $(B, 1, B)$. A B-morphism $u: \xi \to \eta$ is called a B-monomorphism when $\ker u = 0$ or, equivalently, when u is injective; it is called a B-epimorphism when $\operatorname{coker} u = 0$ or, equivalently, when u is surjective.

9. Riemannian and Hermitian Metrics on Vector Bundles

If $x \in \mathbf{R}$, let $\bar{x} = x$, and if $z = x + iy \in \mathbf{C}$, let $\bar{z} = x - iy$. Let F denote either $\mathbf{R}$ or $\mathbf{C}$.

9.1 Definition. Let V be a vector space over F. An inner product on V is a function $\beta: V \times V \to F$, the field of scalars, such that the following axioms hold.

(1) $\beta(ax + a'x', y) = a\beta(x, y) + a'\beta(x', y)$
    $\beta(x, by + b'y') = \bar{b}\beta(x, y) + \bar{b}'\beta(x, y')$
    for each $x, x', y, y' \in V$ and $a, a', b, b' \in F$.
(2) $\beta(x, y) = \overline{\beta(y, x)}$ for each $x, y \in V$.
(3) $\beta(x, x) \geq 0$ in $\mathbf{R}$ and $\beta(x, x) = 0$ if and only if $x = 0$.

With an inner product β on V we can define what it means for $x \in V$ and $y \in V$ to be perpendicular, namely, $\beta(x, y) = 0$. Let W be a subspace of V, and let W^0 be the set of all $y \in V$ with $\beta(x, y) = 0$ for each $x \in W$. Then W^0 is easily proved to be a subspace of V, and $V = W \oplus W^0$.

On $\mathbf{R}^n$ and $\mathbf{C}^n$ there is a natural inner product, the euclidean inner product, which is given by $\beta(x, y) = \sum_{1 \leq i \leq n} x_i \bar{y}_i$. These formulas hold for $\mathbf{R}^\infty$ and $\mathbf{C}^\infty$.

We extend the notion of inner product to vector bundles in the next definition.

9.2 Definition. Let ξ be a real or complex vector bundle over B. A riemannian metric or hermitian metric on ξ is a function $\beta: E(\xi \oplus \xi) \to F$ such that, for each $b \in B$, $\beta|(p^{-1}(b) \times p^{-1}(b))$ is an inner product on $p^{-1}(b)$. The riemannian metric refers to $\mathbf{R} = F$ and the hermitian metric to $\mathbf{C} = F$.

9.3 Example. Let θ^k be the product k-dimensional bundle over B. Then $\beta(b, x, x') = (x|x')$, the euclidean inner product, is a riemannian metric in the real case and a hermitian metric in the complex case.

9.4 Example. For $k \leq m \leq +\infty$, the canonical bundle γ_k^m over $G_k(F^m)$ has a natural riemannian metric derived from the euclidean metric on F^m. For $(V, x, x') \in E(\gamma_k^m \oplus \gamma_k^m)$, let $\beta(V, x, x') = (x|x')$, the euclidean inner product. On each fibre V the function β is clearly an inner product.

Using this construction and a Gauss map, we are able to get the next general result.

9.5 Theorem. *Every real or complex vector bundle with a Gauss map has a riemannian or hermitian metric.*

Proof. Let $g: E(\xi) \to F^\infty$ be a Gauss map. Define $\beta: E(\xi \oplus \xi) \to F$ by the relation $\beta(x, x') = (g(x)|g(x'))$. Then, since g is continuous and a monomorphism on each fibre, β is a riemannian or hermitian metric.

By (5.5) and (9.5), every bundle over a paracompact space has a metric.

Next, we consider the following application of metrics on vector bundles.

9.6 Theorem. *Let $0 \to \xi \overset{u}{\to} \eta \overset{v}{\to} \zeta \to 0$ be a short exact sequence of vector bundles over B; that is, u is a monomorphism, $\mathrm{im}\, u = \ker v$, and v is an epimorphism. Let β be a metric on η. Then there exists a morphism $w: \xi \oplus \zeta \to \eta$ splitting the above exact sequence in the sense that the following diagram is commutative.*

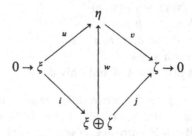

The morphism i is inclusion into the first factor, and j is projection onto the second factor.

Proof. Let ξ' denote $\mathrm{im}\, u$, where $E(\xi') \subset E(\eta)$. Let $E(\zeta')$ be the subset of $x' \in E(\eta)$ such that $\beta(x, x') = 0$ for all $x \in E(\xi')$ with $p_\eta(x) = p_\eta(x')$. Then ζ' is a subbundle of η consisting of vector spaces.

Let $E(\eta)_b$ denote the fibre of η, $E(\xi')_b$ the fibre of ξ', and $E(\zeta')_b$ the fibre of ζ' over $b \in B$.

Let $g: E(\eta) \to E(\xi')$ be the projection of $E(\eta)_b$ onto $E(\xi')_b$ over each $b \in B$. We wish to prove that g is continuous; for this, we consider g locally.

We suppose $u: B \times F^n \to B \times F^m$ is a B-monomorphism, and $\beta(b, x, x')$ is the metric on $B \times F^m$. Then $g: B \times F^m \to B \times F^n$ is given by $g(b, x) = \left(b, \sum_{1 \leq i \leq n} \beta(b, x, u(e_i))e_i \right)$, where $e_i, \ldots, e_n$ is the canonical basis of F^n. Clearly, g is continuous. Since $g: \eta \to \xi'$ is a B-epimorphism, $\ker g$ is a vector bundle, by (8.4). But $\ker g$ equals ζ'.

Clearly, $v|\zeta': \zeta' \to \zeta$ is a B-isomorphism since it is an isomorphism on the fibres. Finally, we define $w|\xi$ equal to the isomorphism $u: \xi \to \xi' \subset \eta$ and $w|\zeta$ equal to the isomorphism $(v|\zeta')^{-1}: \zeta \to \zeta' \subset \eta$. This proves the theorem.

Observe that (9.5) and (9.6) apply to vector bundles over a paracompact space by (5.5).

Exercises

1. Prove that a k-dimensional vector bundle ξ^k is trivial if and only if it has k cross sections $s_1, \ldots, s_k$ such that $s_1(b), \ldots, s_k(b)$ are linearly independent for each $b \in B$.

2. Prove that every metric on a vector bundle ξ with a Gauss map is of the form constructed in (9.5).

3. Define the bundle dimension of a space B, denoted $\dim_b B$, to be the inf of all k such that the inclusion $G_m(\mathbf{R}^{2m}) \to G_m(\mathbf{R}^\infty)$ induces a bijection $[B, G_m(\mathbf{R}^{m+n})] \to [B, G_m(\mathbf{R}^\infty)]$ for all $m, n \geq k$. Let $k = \dim_b B$ in what follows.
   (a) Prove that the natural inclusion $G_k(\mathbf{R}^{2k}) \to G_k(\mathbf{R}^{2k+m})$ induces a bijection $[B, G_k(\mathbf{R}^{2k})] \to [B, G_k(\mathbf{R}^{2k+m})]$.
   (b) Prove that the natural inclusion (see Exercise 2 of Chap. 2) $G_k(\mathbf{R}^{2k}) \to G_m(\mathbf{R}^{k+m})$ induces a bijection $[B, G_k(\mathbf{R}^{2k})] \to [B, G_m(\mathbf{R}^{k+m})]$.
   (c) If ξ^n is an n-dimensional vector bundle over B, a paracompact space, with $n \geq k$, prove that there exists a k-dimensional bundle η such that ξ^n and $\eta^k \oplus \theta^{n-k}$ are B-isomorphic and, moreover, that η is unique up to B-isomorphism.
   (d) If ξ and η are two bundles with $\dim \xi = \dim \eta \geq k$ such that $\xi \oplus \theta^m$ and $\eta \oplus \theta^m$ are isomorphic, prove that ξ and η are isomorphic (over B).
   (e) If ξ is a vector bundle over B, prove that there exists a k-dimensional vector bundle η over B such that $\xi \oplus \eta$ is trivial.

CHAPTER 4

General Fibre Bundles

A fibre bundle is a bundle with an additional structure derived from the action of a topological group on the fibres. In the next chapter the notions of fibre bundle and vector bundle are related. As with vector bundles, a fibre bundle has so much structure that there is a homotopy classification theorem for fibre bundles.

1. Bundles Defined by Transformation Groups

1.1 Definition. A topological group G is a set G together with a group structure and topology on G such that function $(s, t) \mapsto st^{-1}$ is a map $G \times G \to G$.

   The above condition of continuity on $(s, t) \mapsto st^{-1}$ is equivalent to the statement that $(s, t) \mapsto st$ is a map $G \times G \to G$ and $s \mapsto s^{-1}$ is a map $G \to G$.

Examples. The real line $\mathbf{R}$ with addition as group operation, the real line minus zero $\mathbf{R} - \{0\}$ with multiplication as group operation, the full linear groups of nonsingular matrices $GL(n, \mathbf{R})$ and $GL(n, \mathbf{C})$, and the orthogonal, unitary, and symplectic groups $O(n)$, $SO(n)$, $U(n)$, $SU(n)$, and $Sp(n)$ are all topological groups. The orthogonal, unitary, and symplectic groups are considered in detail in Chap. 7.

1.2 Definition. For a topological group G, a right G-space is a space X together with a map $X \times G \to X$. The image of $(x, s) \in X \times G$ under this map is xs. We assume the following axioms.

(1) For each $x \in X$, $s, t \in G$, the relation $x(st) = (xs)t$ holds.
(2) For each $x \in X$, the relation $x1 = x$ holds, where 1 is the identity of G.

A space X is a left G-space provided there is given a map $G \times X \to X$ such that $(st)x = s(tx)$ and $1x = x$ for $s, t \in G$ and $x \in X$.

If X is a left G-space, $xs = s^{-1}x$ defines a right G-space structure on X. Since there is a bijective correspondence between left and right G-space structures, we need study only right G-spaces.

Examples. The space $\mathbf{R}^n$ is a left $GL(n, \mathbf{R})$-space or a left $O(n)$-space, where the G-space action is given by matrix multiplication. The space of orthonormal p-frames in n-space $V_p(\mathbf{R}^n)$ is a right $O(r)$-space for each r, $r \leq p$, where the action of $O(r)$ is to change the first r vectors in the frame by an orthogonal matrix. Scalar multiplication defines a left $(\mathbf{R} - \{0\})$-space structure on $\mathbf{R}^n$.

1.3 Definition. A map $h: X \to Y$ from one G-space to another is called a G-morphism provided $h(xs) = h(x)s$ for all $x \in X$ and $s \in G$.

Let $M_G(X, Y)$ denote the subspace of G-morphisms $X \to Y$. Since the composition of G-morphisms is a G-morphism, the class of G-spaces and G-morphisms forms a category denoted $\mathbf{sp}_G$.

Two elements $x, x' \in X$ in a G-space are called G-equivalent provided there exists $s \in G$ with $xs = x'$. This relation is an equivalence relation, and the set of all xs, $s \in G$, denoted xG, is the equivalence class determined by $x \in X$. Let $X \bmod G$ denote the set of all xG, for $x \in X$, with the quotient topology, that is, the largest topology such that the projection $\pi: X \to X \bmod G$ is continuous. Recall that $\pi(x) = xG$, and, occasionally, π is denoted by π_X. Note that the projection π is an identification map.

1.4 Proposition. *For a G-space X, the map $x \to xs$ is a homeomorphism, and the projection $\pi: X \to X \bmod G$ is an open map.*

Proof. The inverse of $x \mapsto xs$ is the map $x \mapsto xs^{-1}$, and $x \mapsto xs$ is a homeomorphism. If W is an open subset of X, then $\pi^{-1}\pi(W) = \bigcup_{s \in G} Ws$ is an open set, being a union of open sets Ws. Therefore, $\pi(W)$ is open in $X \bmod G$ for each open set W of X.

From the above discussion, it follows that every G-space X determines a bundle $\alpha(X) = (X, \pi, X \bmod G)$. If $h: X \to Y$ is a G-space morphism, we have $h(xG) \subset h(x)G$ for each $x \in X$. The quotient map of h is the map $f: X \bmod G \to Y \bmod G$, where $f(xG) = h(x)G$. Let $\alpha(h)$ denote the bundle morphism $(h, f): \alpha(X) \to \alpha(Y)$.

1.5 Proposition. *The collection of functions $\alpha: \mathbf{sp}_G \to \mathbf{Bun}$ is a functor.*

Proof. Clearly, we have $\alpha(1_X) = (1_X, 1_{X \bmod G})$, and if $h: X \to Y$ and $k: Y \to Z$ are two G-morphisms, then $\alpha(kh) = \alpha(k)\alpha(h)$.

1.6 Definition. A bundle (X, p, B) is called a G-bundle provided (X, p, B) and $\alpha(X)$ are isomorphic for some G-space structure on X by an isomorphism $(1, f)\colon \alpha(X) \to (X, p, B)$, where $f\colon X \bmod G \to B$ is a homeomorphism.

2. Definition and Examples of Principal Bundles

A G-space X has the property that the relation $xs = x$ holds only for $s = 1$ if and only if it has the property that the relation $xs = xt$ holds only for $s = t$.

2.1 Definition. A G-space X is called free provided it has the property that $xs = x$ implies $s = 1$. Let X^* be the subspace of all $(x, xs) \in X \times X$, where $x \in X$, $s \in G$ for a free G-space X. There is a function $\tau\colon X^* \to G$ such that $x\tau(x, x') = x'$ for all $(x, x') \in X^*$. The function $\tau\colon X^* \to G$ such that $x\tau(x, x') = x'$ is called the translation function.

From the definition of the translation function $\tau(x, x')$, it has the following properties:

(1) $\tau(x, x) = 1$.
(2) $\tau(x, x')\tau(x', x'') = \tau(x, x'')$.
(3) $\tau(x', x) = \tau(x, x')^{-1}$ for $x, x', x'' \in X$.

2.2 Definition. A G-space X is called principal provided X is a free G-space with a continuous translation function $\tau\colon X^* \to G$. A principal G-bundle is a G-bundle (X, p, B), where X is a principal G-space.

2.3 Example. The product G-space $B \times G$, where the action of G is given by the relation $(b, t)s = (b, ts)$, is principal. To see this, observe that $((b, t), (b', t')) \in (B \times G)^*$ if and only if $b = b'$, and the translation function has the form $\tau((b, t), (b, t')) = t^{-1}t'$. The corresponding principal G-bundle is the product bundle $(B \times G, p, B)$ (up to natural isomorphism). This is called the product principal G-bundle.

2.4 Example. Let G be a closed subgroup of a topological group Γ. Then G acts on the right of Γ by multiplication in Γ, and $(x, x') \in \Gamma^*$ if and only if $x^{-1}x' \in G$. The translation function for the G-space Γ is $\tau(x, x') = x^{-1}x'$, which is continuous for topological groups. The base space of the corresponding principal G-bundle is the space of left cosets $\Gamma \bmod G$. This example includes some of the examples of fibrations in Chap. 1, Sec. 5.

2.5 Example. Let G be the two-element group $\{+1, -1\}$, and let S^n be the G-space with action given by the relation $x(\pm 1) = \pm x$. Then $(S^n)^*$ is the subspace of $(x, \pm x) \in S^n \times S^n$, and the translation function is $\tau(x, \pm x) = \pm 1$, which is clearly continuous. This principal Z_2-space defines a principal Z_2-bundle with base space RP^n.

Finally, we prove that a principal G-bundle is a bundle with fibre G.

2.6 Proposition. *Let $\xi = (X, p, B)$ be a principal G-bundle. Then ξ is a bundle with fibre G.*

Proof. For $x \in p^{-1}(b)$ we define a bijective map $u: G \to p^{-1}(b)$ by the relation $u(s) = xs$. The inverse function of u is $x' \mapsto \tau(x, x')$, which is continuous, and u is a homeomorphism.

3. Categories of Principal Bundles

3.1 Definition. A morphism $(u, f): (X, p, B) \to (X', p', B')$ between two principal G-spaces is a principal morphism provided $u: X \to X'$ is a morphism between G-spaces.

Since $p^{-1}p(x) = xG$ and since $f(xG) = u(x)G$, the map u determines f. If $B = B'$ and if $f = 1_B$, then u is called a principal B-morphism.

Since the composition of principal morphisms or B-morphisms is a principal morphism or B-morphism, respectively, one can speak of the category **Bun**(G) as consisting of principal G-bundles and morphisms and the subcategory **Bun**$_B(G)$ of **Bun**(G) as consisting of principal G-bundles and morphisms over B. We have two natural structure stripping functors **Bun**$(G) \to$ **Bun** and **Bun**$_B(G) \to$ **Bun**$_B$.

The next theorem is the analogue for principal bundles of 3(2.5).

3.2 Theorem. *Every morphism in **Bun**$_B(G)$ is an isomorphism.*

Proof. Let $u: (X, p, B) \to (X', p', B)$ be a B-morphism of principal G-bundles. First, we prove that u is injective. For this, we let $u(x_1) = u(x_2)$. Since $p(x_1) = p'u(x_2) = p(x_2)$, we have $(x_1, x_2) \in X^*$, and $x_1 s = x_2$ for some $s \in G$. Since $u(x_1) = u(x_2) = u(x_1)s$, we have $s = 1$ and $x_1 = x_2$. Next, we prove that u is surjective. For each $x' \in X'$ we let $x \in X$ be such that $p(x) = p'(x')$. Then $p'(x') = p(x) = p'(u(x))$ and $(u(x), x') \in (X')^*$. Since $u(x)s = x'$ for some $s \in G$, we have $u(xs) = u(x)s = x'$, and u is surjective.

To prove u^{-1} is continuous, let $u(a) = a'$ in X', and let V be an open neighborhood of a in X. By the continuity of the action of G on X, there exist open neighborhoods V_1 of a in X and N of 1 in G such that $V_1 N \subset V$. There is an open neighborhod W of a' in X' such that $\tau'((W \times W) \cap X'^*) \subset N$, where τ' is the translation function of X'. Using the continuity of u, we can replace V_1 by $V_1 \cap u^{-1}(W)$ so that $u(V_1) \subset W$. Now $p(V_1) = U$ is an open neighborhood of $b = p(a) = p(a')$ in B by (1.4), and we replace W by $W \cap (p')^{-1}(U)$ so that $p'(W) = U = p(V_1)$.

For each $x' \in W$, we choose $x \in V_1$ such that $p(x) = p'(x')$. Then we have $u(x), x' \in W$ and $u(x)s = x'$ for some $s \in N$, and $x' = u(x)s = u(xs)$, where $xs \in V_1 N \subset V$. Therefore, for each $x' \in W$, we have $u^{-1}(x') \in V$ and $u^{-1}(W) \subset V$. This proves that u^{-1} is continuous at each $a' \in X'$. The theorem has been proved.

Observe that in the proof of the theorem we used only the fact that X was a free G-space whereas X' must be a principal G-space.

4. Induced Bundles of Principal Bundles

The next proposition says, in effect, that the property of being a G-bundle or principal G-bundle is stable under the operation of taking induced bundles.

4.1 Proposition. *Let X be a G-space with an associated bundle $\xi = (X, p, B)$. For each map $f: B_1 \to B$ the total space X_1 of $f^*(\xi) = (X_1, p_1, B_1)$ has a natural G-space structure, and there is a homeomorphism $g: X_1 \bmod G \to B_1$ making the following diagram commutative.*

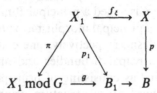

Moreover, the G-space structure on X_1 can be chosen in exactly one way such that f_ξ is a G-morphism. Finally, if (X, p, B) is a principal G-bundle, (X_1, p_1, B_1) is a principal G-bundle.

Proof. We define the action on X_1 by G with the relation $(b_1, x)s = (b_1, xs)$, where $p(xs) = p(x) = f(b_1)$. Then $f_\xi((b_1, x)s) = f_\xi(b_1, xs) = xs = f_\xi(b_1, x)s$, and f_ξ is a G-morphism. Moreover, if f_ξ is a G-morphism, G must act on X_1 by the relation $(b_1, x)s = (b_1, xs)$.

Next we define $g((b_1, x)G) = b_1$. Since $X_1 \bmod G$ has the quotient topology, the function g is continuous. Since p is surjective, the map g is surjective. Since $p(x) = p(x')$ if and only if $x' = xs$, $s \in G$, the map g is injective. If W is an open subset of $X_1 \bmod G$, then $\pi^{-1}(W)$ is an open subset of X, and $g(W) = p_1(\pi^{-1}(W))$ is an open subset of B_1. Note that p_1 is an open map by 2(5.9) and (1.4). Therefore, g is a homeomorphism.

Finally, if $\tau: X^* \to G$ is a translation map for a principal G-space X, then $\tau_1: X_1^* \to G$, defined by $\tau_1((b_1, x), (b_1, x')) = \tau(x, x')$, is a translation map for X_1. Observe that $((b_1, x), (b_1', x')) \in X_1^*$ if and only if $b_1 = b_1'$ and $(x, x') \in X^*$.

4.2 Theorem. *Let $(v, f): \eta \to \xi$ be a morphism of principal G-bundles, and let $\eta \xrightarrow{g} f^*(\xi) \xrightarrow{f_\xi} \xi$ be the canonical factorization given in 2(5.5). Then g is a principal bundle isomorphism over $B(\eta)$, and therefore η and $f^*(\xi)$ are isomorphic principal G-bundles. Finally, $f^*: \mathbf{Bun}_B(G) \to \mathbf{Bun}_{B_1}(G)$ is a functor.*

Proof. We recall $g: X(\eta) \to X(f^*(\xi))$ is given by $g(x) = (p_\eta(x), v(x))$. Then we have $g(xs) = (p_\eta(x), v(x)s) = g(x)s$. Since $f^*(\xi)$ is a principal G-bundle by (4.1),

g is a $B(\eta)$-isomorphism of principal bundles by (3.2). The last result follows from 2(5.6).

Note the similarity of this theorem with 3(3.2). Theorem 3(3.2) plays an important role in the homotopy classification of vector bundles, and Theorem (4.2) is very important for the homotopy classification of principal bundles and fibre bundles.

5. Definition of Fibre Bundles

Let $\xi = (X, p, B)$ be a principal G-bundle, and let F be a left G-space. The relation $(x, y)s = (xs, s^{-1}y)$ defines a right G-space structure on $X \times F$. Let X_F denote the quotient space $(X \times F) \bmod G$, and let $p_F: X_F \to B$ be the factorization of the composition of $X \times F \overset{pr_X}{\to} X \overset{p}{\to} B$ by the projection $X \times F \to X_F$. Explicitly, we have $p_F((x, y)G) = p(x)$ for $(x, y) \in X \times F$.

5.1 Definition. With the above notations, the bundle (X_F, p_F, B), denoted $\xi[F]$, is called the fibre bundle over B with fibre F (viewed as a G-space) and associated principal bundle ξ. The group G is called the structure group of the fibre bundle $\xi[F]$.

Roughly speaking, a principal G-bundle $\xi = (X, p, B)$ consists of a copy of G for each point $b \in B$ all "glued together" by the topology of X. The associated fibre bundle $\xi[F]$ consists of a copy of F for each point of B all "glued together" in a manner prescribed by the topology of the total space X, the action of G on X, and the action of G on F. This gluing is done using the quotient space $X \times F \bmod G$.

5.2 Example. Let ξ be the principal Z_2-bundle $S^1 \to RP^1 = S^1$, and let $F = [-1, +1]$ be the left Z_2-space with action $(\pm 1)t = \pm t$. Then $\xi[F]$ is the fibre bundle consisting of the Moebius band as total space.

The process of going from the principal bundle to the fibre bundle is achieved in this case by "clamping" $[-1, +1]$ onto the two points of S^1 over a given point of $RP^1 = S^1$ and by "sliding" the segment around on the base space to get the total space consisting of the Moebius band. In general, the total space of $\xi[F]$ reflects the "twist" in the topology of the total space X and the "twist" in the action of G on F. In the next proposition we prove that $\xi[F]$ is a bundle with fibre F.

5.3 Proposition. *Let $\xi[F] = \times (X_F, p_F, B)$ be the fibre bundle with associated principal G-bundle $\xi = (X, p, B)$ and fibre F. For each $b \in B$, the fibre F is homeomorphic to $p_F^{-1}(b)$.*

Proof. Let $p(x_0) = b$ for some $x_0 \in X$, and let $f(y) = (x_0, y)G$ be a map $f: F \to X_F$. Since $p_F((x_0, y)G) = p(x_0) = b$, we can view $f: F \to p_F^{-1}(b)$ by the restriction of range from X_F to $p_F^{-1}(b)$.

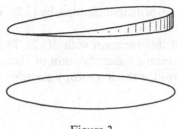

Figure 3

We prove that f has a continuous inverse by considering the map g_1: $p^{-1}(b) \times F \to F$ given by $g_1(x, y) = \tau(x_0, x)y$, where $\tau: X^* \to G$ is the translation map of the principal G-space X. Clearly, we have $g_1(xs, s^{-1}y) = g_1(x, y)$, and by factoring through the restriction of the quotient map $X \times F \to X_F$, we get a map $g: p_F^{-1}(b) \to F$. By the construction, f and g are inverse to each other.

6. Functorial Properties of Fibre Bundles

Let $(u, f): (X, p, B) \to (X', p', B')$ be a principal bundle morphism, and let F be a left G-space. The morphism (u, f) defines a G-morphism $u \times 1_F: X \times F \to X' \times F$, and by passing to quotients, we have a map $u_F: X_F \to X'_F$ such that $(u_F, f): \xi[F] \to \xi'[F]$, where $\xi = (X, p, B)$ and $\xi' = (X', p', B')$.

6.1 Definition. A fibre bundle morphism from $\xi[F]$ to $\xi'[F]$ is a bundle morphism of the form $(u_F, f): \xi[F] \to \xi'[F]$, where $(u, f): \xi \to \xi'$ is a principal bundle morphism. If $B = B'$ and $f = 1_B$, then $u_F: \xi[F] \to \xi'[F]$ is called a fibre bundle morphism over B.

The rest of this chapter is devoted to the study of the category of principal G-bundles and the category of fibre bundles with fibre F and structure group G.

6.2 Proposition. *The functions $\xi \mapsto \xi[F]$ and $(u, f) \mapsto (u_F, f)$ define a functor from the category of principal G-bundles to the category of bundles, admitting the structure of a fibre bundle with fibre F and structure group G.*

Proof. Let $(u, f): \xi \to \xi'$ and $(u', f'): \xi' \to \xi''$ be principal G-bundle morphisms. By applying the quotient space functor to $(u'u) \times 1_F = (u' \times 1_F)(u \times 1_F)$, we get $(u'u)_F = u'_F u_F$. Similarly, we have $(1_X)_F$ as the identity.

A fibre bundle morphism $(u_F, f): \xi[F] \to \xi[F]$ is a fibre bundle isomorphism if and only if $(u, f): \xi \to \xi'$ is a principal bundle isomorphism. This is an example of the fact that the properties of a fibre bundle merely reflect the properties of the associated principal bundle. On the other hand, the properties of a fibre bundle as a bundle reflect also the nature of the action of the group on the fibre; see Exercises 4 and 5.

6.3 Proposition. *Let* $\xi = (X, p, B)$ *be a principal G-bundle and* $\xi[F]$ *a fibre bundle. For each map* $f: B_1 \to B$ *there is a canonical isomorphism* g: $f^*(\xi[F]) \to f^*(\xi)[F]$ *of bundles over* B_1 *such that the natural morphism* $f_{\xi[F]}$: $f^*(\xi[F]) \to \xi[F]$ *factors by* $f^*(\xi[F]) \xrightarrow{g} f^*(\xi)[F] \xrightarrow{(f_\xi)_F} \xi[F]$.

Proof. The total space X_1 of $f^*(\xi[F])$ consists of pairs $(b_1, (x, y)G)$, where $f(b_1) = p_F((x, y)G) = p(x)$ and $f_{\xi[F]}$ is given by the relation $f_{\xi[F]}(b_1, (x, y)G) = (x, y)G$. The total space X_2 of $f^*(\xi)[F]$ consists of pairs $((b_1, x), y)G$, where $f(b_1) = p(x)$ and $(f_\xi)_F$ is given by the relation $(f_\xi)_F(((b_1, x), y)G) = (x, y)G$. We define the isomorphism g by the relation $g(b_1, (x, y)G) = ((b_1, x), y)G$. This isomorphism is the result of applying the quotient space functor to the canonical G-isomorphism $B \times (X \times F) \to (B \times X) \times F$ and observing that $(b_1, xs, s^{-1}y) = (b_1, x, y)$ for $b_1 \in B_1, x \in X, y \in F, s \in G$.

6.4 Corollary. *Let* $\xi[F]$ *be a fibre bundle over* B, *and let* $A \subset B$. *Then* $\xi[F]|A$ *and* $(\xi|A)[F]$ *are canonically A-isomorphic as bundles.*

7. Trivial and Locally Trivial Fibre Bundles

Let ξ be the product principal G-bundle $(B \times G, p, B)$. For each left G-space F, the fibre bundle $\xi[F] = (Y, q, B)$ is B-isomorphic over B to the product bundle $(B \times F, p, B)$. Let $g: Y \to B \times F$ be defined by $g((b, s, y)G) = (b, sy)$. Then g is a B-isomorphism.

7.1 Definition. Two principal G-bundles ξ and η over B are locally isomorphic provided each $b \in B$ has an open neighborhood U such that $\xi|U$ and $\eta|U$ are U-isomorphic (as principal bundles). Two fibre bundles $\xi[F]$ and $\eta[F]$ are locally isomorphic provided ξ and η are locally isomorphic.

7.2 Definition. A principal G-bundle ξ over B is trivial or locally trivial provided ξ is a principal G-bundle that is isomorphic or locally isomorphic to the product principal G-bundle. A fibre bundle $\xi[F]$ is trivial or locally trivial provided ξ is trivial or locally trivial, respectively.

In view of Corollary (6.4), a principal bundle or a fibre bundle that is trivial or locally trivial is trivial or locally trivial as a bundle. Conversely, it is possible for a fibre bundle to be trivial as a bundle but not trivial as a fibre bundle; see Exercise 4.

8. Description of Cross Sections of a Fibre Bundle

The following theorem is of great importance for the classification of principal bundles and fibre bundles, because it yields a criterion for trivality of a
principal bundle and a means of constructing principal bundle morphisms.

8.1 Theorem. *Let* $\xi = (X, p, B)$ *be a principal G-bundle, and let* $\xi[F] =$
(X_F, p_F, B) *be an associated fibre bundle, where F is a left G-space. The cross
sections s of the bundle* $\xi[F]$ *are in bijective correspondence with maps* ϕ:
$X \to F$ *such that* $\phi(xt) = t^{-1}\phi(x)$ *for* $x \in X$, $t \in G$. *The cross section corresponding to* ϕ *is* $s_\phi(xG) = (x, \phi(x))G$ *in* X_F *for each* $xG \in B$.

Proof. Since $(xt, \phi(xt))G = (xt, t^{-1}\phi(x))G = (x, \phi(x))G$ in $X_F = X \times F \mod G$,
the function s_ϕ is well defined. Since s_ϕ is the factorization of $x \mapsto (x, \phi(x))G$
by the quotient map p, the function s_ϕ is continuous. Clearly, the relation
$p_F(s_\phi(xG)) = p_F((x, \phi(x))G) = p(x) = xG$ holds, and the map s_ϕ is a cross
section.

Conversely, let s be a cross section of $\xi[F]$, and let $\phi_s: X \to F$ be defined
by the relation $s(xG) = (x, \phi_s(x))G$ for each $x \in X$. Since $(x, \phi_s(X))G =
(xt, t^{-1}\phi_s(x))G = (xt, \phi_s(xt))G$, and since s is a cross section, the function ϕ_s
satisfies the relation $\phi_s(xt) = t^{-1}\phi_s(x)$ for each $x \in X$, $t \in G$.

Finally, if we prove the continuity of ϕ_s, the theorem will be proved
because $\phi \mapsto s_\phi$ and $s \mapsto \phi_s$ are functions that are inverse to each other.
Let $x_0 \in X$, $y_0 = \phi_s(x_0)$, $b_0 = p(x_0)$, and $s(b_0) = (x_0, y_0)G$. Let W be an open
neighborhood of y_0. By the continuity of the action of G on F, there exist
open neighborhoods W' of y_0 and N of 1 in G such that $NW' \subset W$. Let V be
an open neighborhood of x_0 in X such that $\tau((V \times V) \cap X^*) \subset N$, where τ is
the translation function of ξ. Since s is a map, there exists an open neighborhood U of b_0 such that $s(U) \subset (V \times W') \mod G$. We replace V by $p^{-1}(U) \cap V$.
The relation $s(U) \subset (V \times W') \mod G$ is preserved, and $p(V) = U$. To prove
$\phi_s(V) \subset W$, let $x \in V$ and $b = p(x) \in U$. Then $s(b) = (x', y')G$, where $x' \in V$
and $y' \in W'$. Since $(x', y')G = (x\tau(x, x'), y')G = (x, \tau(x, x')y')G$, we have
$\phi_s(x) = \tau(x, x')y' \in NW' \subset W$ and $\phi_s(V) \subset W$. This proves the theorem.

By observing that $\phi_s: X \to F$ is just a G-morphism with respect to G acting
on the right of X and F, we have two applications of Theorem (8.1) in the
next corollaries.

8.2 Corollary. *Let* $\xi = (X, p, B)$ *and* $\xi' = (X', p', B')$ *be two principal G-bundles. All principal G-bundle morphisms* $\xi \to \xi'$ *are of the form* (ϕ_s, f), *where s is
a cross section of* $\xi[X']$. *Moreover, using the equality* $X_{X'} = X'_X$, *we have
$f = (p'_X)s$.
The following diagram illustrates the situation in this corollary:

$$X_{X'} = X \times X' \bmod G = X'_X$$

$$p'_X \downarrow \qquad\qquad\qquad \downarrow p'_X$$

$$B \xrightarrow{\quad f \quad} B'$$

Proof. By Theorem (8.1) the set of cross sections of $\xi[X']$ are in bijective correspndence with G-morphisms $\phi_s \colon X \to X'$, that is, maps $\phi_s \colon X \to X'$ with $\phi_s(xt) = t^{-1}\phi_s(x) = \phi_s(x)t$.

From the relation $f = (p'_X)s$, we observe that $s(xG) = (x, \phi_s(x))G$ and $p'_X(s(xG)) = p'_X((x, \phi_s(x))G) = p'_X(\phi_s(x))$ for $x \in X$. Since $f(xG) = p'_X(\phi_s(x))$, we have the relation $f = (p'_X)s$. This proves the corollary.

This corollary reduces the problem of the existence of a principal bundle morphism $\xi \to \xi'$ to a more manageable problem of the existence of a cross section. In particular, we can apply the results of Chap. 2, Sec. 7. An application of this corollary is a decisive step in the homotopy classification of locally trivial fibre bundles, and it plays a role similar to that of Gauss maps for vector bundles.

8.3 Corollary. *The following are equivalent statements for a principal G-bundle $\xi = (X, p, B)$.*

(1) *The bundle ξ has a cross section.*
(2) *The bundle ξ is isomorphic to $f^*(\eta)$, where η is the product bundle over a point and f is the unique constant map.*
(3) *The bundle ξ is trivial.*

Proof. Since as bundles $\xi = \xi[G]$, we use (8.1) to prove the equivalence of statements (1) and (2). The set of cross sections of ξ are in bijective correspondence with maps $\phi \colon X \to G$ such that $\phi(xt) = t^{-1}\phi(x) = \phi(x)t$. If $f \colon B \to *$ is the unique constant map, it is induced by ϕ, and by Theorem (4.2), the principal bundles ξ and $f^*(\eta)$ are isomorphic. Clearly, statement (3) implies (1), and statement (2) implies (3) because the induced bundle of a trivial bundle is trivial, by (4.2).

9. Numerable Principal Bundles over $B \times [0, 1]$

This section has the same objectives as Sec. 4 in Chap. 3. Instead of considering all bundles over a paracompact space, as in Chap. 3, we consider numerable bundles over an arbitrary space.

9.1 Definition. An open covering $\{U_i\}_{i \in S}$ of topological space B is numerable provided there exists a (locally finite) partition of unity $\{u_i\}_{i \in S}$ such that $\overline{u_i^{-1}((0, 1])} \subset U_i$ for each $i \in S$.

It is a standard result that a Hausdorff space B is paracompact if and only if each open covering is numerable.

9.2 Definition. A principal G-bundle ξ over a space B is numerable provided there is a numerable over $\{U_i\}_{i \in S}$ of B such that $\xi | U_i$ is trivial for each $i \in S$.

Observe that each numerable principal G-bundle is locally trivial and each locally trivial principal G-bundle over a paracompact space is numerable.

9.3 Proposition. *Let $f: B' \to B$ be a map, and let ξ be a numerable bundle over B. Then $f^*(\xi)$ is a numerable bundle over B'.*

Proof. Suppose that $\xi | U_i$ is trivial, where $\{U_i\}_{i \in S}$ is an open covering of B and $\{u_i\}_{i \in S}$ is a locally finite partition of unity with the closure of $u_i^{-1}(0, 1) \subset U_i$ for each $i \in S$. Then $f^*(\xi) | f^{-1}(U_i)$ is trivial, and $\{u_i f\}_{i \in S}$ is a locally finite partition of unity with the closure of $(u_i f)^{-1}(0, 1] \subset f^{-1}(U_i)$ for each $i \in S$.

The results and the arguments proceed as in Sec. 4 of Chap. 3. Where the proof in this more general situation is completely parallel to that given in Chap. 3, we omit the argument and leave it as an exercise for the reader.

9.4 Lemma. *Let ξ be a principal G-bundle over $B = B_1 \cup B_2$, where $B_1 = A \times [a, c]$ and $B_2 = A \times [c, b]$ for $a < c < b$. If $\xi | B_1$ and $\xi | B_2$ are trivial, then ξ is trivial.*

In order that the argument in 3(4.1) should apply, we have only to observe that the automorphisms $h: B \times G \to B \times G$ are of the form $h(b, s) = (b, g(b)s)$, where $g: B \to G$ is the map given by $(b, g(b)) = h(b, 1)$. This follows from the relation $h(b, s) = h(b, 1)s = (b, g(b)s)$.

The next lemma is the analogue of 3(4.2), but a more delicate argument is required here for the proof which is essentially due to Dold [4].

9.5 Lemma. *Let ξ be a numerable G-bundle over $B \times I$ (where $I = [0, 1]$). Then there exists a numerable covering $\{U_i\}_{i \in S}$ of B such that $\xi | (U_i \times I)$ is trivial for each $i \in S$.*

Proof. Let $\{v_j\}_{j \in T}$ be a (locally finite) partition of unity of $B \times I$ such that $\xi | v_j^{-1}(0, 1]$ is trivial for each $j \in T$. For each r-tuple $\mathbf{k} = (k(1), \ldots, k(r)) \in T^r$, we define

$$v_{\mathbf{k}}(x) = \prod_{1 \leq q \leq r} \min\{v_{k(q)}(x, t)\} \qquad \text{for } t \in \left[\frac{q-1}{r}, \frac{q}{r}\right]$$

Then $v_{\mathbf{k}}: B \to I$ is continuous. Moreover, we have $v_{\mathbf{k}}(x) > 0$ if and only if $x \times [(q-1)/r, q/r] \subset v_{k(q)}^{-1}(0, 1]$ for each $1 \leq q \leq r$. This means, in view of (9.4), that it suffices to prove that the family of open sets $\{v_{\mathbf{k}}^{-1}(0, 1]\}$ for all $\mathbf{k} \in T^r$ and all r is a numerable covering of B, because $\xi | (v_{\mathbf{k}}^{-1}(0, 1] \times I)$ is trivial.

Each $(x,t) \in B \times I$ has a neighborhood which is contained in one set U_i and which intersects only a finite number of sets U_j. Since I is compact, there are for each $x \in B$ a neighborhood N and a natural number r such that

(1) $N \times [(q-1)/r, q/r] \subset U_{k(q)}$ for some $k(q) \in T$ for each $1 \leq q \leq r$.
(2) $N \times I$ intersects only a finite number of $V_j = v_j^{-1}(0,1]$ for $j \in T$.

By property (1), $\{v_k^{-1}(0,1]\}$ is a covering of B, and by property (2), for a given r, the family of $\{v_k\}$ for $\mathbf{k} \in T$ is locally finite.

Now we augment the maps v_k in such a way that we get a locally finite partition of unity. Let $w_r(x)$ denote the sum of all the functions $v_k(x)$ with $\mathbf{k} = (k(1), \ldots, k(s))$ and $s < r$. We define

$$u_k(x) = \max(0, v_k(x) - r w_r(x))$$

For $x \in B$, we have a $\mathbf{k} = (k(1), \ldots, k(r))$ with r minimal with respect to the property that $v_k(x) > 0$. Then $w_r(x) = 0$ and $u_k(x) = v_k(x)$. Consequently, the sets $u_k^{-1}(0,1]$ form an open covering of B. Moreover, let $m > r$ such that $v_k(x) > 1/m$. Then we have $w_m(x) > 1/m$ and $m w_m(y) > 1$ for all y in a neighborhood of x. In this neighborhood all u_k with $\mathbf{k} = (k(1), \ldots, k(s))$ and $s \geq m$ vanish, and consequently the system $\{u_k\}$ is a locally finite partition of unity such that $\xi|(U_k \times I)$ is trivial where $U_k = u_k^{-1}(0,1]$. This proves the proposition.

As with Theorem 3(4.3), the next theorem is the major step in the development of the homotopy properties of numerable principal G-bundles. The method of proof in the next theorem is essentially due to Milnor who introduced it in the setting of microbundles.

9.6 Theorem. *Let $r: B \times I \to B \times I$ be the map $r(b,t) = (b,1)$, and let ξ be a numerable principal G-bundle over $B \times I$. Then there exists a G-morphism $(g, r): \xi \to \xi$.*

Proof. By (9.5) there is a numerable covering $\{U_i\}_{i \in S}$ of B such that $\xi|(U_i \times I)$ is trivial. Moreover, there are maps $u_i: B \to [0,1]$ such that $u_i^{-1}(0,1] \subset U_i$ and such that $\max_{i \in S} u_i(b) = 1$ for each $b \in B$. The maps u_i are easily constructed from the partition of unity. Let $h_i: U_i \times I \times G \to p^{-1}(U_i \times I) \subset E(\xi)$ be a $(U_i \times I)$-isomorphism of principal G-bundles.

We define a morphism $(g_i, r_i): \xi \to \xi$ with the requirements that $r_i(b,t) = (b, \max(u_i(b), t))$, g_i is the identity outside $p^{-1}(U_i \times I)$, and $g_i(h_i(b,t,s)) = h_i(b, \max(u_i(b), t), s)$ for each $(b,t,s) \in U_i \times I \times G$. We choose a well ordering on the set I. For each $b \in B$, there is an open neighborhood $U(b)$ of b such that $U_i \cap U(b)$ is nonempty for $i \in I(b)$, where $I(b)$ is a finite subset of I. We define $r = r_{i(n)} \cdots r_{i(1)}$ on $U(b) \times I$ and $g = g_{i(n)} \cdots g_{i(1)}$ on $p^{-1}(U(b) \times I)$, where $I = \{i(1), \ldots, i(n)\}$ and $i(1) < i(2) < \cdots < i(n)$. For $i \notin I(b)$ the map r_i is the identity on $U(b) \times I$, and g_i is the identity on $p^{-1}(U(b) \times I)$. The maps r and

q are infinite compositions of maps where all but a finite number of terms equal the identity near a point.

9.7 Corollary. *With the notations of* (9.6), *the principal G-bundles ξ and $r^*(\xi)$ are isomorphic over $B \times I$.*

Let $\xi = (X, p, B)$ be a principal G-bundle, and let W be a locally compact space. Then the relation $(x, w)s = (xs, w)$ defines a right G-space structure on $X \times W$. The translation map τ_1 for $X \times W$ is defined by the relation $\tau_1((x, w), (x', w)) = \tau(x, x')$, where τ is the translation map for $X \bmod G$ and $X \times W \to (X \times W) \bmod G$ are identification maps, and the function ϕ: $(X \times W) \bmod G \to (X \bmod G) \times W$, where $\phi((x, w)G) = (xG, w)$, is a homeomorphism. We are most interested in the above construction for $W = [0, 1] = I$. The following result is a corollary of (9.6) and (9.7).

9.8 Theorem. *Let ξ be a numerable principal G-bundle over $B \times I$. Then the bundles ξ, $(\xi|(B \times 1)) \times I$, and $(\xi|(B \times 0)) \times I$ are G-isomorphic. If $\varepsilon_i: B \to B \times I$ is the map $\varepsilon_i(b) = (b, i)$ for $i = 0$, 1, then $\varepsilon_0^*(\xi)$ and $\varepsilon_1^*(\xi)$ are B-isomorphic bundles.*

Proof. Observe that $r^*(\xi)$ and $(\xi|(B \times 1)) \times I$ are $(B \times I)$-isomorphic principal G-bundles, where $r(b, t) = (b, 1)$. By (9.7), $r^*(\xi)$ and ξ are isomorphic bundles, and consequently ξ and $(\xi|(B \times 1)) \times I$ are isomorphic principal G-bundles. Similarly, ξ and $(\xi|(B \times 0)) \times I$ are isomorphic.

For the last statement, observe that $r\varepsilon_0 = \varepsilon_1$, and the bundles $r^*(\xi)|(B \times 0)$ and $\xi|(B \times 0)$ are G-isomorphic. Therefore, $\varepsilon_1^*(\xi) = \varepsilon_0^* r^*(\xi)$ and $\varepsilon_0^*(\xi)$ are isomorphic principal G-bundles. This proves the theorem.

As a corollary we have the following result.

9.9 Theorem. *Let ξ be a numerable principal G-bundle over B, and let $f_t: B' \to B$ be a homotopy. Then the principal G-bundles $f_0^*(\xi)$ and $f_1^*(\xi)$ are isomorphic over B'.*

Proof. Consider the homotopy $f: B' \times I \to B$ and the maps $\varepsilon_i: B' \to B' \times I$, where $\varepsilon_i(b) = (b, i)$ for $i = 0$, 1. Clearly, we have $f_i = f\varepsilon_i$, and $f^*(\xi)$ is isomorphic to $\varepsilon_i^* f^*(\xi)$. By (9.8) we have the result.

10. The Cofunctor k_G

For each space B, let $k_G(B)$ denote the set of isomorphism classes of numerable principal G-bundles over B. Let $\{\xi\}$ denote the isomorphism class of the principal G-bundle ξ over B. For a homotopy class $[f]: X \to Y$ we define a function $k_G([f]): k_G(Y) \to k_G(X)$ by the relation $k_G([f])\{\xi\} =$

$\{f^*(\xi)\}$. By Theorem (4.2) the element $\{f^*(\xi)\}$ is independent of the representative ξ of $\{\xi\}$, and by Theorem (9.9) it is independent of the representative f in $[f]$. Consequently, $k_G([f]): k_G(Y) \to k_G(X)$ is a function.

Let **H** denote the category of all spaces and homotopy classes of maps.

10.1 Theorem. *The collection of functions k_G: **H** → **ens** is a cofunctor.*

Proof. Let $f: X \to Y$ and $g: Y \to Z$ be two maps, and let ξ be a numerable principal G-bundle over Z. Then the bundles $(gf)^*(\xi)$ and $f^*(g^*(\xi))$ are isomorphic over X by the construction used in 2(5.7). Therefore, we have $k_G([g][f]) = k_G([f])k_G([g])$. Since ξ and $1_Z^*(\xi)$ are isomorphic over Z, the function $k_G(1_Z)$ equals the identity on $k_G(Z)$.

From general properties of cofunctors we have the next corollary.

10.2 Corollary. *If $f: X \to Y$ is a homotopy equivalence, $k_G([f]): k_G(Y) \to k_G(X)$ is a bijection.*

10.3 Corollary. *If X is contractible, each numerable principal G-bundle over X is trivial.*

Proof. Observe that $k_G(*)$ has only one point, the class of the trivial bundle. Now use (10.2).

Let $\omega = (E_0, p_0, B_0)$ be a fixed numerable principal G-bundle. For each space X we define a function $\phi_\omega(X): [X, B_0] \to k_G(X)$ by the relation $\phi_\omega(X)[u] = \{u^*(\omega)\}$. By (9.9), $\phi_\omega(X)$ is a function.

10.4 Proposition. *The set of functions $\phi_\omega: [-, B_0] \to k_G$ is a morphism of cofunctors defined* **H** → **ens**.

Proof. Let $f: X \to Y$ be a map, and let $[u]$ be an element of $[Y, B_0]$. We compute

$$\phi_\omega(X)[[f], B_0][u] = \phi_\omega(X)([u][f]) = \{(uf)^*(\omega)\}$$

and

$$k_G([f])\phi_\omega(Y)[u] = k_G([f])\{u^*(\omega)\} = \{f^*(u^*(\omega))\}$$

Therefore, we have $\phi_\omega(X)[[f], B_0] = k_G([f])\phi_\omega(Y)$, and ϕ_ω is a morphism of cofunctors.

10.5 Definition. A principal G-bundle $\omega = (E_0, p_0, B_0)$ is universal provided ω is numerable and $\phi_\omega: [-, B_0] \to k_G$ is an isomorphism. The space B_0 is called a classifying space of G.

From general principles, $\phi_\omega: [-, B_0] \to k_G$ is an isomorphism of cofunctors if and only if each function $\phi_\omega(X): [X, B_0] \to k_G(X)$ is a bijection.

10.6 Proposition. *A numerable principal G-bundle* $\omega = (E_0, p_0, B_0)$ *is universal if and only if the following are true.*

(1) *For each numerable principal G-bundle* ξ *over* X *there exists a map* f: $X \to B_0$ *such that* ξ *and* $f^*(\omega)$ *are isomorphic over* X.
(2) *If* f, g: $X \to B_0$ *are two maps such that* $f^*(\omega)$ *and* $g^*(\omega)$ *are isomorphic over* X, *then* f *and* g *are homotopic.*

Proof. Observe that condition (1) says that $\phi_\omega(X)$ is surjective and (2) says that $\phi_\omega(X)$ is injective.

This criterion for a bundle to be universal leads to the following definition.

10.7 Definition. A bundle ω is called *n-universal*, or universal for dimensions $\leq n$, provided $\phi_\omega(X)$ is a bijection for each CW-complex X with $\dim X \leq n$.

11. The Milnor Construction

In this section, we consider the very simple elegant construction for a universal G-bundle and prove that the conditions of (10.6) are satisfied.

11.1 Definition of the *G*-space E_G. For a group G, there is an infinite join

$$E_G = G * G * \cdots * G * \cdots$$

An element of E_G is denoted $\langle x, t \rangle$ and written

$$\langle x, t \rangle = (t_0 x_0, t_1 x_1, \ldots, t_k x_k, \ldots)$$

where each $x_i \in G$ and $t_i \in [0, 1]$ such that only a finite number of $t_i \neq 0$ and $\sum_{0 \leq i} t_i = 1$. In the set E_G there is the identification $\langle x, t \rangle = \langle x', t' \rangle$ provided $t_i = t_i'$ for each i, and $x_i = x_i'$ for all i with $t_i = t_i' > 0$. Observe that if $t_i = t_i' = 0$ then x_i and x_i' may be different but $\langle x, t \rangle = \langle x', t' \rangle$ in the set E_G. We define a right action of G on E_G by the relation $\langle x, t \rangle y = \langle xy, t \rangle$ or $(t_0 x_0, t_1 x_1, \ldots) y = (t_0 x_0 y, t_1 x_1 y, \ldots)$ for $y \in G$.

Now we put a topology on E_G in such a way that E_G is a G-space. For this we consider two families of functions t_i: $E_G \to [0, 1]$ for $0 \leq i$, which assigns to the element $(t_0 x_0, t_1 x_1, \ldots)$ the component $t_i \in [0, 1]$, and x_i: $t_i^{-1}(0, 1] \to G$ for $0 \leq i$, which assigns to the element $(t_0 x_0, t_1 x_1, \ldots)$ the component $x_i \in G$. Observe that x_i cannot be uniquely defined outside $t_i^{-1}(0, 1]$ in a natural way. For $a \in E_G$ and $y \in G$ there are the following relations between the action of G and the functions x_i and t_i:

$$x_i(ay) = x_i(a)y \quad \text{and} \quad t_i(ay) = t_i(a)$$

The set E_G is made into a space by requiring it to have the smallest topology such that each of the functions t_i: $E_G \to [0, 1]$ and x_i: $t_i^{-1}(0, 1] \to G$ is continuous, where $t_i^{-1}(0,1]$ has the subspace topology. From the relations $t_i(ay) = t_i(a)$ and $x_i(ay) = x_i(a)y$ it is clear that E_G is a G-space because the G-set structure map $E_G \times G \to E_G$ is continuous.

We denote the quotient space E_G mod G by B_G and the resulting bundle by $\omega_G = (E_G, p, B_G)$. This is the Milnor construction.

11.2 Theorem. *The G-bundle ω_G is a numerable principal G-bundle.*

Proof. First, we prove that ω_G is a principal G-bundle. The translation function $\tau(a, a')$ is given by the continuous function $x_i(a)x_i(a')^{-1}$ on the open set $(t_i^{-1}(0,1] \times t_i^{-1}(0,1]) \cap E_G^*$, where E_G^* is the subspace of pairs (a, a') with $p(a) = p(a')$ in B_G. Finally, we observe that the open sets $t_i^{-1}(0,1] \times t_i^{-1}(0,1]$ for $0 \leq i$ cover E_G^*. Consequently, ω_G is a principal G-bundle.

Since $t_i(ay) = t_i(a)$ for $y \in G$, we can define a unique map $u_i: B_G \to [0,1]$ with the property that $u_i p = t_i$ on E_G. Now we show that ω_G is trivial over each $V_i = u_i^{-1}(0,1] = p(t_i^{-1}(0,1])$ by defining a cross section s_i of ω_G over V_i. For this we use the map $s_i': t_i^{-1}(0,1] \to t_i^{-1}(0,1]$ defined by $s_i'(a) = ax_i(a)^{-1}$. This map has the property that $s_i'(ay) = ay(x_i(ay))^{-1} = ay(x_i(a)y)^{-1} = ayy^{-1}x_i(a)^{-1} = ax_i(a)^{-1} = s_i'(a)$. By passing to quotients, the map s_i' defines a map $s_i: V_i \to t_i^{-1}(0,1] \subset E_G$ such that $s_i' = s_i p$. Since $p(a) = p(s_i'(a))$ for each $a \in t_i^{-1}(0,1]$, we have $p s_i(b) = b$ for $b \in V_i$. By (8.3) the existence of a cross section implies that $\omega_G | V_i$ is a trivial principal G-bundle.

Finally, to show that ω_G is numerable, we shall construct a (locally finite) partition of unity $\{v_i\}_{j \leq i}$ on B_G with $v_i^{-1}(0,1] \subset V_i = u_i^{-1}(0,1]$. We define

$$w_i(b) = \max(0, u_i(b) - \sum_{j < i} u_j(b))$$

and $w_i: B_G \to [0,1]$ is a map with $w_i^{-1}(0,1] \subset V_i$. For $b \in B_G$, let w be the smallest i such that $u_i(b) \neq 0$, and let n be the largest. Then we have $\sum_{m \leq i \leq n} u_i(b) = 1$. Then $u_m(b) = w_m(b)$, and B_G is covered by the open sets $w_i^{-1}(0,1]$. Since $u_i(b) = 0$ for $n < i$, we have $w_i(b') = 0$ for all b' with $\sum_{0 \leq i \leq n} u_i(b') > \frac{1}{2}$. This is an open neighborhood $N_n(b)$ of b such that $N_n(b) \cap w_i^{-1}(0,1]$ is empty for $i > n$. The open covering $\{w_i^{-1}(0,1]\}$ is locally finite. Now we replace w_i with $v_i = w_i / \sum_j w_j$. This proves the theorem.

Observe that the spaces E_G and B_G are filtered by subspaces

$$\cdots \subset E_G(n) \subset E_G(n+1) \subset \cdots \subset E_G$$

$$\cdots \subset B_G(n) \subset B_G(n+1) \subset \cdots \subset B_G$$

where $p(E_G(n)) = B_G(n)$ and $(t_0 x_0, t_1 x_1, \dots) \in E_G(n)$, provided $t_i = 0$ for $i > n$.

11.3 Example $G = Z_2$. The space $E_G(n)$ is just the n-sphere S^n up to homeomorphism, and the action of Z_2 on $E_G(n) = S^n$ is by the identity and the antipodal map. The space $B_G(n)$ is RP^n. The inclusions $E_G(n) \subset E_G(n + 1)$ and $B_G(n) \subset B_G(n + 1)$ are just the natural inclusions $S^n \subset S^{n+1}$ and $RP^n \subset RP^{n+1}$. The space E_G is S^∞ and B_G is RP^∞.

The bundle (S^{n+1}, p, RP^{n+1}) is universal for dimensions $\leqq n$.

11.4 Example $G = S^1$. The space $E_G(n)$ is just the $(2n + 1)$-sphere S^{2n+1} up to a homeomorphism, and the action of S^1 on $E_G(n) = S^{2n+1}$ is by the relation $(z_0, z_1, \ldots, z_n)e^{i\theta} = (e^{i\theta}z_0, e^{i\theta}z_1, \ldots, e^{i\theta}z_n)$ for each $e^{i\theta} \in S^1$. The space $B_G(n)$ is CP^n, complex n-dimensional projective space. The inclusions $E_G(n) \subset E_G(n + 1)$ and $B_G(n) \subset B_G(n + 1)$ are just the natural inclusions $S^{2n+1} \subset S^{2n+3}$ and $CP^n \subset CP^{n+1}$. The space E_G is S^∞, and B_G is CP^∞.

The bundle (S^{2n+1}, p, CP^n) is universal for dimensions $\leqq 2n$.

12. Homotopy Classification of Numerable Principal G-Bundles

In this section, we prove that the bundle ω_G which comes from the Milnor construction is a universal principal G-bundle.

12.1 Proposition. *Let ξ be a numerable principal G-bundle over a space B. Then there exists a countable partition of unity $\{u_n\}_{0 \leqq n}$ such that $\xi | u_n^{-1}(0, 1]$ is trivial for each natural number n.*

Proof. Let $\{v_i\}_{i \in T}$ be a partition of unity on B such that ξ is trivial over each $v_i^{-1}(0, 1]$ for $i \in T$. For each $b \in B$ we denote the finite set of $i \in T$ with $v_i(b) > 0$ by $S(b)$, and for each finite subset S of T we denote by $W(S)$ the open set of $b \in B$ such that $v_i(b) > v_j(b)$ for each $i \in S$ and $j \in T - S$. Let $u_S: B \to [0, 1]$ be the continuous map given by the

$$u_S(b) = \max\left[0, \min_{i \in S, j \in T-S} (v_i(b) - v_j(b))\right]$$

Then we have $W(S) = u_S^{-1}(0, 1]$.

Let Card S denote the number of elements in S. If Card $S =$ Card S', then $W(S) \cap W(S')$ is empty. To see this, let $i \in S - S'$ and $j \in S' - S$. For $b \in W(S)$ we have $v_i(b) > v_j(b)$, and for $b \in W(S')$ we have $v_j(b) > v_i(b)$. The relations cannot hold simultaneously.

Finally, we denote

$$W_m = \bigcup_{m = \text{Card } S} W(S) \quad \text{and} \quad w_m(b) = \sum_{m = \text{Card } S} u_S(b)$$

Then we have $w_m^{-1}(0, 1] = W_m$ and $u_m(b) = w_m(b) / \sum_{0 \leqq n} w_n(b)$. The family $\{u_n\}$ is the desired partition of unity since $u_n^{-1}(0, 1] = W_n$. Observe that $\xi | W_n$ is trivial because $\xi | W(S)$ is trivial and W_n is a disjoint union.

In the next theorem we prove that condition (1) of the criterion (10.6) for ω_G to be a universal principal G-bundle holds.

12.2 Theorem. *For each numerable principal G-bundle ξ over a space B there exists a map $f: B \to B_G$ such that ξ and $f^*(\omega_G)$ are B-isomorphic principal G-bundles.*

Proof. It suffices to define a G-morphism $(g, f): \xi \to \omega_G$, for then, by (4.2), ξ and $f^*(\omega_G)$ are isomorphic. By (12.1) we can assume that there is a countable partition of unity $\{u_n\}_{0 \leq n}$ on B such that $\xi | u_n^{-1}(0, 1]$ is trivial for $n \geq 0$. Let U_n denote $u_n^{-1}(0, 1]$, and let $h_n: U_n \times G \to E(\xi | U_n) \subset E(\xi)$ be an isomorphism defining the locally trivial character of ξ.

We define $g: E(\xi) \to E_G$ by the relation

$$g(z) = (u_0 p(z)(q_0 h_0^{-1}(z)), \ldots, u_n p(z)(q_n h_n^{-1}(z)), \ldots)$$

where $q_n: U_n \times G \to G$ is the projection on the second factor. Since for z, where $h_n^{-1}(z)$ is undefined, we have $u_n(p(z)) = 0$, the map g is well defined and the relation $g(zs) = g(z)s$ holds for each $s \in G$ since $h_n(zs) = h_n(z)s$. The map g induces $f: B \to B_G$ and $(g, f): \xi \to \omega_G$ is a bundle morphism. This proves the theorem.

We consider various maps $B_G \to B_G$ determined by maps $E_G \to E_G$. Let E_G^{od} denote the subspace of all $(x, t) \in E_G$ with $t_{2i+1} = 0$ for all $i \geq 0$, and let E_G^{ev} denote the subspace of all $(x, t) \in E_G$ with $t_{2i} = 0$ for all $i \geq 0$. Let B_G^{od} be the subspace $p(E_G^{\text{od}})$ and B_G^{ev} the subspace $p(E_G^{\text{ev}})$.

We use the following linear functions.

$$\alpha_n: [1 - (\tfrac{1}{2})^n, 1 - (\tfrac{1}{2})^{n+1}] \to [0, 1]$$

given by $\alpha_n(t) = 2^{n+1} t - 2^{n+1} + 2$. Clearly, we have $\alpha_n(1 - (\tfrac{1}{2})^n) = 0$ and $\alpha_n(1 - (\tfrac{1}{2})^{n+1}) = 1$. We define a homotopy $h_s^{\text{od}}: E_G \to E_G$ such that $h_s^{\text{od}}(x, t)y = h_s^{\text{od}}(xy, t)$ by the following relation $h_s^{\text{od}}(x, t) = (x', t')$ where for $s \in I_n$

$$x_i' = \begin{cases} x_i & 0 \leq i \leq n \\ x_{n+j} & i = n + 2j - 1 \qquad \text{for } 0 < j < \infty \\ & = n + 2j \end{cases}$$

$$t_i' = \begin{cases} t_i & 0 \leq i \leq n \\ \alpha_n(s) t_{n-j} & i = n + 2j - 1 \\ (1 - \alpha_n(s)) t_{n-j} & i = n + 2j \end{cases}$$

and for $s = 1$ the relation $(x', t') = (x, t)$. One can check that this is well defined for $s = 1 - 2^{-n} \in I_n \cap I_{n-1}$. The homotopy is continuous because it is continuous on the locally finite open covering of $v_i^{-1}(0, 1]$, where $\{v_i\}$ is the partition of unity. The maps $h_s^{\text{od}}: E_G \to E_G$ induce $g_s^{\text{od}}: B_G \to B_G$ such that $(h_s^{\text{od}}, g_s^{\text{od}}): \omega_G \to \omega_G$ is a homotopy of bundle morphisms. Finally, we observe that $h_0^{od}(E_G) = E_G^{od}$ and $g_0^{od}(B_G) = B_G^{od}$. We have $(h_s^{ev}, g_s^{ev}): \omega_G \to \omega_G$ by a similar construction, where $h_0^{ev}(E_G) = E_G^{ev}$ and $g_0^{ev}(B_G) = B_G^{ev}$.

In summary we have the following proposition.

12.3 Proposition. *There are two homotopies* (h_s^{od}, g_s^{od}), $(h_s^{ev}, g_s^{ev}): \omega_G \to \omega_G$ *by G-bundle morphisms of the identity* $1_G: \omega_G \to \omega_G$, *such that* $h_0^{od}(E_G) = E_G^{od}$, $g_0^{od}(B_G) = B_G^{od}$, $h_0^{ev}(E_G) = E_G^{ev}$, *and* $g_0^{ev}(B_G) = B_G^{ev}$. *The bundles* ω_G, $(g_0^{od})^*\omega_G$, *and* $(g_0^{ev})^*\omega_G$ *are all isomorphic.*

Proof. Only the last statement has to be checked. It follows from $(h_1^{od}, g_1^{od}) = (h_1^{ev}, g_1^{ev}) = 1$ and Theorem (9.9).

Now we are in a position to check condition (2) of (10.6) for the bundle ω_G.

12.4 Theorem. *Let* $f_0, f_1: X \to B_G$ *be two maps such that* $f_0^*(\omega_G)$ *and* $f_1^*(\omega_G)$ *are isomorphic. Then* f_0 *and* f_1 *are homotopic.*

Proof. Let ξ be any numerable principal G-bundle isomorphic to $f_0^*(\omega_G)$ and $f_1^*(\omega_G)$. Then f_0 is homotopic to $g_0^{od}f_0$ by (12.3), and f_1 is homotopic to $g_0^{ev}f_1$ by (12.3). Consequently, by changing f_0 and f_1 up to homotopy, we can assume that $f_0(X) \subset B_G^{od}$ and $f_1(X) \subset B_G^{ev}$.

Now we define the G-morphism $(k, f): \xi \times [0, 1] \to \omega_G$ with $f|X \times 0 = f_0$ and $f|X \times 1 = f_1$. We have

$$k(z, 0) = (t_0(z)x_0(z), 0, t_2(z)x_2(z), 0, \ldots)$$

$$k(z, 1) = (0, t_1(z)x_1(z), 0, t_3(z)x_3(z), \ldots)$$

and we prolong to $k: E(\xi) \times I \to E_G$ by the function $k(z, s) = ((1 - s) \times t_0(z)x_0(z), st_1(z)x_1(z), (1 - s)t_2(z)x_2(z), st_3(z)x_3(z), \ldots)$. Since clearly for $y \in G$ we have $k(zy, s) = k(z, s)y$, the map $k(z, s)$ induces a map $f: X \times I \to B_G$ such that $f(b, 0) = f_0(b)$ and $f(b, 1) = f_1(b)$. This proves the theorem.

12.5 Summary. The bundle ω_G which arises from the Milnor construction is universal.

13. Homotopy Classification of Principal G-Bundles over CW-Complexes

Since all CW-complexes are paracompact (see Miyazaki [1]) and since all open coverings of a paracompact space are numerable, the above results on the homotopy classification of principal G-bundles apply to all locally trivial principal G-bundles over a CW-complex. Using the results of Chap. 2, Sec. 6, we derive the following more precise result.

13.1 Theorem. *Let B be a CW-complex, and let $\omega = (X_0, p_0, B_0)$ be a locally trivial principal G-bundle such that $\pi_i(X_0) = 0$ for $i \leq \dim B$. Then the function $\phi_\omega(B): [B, B_0] \to k_G(B)$ defined in the paragraph before (10.1) is a bijection.*

Proof. First, we prove that $\phi_\omega(B)$ is surjective. Let ξ be a locally trivial principal G-bundle over B. By Theorem 2 (7.1) under hypothesis (H1), the fibre bundle $\xi[X_0]$ has a cross section. By Theorem (8.2) we have a principal bundle morphism $(u, f): \xi \to \omega$, and by Theorem (4.2) we have $\{\xi\} = \{f^*(\omega)\} = \phi_\omega(B)([f])$.

Second, we prove that $\phi_\omega(B)$ is injective. Let $f, g: B \to B_0$ be maps such that $\{f^*(\omega)\} = \{g^*(\omega)\}$. By (8.2) the fibre bundle $(f^*(\omega) \times I)[X_0]$ has a cross section s over $B \times \{0, 1\}$, resulting from the principal bundle morphisms which are the compositions of $(f^*(\omega) \times I)|(B \times 0) \to f^*(\omega) \to \omega$ and $(f^*(\omega) \times I)|(B \times 1) \to g^*(\omega) \to \omega$. By Theorem 2 (7.1) under hypothesis (H1), this cross section s prolongs to a cross section s^* of $(f^*(\omega) \times I)[X_0]$ over $B \times I$. By (8.2), we have a principal bundle morphism $(w, h): f^*(\omega) \times I \to \omega$ such that $h(b, 0) = f(b)$ and $h(b, 1) = g(b)$ for $b \in B$. This proves the theorem.

Exercises

1. Let X be a G-space, and Y a Hausdorff G-space. Prove that $M_G(X, Y)$ is a closed subset of Map (X, Y).

2. Let X be a G-space, where G is a finite group. Prove that $X \to X$ mod G is a closed map.

3. Let H and G be two closed subgroups of a topological group Γ. If H is a normal subgroup of G, prove that Γ mod H has a principal G/H structure with an associated principal bundle (Γ mod H, p, Γ mod G), where p is the quotient of the identity $\Gamma \to \Gamma$.

4. Let ξ be a principal G-bundle, and let G act trivially on the left of F, that is, $sy = y$ for each $s \in G$, $y \in F$. Prove that $\xi[F]$ is a trivial bundle with fibre F.

5. Let ξ be a principal G-bundle, and let G act on the left of F in such a way that for some $y_0 \in F$ the result is $ty_0 = y_0$ for each $t \in G$. Prove directly that $\xi[F]$ has a cross section, and then prove the result, using Theorem (8.1).

6. In Theorem (8.1), to what extent are the functions $s \to \phi_s$ and $\phi \to s_\phi$ continuous if the set of cross sections and the set of G-morphisms have the subspace topology induced by the compact-open topology?

7. Let G be a discrete group. Formulate the property that X is a G-space and a principal G-space in terms of the topology of X and the action of G on X.

8. Find a universal bundle for the group $\mathbf{Z}$ whose corresponding classifying space is S^1.

9. Find a universal bundle and classifying space for Z_n such that the classifying space is a CW-complex.

10. Let $\xi_i = (X_i, p_i, B_i)$ be a principal G_i-bundle, $i = 1, 2$. Then $\xi_1 \times \xi_2 = (X_1 \times X_2,$ $p_1 \times p_2, B_1 \times B_2)$ has the structure of a principal $G_1 \times G_2$-bundle. If ξ_i is a universal G_i-bundle, $i = 1, 2$, then $\xi_1 \times \xi_2$ is a universal $G_1 \times G_2$-bundle.

    Apply this theorem to prove the existence of a universal bundle with a CW-complex as classifying space for each finitely generated abelian group (finite direct sum of cyclic groups).

11. If $u: G \to H$ is a continuous group morphism, define a map $B(u): B_G \to B_H$. Under what circumstances do you have a functor?

12. If X is an n-connected CW-complex and if Y is an m-connected CW-complex, prove that $X * Y$ is $(n + m + 1)$-connected.

13. Let $\xi = (X, p, B)$ be a numerable principal G-bundle. Prove that ξ is universal if and only if X is contractible.
    Hint: See the article of Dold on partitions of unity, Theorem (7.5).

CHAPTER 5

Local Coordinate Description of Fibre Bundles

In the first section, we show that, up to isomorphism, vector bundles are just locally trivial fibre bundles with a finite-dimensional vector space V as fibre and $\mathbf{GL}(V)$, the group of automorphisms of V, as a structure group. This is done by examining how trivial bundles are pieced together, using systems of transition functions to define a general locally trivial fibre bundle. We can apply this analysis to prove a theorem which says that any continuous functorial operation on vector spaces determines an operation on vector bundles. This allows construction of tensor products, exterior products, etc., of vector bundles.

1. Automorphisms of Trivial Fibre Bundles

In 3(2.3) we saw that the B-morphisms of the trivial vector bundles $u: B \times F^n \to B \times F^m$ have the form $u(b, x) = (b, f(b)u)$, where $f: B \to \mathbf{L}(F^n, F^m)$ is a map. Moreover, u is an automorphism if and only if $n = m$ and $f(b): F^n \to F^n$ is a linear isomorphism for each $b \in B$. A similar result holds for trivial fibre bundles.

1.1 Theorem. *Let $\xi = (B \times G, p, B)$ be the product principal G-bundle. Then the B-automorphisms $\xi \to \xi$ over B are in bijective correspondence with maps $B \to G$. More precisely, the B-automorphisms of ξ have the form $h_g(b, s) = (b, g(b)s)$, where $g: B \to G$ is a map.*

Proof. From the relation $h_g(b, st) = (b, g(b)st) = (b, g(b)s)t = h_g(b, s)t$, it follows that h_h is an automorphism with inverse morphism $h_g^{-1} = h_{g'}$, where $g'(b) = g(b)^{-1}$ for each $b \in B$. Conversery, let $h: \xi \to \xi$ be a B-automorphism.

Since $ph = p$, we have $h(b, s) = h(b, f(b, s))$ for some map $f: B \times G \to G$. For $g(b) = f(b, 1)$, we have $h(b, s) = h(b, 1)s = (b, g(b))s = (b, g(b)s) = h_g(b, s)$. This proves the theorem.

1.2 Corollary. *Let F be a left G-space. The fibre bundle automorphisms $\xi[F] \to \xi[F] = (B \times F, p, B)$ are all of the form $h_g(b, y) = (b, g(b)y)$, where $g: B \to G$ is a map.*

Proof. By (1.1), fibre bundle automorphisms are quotients of $(b, s, y) \mapsto (b, g(b)s, y)$. Since $(b, g(b)s, y) \bmod G = (b, g(b)y)$, the fibre bundle automorphisms are of the form $(b, y) \mapsto (b, g(b)y) = h_g(b, y)$.

Note that for $h_g(b, y) = (b, g(b)y)$ and $h_{g'}(b, y) = (b, g'(b)y)$ we have $h_{g'} h_g = h_{g'g}$, and h_g is the identity if and only if $g(b) = 1$ for each $b \in B$.

2. Charts and Transition Functions

Let G be a group, and let Y be a left G-space. In this section, all principal bundles are G-bundles, and all fibre bundles have fibre Y. For a space B, let $\theta(B)$ denote the product fibre bundle $(B \times Y, p, B)$. We view the total space of a restricted bundle $\eta|A$ as a subspace of the total space of η.

Since we have yet to relate formally the concepts of vector bundle and fibre bundle, the above definitions and results are stated for both concepts. A result of the following discussion is the relation between vector bundles and fibre bundles.

2.1 Definition. Let η be a fibre bundle over B, and let U be an open subset of B. A chart of η over U is an isomorphism $h: \theta(U) \to \eta|U$.

A chart of a k-dimensional vector bundle η is a U-vector bundle isomorphism $U \times F^k \to \eta|U$.

If $h: \theta(U) \to \eta|U$ is a chart over U, and if V is an open subset of U, then h restricts to a chart $\theta(V) \to (\eta|U)|V = \eta|V$ of η over V.

2.2 Proposition. *For two charts $h_1, h_2: \theta(U) \to \eta|U$ of η over U, there is a map $g: U \to G$ such that $h_1(b, y) = h_2(b, g(b)y)$ for each $(b, y) \in U \times Y$. Moreover, this g is unique. For two charts of n-dimensional vector bundles, the map g is defined $U \to \mathbf{GL}(n, F)$.*

Proof. By (1.2), the automorphism $h_2^{-1} h_1: \theta(U) \to \theta(U)$ has the form $h_2^{-1} h_1(b, y) = (b, g(b)y)$, and we have $h_1(b, y) = h_2(b, g(b)y)$.

2.3 Definition. An atlas of charts for a fibre (vector) bundle η is a family of pairs $\{(h_i, V_i)\}$ for $i \in I$ such that h_i is a chart of η over V_i and the family of open sets $\{V_i\}$ covers B. An atlas is complete provided it includes all charts of the bundle.

A fibre bundle is locally trivial if and only if it has an atlas of charts. Then it has a unique complete atlas. The covering associated with the atlas $\{(h_i, V_i)\}$ is the open covering $\{V_i\}$. A vector bundle is defined in terms having an atlas of charts [see 3(1.1)].

Let $\{(h_i, V_i)\}$ with $i \in I$ be an atlas for a fibre (vector) bundle η. We restrict h_i and h_j to $V_i \cap V_j$ and apply Proposition (2.2). There exists a unique map $g_{i,j}: V_i \cap V_j \to G$ such that $h_j(b, y) = h_i(b, g_{i,j}(b)y)$ for $(b, y) \in (V_i \cap V_j) \times Y$. The functions $g_{i,j}$ on $V_i \cap V_j$ have the following properties:

(T1) For each $b \in V_i \cap V_j \cap V_k$, the relation $g_{i,k}(b) = g_{i,j}(b)g_{j,k}(b)$ holds.
(T2) For each $b \in V_i$, the map $g_{i,i}(b)$ is equal to the identity in G.
(T3) For each $b \in V_i \cap V_j$, the relation $g_{i,j}(b) = g_{j,i}(b)$ holds.

Properties (T1) to (T3) follow from the fact that $g_{i,j}$ is the only map satisfying the relation $h_j(b, y) = h_i(b, g_{i,j}(b)y)$.

2.4 Definition. A system of transition functions on a space B relative to an open covering $\{V_i\}$ with $i \in I$ of B is a family of maps $g_{i,j}: V_i \cap V_j \to G$ for each $i, j \in I$ such that propertyy (T1) is satisfied.

Since $g_{i,i}(b)g_{i,i}(b) = g_{i,i}(b)$ for $b \in V_i$, it follows that (T2), and similarly (T3), is satisfied for a system of transition functions.

From the above discussion, there is a natural system of transition functions $\{g_{i,j}\}$ associated with each atlas $\{(h_i, V_i)\}$ of a fibre (vector) bundle η, namely, those functions defined by the relations $h_j(b, y) = h_i(b, g_{i,j}(b)y)$.

2.5 Proposition. *Let* $\{(h_i, V_i)\}$ *and* $\{h'_i, V_i)\}$ *be two atlases for a fibre (vector) bundle* η *with the same associated covering* $\{V_i\}$ *for* $i \in I$ *and with systems of transition functions* $\{g_{i,j}\}$ *and* $\{g'_{i,j}\}$. *Let* $r_i: V_i \to G$ *be the unique maps such that* $h'_i(b, y) = h_i(b, r_i(b)y)$ *for* $b \in V_i$ *and* $y \in Y$. *Then* $g'_{i,j}(b) = r_i(b)^{-1}g_{i,j}(b)r_j(b)$ *for each* $b \in V_i \cap V_j$.

Proof. We compute $h'_j(b, y) = h_j(b, r_j(b)y) = h_i(b, g_{i,j}(b)r_j(b)y)$ and $h'_i(b, g'_{i,j}(b)y) = h_i(b, r_i(b)g'_{i,j}(b)y)$. Since $h'_j(b, y) = h'_i(b, g'_{i,j}(b)y)$, we have $r_i(b)g'_{i,j}(b) = g_{i,j}(b)r_j(b)$ or $g'_{i,j}(b) = r_i(b)^{-1}g_{i,j}(b)r_j(b)$.

This proposition leads to the next definition.

2.6 Definition. Two systems of transition functions $\{g_{i,j}\}$ and $\{g'_{i,j}\}$ relative to the same open covering $\{V_i\}$ of a space B are equivalent provided there exist maps $r_i: V_i \to G$ satisfying the relation (E): $g'_{i,j}(b) = r_i(b)^{-1}g_{i,j}(b)r_j(b)$ for each $b \in V_i \cap V_j$.

The reader can easily verify that this relation is an equivalence relation.

2.7 Theorem. *Let* η *and* η' *be two fibre bundles with fibre* F *and structure group* G *over a space* B *or two vector bundles of dimension* k *with* $G = \mathrm{GL}(k, F)$. *Let* $\{(V_i, h_i)\}$ *be an atlas of* η *with transition functions* $\{g'_{i,j}\}$, *and let*

$\{(V_i, h_i')\}$ be an atlas of η' with transition functions $\{g_{i,j}'\}$. Then η and η' are isomorphic over B if and only if $\{g_{i,j}\}$ and $\{g_{i,j}'\}$ are equivalent systems of transition functions.

Proof. Let $f: \eta \to \eta'$ be an isomorphism of fibre bundles or vector bundles. From the relation $h_j(b, y) = h_i(b, g_{i,j}(b)y)$, it follows that $fh_j(b, y) = fh_i(b, g_{i,j}(b)y)$, and $\{(V_i, fh_i)\}$ is an atlas for η' with transition functions $\{g_{i,j}\}$. By applying Proposition (2.5) to the atlases $\{V_i, h_i'\}$ and $\{(V_i, fh_i)\}$, we find that $\{g_{i,j}'\}$ and $\{g_{i,j}\}$ are equivalent sets of transition functions.

Conversely, let $g_{i,j}'(b) = r_i(b)^{-1} g_{i,j}(b) r_j(b)$ for each $b \in V_i \cap V_j$. We define f_i: $V_i \times Y \to V_i \times Y$ by $f_i(b, y) = (b, r_i(b)^{-1}y)$, and we define $f: \eta \to \eta'$ by requiring that $f|(\eta|V_i) = h_i' f_i h_i^{-1}$ or $h_i' f_i = fh_i$ on $V_i \times Y$. For $b \in V_i \cap V_j$ the two definitions of f lead to the same map. To see this, we choose $(b, y) \in (V_i \cap V_j) \times Y$ and prove that $h_j' f_j(b, y) = fh_j(b, y)$ implies $h_i' f_i(b, y) = fh_i(b, y)$. For this we make the following calculation: $h_j' f_j(b, y) = h_j'(b, r_j(b)^{-1}y) = h_i'(b, g_{i,j}'(b) r_j(b)^{-1}y) = h_i'(b, r_i^{-1}(b) g_{i,j}(b)y) = h_i' f_i(b, g_{i,j}(b)y)$. Using $fh_j(b, y) = fh_i(b, g_{i,j}(b)y)$, we have $fh_i(b, g_{i,j}(b)y) = h_i' f_i(b, g_{i,j}(b)y)$ or $fh_i = h_i' f_i$. Therefore, f is a well-defined map that is locally, and consequently globally, an isomorphism.

2.8 Remark. Isomorphism classes of k-dimensional vector bundles and fibre bundles with fibre F^k and group $\mathbf{GL}(k, F)$ are determined by transition functions that have values in $\mathbf{GL}(k, F)$. In the next section we see that a bijection can be constructed between these two sets of isomorphism classes.

The discussion up to this point has been relative to a given covering $\{V_i\}$ such that $\eta|V_i$ is trivial. If $\{V_j'\}$ is a second open covering such that $\eta|V_j'$ is trivial, by working with $\{V_i \cap V_j'\}$ we get an open covering for which charts over V_i and V_j can be compared.

3. Construction of Bundles with Given Transition Functions

3.1 Remark. Let $\eta = \xi[Y]$ be a fibre bundle with fibre Y, structure group G, and base space B. If $\{(V_i, k_i)\}$ is an atlas of η with transition functions $\{g_{i,j}\}$, there is an atlas $\{(V_i, h_i)\}$ of ξ with transition functions $\{g_{i,j}\}$, where $h_i([Y]) = k_i$.

3.2 Theorem. *Let $\{V_i\}$ with $i \in I$ be an open covering of a space B, let G be a topological group, let Y be a left G-space, and let $\{g_{i,j}\}$ be a system of transition functions associated with the open covering $\{V_i\}$. Then there exist a fibre bundle $\eta = \xi[Y]$ and an atlas $\{(V_i, h_i)\}$ for η such that the set of transition functions of this atlas is $\{g_{i,j}\}$. Moreover, if $Y = F^k$ and if G is a closed subgroup of $\mathbf{GL}(k, F)$, then η admits the structure of a vector bundle. Finally, η is unique up to B-isomorphism.*

Proof. By Theorem (2.7), if η exists, it is unique. We begin by constructing ξ. For this, let Z be the sum space (i.e., coproduct or disjoint union) of the family $\{V_i \times G\}$ for $i \in I$. An element of Z is of the form (b, s, i), where $b \in V_i$ and $s \in G$. The inclusion maps $q_i: V_i \times G \to Z$ are defined by the relation $q_i(b, s) = (b, s, i)$, and Z has the largest topology such that all the q_i are continuous.

On the space Z, we define an equivalence relation R by the requirement that (b, s, i) and (b', s', j) are R-related provided $b = b'$ and $s' = g_{j,i}(b)s$. From properties (T1) to (T3) for transition functions we see that R is an equivalence relation. Let X be the quotient space Z mod R, let $q: Z \to X$ be the canonical quotient map, and let $h_i = qq_i$ for each $i \in I$. We denote by $\langle b, s, i \rangle$ the class of (b, s, i) in the space X.

We define $p: X \to B$ by $p(\langle b, s, i \rangle) = b$. Since p is a quotient of a projection, it is an open map. For $b \in V_i$, we have $ph_i(b, s) = b$, and $h_i: V_i \times G \to X$ is a continuous injection.

The group acts on Z by the requirement that $(b, s, i)t = (b, st, i)$. If (b, s, i) and (b, s', j) are R-related, then (b, st, i) and $(b, s't, j)$ are R-related. Therefore, X becomes a G-space under the action of G defined by $\langle b, s, i \rangle t = \langle b, st, i \rangle$. Clearly, we have $p(x) = p(x')$ if and only if $xt = x'$ for some $t \in G$, and $xt = x$ implies that $t = 1$. The translation function $\tau(\langle b, s_1, i \rangle, \langle b, s_2, j \rangle) = \tau(\langle b, s_1, i \rangle, \langle b, g_{i,j}(b)s_2, i \rangle) = s_1^{-1} g_{i,j}(b)s_2$ is continuous. Consequently, (X, p, B) is a principal G-bundle since p is an open map.

Since $h_i(b, s)t = \langle b, s, i \rangle t = \langle b, st, i \rangle = h_i(b, st)$, the maps $h_i: V_i \times G \to \xi | V_i$ are G-isomorphisms. From the relation $h_i(b, g_{i,j}(b)s) = \langle b, g_{i,j}(b)s, i \rangle = \langle b, s, j \rangle = h_j(b, s)$, the set of functions $\{g_{i,j}\}$ is the set of transition functions for the atlas $\{(V_i, h_i)\}$.

Finally, to define a vector bundle structure on $\eta = \xi[F^k]$, we require $a(x, y) \bmod G + a'(x, y') \bmod G = (x, ay + a'y') \bmod G$. If $k_i: V_i \times F^k \to p^{-1}(V_i)$ is the chart given by the relation $k_i(b, y) = ((b, 1, i)y) \bmod G$, then k_i is a V_i-isomorphism of vector bundles.

3.3 Remark. There is a bijection between isomorphism classes of fibre bundles with fibre F^k and group $\mathbf{GL}(k, F)$ and isomorphism classes of k-dimensional vector bundles. In both cases these isomorphism classes are determined by an equivalence class of transition functions associated with a complete atlas of a member of the isomorphism class. Moreover, if ξ is a principal $\mathbf{GL}(k, F)$-bundle, then $\xi[F^k]$ is a vector bundle where the vector space operations on F^k determine the vector space operations on the fibres of $\xi[F^k]$.

4. Transition Functions and Induced Bundles

In this next proposition we calculate the transition functions of induced bundles.

4.1 Proposition. *Let* $\eta = \xi[Y]$ *be a fibre bundle over* B *with group* G *and fibre* Y, *let* $f: B_1 \to B$ *be a map, and let* $\{(V_i, h_i)\}$ *be an atlas for* η *with transition functions* $\{g_{i,j}\}$. *Then* $\{(f^{-1}(V_i), f^*(h_i))\}$ *is an atlas for* $f^*(\eta) = f^*(\xi)[Y]$ *with transition functions* $\{g_{i,j}f\}$.

Proof. By 4(4.2) and 4(6.3) the morphism $f^*(h_i): \theta(f^{-1}(V_i)) \to f^*(\eta)|f^{-1}(V_i)$ is an isomorphism given by $f^*(h_i)(b_1, y) = (b_1, h_i(f(b_1), y))$. If $h_i(b, g_{i,j}(b)y) = h_j(b, y)$, we have $f^*(h_i)(b_1, g_{i,j}(f(b_1)))y) = f^*(h_j)(b_1, y)$, and $\{g_{i,j}f\}$ is the set of transition functions for $f^*(\eta)$.

5. Local Representation of Vector Bundle Morphisms

Let $\xi = (X, p, B)$, $\eta = (X', p', B)$, and $\zeta = (X'', p'', B)$ be three vector bundles over B with atlases $\{(U_a, h_a)\}$ for $a \in A$, $\{(V_i, h'_i)\}$ for $i \in I$, and $\{(W_r, h''_r)\}$ for $r \in R$, respectively. Let $\{g_{a,b}\}$ for $a, b \in A$, $\{g'_{i,j}\}$ for $i, j \in I$, and $\{g''_{r,s}\}$ for r, $s \in R$ be the transition functions for ξ, η, and ζ, respectively.

Let $u: \xi \to \eta$ be a vector bundle morphism. Over the open set $U_a \cap V_i$, there are the following vector bundle morphisms:

$$(U_a \cap V_i) \times F^n \xrightarrow{h_a} \xi|(U_a \cap V_i) \xrightarrow{u} \eta|(U_a \cap V_i) \xrightarrow{(h'_i)^{-1}} (U_a \cap V_i) \times F^m$$

The composition of these morphisms has the form $(z, x) \mapsto (z, u_{i,a}(z)x)$, where $u_{i,a}: U_a \cap V_i \to L(F^n, F^m)$ is a map.

5.1 Proposition. *With the above notations, there is a bijection between the set of vector bundle morphisms* $u: \xi \to \eta$ *and sets of maps* $\{u_{i,a}\}$, *where* $i \in I$, $a \in A$, *and* $u_{i,a}: U_a \cap V_i \to L(F^n, F^m)$ *such that for* $z \in U_a \cap U_b \cap V_i \cap V_j$

$$(C): u_{j,b}(z) = g'_{j,i}(z)u_{i,a}(z)g_{a,b}(z)$$

Moreover, the maps $u_{i,a}$ *corresponding to* u *are given by* $h'^{-1}_i uh_a(z, x) = (z, u_{i,a}(z)x)$.

Proof. For given $u: \xi \to \eta$ we begin by verifying the relation (C) of compatibility. For this, we calculate

$$h'_j(z, u_{j,b}(z)x) = uh_b(z, x) = uh_a(z, g_{a,b}(z)x) = h'_i(z, u_{i,a}(z)g_{a,b}(z)x)$$

$$= h'_j(z, g'_{j,i}(z)u_{i,a}(z)g_{a,b}(z)x).$$

Since h'_i is an isomorphism, we have the desired relation.

Conversely, for each family $\{u_{i,a}\}$ of maps satisfying (C), a corresponding morphism $u: \xi \to \eta$ is defined uniquely by the relation $uh_a(z, x) = h'_i(z, u_{i,a}(z)x)$. Observe that the images of h_a cover X and the compatibility relation (C) says that u is uniquely determined where the images of h_a and of h_b intersect. This proves the proposition.

For $\xi = \eta$ and the same local charts for each vector bundle, we observe that $u = 1$ if and only if $u_{i,a}(z) = 1$ for each $z \in U_i \cap V_a$ and $i \in I$, $a \in A$. We speak of the family $\{u_{i,a}\}$ representing $u: \xi \to \eta$ with respect to the charts $\{(U_a, h_a)\}$ of ξ and $\{(V_i, h_i')\}$ of η.

5.2 Proposition. *With the above notations, let $\{u_{i,a}\}$ represent the morphism $u: \xi \to \eta$, let $\{v_{r,i}\}$ represent the morphism $v: \eta \to \zeta$, and let $\{w_{r,a}\}$ represent the morphism $uv: \xi \to \zeta$. Then for $z \in U_a \cap V_i \cap W_r$ we have the relation $w_{r,a}(z) = v_{r,i}(z)u_{i,a}(z)$.*

Proof. For $z \in U_a \cap V_i \cap W_r$, we have $(z, w_{r,a}(z)x) = (h_r'')^{-1}vuh_a(z, x) = ((h_r'')^{-1}vh_i')((h_i')^{-1}uh_a)(z, x) = ((h_r'')^{-1}vh_i')(z, u_{i,a}(z)x) = (z, v_{r,i}(z)u_{i,a}(z)x)$. This proves the desired result.

In (5.2), the $w_{r,a}: U_a \cap W_r \to L(F^n, F^q)$ are determined by the relation $w_{r,a}(z) = v_{r,i}(z)u_{i,a}(z)$ for $z \in U_a \cap V_i \cap W_r$.

The result 3(2.5) is an immediate corollary of the analsis in (5.1) and (5.2).

6. Operations on Vector Bundles

We wish to prove that every (continuous) operation on vector spaces defines a corresponding operation on vector bundles. This will allow us to speak of the direct sum $\xi \oplus \eta$ of two vector bundles, which is the Whitney sum; the tensor product $\xi \otimes \eta$; the vector bundle $\operatorname{Hom}(\xi, \eta)$; and the rth exterior power $\Lambda^r \xi$, to mention a few examples.

Recall the notation $\mathbf{VB}_B$ for the category of all vector bundles over a space B. For B equal to a point, the category $\mathbf{VB}_B$ can be viewed as the category of vector space. Let $\mathbf{VB}_B(p, q)$ denote the product category consisting of p copies of $\mathbf{VB}_B$ and q copies of $\mathbf{VB}_B^*$, the dual category of $\mathbf{VB}_B$. In the next definition we make precise the definition of a continuous operation.

6.1 Definition. Let 0 denote the one-point space. A functor $F: \mathbf{VB}_0(p, q) \to \mathbf{VB}_0$ is called continuous provided for each family of maps $u_i: Z \to L(V_i, W_i)$, where $1 \le i \le p + q$, the function $z \mapsto F(u_1, \ldots, u_{p+q})$ is a map

$$Z \to L(F(V_1, \ldots, V_p, W_{p+1}, \ldots, W_{p+q}), F(W_1, \ldots, W_p, V_{p+1}, \ldots, V_{p+q}))$$

The following functors are continuous: $V \oplus W$, $V \otimes W$, and $\operatorname{Hom}(V, W)$.

In the next theorem we see that a continuous functor on $\mathbf{VB}_0$ defines a functor on each $\mathbf{VB}_B$.

6.2 Theorem. *For each continuous functor $F: \mathbf{VB}_0(p, q) \to \mathbf{VB}_0$ there exists a family of functors $F_B: \mathbf{VB}_B(p, q) \to \mathbf{VB}_B$, one for each space B, such that $F_{B_1}(f^*(\xi_1), \ldots, f^*(\xi_{p+q}))$ and $f^*F_B(\xi_1, \ldots, \xi_{p+q})$ are B_1-isomorphic bundles for each map $f: B_1 \to B$. Moreover, it is required that $F = F_0$.*

Proof. We carry out the proof for the case $p = q = 1$. Let ξ and ξ' be two vector bundles over B with local coordinate charts $\{(U_a, h_a)\}$ and $\{(U_a, k_a)\}$ with transition functions $\{g_{a,b}\}$ and $\{f_{a,b}\}$ for $a, b \in A$. We define $F_B(\xi, \xi')$ to be a vector bundle with the (continuous) transition functions $\{F_B(g_{a,b}, f_{b,a})\}$ for $a, b \in A$.

If $u: \xi \to \eta$ and $u': \eta' \to \xi'$ are morphisms, they are represented by $\{u_{i,a}\}$ and $\{u'_{i,a}\}$, respectively, where $\{(V_i, h'_i)\}$ and $\{(V_i, k'_i)\}$ are atlases of η and η', respectively. Then the family of maps $F(u_{i,a}, u'_{i,a})$ defines a morphism $F_B(\xi, \xi') \to F_B(\eta, \eta')$, by (5.1), which is denoted $F_B(u, u')$. Since F is a functor, it preserves the compatibility condition (C) of Proposition (5.1). Clearly, $F_B(1, 1) = 1$ by the remark following Proposition (5.1). With Proposition (5.2), we see that $F_B(vu, u'v') = F_B(v, v')F_B(u, u')$ by applying F to the local morphisms representing u, u', v, and v', where $v: \eta \to \zeta$ and $v': \zeta' \to \eta'$ are morphisms.

Finally, for a map $f: B_1 \to B$ we find that $F_{B_1}(f^*(\xi), f^*(\xi'))$ and $f^*(F_B(\xi, \xi'))$ are B_1-isomorphic. For this, observe that $\{F(g_{a,b}f, f_{b,a}f)\}$ for a, $b \in A$ is a set of transition functions for both $f^*(F_B(\xi, \xi'))$ and $F_{B_1}(f^*(\xi), f^*(\xi'))$ with respect to the open covering $\{f^{-1}(U_a)\}$, where $a \in A$ by Proposition (4.1).

In the next theorem we investigate to what extent the functors F_B are uniquely determined by F.

6.3 Theorem. *Let F, $G: \mathbf{VB}_0(p, q) \to \mathbf{VB}_0$ be two continuous functors, and let $\phi: F \to G$ be a morphism of functors. Then for each space B there exists a morphism $\phi_B: F_B \to G_B$ of functors defined in (6.2), where $\phi_B(\xi_1, \ldots, \xi_{p+q})$: $F_B(\xi_1, \ldots, \xi_{p+q}) \to G_B(\xi_1, \ldots, \xi_{p+q})$ restricted to the fibre over $z \in B$ is just*

$$\phi(\xi_{1,z}, \ldots, \xi_{p+q,z}): F(\xi_{1,z}, \ldots, \xi_{p+q,z}) \to G(\xi_{1,z}, \ldots, \xi_{p+q,z})$$

With respect to this property, ϕ_B is unique.

Proof. For product vector bundles ξ_i of dimension $r(i)$, we have $\phi_B(\xi_1, \ldots, \xi_{p+q}) = 1_B \times \phi(F^{r(1)}, \ldots, F^{r(p+q)})$. Since every vector bundle is locally trivial, the $\phi_B(\xi_1, \ldots, \xi_{p+q})$ are well-defined morphisms by the above requirement. The uniqueness is clear.

6.4 Corollary. *If $\psi: G \to H$ is a second morphism in (6.3), then $(\psi\phi)_B = \psi_B\phi_B$. If $\phi: F \to F$ is the identity, $\phi_B: F_B \to F_B$ is an isomorphism. If $\phi: F \to G$ is an isomorphism, then $\phi_B: F_B \to G_B$ is an isomorphism.*

Proof. This results from the uniqueness statement in (6.3).

6.5 Remark. The functors F_B in (6.2) are unique up to isomorphism. Now we discuss the following examples.

6.6 Example. If ξ and η are two vector bundles over B, the Whitney sum $\xi \oplus \eta$ is the prolongation to vector bundles of the direct sum functor. This follows from the fact that the fibre of $\xi \oplus \eta$ over a point of B is the direct sum of the fibres of ξ and of η. The usual properties of direct sums of vector spaces prolong to Whitney sums of vector bundles, by (6.3) and (6.4). There are the following isomorphisms, for example:

$$\xi \oplus \eta \cong \eta \oplus \xi \quad \text{and} \quad \xi \oplus (\eta \oplus \zeta) \cong (\xi \oplus \eta) \oplus \zeta$$

6.7 Example. The tensor product functor is continuous and, in view of (6.2), we may define the tensor product $\xi \otimes \eta$ of two vector bundles over B. If $u: \xi \oplus \eta \to \zeta$ is a bundle map that is bilinear on each fibre $b \in B$, then u defines a vector bundle morphism $v: \xi \otimes \eta \to \zeta$ which is the usual factorization of bilinear maps through the tensor product on each fibre. Using (6.3) and (6.4), we have the following isomorphisms:

$$\xi \otimes \eta \cong \eta \otimes \xi \quad \xi \otimes (\eta \otimes \zeta) \cong (\xi \otimes \eta) \otimes \zeta$$
$$\xi \otimes \theta^1 \cong \xi \quad \text{and} \quad \xi \otimes (\eta \oplus \zeta) \cong (\xi \otimes \eta) \oplus (\xi \otimes \zeta)$$

where ξ, η, ζ are vector bundles over B and θ^1 is the trivial line bundle over B. This discussion holds for real and complex vector bundles.

6.8 Example. The homomorphism functor $\text{Hom}(V, W)$ is continuous, and in view of (6.2), we may define the homomorphism vector bundle $\text{Hom}(\xi, \eta)$. The fibre over $b \in B$ is the vector space of homomorphisms $\xi_b \to \eta_b$, where ξ_b and η_b are the fibres of ξ and η over $b \in B$, respectively. A cross section s of $\text{Hom}(\xi, \eta)$ is just a vector bundle morphism $u: \xi \to \eta$. The continuity of s and the continuity of u are equivalent to each other.

6.9 Example. The rth exterior power functor $\Lambda^r V$ is continuous, and, in view of (6.2), we may define the rth exterior power of $\Lambda^r \xi$ of a vector bundle ξ. In the next proposition we relate the functors in (6.6), (6.7), and (6.9).

6.10 Proposition. *Let* $\xi = \lambda_1 \oplus \cdots \oplus \lambda_n$ *be the Whitney sum of n line bundles. Then for $r \leq n$ there is an isomorphism*

$$\Lambda^r(\lambda_1 \oplus \cdots \oplus \lambda_n) \cong \sum_{i(1) < \cdots < i(r)} (\lambda_{i(1)} \otimes \cdots \otimes \lambda_{i(r)})$$

Proof. The above relation holds for vector spaces, and by (6.4) it holds for vector bundles.

7. Transition Functions for Bundles with Metrics

The following subgroups of the full linear group will play a very important role in subsequent developments. For $x \in \mathbf{R}$, the real numbers let $\bar{x} = x$; for $x \in \mathbf{C}$, the complex numbers, let $\bar{x} = x_1 - ix_2$, where $x = x_1 + ix_2$; and for

$x \in \mathbf{H}$, the quaternions of Hamilton, let $\bar{x} = x_0 - ix_1 - jx_2 - kx_3$, where $x = x_0 + ix_1 + jx_2 + kx_3$. Let F equal $\mathbf{R}$, $\mathbf{C}$, or $\mathbf{H}$. On the vector space F^n we define the inner product $(x|y) = x_1\bar{y}_1 + \cdots + x_n\bar{y}_n$ and the norm $\|x\| = (x|x)^{1/2}$ for $x = (x_1, \ldots, x_n)$, $y = (y_1, \ldots, y_n) \in F^n$.

7.1 Definition. The orthogonal group in k dimensions, denoted $O(k)$, is the subgroup of $u \in \mathbf{GL}(k, \mathbf{R})$ such that $(u(x)|u(y)) = (x|y)$ for each x, $y \in \mathbf{R}^k$. The unitary group in k dimensions, denoted $U(k)$, is the subgroup of $u \in \mathbf{GL}(k, \mathbf{C})$ such that $(u(x)|u(y)) = (x|y)$ for each x, $y \in \mathbf{C}^k$. The symplectic group in k dimensions, denoted $Sp(k)$, is the subgroup of $u \in \mathbf{GL}(k, \mathbf{H})$ such that $(u(x)|u(y)) = (x|y)$ for each x, $y \in \mathbf{H}^k$.

In each case, $O(k)$, $U(k)$, and $Sp(k)$ are closed and bounded subsets of the space of matrices. Therefore, they are compact (topological) groups. There are important subgroups of $O(k)$ and $U(k)$.

7.2 Definition. The special orthogonal group in k dimensions, denoted $SO(k)$, is the closed subgroup of $u \in O(k)$ with det $u = +1$. The special unitary group in k dimensions, denoted $SU(k)$, is the closed subgroup of $u \in U(k)$ with det $u = +1$.

Further properties of $O(k)$, $SO(k)$, $U(k)$, $SU(k)$, and $Sp(k)$ are developed in later chapters. These groups are referred to as the classical groups.

In the next proposition, we see that a standard orthonormalization process can be applied to cross sections of vector bundles with a metric.

7.3 Proposition. *Let ξ be a vector bundle over B with a metric β, and let $s_1, \ldots, s_m$ be cross sections of ξ such that $s_1(b), \ldots, s_m(b)$ are linearly independent. Then there exist cross sections $s_1^*, \ldots, s_m^*$ of ξ which are linear combinations of $s_1, \ldots, s_m$ with continuous scalar-valued functions as scalars such that $\beta(s_i^*, s_j^*) = \delta_{i,j}$.*

Proof. Let $s_1^*(b)$ equal $s_1(b)$ divided by its length $\beta(s_1(b), s_1(b))^{1/2}$ (which is >0) at each $b \in B$. If the $s_1^*, \ldots, s_{k-1}^*$ have been chosen with $\beta(s_i^*, s_j^*) = \delta_{i,j}$, $1 \leqq i, j \leqq k - 1$, we define $s_k^*(b)$ to be $s_k(b) - \sum_{1 \leqq j \leqq k-1} (s_k(b)|s_j^*(b))s_j^*(b)$ divided by its length. In this way we define $s_1^*, \ldots, s_m^*$ with the desired properties.

The following theorem says that vector bundles over a paracompact space have the orthogonal (unitary or symplectic) group as the structure group.

7.4 Theorem. *Let ξ be a vector bundle over B with a metric β. Then ξ has an atlas $\{(V_i, h_i^*)\}$ of charts such that $(x|y) = \beta(h_i^*(b, x), h_i^*(b, y))$ for each x, $y \in F^n$ and $b \in V_i$. The transition functors $\{g_{i,j}\}$ of this atlas have their values in $O(n)$, the real case with $F = \mathbf{R}$; $U(n)$, the complex case with $F = \mathbf{C}$; and $Sp(n)$, the quaternionic case with $F = \mathbf{H}$.*

Proof. Let $\{(V_i, h_i)\}$ be an atlas of ξ. Then there are cross sections $h_i(b, e_j) = s_j(b)$ for $1 \leq j \leq n$ of ξ over V_i which are a base of the fibre over each $b \in V_i$. By Proposition (7.3) there are n cross sections $s_1^*, \ldots, s_n^*$ of ξ over V_i such that $\beta(s_i^*, s_j^*)$ equals 1 for $i = j$ and 0 for $i \neq j$ at each $b \in V_i$. We define $h_i^*: V_i \times F^n \to \xi \mid V_i$ by $h_i^*(b, a_1, \ldots, a_n) = a_1 s_1^*(b) + \cdots + a_n s_n^*(b)$. Then h_i^* is a chart of ξ over V_i, and $\{(V_i, h_i^*)\}$ is an atlas with the desired property.

For the last statement, we recall that $h_i(b, g_{i,j}(b)x) = h_j(b, x)$ for $b \in V_i \cap V_j$ and $x \in F^n$. For $b \in V_i \cap V_j$ we have $(x \mid y) = \beta(h_i^*(b, x), h_j^*(b, y)) = \beta(h_i^*(b, g_{i,j}(b)x), h_j^*(b, g_{i,j}(b)x)) = (g_{i,j}(b)x \mid g_{i,j}(b)y)$. Consequently, the last statement follows for $F = \mathbf{R}, \mathbf{C},$ or $\mathbf{H}$.

7.5 Remark. By 3(9.5) every vector bundle over a paracompact space has a metric β, and therefore Theorem (7.4) applies. This theorem applies to the real, complex, and quaternionic cases.

7.6 Definition. Let ξ be a vector bundle. Then the conjugate bundle to ξ, denoted $\bar{\xi}$, has the same underlying structure (E, p, B) and addition morphism $E \oplus E \to E$. The scalar multiplication map is given by ax equal to $\bar{a}x$ in ξ.

For a real vector bundle, $\xi = \bar{\xi}$.

7.7 Definition. A line bundle is a one-dimensional vector bundle.

7.8 Theorem. *Let ξ be a real or complex line bundle with a metric (riemannian or hermitian). Then $\xi \otimes \bar{\xi}$ is a trivial line bundle.*

Proof. A metric $\beta: \xi \oplus \xi \to F$, where $F = \mathbf{R}$ or $\mathbf{C}$, defines a vector bundle morphism $u: \xi \otimes \bar{\xi} \to B \times F$. Since u is a surjective morphism of one-dimensional vector bundles, u is an isomorphism by 3(2.5). This proves the theorem.

7.9 Remark. Theorem (7.8) holds over every paracompact space.

Exercises

1. In Theorem (1.1) and Corollary (1.2) determine to what extent $g \mapsto h_g$ is a homeomorphism of Map (B, G) onto $\mathrm{Hom}_G(B \times G, B \times G)$, where $\mathrm{Hom}_G(B \times G, B \times G)$ has the subspace topology from Map $(B \times G, B \times G)$.

2. Let η be a locally trivial fibre bundle with group G and transition functions $\{g_{i,j}\}$, for $i, j \in I$, defined with respect to an open covering $\{V_i\}$ of the base space. Prove that η is trivial if and only if there exist maps $r_i: V_i \to G$ with $g_{i,j}(b) = r_i(b)^{-1} r_j(b)$ for $b \in V_i \cap V_j$.

3. For the canonical principal bundles $p: S^n \to RP^n$ and $p: S^{2n+1} \to CP^n$ determine an atlas and compute the transition functions.

4. Let U and V be two open subsets of a space B such that $B = U \cup V$ and $U \cap V = W_1 \cup \cdots \cup W_k$, the connected components of $U \cap V$. Determine, up to isomorphism, all principal G-bundles ξ over B for G, a discrete group such that $\xi|U$ and $\xi|V$ are trivial.

5. Let ξ be a k-dimensional vector bundle with atlas $\{(V_i, h_i)\}$ and transition functions $\{g_{i,j}\}$, where $g_{i,j}(b) \in O(k)$ for each $b \in V_i \cap V_j$. Prove that ξ has a metric β and, moreover, that the metric β is unique under the requirement that each h_i is metric preserving.

CHAPTER 6
Change of Structure Group in Fibre Bundles

In this chapter we consider the relation between principal H-bundles and principal G-bundles, where H is a closed subgroup of G. We do this for general principal bundles and then describe the relation, using the classifying spaces and the local coordinate description. This is a generalization of the process in Chap. 5, Sec. 7.

1. Fibre Bundles with Homogeneous Spaces as Fibres

Let $\xi = (X, p, B)$ be a principal G-bundle, and let H be a closed subgroup of G. Then the relation on X defined by the action of the group H is compatible with the projection $p: X \to B$. Therefore, there is a bundle $\xi \bmod H = (X \bmod H, q, B)$, where q is the result of factoring p by the canonical map $X \to X \bmod H$.

1.1 Theorem. *Let $\xi = (X, p, B)$ be a principal G-bundle, and let H be a closed subgroup of G. Then there is a canonical B-isomorphism of bundles $\xi \bmod H \to \xi[G \bmod H]$, where the fibre $G \bmod H$ is the homogeneous space of right cosets of H in G.*

Proof. Let $h: X \bmod H \to X_{G \bmod H}$ be defined by the relation $h(xH) = (x, eH)G$. For $v \in H$, we have $h(xH) = h(xvH) = (xv, eH)G = (xv, v^{-1}eH)G = (x, eH)G$, and h is a well-defined function. Since $X \bmod H$ has the quotient topology, h is continuous. For each $u \in G$, we have $(x, uH)G = (xu, eH)G$, and h is surjective. For $h(xH) = h(x'H)$, we have $(x, eH)G = (x', eH)G$ and $x' = xv$ for $v \in H$. Consequently, h is injective.

Finally, we prove that h^{-1} is continuous. The functions $g_1: X \times G \to X$ and $g_2: X \times (G \bmod H) \to X \bmod H$ given by the relations $g_1(x, u) = xu$ and $g_2(x, uH) = xuH$ are continuous. Since $g_2(x, uH) = g_2(xv, v^{-1}uH)$, the map g_2 induces a map $(X \times (G \bmod H)) \bmod G \to X \bmod H$ which is h^{-1}. This proves the theorem.

1.2 Corollary. *The principal G-bundle ξ and the fibre bundle $\xi[G]$, where G acts on G by left multiplication, are isomorphic as bundles over B.*

The corollary can be seen easily by a direct argument.

2. Prolongation and Restriction of Principal Bundles

2.1 Definition. For a closed subgroup H of G, let $\xi = (X, p, B)$ be a principal G-bundle, and $\eta = (Y, q, B)$ a principal H-bundle. Let $f: Y \to f(Y) \subset X$ be a homeomorphism onto a closed subset $f(Y)$ such that $f(ys) = f(y)s$ for $y \in Y$ and $s \in H$. Then η is called a restriction of ξ, and ξ is a prolongation of η.

In other words, η is the result of restricting the structure group G of ξ to H, and ξ is the result of prolonging the structure group H of η to G. In the next two theorems the possibility of restriction and prolongation is discussed.

2.2 Theorem. *Let H be a closed subgroup of G. Every principal H-bundle $\eta = (Y, q, B)$ has a prolongation $\xi = (X, p, B)$ with structure group G. Moreover, if η is trivial, locally trivial, or numerable, ξ is trivial, locally, trivial, or numerable, respectively.*

Proof. Since H acts on the left of G by multiplication in the group, we can form $\eta[G] = \xi$, where $\xi = (X, p, B)$ and $X = (Y \times G) \bmod H$. Then G acts on the right of X by the relation $xt = ((y, s)H)t = (y, st)H$, which is compatible with $(hr, r^{-1}s)H = (y, s)H$ for each $r \in H$. Moreover, the relation $f(y) = (y, e)H$ defines a homeomorphism $f: Y \to f(Y) \subset X$. Since the set $Y \times (G - H)$ is open in $Y \times G$, its projection $X - f(Y)$ is open in X and $f(Y)$ is closed in X. The action of H is preserved by f because $f(yr) = (yr, e)H = (yr, r^{-1}e)Hr = (y, e)Hr = f(y)r$ for $r \in H$. The translation function τ for ξ is given by the relation $\tau((y, s)H, (y', s')H) = s^{-1}\tau_1(y, y')s'$, which is continuous by properties of quotient topologies. Clearly, the prolongation of a trivial bundle is trivial. The last statement of the theorem follows from 4(6.4).

2.3 Theorem. *Let H be a closed subgroup of G. A principal G-bundle $\xi = (X, p, B)$ has a restriction to a principal H-bundle $\eta = (Y, q, B)$ if and only if $\xi \bmod H$ (or $\xi[G \bmod H]$) has a cross section. Moreover, if ξ is trivial or locally trivial, and if the principal H-bundle associated with G is trivial or locally trivial, then η is trivial or locally trivial, respectively. If ξ is numerable and if the principal H-bundle associated with G is locally trivial, then η is numerable.*

Proof. Let $f: Y \to X$ be a map defining a restriction of principal bundles. Then the composition of $f: Y \to X$ and the quotient map $X \to X \bmod H$ is a map σ^* such that $\sigma^*(ys) = \sigma^*(y)s$ for each $s \in H$. This map σ^* defines a cross section of $\xi \bmod H$ (or $\xi[G \bmod H]$).

Conversely, a cross section σ of $\xi[G \bmod H]$ defines a map $g: X \to G \bmod H$ such that $g(xs) = s^{-1}g(x) = g(x)s$ for each $s \in G$ by Theorem 4(8.1). Let Y be the closed subset $g^{-1}(eH)$ of X, let q be the restriction $p|Y$, and let $\eta = (Y, q, B)$. Let $y_1, y_2 \in Y$ such that $q(y_1) = q(y_2)$. For some $s \in G$ we have $y_2 = y_1 s$, $eH = g(y_2) = g(y_1 s) = s^{-1}g(y_1) = s^{-1}eH$, and $s \in H$. Finally, the restriction of the translation function of the principal G-space X is the translation function of the principal H-space Y.

If ξ is trivial, we have H-morphisms $Y \to X$, $X \to G$, and $G \to H$ which compose to an H-morphism $Y \to H$, and η is trivial by 4(8.3). For the statement concerning local triviality we can use 4(6.4).

2.4 Corollary. *Let B be a CW-complex, and let $\pi_i(G \bmod H) = 0$ for each $i < \dim B$. Then every principal G-bundle has a restriction to a principal H-bundle.*

Proof. The corollary follows from (2.3) by using Theorem 2(7.1) under hypothesis (H1).

3. Restriction and Prolongation of Structure Group for Fibre Bundles

The next theorem says that fibre bundles remain unchanged as bundles under restriction or prolongation of the structure group.

3.1 Theorem. *Let H be a closed subgroup of G, let F be a left G-space that is also a left H-space, and let ξ be a principal G-bundle with η as restriction to the subgroup H. Then there is a natural isomorphism $g: \eta[F] \to \xi[F]$ of bundles over B.*

Proof. Let $f: Y \to X$ be the map defining the restriction of the principal bundles. We define $g: Y_F \to X_F$ by the relation $g((y, z)H) = (f(y), z)G$. Then g is a map since it is the quotient of $f \times 1_F$.

Since every element of X_F has the form $(x, z)G$, where $x \in f(Y)$, the map g is surjective. Since the relation on $X \times F$ determined by the action of G induces the relation on $f(Y) \times F$ determined by the action of H, the map g is injective.

Finally, we must show that g is an open map. Let $g((y_0, z_0)H) = (f(y_0), z_0)G$ and $x_0 = f(y_0)$. Let W be an open set containing $(y_0, z_0)H$ in Y_F, let $q_1: Y \times F \to Y_F$ and $p_1: X \times F \to X_F$ denote the natural projections, and let W_1 be the open neighborhood of $y_0 \in Y$ and W_2 of $z_0 \in F$ such that $q_1(W_1 \times W_2) \subset W$. There exists a symmetric open neighborhood N of $1 \in G$ and a neighborhood V of z_0 in F such that $NV \subset W_2$. Let U be an open neighborhood of x_0 in X such that $p(U) \subset q(W_1)$ and $r((U \times U) \cap X^*) \subset N$. We replace W_1 by $W_1 \cap q^{-1}(p(U)) \cap f^{-1}(U)$ and U by $p^{-1}(q(W)) \cap U$. Then we have $p(U) = q(W_1)$ and $f(W_1) \subset U$. For each $(x, z) \in U \times V$, there exists $y \in W_1$ such that $f(y) \in U$ and $f(y) = xs$. This means that $s \in N$, and $g((y, s^{-1}z)H) = (f(y), s^{-1}z)G = (x, z)G$. Therefore, we have $g(W) \supset g(q_1(W_1 \times W_2)) \supset p_1(U \times V)$, and $g(W)$ is an open set. This proves the theorem.

3.2 Corollary. *With the notations of* (3.1), *the bundles* $\eta[G]$ *and* ξ *are isomorphic as bundles over* B, *and there is a map* $g: Y_G \to X_G = X$ *that is a G-space isomorphism.*

Proof. We need prove only that g commutes with the action of G. Since $g((y, s)H)t = ((f(y), s)G)t = (f(y), st)G = g((y, st)H)$, it follows that g is a G-morphism.

4. Local Coordinate Description of Change of Structure Group

Theorem (2.3) has the following interpretation for locally trivial bundles and their local coordinate transformations.

4.1 Theorem. *Let H be a closed subgroup of G, let $\xi = (X, p, B)$ be a locally trivial principal G-bundle, and let $\{h_i, V_i\}$ for $i \in I$ be an atlas of charts for ξ with transition functions $\{g_{i,j}\}$. The bundle ξ has a restriction to a principal H-bundle if and only if there exist maps $r_i: V_i \to G$ such that the transition functions $g'_{i,j}(b) = r_i(b)^{-1} g_{i,j}(b) r_j(b)$ have values in H for each $b \in V_i \cap V_j$.*

Proof. By Corollary (3.2) and Theorem 5(2.7) the maps r_i must exist because ξ and $\eta[G]$ are G-isomorphic. For the converse, by Theorem 5(3.2), the transition functions describe the principal H-bundle $\eta = (Y, q, B)$ and the principal G-bundle ξ. The natural inclusion $V_i \times H \to V_i \times G$ defines a map $f: Y \to X$ which defines the restriction of ξ to η.

Theorem (4.1) says that if the transition functions have values in a subgroup H of G, where G is the structure group of a fibre bundle $\xi[F]$, this fibre bundle is isomorphic to $\eta[F]$, where H is the structure group of η.

The restriction process in this form has been carried through for vector bundles over a paracompact space; see Chap. 5, Sec. 7.

5. Classifying Spaces and the Reduction of Structure Group

Let H be a closed subgroup of G, let $\omega_H = (Y_0, q_0, B_H)$ be a universal bundle for H, and let $\omega_G = (X_0, p_0, B_G)$ be a universal bundle for G. By Corollary (3.2), $\omega_H[G]$ is a numerable principal G-bundle over B_H. By the classification theorem 4(12.2), there is a principal G-bundle morphism $(h_0, f_0)\colon \omega_H[G] \to \omega_G$, where $f_0^*(\omega_G)$ and $\omega_H[G]$ are isomorphic over B_H.

5.1 Theorem. *With the above notations, let $\xi = (X, p, B)$ be a numerable principal G-bundle over B with classifying map $f\colon B \to B_G$; that is, $f^*(\omega_G)$ and ξ are B-isomorphic. Then the restrictions $\eta = (Y, q, B)$ of ξ are in bijective correspondence with homotopy classes of maps $g\colon B \to B_H$ such that $f_0 \circ g$ and f are homotopic. We have the following diagram:*

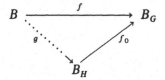

Proof. Let η be a restriction of ξ. Then there is a unique homotopy class determined by a map $g\colon B \to B_H$ such that $g^*(\omega_H)$ and η are isomorphic. Consequently, $\eta[G]$ and $g^*(\omega_H[G])$ are isomorphic. From the above discussion we know that $\eta[G]$ is B-isomorphic to $g^*(f_0^*(\omega_G))$, and ξ is B-isomorphic to $f^*(\omega_G)$. Since, by Corollary (3.2), $\eta[G]$ and ξ are isomorphic, the classification theorem 4(11.2) implies that f and $f_0 g$ are homotopic.

Conversely, if g exists, let η denote $g^*(\omega_H)$, where $\eta = (Y, q, B)$. We have an H-morphism $h'\colon Y_0 \to X_1$, where $f_0^*(\omega_G) = (X_1, p, B_H)$ which is the composition of the quotient mod H of the inclusion $Y_0 \times H \to Y_0 \times G$ and the isomorphism $\omega_H[G] \to f_0^*(\omega_G)$. Under the induced bundle functor g^*, this defines an H-morphism $h'' = g^*(h')$ of Y into the total space of $g^*(f_0^*(\omega_G))$. The H-morphism h'' when composed with the isomorphism $g^*(f_0^*(\omega_G)) \to \xi$ over B defines a restriction morphism $h\colon Y \to X$ which commutes with the action of H. This proves the theorem.

In the case of the universal bundle ω_G defined by the Milnor construction, an inclusion $H \subset G$ defines a natural inclusion $E_H \subset E_G$. This inclusion induces a natural map $B_H \to B_G$ and morphism $\omega_H \to \omega_G$.

Exercises

1. Prove that a principal G-bundle ξ has a restriction to the subgroup 1 if and only if ξ is trivial.

2. For a subgroup H of G, prove that every principal G-bundle restricts to a principal H-bundle if and only if the map $f_0\colon B_H \to B_G$ has a right (homotopy) inverse g_0.

$B_G \to B_H$, that is, $f_0 g_0 \simeq 1$. Give a homotopy criterion for the restriction always to be unique when g_0 exists.

3. Define restriction and prolongation of principal bundles for a (continuous) group morphism $h: H \to G$. In what sense do (2.2), (2.3), (3.1), (4.1), and (5.1) have analogues for this more general situation?

CHAPTER 7

The Gauge Group of a Principal Bundle

The gauge group of a principal bundle is simply its automorphism group with a topology coming from the mapping space topology. The mapping space topology (called the compact open topology) will not be considered in detail but will be outlined in the first section.

The classifying space $B\operatorname{Aut}_B(P)$ of the gauge $\operatorname{Aut}_B(P)$ of a principal bundle P over B with structure group G enables us to describe the mapping space $\operatorname{Map}(B, BG)$, whose connected components indexed by $[B, BG]$ are just the isomorphism classes of principal G-bundles over B. The fact that $[B, BG]$ is naturally the set of isomorphism classes of principal bundles is the homotopy classification of principal bundles, see chapter 3. This chapter can be viewed as an extension of chapter 3 and as background for the applications of the gauge group to differential geometry and mathematical physics.

1. Definition of the Gauge Group

1.1 Definition. Let $p: P \to B$ be a principal G-bundle. The gauge $\operatorname{Aut}_B(P)$ of P is the subspace of $u \in \operatorname{Map}(P, P)$ such that

$$pu = p \qquad \text{and} \qquad u(xs) = u(x)s \quad \text{for } s \in G, x \in P.$$

Thus, an element of the gauge group is G-equivariant and preserves the projection to B, and this is why we denote it by $\operatorname{Aut}_B(P)$. From 3(3.2) we know that if $u \in \operatorname{Aut}_B(P)$, then the inverse $u^{-1}: P \to P$ is defined and $u^{-1} \in \operatorname{Aut}_B(B)$.

We give $\operatorname{Aut}_B(P)$ the subspace topology from the mapping space $\operatorname{Map}(P, P)$. A general reference for mapping spaces is Bourbaki, *General to-*

pology, Chapter 10. From general mapping space theory we know that composition in $\text{Aut}_B(P)$ is continuous, but there remains the question as to whether or not $u \mapsto u^{-1}$ defined $\text{Aut}_B(P) \to \text{Aut}_B(P)$ is continuous. We will leave this question unanswered for the moment, for it will be resolved in the context of other descriptions of the gauge group.

1.2 Remark. Let P denote a principal G-bundle over B. To each $u \in \text{Aut}_B(P)$, we assign the continuous function $\phi_u \colon P \to G$ defined by the relation $u(x) = x\phi_u(x)$ or by the formula $\phi_u(x) = \tau(x, u(x))$ where $\tau \colon P \times_B P \to G$ is the translation function of the principal bundle P. Next, observe that the relation $u(xs) = u(x)s$ is equivalent to the functional relation $\phi_u(xs) = s^{-1}\phi_u(x)s$ for all $x \in P$, $s \in G$.

1.3 Notation. For a topological group G we denote by $\text{Ad}(G)$ the right G-space G with right adjoint G-action given by $x \cdot s = s^{-1}xs$. Here the operation $s^{-1}xs$ is multiplication in G. In addition, $\text{Ad}(G)$ is a G-group because the topological group structure on G is preserved by the right G-action. For X and Y, two right G-spaces, we denote by $\text{Map}_G(X, Y)$ the subspace of all $f \in \text{Map}(X, Y)$ satisfying $f(xs) = f(x)s$ for all $s \in G$, $x \in X$.

In particular, we have that $\text{Aut}_B(P)$ is a subspace of $\text{Map}_G(P, P)$.

1.4 Remark. The function which assigns to an automorphism $u \in \text{Aut}_B(P)$ the function $\phi_u \in \text{Map}_G(P, \text{Ad}(G))$ given by $\phi_u(x) = \tau(x, u(x))$ is a continuous bijection. If $\phi \in \text{Map}_G(P, \text{Ad}(G))$, then we form

$$u(x) = x\phi(x) \in \text{Aut}_G(P)$$

and this gauge transformation u satisfies $\phi = \phi_u$. This formula shows that this mapping $\text{Aut}_B(P) \to \text{Map}_G(P, \text{Ad}(G))$ is a homeomorphism. In this context, we see that we have a topological group where for $\phi, \phi' \in \text{Map}_G(P, \text{Ad}(G))$ the product is given by the relation $\phi\phi'(x) = \phi(x)\phi'(x)$ in G and the inverse by $\phi^{-1}(x) = \phi(x)^{-1}$ in G.

Now we have a third description in terms of cross sections of a fibre bundle.

1.5 Notation. For a bundle $p \colon E \to B$ we denote by $\Gamma(E/B)$ the subspace of cross sections $s \in \text{Map}(B, E)$. For a principal G-bundle P over B we denote by $\text{Ad}(P)$ the fibre bundle with fibre $\text{Ad}(G)$ sometimes denoted $P[\text{Ad}(G)]$ or $P \times^G \text{Ad}(G)$. The fibre bundle $\text{Ad}(P)$ is a bundle of groups since the group structure on $\text{Ad}(G)$ is G-equivariant, and the space $\Gamma[\text{Ad}(P)/B]$ is a topological group.

1.6 Remark. In 3(8.1) cross sections of a fibre bundle are described by mappings of the principal bundle into the fibre, and thus we have the following bijection $\text{Map}_G(P, \text{Ad}(G)) \to \Gamma[\text{Ad}(P)/B]$ given by the function which assigns to the G-map $\phi: P \to \mathrm{Ad}(G)$ the cross section $s_\phi: B \to \mathrm{Ad}(P)$ where $s_\phi(b) = (x, \phi(x)) \bmod G$ for $x \in P_b$ an arbitrary element of the fibre P_b over $b \in B$. In chapter 3 we did not take up the question of the continuity of this bijection or its inverse, and we will not need it here since we use only the first two descriptions of the gauge group.

Finally, we close with the following calculations of the gauge group.

1.7 Proposition. *Let P be a principal G-bundle. If either P is trivial or if G is abelian, then the topological group $\mathrm{Aut}_B(P)$ is isomorphic to the topological group $\mathrm{Map}(B, G)$.*

Proof. In the first case $\mathrm{Map}_G(B \times G, G) = \mathrm{Map}(B, G)$ since $f(b, s)$ with $f(b, ss') = f(b, s)s'$ is of the form $f(b, s) = f(b, 1)s$ for $s \in G$, $b \in B$, and in the second case, $\mathrm{Map}_G(P, \mathrm{Ad}(G)) = \mathrm{Map}(B, G)$ because $f(xs) = f(x)$ since the action of G on $\mathrm{Ad}(G)$ is trivial. This proves the preposition.

2. The Universal Standard Principal Bundle of the Gauge Group

2.1 Notation. Let $p: P \to B$ be a principal G-bundle, and let $E_G \to B_G$ denote a universal bundle for G.

(1) Let $\mathrm{Map}_G(P, E_G)$ denote the subspace of $w \in \mathrm{Map}_G(P, E_G)$ such that $w(xs) = w(x)s$ for all $x \in P$, $s \in G$.
(2) Let $\mathrm{Map}_P(B, B_G)$ denote the subspace of $f \in \mathrm{Map}(B, B_G)$ such that P and $f^*(E_G)$ are isomorphic.
(3) Let $p_P: \mathrm{Map}_G(P, E_G) \to \mathrm{Map}_P(B, E_G)$ denote the function which assigns to $w \in \mathrm{Map}_G(P, E_G)$ the quotient map $f \in \mathrm{Map}_G(P, B_G)$ on the base spaces of the bundles. The diagram related to this situation is the following

$$
\begin{array}{ccc}
P & \xrightarrow{\ w\ } & E_G \\
\downarrow & & \downarrow \\
B & \xrightarrow{\ f\ } & B_G.
\end{array}
$$

2.2. Remark. The gauge group $\mathrm{Aut}_B(P)$ acts on the right of $\mathrm{Map}_G(E, E_G)$ by composition of bundle morphisms, that is, if $u \in \mathrm{Aut}_B(P)$ and $w \in \mathrm{Map}_G(P, E_G)$, then we have $wu \in \mathrm{Map}_G(P, E_G)$. Moreover, if $w, w' \in \mathrm{Map}_G(P, E_G)$ such that $p_P(w) = p_P(w')$, then we have $\tau(w, w') = w^{-1}w' \in \mathrm{Aut}_B(P)$ with $w\tau(w, w') = w'$ which is a translation function. This function

$$
\tau: \mathrm{Map}_G(P, E_G) \times_{\mathrm{Map}_P(B, B_G)} \mathrm{Map}_G(P, E_G) \to \mathrm{Aut}_B(P)
$$

is continuous being composition and inverse functions. In particular, we have a principal bundle for the gauge group $\mathrm{Aut}_B(P)$.

2.3 Definition. Let $p: P \to B$ be a principal G-bundle. The standard principal bundle for the gauge group $\text{Aut}_B(P)$ is

$$p_P: \text{Map}_G(P, E_G) \to \text{Map}_P(B, B_G).$$

Now we consider this construction for the product principal bundle $\text{pr}_1: B \times G \to B$. Then the function which assigns to a G-map $w: B \times G \to E_G$ the map $w': B \to E_G$ given by $w'(b) = w(b, 1)$ is a homeomorphism $\text{Map}_G(B \times G, E_G) \to \text{Map}(B, E_G)$; with this notation we have $p_P(w) = p_G w'$ where $p_G: E_G \to B_G$ is the projection in the universal bundle. This homeomorphism is right $\text{Map}(B, G)$ equivariant where $\text{Map}(B, G)$ is $\text{Aut}_B(P)$ for the product bundle $B \times G$ over B. The result is the following.

2.4 Remark. The standard principal bundle for the gauge group $\text{Aut}_B(B \times G) = \text{Map}(B, G)$ of the product bundle is isomorphic to

$$p_G: \text{Map}(B, E_G) \to \text{Map}_0(B, B_G)$$

where p_G denotes composition by p_G on the left and $\text{Map}_0(B, B_G)$ denotes the subspace of null homotopic maps $B \to B_G$.

Note that for this last statement we make use of the homotopy classification of bundles to assert that $\text{Map}_{B \times G}(B, B_G)$ is the subspace of null homotopic maps. For the homotopy theory to be valid we must consider only locally trivial bundles.

2.5 Convention. Locally trivial bundles are always trivial bundles over the open sets of a numerical covering, i.e. a covering with a subordinate partition of unity.

3. The Standard Principal Bundle as a Universal Bundle

3.1 Definition. Let $p: P \to B$ be a principal G-bundle. Then the product of P with Y is the principal G-bundle $q: P \times Y \to B \times Y$ where $P \times Y$ has the action of G given by

$$(x, y)s = (xs, y) \qquad \text{for } s \in G, \ x \in X, \ y \in Y.$$

Note that if P is a locally trivial bundle, then so $P \times Y \to B \times Y$ the product with a space Y.

3.2 Remark. Let P be a locally trivial bundle over a space $B_0 \times [0, 1]$. Form the restriction $P|B_0 \times \{0\} = P_0$. Then the first step in the homotopy classification asserts that P is isomorphic to $P_0 \times [0, 1]$ over $B_0 \times [0, 1]$, see 3(9.8).

The next proposition is a direct corollary of the homotopy classification theory of principal bundles.

3.3 Proposition. *Let P be a locally trivial principal G-bundle. Then the space* $\mathrm{Map}_G(P, E_G)$ *is nonempty and path connected, and* $\mathrm{Map}_P(B, B_G)$ *is the path component of* $f\colon B \to B_G$ *with* $f^*(E_G)$ *isomorphic to* P.

Proof. The fact that the space is nonempty is equivalent to the assertion that P is isomorphic to $f^*(E_G)$ for some map $f\colon B \to B_G$, see 3(12.3), which in turn is part of a bundle morphism $u\colon P \to E_G$ inducing f. Thus, $u \in \mathrm{Map}_G(P, E_G)$ and the space is nonempty. To see that it is path connected, we consider two elements $u,\, u' \in \mathrm{Map}_G(P, E_G)$ inducing maps $f,\, f'\colon B \to B_G$ where P is isomorphic to both $f^*(E_G)$ and $f'^*(E_G)$. Hence, f and f' are homotopic with a morphism of bundles $(w, h)\colon P \times [0,1] \to E_G$ such that $w(x, 0) = u(x)$ and $w(x, 1) = u'(x)$, see 3(12.4). Thus, w defines a path from u to u' in $\mathrm{Map}_G(P, E_G)$, and proves the last statement. This proves the proposition.

Now we can prove more if we have the "exponential law" for mapping space, namely a homeomorphism

$$\mathrm{Map}(P \times Y, Z) \to \mathrm{Map}(Y, \mathrm{Map}(P, Z))$$

given by assigning to $f(x, y)$ the map $y \mapsto f(y)(x) = f(x, y)$. The exponential law holds for P a locally compact space or in the general simplicial setting.

3.4 Theorem. *Let P be a principal bundle over B such that P is a locally compact space. Then $\mathrm{Map}_G(P, E_G)$ is a contractible space.*

Proof. Applying the exponential law for any space Y we have a homeomorphism $\mathrm{Map}_G(P \times Y, E_G) \to \mathrm{Map}(Y, \mathrm{Map}_G(P, E_G))$ between two path connected spaces. We apply this to $Y = \mathrm{Map}_G(P, E_G)$ where the identity and the constant are connected by a path, that is, by a contracting homotopy. This proves the theorem.

3.5 Corollary. *For a principal bundle P as in (3.4) the classifying space of the gauge group $\mathrm{Aut}_B(P)$ is $\mathrm{Map}_P(B, B_G)$ with universal bundle the standard principal bundle for the gauge group*

$$\mathrm{Map}_G(P, E_G) \to \mathrm{Map}_P(B, B_G) = B\,\mathrm{Aut}_B(P).$$

4. Abelian Gauge Groups and the Künneth Formula

We begin with two general results in algebraic topology and tie these results to gauge group, see Spanier, *Algebraic topology*.

(4.1) Definition. A space X is called a $K(\Pi, n)$, or Eilenberg-MacLane space, provided the homotopy groups of this space are given by

$$\pi_i(X) = \begin{cases} \Pi & \text{for } i = n \\ 0 & \text{for } i \neq n \end{cases}.$$

In other words, there is one nonvanishing homotopy group and it is degree n. This means that Π must be abelian for $n > 1$, but Π can be any group for $n = 1$, and for $n = 0$ we can choose Π to be any set. Since a $K(\Pi, n)$ is path connected, we can use any base point for $n > 0$.

4.2 Remark. For any abelian group Π and any $n \geq 0$ there exists an abelian topological group A which is a $K(\Pi, n)$ as a space. Moreover, the classifying space $B(A) = BK(\Pi, n) = K(\Pi, n + 1)$, which gives an inductive construction of $K(\Pi, n)$ as an abelian group using a classifying space construction which preserves products, and the loop space $\Omega K(\Pi, n) = K(\Pi, n - 1)$ from the homotopy exact triangle for a fibre sequence. Note that we have used the symbol $K(\Pi, n)$ for both the space and the type of space that it is.

4.3 Remark. Any abelian topological group A is a product of $K(\Pi, n)$ spaces as a commutative H-space up to homotopy, and indeed up to higher homotopies. This is a theorem of J. C. Moore which was proved in the simplicial set/group context, see H. Cartan Seminaire 1954–55. This is also a reference for (4.2).

4.4 Remark. The first nonzero reduced cohomology group of a $K(\Pi, n)$ is in degree n where $H^n(K(\Pi, n), \Pi) = \text{Hom}(\Pi, \Pi)$ which has a canonical element ι_n corresponding to the identity on Π. Using ι_n, we define a natural morphism of abelian group valued functors on the category of spaces and homotopy classes of maps

$$[X, K(\Pi, n)] \to H^n(X, \Pi)$$

by assigning to a homotopy class of maps $[f] : X \to K(\Pi, n)$ the element $H^n(f)(\iota_n) \in H^n(X, \Pi)$. On the subcategory of spaces with the homotopy type of a CW-complex or on the category of simplicial sets this morphism of functors is an isomorphism of functors.

4.5 Remark. From bundle theory using the fact that

$$K(\Pi, n) = BK(\Pi, n - 1),$$

we see that the homotopy set $[X, K(\Pi, n)]$, interpreted as cohomology in the previous remark, can also be interpreted as isomorphism classes of principal $K(\Pi, n - 1)$-bundles over X. In line with the topic of this chapter we study the gauge group of one of these bundles. Since the structure is abelian, by (1.7), the gauge group is independent of which principal bundle over X and isomorphic to the mapping space $\text{Map}(X, K(\Pi, n - 1))$.

This leads us to the study of mapping spaces into $K(\Pi, n)$ spaces which begins with the connected components $[X, K(\Pi, n)]$ of $\text{Map}(X, K(\Pi, n))$. An application of the theorem of J. C. Moore cited in (4.3) applied to the abelian group structure defined by multiplication of the abelian group valued functions gives that this space is a product of $K(\Pi, q)$ spaces. The groups arising are given in the following theorem of René Thom.

4.6 Theorem. *The space* $\mathrm{Map}(X, K(\Pi, n))$ *is homotopy equivalent to the product* $\prod\limits_{0 \le q \le n} K(H^{n-q}(X, \Pi), q)$.

Proof. We give two proofs of this theorem. The first is by induction on n. The case $n = 0$ is clear since $K(\Pi, 0) = \Pi$ is a discrete space. Next we apply the inductive hypothesis for $n - 1$ to describe using (3.5) and (4.2). The connected components of $\mathrm{Map}(X, Y)$ are indexed by $[X, Y]$ the homotopy classes of maps $X \to Y$

$$\mathrm{Map}(X, K(\Pi, n)) = \prod_{[X, K(\Pi, n)]} B(\mathrm{Map}(X, K(\Pi, n - 1)))$$

where each connected component is the classifying space

$$B(\mathrm{Map}(X, K(\Pi, n - 1)))$$

of the gauge group of the $K(\Pi, n - 1)$ bundles over X. Now we use the inductive hypothesis to rewrite this expression for $\mathrm{Map}(X, K(\Pi, n))$

$$\mathrm{Map}(X, K(\Pi, n)) = K(H^n(X, \Pi), 0) \times B\left(\prod_{0 \le p \le n-1} K(H^{n-p-1}(X, \Pi), p) \right)$$

$$= \prod_{0 \le q \le n} K(H^{n-q}(X, \Pi), q) \qquad \text{for } p + 1 = q.$$

This completes the inductive step and proves the theorem.

For another proof we start with the evaluation map $e(f) = f(*)$ at a base point $* \in X$. This is a fibration $e \colon \mathrm{Map}(X, K(\Pi, n)) \to K(\Pi, n)$ with fibre $\mathrm{Map}_0(X, K(\Pi, n))$. Now the homotopy groups $\pi_q(\mathrm{Map}_0(X, K(\Pi, n))$ can be calculated using loop space properties of $K(\Pi, n)$ spaces and the homotopy groups

$$\pi_q(\mathrm{Map}_0(X, K(\Pi, n)) = \pi_0(\Omega^q \mathrm{Map}_0(X, K(\Pi, n)))$$

$$= [X, \Omega^q K(\Pi, n)]_0$$

$$= [X, K(\Pi, n - q)]_0$$

$$= \bar{H}^{n-q}(X, \Pi) \qquad \text{by (4.4)},$$

and this is $\pi_q(\mathrm{Map}(X, K(\Pi, n))$ for $q < n$ and for $q = n$ we get an extra factor of Π from $\pi_n(K(\Pi, n)) = \Pi$ in the fibre sequence. After knowing that the bundle is trivial, we have another proof of the theorem.

Now we consider the relation of this decomposition of $\mathrm{Map}(Y, K(\Pi, n))$ to a Künneth formula for cohomology.

4.7 Theorem. *Let X and Y be two spaces with the homotopy type of CW-complexes. Then we have the following direct sum decomposition of the cohomology*

$$H^n(X \times Y, \Pi) = \bigoplus_{0 \leq q \leq n} H^q(X, H^{n-q}(Y, \Pi)).$$

Proof. We use the description of cohomology as homotopy classes of maps as given in (4.4) to calculate

$$
\begin{aligned}
H^n(X \times Y, \Pi) &= [X \times Y, K(\Pi, n)] \\
&= [X, \mathrm{Map}(Y, K(\Pi, n)] \\
&= \prod_{0 \leq q \leq n} [X, K(H^{n-q}(Y, \Pi), q)] \\
&= \bigoplus_{0 \leq q \leq n} H^q(X, H^{n-q}(Y, \Pi)).
\end{aligned}
$$

The proves the theorem on the Künneth formula.

Observe that the Künneth formula is equivalent to the mapping space decomposition of (4.6), and from the Künneth formula we obtain another proof of the decomposition (4.6).

CHAPTER 8

Calculations Involving the Classical Groups

In this chapter we consider fibre bundles with the classical groups as fibre. By studying these bundles, we are able to calculate homotopy groups of the Stiefel varieties and classical groups. In the course of the development, we derive a homotopy classification for general fibre bundles over a suspension. Using this classification theorem and the homotopy groups of the classical groups, we are able to classify all vector bundles over a sphere of dimension ≤ 4.

1. Stiefel Varieties and the Classical Groups

In 5(7.1) and 5(7.2) we defined the five classical groups $SO(k)$, $O(k)$, $SU(k)$, $U(k)$, and $Sp(k)$. For discussing the common properties of these groups it is convenient to unify the notation connected with them. Let F denote $\mathbf{R}$, $\mathbf{C}$, or $\mathbf{H}$. Let $U_F(k)$ denote $O(k)$ for $F = \mathbf{R}$, $U(k)$ for $F = \mathbf{C}$, and $Sp(k)$ for $F = \mathbf{H}$. Let $SU_F(k)$ denote $SO(k)$ for $F = \mathbf{R}$, and $SU(k)$ for $F = \mathbf{C}$. Consequently, $U_F(k)$ is the group of linear transformations $u: F^k \to F^k$ such that $(u(x)|u(y)) = (x|y)$ for $x, y \in F^k$, and $SU_F(k)$ is the closed subgroup of $u \in U_F(k)$ such that $\det u = +1$.

As a vector subspace, we can consider $F^k \subset F^{k+1}$, where a vector in F^k has zero in its $(k + 1)$-coordinate when viewed in F^{k+1}. Consequently, there exist natural inclusions $U_F(k) \subset U_F(k + 1)$ and $SU_F(k) \subset SU_F(k + 1)$ because $(x|y)$ is the same for $x, y \in F^k$ or x, y viewed in F^{k+1}. The groups $U_F(k)$ and $SU_F(k)$ are the subgroups of elements u, where $u \in U_F(k + 1)$ and $u \in SU_F(k + 1)$, respectively, such that $u(e_{k+1}) = e_{k+1}$. As usual, we denote $e_i = $
(i)
$(0, \ldots, 0, 1, 0, \ldots, 0)$.

With the inclusions $F^1 \subset F^2 \subset \cdots \subset F^k \subset F^{k+1} \subset \cdots$, we are able to give $F^\infty = \bigcup_{1 \leq k} F^k$ the inductive topology [see 1(1.1)]. The vector space F^∞ has the inner product induced by the inner product on each F^k.

1.1 Definition. The infinite classical groups are the union $\bigcup_{1 \leq k} U_F(k)$, denoted U_F (or $U_F(\infty)$), with the inductive topology and the union $\bigcup_{1 \leq k} SU_F(k)$, denoted SU_F (or $SU_F(\infty)$), with the inductive topology.

In special cases, we use the notation $U_F = 0$ and $SU_F = SO$ for $F = \mathbf{R}$, $U_F = U$ and $SU_F = SU$ for $F = \mathbf{C}$, and $U_F = Sp$ for $F = \mathbf{H}$.

We recall that $V_k(F^n)$ is the subspace of F^{kn} consisting of k-tuples $(v_1, \ldots, v_k)$, where $(v_i | v_j) = \delta_{i,j}$ for $1 \leq i, j \leq k$. Here we may have $n = \infty$, in which case $V_k(F^\infty)$ has the inductive topology defined by the following subsets:

$$V_k(F^k) \subset V_k(F^{k+1}) \subset \cdots \subset V_k(F^n) \subset \cdots \subset V_k(F^\infty) = \bigcup_{k \leq n} V_k(F^n)$$

We define a map η_k^n or just $\eta: U_F(n) \to V_k(F^n)$ by the relation $\eta(u) = (u(e_1), \ldots, u(e_k))$. The map η is a continuous surjection. Moreover, the following diagram is commutative.

$$
\begin{array}{ccc}
U_F(n) & \xrightarrow{\ \eta_k^n\ } & V_k(F^n) \\
\big\downarrow & & \big\downarrow \\
U_F(n+1) & \xrightarrow{\ \eta_k^{n+1}\ } & V_k(F^{n+1})
\end{array}
$$

The vertical maps are inclusions.

There are two embeddings of $U_F(m)$ in $U_F(n)$ for $m \leq n$. First, $U_F(m)$ acts on $e_1, \ldots, e_m$, yielding an embeading denoted $U_F(m) \subset U_F(n)$; second, $U_F(m)$ acts on $e_{n-m+1}, \ldots, e_n$, yielding an embedding denoted $1_{n-m} \times U_F(m) \subset U_F(n)$. The first embedding was used for Definition (1.1).

1.2 Proposition. *For* u, $v \in U_F(n)$ *we have* $\eta_k^n(u) = \eta_k^n(v)$ *if and only if* $u = vw$, *where* $w \in 1_k \times U_F(n - k)$. *Moreover,* $(\eta_k^n)^{-1}\eta_k^n(u)$ *equals the coset* $u(1_k \times U_F(n - k))$.

Proof. Clearly, $\eta_k^n(u) = \eta_k^n(v)$ if and only if $u(e_i) = v(e_i)$ for $1 \leq i \leq k$. This is equivalent to $v^{-1}u(e_i) = e_i$ for $1 \leq i \leq k$, or $v^{-1}u \in 1_k \times U_F(n - k)$. The second statement is clear.

If $U_F(n) \bmod U_F(n - k)$ or $U_F(n)/U_F(n - k)$ denotes the homogeneous space of right cosets modulo $1_k \times U_F(n - k)$, then η_k^n defines a continuous bijection.

$$\theta_k^n: U_F(n) \bmod U_F(n - k) \to V_k(F^n)$$

The following theorem results from (1.2) and the fact that the spaces involved are compact.

1.3 Theorem. *The map θ_k^n given by the relation $\theta_k^n((u)(1_k \times U_F(n-k)) = (u(e_1), \ldots, u(e_k))$ is a homeomorphism $U_F(n) \bmod U_F(n-k) \to V_k(F^n)$.*

Moreover, where is the following commutative diagram:

$$
\begin{array}{ccc}
U_F(n) \bmod U_F(n-k) & \xrightarrow{\;\theta_k^n\;} & V_k(F^n) \\
\downarrow & & \downarrow \\
U_F(n+1) \bmod U_F(n+1-k) & \xrightarrow{\;\theta_k^{n+1}\;} & V_k(F^{n+1})
\end{array}
$$

The vertical maps are inclusions or are induced by inclusion.

By taking unions we have for $n = +\infty$ a homeomorphism θ_k: $U_F \bmod (1_k \times U_F) \to V_k(F^\infty)$, where $U_F \bmod (1_k \times U_F)$ has the inductive topology which, as the reader can easily verify, equals the quotient topology.

1.4 Proposition. *For $k \neq n$, and for $F = \mathbf{R}$ or $\mathbf{C}$, the map induced by the inclusion $SU_F(n) \bmod SU_F(n-k) \to U_F(n) \bmod U_F(n-k)$ is a homeomorphism.*

Proof. The relation $u = vw$ for $w \in SU_F(n-k)$ or $U_F(n-k)$ as given in (1.2) determines whether or not u and v determine the same right coset. If $\det u = \det v = 1$, we have $\det w = 1$, and the above map is an injection. For each u there exist v and w as above, with $\det v = 1$, and the above map is a surjection. Since the spaces are compact, the map is a homeomorphism.

1.5 Examples. We have the following identifications, using the above results and some elementary considerations.

(1) For $k = n$,

$$O(n) = V_n(\mathbf{R}^n) \qquad U(n) = V_n(\mathbf{C}^n) \qquad \text{and} \qquad Sp(n) = V_n(\mathbf{H}^n)$$

(2) For $k < n$,

$$V_k(\mathbf{R}^n) = \frac{O(n)}{O(n-k)} = \frac{SO(n)}{SO(n-k)}$$

$$V_k(\mathbf{C}^n) = \frac{U(n)}{U(n-k)} = \frac{SU(n)}{SU(n-k)}$$

$$V_k(\mathbf{H}^n) = \frac{Sp(n)}{Sp(n-k)}$$

(3) For $k = 1$,

$$V_1(\mathbf{R}^n) = S^{n-1} = \frac{O(n)}{O(n-1)} = \frac{SO(n)}{SO(n-1)}$$

$$V_1(\mathbf{C}^n) = S^{2n-1} = \frac{U(n)}{U(n-1)} = \frac{SU(n)}{SU(n-1)}$$

$$V_1(\mathbf{H}^n) = S^{4n-1} = \frac{Sp(n)}{Sp(n-1)}$$

(4) For $n = 1$ or 2,

$$SO(1) = SU(1) = 1 \qquad O(1) = Z_2$$

$$U(1) = SO(2) = S^1 \qquad \text{the circle group}$$

$$Sp(1) = SU(2) = S^3 \qquad \text{the multiplicative group of}$$
$$\text{quaternions of length } 1$$

(5) For $k = n - 1$,

$$SO(n) = V_{n-1}(\mathbf{R}^n) \qquad \text{and} \qquad SU(n) = V_{n-1}(\mathbf{C}^n)$$

(6) For the relation between SU_F and U_F, there are two exact sequences

$$SO(n) \to O(n) \overset{\det}{\to} Z_2 \qquad SU(n) \to U(n) \overset{\det}{\to} S^1$$

2. Grassmann Manifolds and the Classical Groups

For $(v_1, \ldots, v_k) \in V_k(F^n)$ we denote by $\langle v_1, \ldots, v_k \rangle$ or $p(v_1, \ldots, v_k)$ the k-dimensional subspace of F^n with base $v_1, \ldots, v_k$. As before, we let $G_k(F^n)$ denote the set of k-dimensional subspaces of F^n with the largest topology such that $p: V_k(F^n) \to G_k(F^n)$ is continuous. Since every subspace of F^n has an orthonormal base, p is surjective.

We consider the continuous surjection

$$p\theta_k^n: U_F(n)/U_F(n - k) \to G_k(F^n)$$

2.1 Proposition. *For $u \bmod U_F(n - k)$ and $v \bmod U_F(n - k)$, we have $p\theta_k^n(u) = p\theta_k^n(v)$ if and only if $v = us_1 s_2$, where $s_1 \in 1_k \times U_F(n - k)$ and $s_2 \in U_F(k)$.*

Proof. Let $\theta_k^n(u) = (u_1, \ldots, u_k) \in V_k(F^n)$ and $\theta_k^n(v) = (v_1, \ldots, v_k) \in V_k(F^n)$. We have $\langle u_1, \ldots, u_k \rangle = \langle v_1, \ldots, v_k \rangle$ if and only if $u_i = s(v_i)$, where $s \in U_F(k)$. Since $u_i = u(e_i)$ and $v_i = v(e_i)$, by (1.2) we have $u_i = us_1(e_i)$ for $1 \leq i \leq k$, and $s(v_i) = us_1(e_i)$ for $k + 1 \leq i \leq n$, where $s_1 \in 1_k \times U_F(n - k)$. Finally, the relation $u_i = s(v_i)$ is equivalent to $us_1 = vs$ or $v = us_1 s^{-1}$.

Consequently, $p\theta_k^n$ defines a continuous bijection $\psi_k^n: U_F(n)/(U_F(k) \times U_F(n - k)) \to G_k(F^n)$. Since the above spaces are compact, we have the next theorem, in view of (2.1).

2.2 Theorem. *The function $\psi_k^n(u \bmod U_F(k) \times U_F(n - k)) = \langle u(e_1), \ldots, u(e_k) \rangle$ defines a homeomorphism*

$$\psi_k^n: U_F(n)/(U_F(k) \times U_F(n - k)) \to G_k(F^n)$$

Moreover, the following diagrams are commutative.

$$U_F(n)/U_F(n-k) \xrightarrow{\theta_k^n} V_k(F^n)$$

$$\downarrow \qquad\qquad\qquad\qquad \downarrow p$$

$$U_F(n)/(U_F(k) \times U_F(n-k)) \xrightarrow{\psi_k^n} G_k(F^n)$$

The vertical maps are quotients in the above diagram.

$$U_F(n)/(U_F(k) \times U_F(n-k)) \xrightarrow{\psi_k^n} G_k(F^n)$$

$$\downarrow \qquad\qquad\qquad\qquad \downarrow$$

$$U_F(n+1)/(U_F(k) \times U_F(n+1-k)) \xrightarrow{\psi_k^{n+1}} G_k(F^{n+1})$$

The vertical maps are inclusions in the above diagram.

2.3 Remark. For $F = \mathbf{R}$ or $\mathbf{C}$ we leave it to the reader to discuss the space $SG_k(F^n)$ of oriented k-subspaces, that is, subspaces together with an equivalence class of orthonormal basis that differ by the action of $SU_F(k)$. As above, one defines a homeomorphism

$$\psi_k^n: U_F(n)/(SU_F(k) \times U_F(n-k)) \to SG_k(F^n)$$

The next theorem follows from the fact that homogeneous spaces of compact groups define principal bundles.

2.4 Theorem. *The bundle $(V_k(F^n), p, G_k(F^n))$ is a principal $U_F(k)$-bundle, and the bundle $(V_k(F^n), p, SG_k(F^n))$ is a principal $SU_F(k)$-bundle.*

A map $SG_k(F^n) \to G_k(F^n)$ results by "forgetting" orientation. Over each point of $G_k(\mathbf{R}^n)$ are precisely two points representing the two orientations of a subspace, and over each point of $G_k(\mathbf{C}^n)$ is a circle.

2.5 Examples. We have $SG_1(\mathbf{R}^n) = S^{n-1}$, $G_1(\mathbf{R}^n) = RP^{n-1}$, real projective $(n-1)$-space; $SG_1(\mathbf{C}^n) = S^{2n-1}$, $G_1(\mathbf{C}^n) = CP^{n-1}$, complex projective $(n-1)$-space; and $G_1(H^n) = HP^{n-1}$, quaternionic projective $(n-1)$-space.

3. Local Triviality of Projections from Stiefel Varieties

We begin by considering the action of $U_F(k)$ on the right of $V_k(F^n)$. For $a = [a_{i,j}] \in U_F(k)$ this action is given by the relation $(v_1,\dots,v_k)a = (v_1',\dots,v_k')$, where $v_j' = \sum_{1 \le i \le k} a_{i,j}v_i$. Then for $(v_1,\dots,v_k)$ and $(v_1',\dots,v_k')$, two elements of $V_k(F^n)$, it is clear by elementary linear algebra that $\langle v_1,\dots,v_k \rangle = \langle v_1',\dots,v_k' \rangle$ if and only if there exists $a \in U_F(k)$ with $(v_1,\dots,v_k)a = (v_1',\dots,v_k')$. Moreover, a is unique if it exists.

To justify our definition of orthogonal projection in (3.2), we use the following result.

3.1 Proposition. *For each* $x \in F^n$ *and* $(v_1, \ldots, v_k)a = (v_1', \ldots, v_k')$, *the relation* $\sum_{1 \leq i \leq k} (x|v_i)v_i = \sum_{1 \leq i \leq k} (x|v_i')v_i'$ *holds.*

Proof. We calculate $\sum_j (x|v_j')v_j' = \sum_{i,j,m} (x|a_{i,j}v_i)a_{m,j}v_m = \sum_{i,m} \left(\sum_j \bar{a}_{i,j}a_{m,j} \right)(x|v_i)v_m = \sum_{i,m} \delta_{i,m}(x|v_i)v_m = \sum_i (x|v_i)v_i$. This proves the proposition.

Let $\pi: V_k(F^n) \times F^n \to F^n$ be defined by the relation $\pi'((v_1, \ldots, v_k),$ $= \sum_{1 \leq i \leq k} (x|v_i)v_i$. By (3.1), $(v_i) \mapsto \pi'(v_i)(x)$ is constant on the fibre over a point of $G_k(F^n)$ of the projection $p: V_k(F^n) \to G_k(F^n)$. Therefore, since F^n is locally compact, π' defines a map $\pi: G_k(F^n) \times F^n \to F^n$ which we write $\pi_W(x)$ for $W \in G_k(F^n)$ and $x \in F^n$. If $W = \langle v_1, \ldots, v_k \rangle$, then $\pi_W(x)$ equals $\pi'(v_i)(x)$, and we have $\langle v_i | x - \pi_W(x) \rangle = 0$ for each i with $1 \leq i \leq k$. Let $v_W(x)$ denote $x - \pi_W(x)$. With these notations we state the following definition.

3.2 Definition. The orthogonal projection of $x \in F^n$ on $W \in G_k(F^n)$ is $\pi_W(x)$, and the normal projection of $x \in F^n$ to $W \in G_k(F)$ is $v_W(x)$.

We define a function $\sigma': V_k(F^n) \times V_k(F^n) \to \mathbf{R}$ by the relation $\sigma'((v_1, \ldots, v_k), (v_1', \ldots, v_k')) = \|\det[(v_i|v_j')]\|$. Since the determinant of an element of $U_F(k)$ is a unit, we have $\sigma'((v_1, \ldots, v_k)a, (v_1', \ldots, v_k')b) = \sigma'((v_1, \ldots, v_k), (v_1', \ldots, v_k'))$. Consequently, σ' defines a map $\sigma: G_k(F^n) \times G_k(F^n) \to \mathbf{R}$, where $\sigma(W, W') \geq 0$. In the next proposition we determine what it means for $\sigma(W, W') > 0$. For a subset $S \subset F^n$, let S^* denote the subspace of all $v \in F^n$ with $(v|S) = 0$.

3.3 Proposition. *Let* $W, W' \in G_k(F^n)$. *Then the following statements are equivalent.*

(1) $\sigma(W, W') > 0$.
(2) $\pi_{W'}(W) = W'$.　　(2') $\pi_W(W') = W$.
(3) $W \cap W'^* = 0$.　　(3') $W' \cap W^* = 0$.

Two k-subspaces W and W' are called related provided they satisfy the above five conditions.

Proof. For reasons of symmetry, it suffices to prove that (1), (2), and (3) are equivalent. The fact that $\sigma(W, W') > 0$ is equivalent to the statement that $\pi_{W'}(v_1), \ldots, \pi_{W'}(v_k)$ is a base of W' where $W = \langle v_1, \ldots, v_k \rangle$. This statement, in turn, is equivalent to (2). For $\pi_{W'}(W) = W'$ we have $v_{W'}(x) \neq x$ for each $x \in W$ and $(x|W') \neq 0$. This is a restatement of (3). This proves the proposition.

3.4 Corollary. *Let* W *and* W' *be two related elements of* $G_k(F^n)$. *Then there exist neighborhoods* N *of* W *and* N' *of* W' *such that* V *and* V' *are related for each pair* $(V, V') \in N \times N'$.

3.5 Definition. Let $\tilde{V}_k(F^n)$ be the subspace of $(F^n)^k$ consisting of linearly independent k-tuples in F^n. The Gram-Schmidt map $GS: \tilde{V}_k(F^n) \to V_k(F^n)$ is given by $GS(y_1,\ldots,y_k) = (v_1,\ldots,v_k)$, where v_1 equals y_1 divided by $\|y_1\|$ and v_j equals $y_j - \sum_{i<j} (y_j|v_i)v_i$ divided by

$$\left\| y_j - \sum_{i<j} (y_j|v_i)v_i \right\|$$

For $W \in G_k(F^n)$, let $O(W)$ denote the open set of all W' related to W. For $W = \langle v_1,\ldots,v_k \rangle$ we define a linear inner product preserving map

$$f_{(v_1,\ldots,v_k),W'}: W \to F^n$$

where $f_{(v_1,\ldots,v_k)W'}(W) = W'$, and the map f is continuous in $(v_1,\ldots,v_k) \in V_k(F^n)$ and $W' \in O(W)$. This s done by requiring $f_{(v_1,\ldots,v_k),W'}(v_i) = v_i'$, where $(v_1',\ldots,v_k') = GS(\pi_{W'}(v_1),\ldots,\pi_{W'}(v_k))$. The above definition is well defined in view of (3.3). With these notations we prove the following theorem.

3.6 Theorem. *The projection $p: V_k(F^n) \to G_k(F^n)$ is locally trivial, and $(V_k(F^n), p, G_k(F^n))$ is a locally trivial principal bundle with fibre $V_k(F^k) = U_F(k)$.*

Proof. For each subset of k elements $H \subset \{1,2,\ldots,n\}$ or $\{1,2,\ldots\}$ for $n = +\infty$, we define O_H to be the open set $O(F_H)$, where $F_H = \sum_{i \in H} Fe_i \subset F^n$. We define $j: F^k \to F_H$ to be the natural inclusion where $j(e_1) = e_{i(1)}$ and $i(1)$ is the minimum of $i \in H$, etc. Then we have the commutative diagram where f is an O_H-isomorphism:

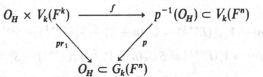

We define $f(W,v_1,\ldots,v_k) = (v_1',\ldots,v_k')$, where $v_i' = f_{(j(e_1),\ldots,j(e_k)),W}(v_i)$, with $1 \leq i \leq k$. This proves the first statement of the theorem since the O_H cover $G_k(F^n)$. For the second, observe that f commutes with the action of $U_F(k)$ on the right.

3.7 Corollary. *The projection $p: V_k(F^n) \to G_k(F^n)$ is a fibre map.*

Proof. For this, we use (3.6) and 1(5.2).

3.8 Theorem. *The map $q: V_{k+1}(F^n) \to V_k(F^n)$ defined by $q(v_1,\ldots,v_{k+1}) = (v_1,\ldots,v_k)$ is locally trivial with fibre $V_1(F^{n-k}) = S^{n-k-1}$.*

Proof. Let $H \subset \{1,\ldots,n\}$ be a subset of k elements, and let $H' = \{1,\ldots,n\} - H$. Let $F_{H'} = \sum_{i \in H'} Fe_i$ as in (3.6). Let O_H^* denote the open subset $p^{-1}(O_H)$, where O_H is defined as in (3.6). Then we have the following commutative diagram, where g is an O_H^*-isomorphism:

$$O_H^* \times V_1(F_{H'}) \xrightarrow{\quad g \quad} q^{-1}(O_H^*) \subset V_{k+1}(F_n)$$

$$pr_1 \searrow \qquad \swarrow q$$

$$O_H^* \subset V_k(F_n)$$

We define $g(v_1, \ldots, v_k, x) = (v_1, \ldots, v_k, v_{k+1})$, where $v_{k+1} = f_{(v_1, \ldots, v_k), W*}(x)$. This proves the theorem.

Observe in (3.8) that the fibre is a sphere, $V_1(F^{n-k}) = S^{n-k-1}$ for $F = \mathbf{R}$, $S^{2n-2k-1}$ for $F = \mathbf{C}$, and $S^{4n-4k-1}$ for $F = \mathbf{H}$.

3.9. Corollary. *The projection* $q: V_{k+1}(F^n) \to V_k(F^n)$ *is a fibre map.*

Proof. For this, we use (3.8) and 1(5.2).

3.10 Remark. Theorems (3.6) and (3.8) are usually proved using deep properties of Lie groups. Our proofs are elementary and geometric in character.

4. Stability of the Homotopy Groups of the Classical Groups

Let F denote $\mathbf{R}$, $\mathbf{C}$, or $\mathbf{H}$, and let c be a corresponding integer, where c equals the dimension of the real vector space F over $\mathbf{R}$. We make use of the following fibre maps p.

$$U_F(n) \to V_{n+1}(F^{n+1}) = U_F(n+1) \xrightarrow{p} V_1(F^{n+1}) = S^{(n+1)c-1}$$

$$SU_F(n) \to V_n(F^{n+1}) = SU_F(n+1) \xrightarrow{p} V_1(F^{n+1}) = S^{(n+1)c-1}$$

The above maps p (the second is defined only for $F = \mathbf{R}$ or $\mathbf{C}$) are compositions of the fibre maps $V_{k+1}(F^{n+1}) \to V_k(F^{n+1})$ considered in (3.8). Now the next theorem is an easy consequence of the homotopy sequence of a fibre map.

4.1 Theorem. *The natural inclusions* $U_F(n) \to U_F(n+q)$ *and* $SU_F(n) \to SU_F(n+q)$ *induce morphisms* $\pi_i(U_F(n)) \to \pi_i(U_F(n+q))$ *and* $\pi_i(SU_F(n)) \to \pi_i(SU_F(n+q))$ *which are isomorphisms for* $i \leq c(n+1) - 3$ *and epimorphisms for* $i \leq c(n+1) - 2$.

Proof. If we prove the statement for $q = 1$, then by factoring the above inclusion $U_F(n) \to U_F(n+1) \to \cdots \to U_F(n+q)$, we get our result. For $q = +\infty$, we use 1(4.3) and observe the hypotheses are easily satisfied.

For $q = 1$ we use the above fibre maps and the homotopy sequence [see 1(5.3)]. We have

$$\pi_{i+1}(S^{(n+1)c-1}) \to \pi_i(U_F(n)) \to \pi_i(U_F(n+1)) \to \pi_i(S^{(n+1)c-1})$$

and a similar sequence for $SU_F(n) \to SU_F(n + 1)$. Our morphism is an epimorphism if $\pi_i(S^{(n+1)c-1}) = 0$ or $i < (n + 1)c - 1$ and an isomorphism if, in addition, $\pi_{i+1}(S^{(n+1)c-1}) = 0$ or $i + 1 < (n + 1)c - 1$. This proves the theorem.

4.2 Remark. For fixed i, $i \leq (n + 1)c - 3$ determines the stable range of n, where

$$\pi_i(U_F(n)) \to \pi_i(U_F) \qquad \text{or} \qquad \pi_i(SU_F(n)) \to \pi_i(SU_F)$$

are isomorphisms. In the real case, this range is $i \leq n - 2$; in the complex case, this range is $i \leq 2n - 1$; and in the quaternionic case, this range is $i \leq 4n + 1$.

5. Vanishing of Lower Homotopy Groups of Stiefel Varieties

In this section we make use of the following fibre map p.

$$U_F(m) \to U_F(k + m) \to V_k(F^{k+m})$$

Again p is the composition of fibre maps $V_{k+1}(F^n) \to V_k(F^n)$ considered in (3.8). The next theorem is an easy consequence of the homotopy sequence of a fibre map.

5.1 Theorem. $\pi_i(V_k(F^{k+m})) = 0$ for $i \leq (m + 1)c - 2$.

Proof. Applying the homotopy sequence 1(5.3) of a fibre map, we have the following exact sequence:

$$\cdots \to \pi_i(U_F(m)) \xrightarrow{\alpha} \pi_i(U_F(m + k)) \to \pi_i(V_k(F^{k+m})) \to \pi_{i-1}(U_F(m)) \xrightarrow{\beta} \cdots$$

By Theorem (4.1), if $i \leq c(m + 1) - 2$, then α is an epimorphism and β is an isomorphism, and $\pi_i(V_k(F^{k+m})) = 0$. This proves the theorem.

Observe that $\pi_i(V_k(F^\infty)) = 0$ for all i. In the real case, $\pi_i(V_k(\mathbf{R}^{k+m})) = 0$ for $i \leq m - 1$; in the complex case, $\pi_i(V_k(\mathbf{C}^{k+m})) = 0$ for $i \leq 2m$; and in the quaternionic case, $\pi_i(V_k(\mathbf{H}^{k+m})) = 0$ for $i \leq 4m + 2$.

6. Universal Bundles and Classifying Spaces for the Classical Groups

Putting together Theorems (5.1) and (3.6) and definition 4(10.7), we have the next theorem immediately.

6.1 Theorem. For $U_F(k)$, the principal bundle $V_k(F^{k+m}) \to G_k(F^{k+m})$ is universal in dimensions $\leq c(m + 1) - 2$, and $V_k(F^\infty) \to G_k(F^\infty)$ is a universal bundle.

For $SU_F(k)$, the principal bundle $V_k(F^{k+m}) \to SG_k(F^{k+m})$ is universal in dimensions $\leq c(m + 1) - 2$, and $V_k(F^\infty) \to SG_k(F^\infty)$ is a universal bundle.

We suggest that the reader compare $V_k(F^{mk})$ and the join $U_F(k) * \cdots * U_F(k)$. In this way the universal bundle $V_k(F^\infty) \to G_k(F^\infty)$ can be mapped naturally into the bundle coming from the Milnor construction.

7. Universal Vector Bundles

In Theorem 3(7.2) vector bundles over paracompact spaces were classified using the universal principal bundle $\alpha_k^n = (E, p, G_k(F^n))$ when $n = +\infty$. By (6.1), $\alpha_k^n = (V_k(F^n), p, G_k(F^n))$ is the universal $U_F(k)$-bundle in dimensions $\leq c(n - k + 1) - 2$. By 5(3.3), the fibre bundle $\alpha_k^n[F^k]$ is a vector bundle. We compare $\alpha_k^n[F^k]$ and γ_{kn} using the bundle map $f: \alpha_k^n[F^k] \to \gamma_{kn}$, where

$$f((v_1, \ldots, v_k, y) \bmod U_F(k)) = (\langle v_1, \ldots, v_k \rangle, y_1 v_1 + \cdots + y_k v_k)$$

For $a \in U_F(k)$, let $(v_1, \ldots, v_k)a = (v_1', \ldots, v_k')$ and $ay = y'$. Then we have $\sum_m y_m' v_m = \sum_{m,j} a_{m,j} y_j v_m = \sum_j y_j v_j'$, and u is a well-defined bundle morphism.

7.1 Proposition. *The bundle morphism $f: \alpha_k^n[F^k] \to \gamma_{kn}$ is a vector bundle isomorphism.*

Proof. From the definition of the vector bundle structure on $\alpha_k^n[F^k]$ [see 5(3.3)], and the definition of f, it follows that f is a vector bundle morphism which is an isomorphism on each fibre. By 3(2.5), f is an isomorphism. ∎

Theorem (6.1) has an interpretation for vector bundles in view of (7.1) and Remark 5(3.3).

7.2 Theorem. *Let X be an n-dimensional CW-complex. The function that assigns to each homotopy class $[f]: X \to G_k(F^{k+m})$ the isomorphism class of the k-dimensional vector bundle $f^*(\gamma_k^{k+m})$ over X is a bijection when $n \leq c(m + 1) - 2$.*

7.3 Remark. We can reverse the process in (7.1) and get a description of the associated principal $U(k)$-bundle, denoted $\Pr \xi$, for any vector bundle ξ. The total space of $\Pr \xi$ is the subspace of the total space of $\xi \oplus \overset{(k)}{\cdots} \oplus \xi$ consisting of orthonormal k-tuples $(v_1, \ldots, v_k)$ of vectors (of necessity all in the same fibre). Then $U(k)$ acts on the total space of $\Pr \xi$ by $(v_1, \ldots, v_k)a = (v_1', \ldots, v_k')$, where $v_j' = \sum_i a_{i,j} v_i$. If ξ is trivial, $\Pr \xi$ is a trivial principal $U(k)$-bundle, and the operation $\xi \mapsto \Pr \xi$ commutes with restriction. Therefore, $\Pr \xi$ is a principal $U(k)$-bundle, and, as above, it has the appropriate transition functions.

The reader can investigate to what extent $\Pr$ is a functor. The process described above can be thought of as the analogue for vector bundles of

$F^k \mapsto V_k(F^k)$ for vector spaces. We could take the operation $F^k \mapsto V_m(F^k)$ and define the bundle of m-frames in a k-dimensional vector bundle or $F^k \mapsto V_m(F^k)$ and define the bundle of m-dimensional subspaces in a k-dimensional vector bundle. We explore these constructions further in Chap. 15.

7.4 Example. For $\tau(S^n)$, $\mathrm{Pr}\,\tau(S^n)$ is just $V_{k+1}(\mathbf{R}^{n+1}) \to V_1(\mathbf{R}^{n+1})$. Similarly, for k-frames on $S^n = V_1(\mathbf{R}^{n+1})$ the bundle $V_{k+1}(\mathbf{R}^{n+1}) \to V_1(\mathbf{R}^{n+1})$ is used.

8. Description of all Locally Trivial Fibre Bundles over Suspensions

We view the nonreduced suspension SX of a space X as the quotient of $X \times [-1, +1]$ with $X \times \{-1\}$ identified to one point, $X \times \{+1\}$ identified to a second point, and no further identifications. For $J \subset [-1, +1]$, we denote by XJ the image of $X \times J$ in SX. For $x \in X$, $t \in [-1, +1]$, let $\langle x, t \rangle$ denote the point in SX corresponding to $(x, t) \in X \times [-1, +1]$. Let SX_+ denote $X[0, 1]$ and SX_- denote $X[-1, 0]$. Both the upper cone SX_+ and the lower cone SX_- are contractible.

8.1 Definition. A bundle ξ over a suspension SB is regular provided, for some $\varepsilon > 0$, $\xi | B(-\varepsilon, +1]$ and $\xi | B[-1, +\varepsilon)$ are trivial.

Every numerable principal G-bundle is regular over SB.

If $\{(B(-\varepsilon, +1], h_1), (B[-1, +\varepsilon), h_2)\}$ is an atlas for a regular principal G-bundle ξ over a suspension SB, there is just one transition function $g_{1,2}$: $B(-\varepsilon, +\varepsilon) \to G$ to consider, where for $\langle b, t \rangle \in B(-\varepsilon, +\varepsilon)$ and $s \in G$ there is the relation $h_2(\langle b, t \rangle, s) = h_1(\langle b, t \rangle, g_{1,2}\langle b, t \rangle s)$. We restrict $g_{1,2}$ to $B = B[0] \subset SB$ and denote the resulting map by c_ξ. In the next theorem, we determine the relation between the isomorphism class of ξ and the map c_ξ.

8.2 Theorem. *Two regular principal G-bundles ξ and ξ' over SB are isomorphic if and only if there exists $s \in G$ such that $c_{\xi'}$ and $sc_\xi s^{-1}$ are homotopic.*

Proof. If ξ and ξ' are isomorphic, there exist $\varepsilon > 0$ and maps $r_1: B(-\varepsilon, +1] \to G$ and $r_2: B[-1, +\varepsilon) \to G$ such that $c_{\xi'}(b) = r_1(b)c_\xi(b)r_2(b)^{-1}$, where $r_k(b) = r_k(b, 0)$ for $b \in B$. This statement follows from Theorem 5(2.7). By changing the charts by the constant action of G, we can assume that $c_\xi(b_0) = c_{\xi'}(b_0) = 1$. Consequently, we have $r_1(b_0) = r_2(b_0) = s \in G$. Since r_1 and r_2 are defined on $B(-\varepsilon, +1]$ and $B[-1, +\varepsilon)$, which are contractible, there exist homotopies $h_{i,t}: B \to G$ for $i = 1, 2$ with $h_{i,t}(b_0) = h_{i,1}(b) = s$ for each $t \in [0, 1]$ and $b \in B$ and $h_{i,0}(b) = r_i(b)$ for each $b \in B$. Then $h_{1,t}(b)c_\xi(b)h_{2,t}(b)^{-1}$ is a homotopy from $c_{\xi'}$ to $sc_\xi s^{-1}$.

Conversely, by using $r_i(b, t) = s$ on the appropriate domain, we can assume c_ξ and $c_{\xi'}$ are homotopic. Therefore, the map $c_{\xi'}c_\xi^{-1}$ is homotopic to the map constantly equal to $1 \in G$. Therefore, we have $r_1: B(-\varepsilon, 1] \to G$ such that

$r_1 | B(-\varepsilon, +\varepsilon) = g'_{1,2} g_{1,2}^{-1}$. If we define r_2 to be constantly equal to $1 \in G$ with domain $B[-1, +\varepsilon)$, we have $g'_{1,2}(x) = r_1(x)g_{1,2}(x)r_2(x)^{-1}$ for $x \in B(-\varepsilon, +\varepsilon)$. Consequently, by Theorem 5(2.7), ξ and ξ' are isomorphic.

8.3 Corollary. *Let G be a path connected group. The function that assigns to each isomorphism class of regular principal G-bundles ξ over SB the homotopy class $[c_\xi] \in [B, G]$ is a bijection.*

Proof. In Theorem (8.2), observe that if $u(t)$ is a path from 1 to s then $u(t)c_\xi u(t)^{-1}$ is a homotopy from c_ξ to $sc_\xi s^{-1}$. This remark and Theorem (8.2) demonstrate the corollary.

8.4 Corollary. *Let G be a path connected group. The isomorphism classes of locally trivial principal G-bundles ξ over S^n are classified by elements $[c_\xi]$ of $\pi_{n-1}(G)$.*

Proof. Note that $S^n = S(S^{n-1})$.

8.5 Definition. A map $c_\xi: B \to G$ associated with the principal G-bundle ξ over SB as in (8.1) is called a characteristic map of ξ.

The above discussion could be carried out for vector bundles of dimension k, using $G = U_F(k)$.

9. Characteristic Map of the Tangent Bundle over S^n

9.1 Notation. Let $a, b \in S^{n-1}$ and $a \neq -b$. Let $R(b, a)$ denote the rotation in $SO(n)$ with the two defining properties:

(R1) For each y with $(a|y) = (b|y) = 0$ there is the relation $R(b, a)y = y$.
(R2) $R(b, a)a = b$, where the rotation is along the shortest great circle from a to b.

The following formula holds:

$$R(b, a)x = x - \frac{((a + b)|x)}{1 + (a|b)}(a + b) + 2(a|x)b$$

By a straightforward calculation, the reader can verify (R1), (R2), and the relation $\| R(b, a)e_i \| = 1$ for $0 \leq i \leq n - 1$.

9.2 Principal Bundle of $\tau(S^n)$. By (7.4) the principal $SO(n)$ bundle associated with $\tau(S^n)$ is $SO(n + 1) \xrightarrow{p} S^n$, where $p(u) = u(e_n)$. The fibre is $SO(n)$ with natural inclusion. Equivalently, the calculation could be made for $O(n + 1) \xrightarrow{p} S^n$ with fibre $O(n)$.

9.3 Notation. We define $\phi: S^n - \{-e_n\} \to SO(n+1)$ by the relation $\phi(x) = R(x, e_n)$, where R is defined in (9.1). Let $r(x) = \phi(e_{n-1})^2(x)$. We can calculate $\phi(e_{n-1})(x_0, \ldots, x_n) = (x_0, \ldots, x_{n-2}, x_n, -x_{n-1})$, and consequently we have $r(x_0, \ldots, x_n) = (x_0, \ldots, x_{n-2}, -x_{n-1}, -x_n)$ and $r^2 = 1$.

9.4 Local Charts for $SO(n+1) \to S^n$. We define maps $h_1: (S^n - \{-e_n\}) \times SO(n) \to SO(n+1)$ and $h_2: (S^n - \{e_n\}) \times SO(n) \to SO(n+1)$ by the relations $h_1(x, u) = \phi(x)u$ and $h_2(x, u) = r\phi(r(x))u$, where the following diagrams are commutative.

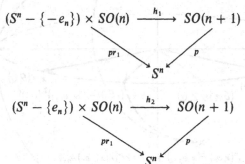

Observe that $p(\phi(x)u) = \phi(x)u(e_n) = \phi(x)e_n = x$ and $ph_2(x, u) = r\phi(r(x))u(e_n) = r\phi(r(x))e_n = r(r(x)) = x$ by (9.3). Clearly, h_1 and h_2 commute with the action of $SO(n)$ on the right. The transition map $g_{1,2}: S^n - \{e_n, -e_n\} \to SO(n)$ satisfies the relation $h_1(x, g_{1,2}(x)u) = h_2(x, u)$. Therefore, we have $g_{1,2}(x) = (\phi(x))^{-1}r\phi(r(x))$, and, restricting to S^{n-1}, we have $c_n(x) = (\phi(x))^{-1}r\phi(r(x))$, where c_n denotes the characteristic map of the principal bundle $SO(n+1) \to S^n$.

9.5 Theorem. *The characteristic map of $\tau(S^n)$ has the form $c_n(x) = R(x, e_{n-1})^2$ for $x \in S^{n-1}$.*

Proof. First, we compute $c_n(e_{n-1}) = (\phi(e_{n-1}))^{-1}r\phi(-e_{n-1}) = (\phi(e_{n-1}))^{-1}\phi(e_{n-1}) = R(e_{n-1}, e_{n-1})^2$. For $x \neq e_{n-1}$, we parametrize the intersection of $Re_n \oplus Re_{n-1} \oplus Rx$ with S^n by spherical coordinates (θ, ϕ); see Fig. 4. Then $e_n = (0, \pi/2)$, $e_{n-1} = (0, 0)$, and $x = (\theta, 0)$. If $y \in S^{n-1}$ is such that $(y|x) = (y|e_{n-1}) = 0$, then $c_n(x)y = y = R(x, e_{n-1})^2y$, and so we consider what happens in the 2-sphere $(Re_n \oplus Re_{n-1} \oplus Rx) \cap S^n$. Since $c_n(x)e_n = e_n = R(x, e_{n-1})^2e_n$, we need prove only that $R(x, e_n)R(x, e_{n-1})^2e_{n-1} = rR(r(x), e_n)e_{n-1}$.

First, we do the calculation for $0 \leq \theta \leq \pi/2$ as in Fig. 4. We have $R(x, e_{n-1})^2(0, 0) = (2\theta, 0)$. Since $R(x, e_n)$ is a $90°$ rotation from e_n to x around the axis $(\theta + \pi/2, 0)$ and $2\theta \leq \theta + \pi/2$, we have $R(x, e_n)(2\theta, 0) = (\theta + \pi/2, \theta - \pi/2)$. For the other side of the relation, we observe first that $r(\theta, \phi) = (\pi - \theta, -\phi)$ for $0 \leq \theta \leq \pi$. Then $R(r(x), e_n)$ is a $90°$ rotation from e_n to $(\pi - \theta, 0)$ around the axis $(\pi/2 - \theta, 0)$, and we have $R(r(x), e_n)(0, 0) = (\pi/2 - \theta, \pi/2 - \theta)$. Finally, we have $rR(r(x), e_n)(0, 0) = r(\pi/2 - \theta, \pi/2 - \theta) = (\pi/2 + \theta, \theta - \pi/2)$.

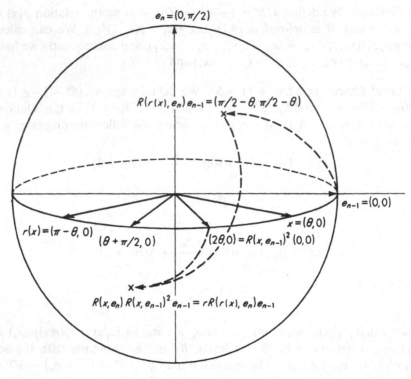

Figure 4

Second, we do the calculation for $\pi/2 \leq \theta \leq \pi$. We have $R(x, e_{n-1})^2(0,0) =$ $(2\theta - 2\pi, 0)$. Since $R(x, e_n)$ is a $90°$ rotation from e_n to x around the axis $(\theta - 3\pi/2, 0)$ and $\theta - 3\pi/2 \leq 2\theta - 2\pi$, we have $R(x, e_n)(2\theta - 2\pi, 0) =$ $(\theta - 3\pi/2, \theta - \pi/2)$. For the other side of the relation, we observe also that $r(\theta, \phi) = (-\pi - \theta, -\phi)$ for $-\pi \leq \theta \leq 0$. Then $R(r(x), e_n)$ is a $90°$ rotation from e_n to $(\pi - \theta, 0)$ around the axis $(\pi/2 - \theta, 0)$, and we have $R(r(x), e_n)(0,0) =$ $(\pi/2 - \theta, \pi/2 - \theta)$. Finally, we have $rR(r(x), e_n)(0,0) = r(\pi/2 - \theta, \pi/2 - \theta) =$ $(\theta - 3\pi/2, \theta - \pi/2)$. This proves the theorem since a rotation is determined by its image on two points x and y with $x \neq y$ and $x \neq -y$.

9.6 Corollary. *The relation $c_{r(S^n)}(-x) = c_{r(S^n)}(x)$ is satisfied for each $x \in S^{n-1}$.*

Proof. We must prove that $R(-x, e_{n-1})^2 = R(x, e_{n-1})^2$. In the (e_{n-1}, x)-plane, where e_{n-1} is represented by 0 and x by $\theta, 0 \leq \theta - \pi$, we have $-x$ represented by $\theta - \pi$. Then $R(x, e_{n-1})^2$ is a rotation by 2θ, and $R(x, e_{n-1})^2$ is a rotation by $2\theta - 2\pi$ or that which is the same as 2θ.

9.7 Representations of the Characteristic Map. Let $\alpha: S^{n-1} \to O(n)$ be the map defined by the requirement that $\alpha(x)$ be a reflection through the hyperplane orthogonal to x. For $y \in S^{n-1}$, we have $y = (y|x)x + (y - (y|x)x)$, and consequently $\alpha(x)y = y - 2(y|x)x$. Let $\beta: S^{n-1} \to O(n)$ be the map defined by the requirement that $\beta(x)$ is a "rotation" by $180°$ around the axis x. For $y \in S^{n-1}$, we have $\beta(x)y = 2(y|x)x - y$. Note that $\alpha(x)y = -\beta(x)y$, and $\alpha(x)y = y$ or $\beta(x)y = -y$ if and only if $(x|y) = 0$. We summarize further properties of α and β in the next proposition.

9.8 Proposition. *The relation* $R(x_1, x_2)^2 = \alpha(x_1)\alpha(x_2) = \beta(x_1)\beta(x_2)$ *holds.*

Proof. Clearly, $\alpha(x_1)\alpha(x_2) = \beta(x_1)\beta(x_2)$. Moreover, if y is orthogonal to x_1 and x_2, we have $R(x_1, x_2)^2 y = \alpha(x_1)\alpha(x_2)y = \beta(x_1)\beta(x_2)y = y$. It suffices to show that $R(x_1, x_2)^2 x_2 = \beta(x_1)\beta(x_2)x_2 = \beta(x_1)x_2$ equal the image of x_2 by a rotation of $180°$ around x_1. But this is the image of x_2 under a rotation through x_1 by an angle twice the angle between x_1 and x_2.

9.9 Corollary. *The relation* $c_n(x) = R(x, e_{n-1})^2 = \alpha(x)\alpha(e_{n-1}) = \beta(x)\beta(e_{n-1})$ *holds.*

10. Homotopy Properties of Characteristic Maps

For a list of properties of the degree of a map $S^n \to S^n$, see Eilenberg and Steenrod [1].

10.1 Theorem. *Let* $p: SO(n) \to S^{n-1}$ *be the projection* $p(u) = u(e_{n-1})$, *and let* $c_{n-1}: S^{n-1} \to SO(n)$ *be the characteristic map. Then the degree of* pc_{n-1} *is* $(+1) + (-1)^n$.

Proof. Let $H_{\pm}^{n-1}$ be the closed subspace of $x \in S^{n-1}$ with $\pm x_{n-1} \geq 0$. Then $pc_{n-1}: H_{\pm}^{n-1} - S^{n-2} \to S^{n-1} - \{-e_{n-1}\}$ is a homeomorphism, and pc_{n-1} defines a map $f_1: H_+^{n-1}/S^{n-2} \to S^{n-1}$ of degree $+1$. Since $c_{n-1}(x) = c_{n-1}(-x)$, we have a homeomorphism $pc_{n-1}: H_+^{n-1}/S^{n-2} \to S^{n-1}$, and pc_{n-1} defines a map $f_2: H_-^{n-1}/S^{n-2} \to S^{n-1}$ of degree equal to the degree of $x \mapsto -x$, which is $(-1)^n$ on S^{n-1}.

We have the following commutative diagram.

$$
\begin{array}{ccc}
S^{n-1} & \xrightarrow{\;pc_{n-1}\;} & S^{n-1} \\
& \searrow & \Big\uparrow {\scriptstyle f_1 \vee f_2} \\
& & S^{n-1}/S^{n-2} = S^{n-1} \vee S^{n-1}
\end{array}
$$

Consequently, the relation $\deg pc_{n-1} = +1 + (-1)^n$ holds, and this proves the theorem.

Therefore, $\deg pc_{n-1}$ is 2 for n even and 0 for n odd. For a map $f: S^{n-1} \times S^{n-1} \to S^{n-1}$ there is a bidegree $(a, b) \in \mathbf{Z} \times \mathbf{Z}$ associated with it, where a is the degree of $x \to f(x, y_0)$ and b is the degree of $y \to f(x_0, y)$. The pair (a, b) is independent of the pair (x_0, y_0).

10.2 Corollary. *The bidegree of* $(x, y) \mapsto \alpha(x)y$ *of* $S^{n-1} \times S^{n-1} \to S^{n-1}$ *is* $(+1 + (-1)^n, -1)$.

Proof. By (10.1) $x \mapsto \alpha(x)\alpha(e_{n-1})y$ has degree $1 + (-1)^n$ and is independent of $\alpha(e_{n-1})y \in S^{n-1}$. Moreover, the map $y \mapsto \alpha(x)y$ is a reflection through a hyperplane. It has degree -1.

10.3 Corollary. *The bidegree of* $(x, y) \mapsto \beta(x)y$ *of* $S^{n-1} \times S^{n-1} \to S^{n-1}$ *is* $(+1 + (-1)^n, -(-1)^n)$.

Proof. Since $\beta(x)y = -\alpha(x)y$, the degree of $x \mapsto -x$ which is $(-1)^n$ multiplied by the bidegree of $(x, y) \mapsto \alpha(x)y$ yields the bidegree of $\beta(x)y$.

The following theorem is very useful in making calculations of homotopy groups.

10.4 Theorem. *Let* $\xi[F] = (E_F, p_F, S^n)$ *be a fibre bundle with structure group* G *and fibre* F. *Let* $c_\xi \colon S^{n-1} \to G$ *be the characteristic map of* ξ, *and for* $y_0 \in F$, *let* $c_F \colon S^{n-1} \to F$ *be the map defined by* $c_F(x) = c_\xi(x)y_0$. *Then* $\mathrm{im}\, \partial = \ker \alpha$ *is generated by* $[c_F]$ *in the exact sequence*

$$\mathbf{Z} = \pi_n(S^n) \overset{\partial}{\to} \pi_{n-1}(F, y_0) \overset{\alpha}{\to} \pi_{n-1}(E_F) \to 0$$

Proof. We need only show that $\partial([f]) = [c_F]$, where $\deg f = \pm 1$. Let $f(x) = R(x, e_n)^2(e_n)$, where we consider $f \colon (H^+, S^{n-1}) \to (S^n, \{-e_n\})$, and the induced map $f \colon H^+/S^{n-1} \to S^n$ has degree $+1$. For $x \in H^+$ with $x \neq e_n$, let $h(x)$ be the point in S^{n-1} where the great circle from e_n to x intersects S^{n-1}.

Let $h_1 \colon H^- \times F \to p_F^{-1}(H^-)$ and $h_2 \colon H^+ \times F \to p_F^{-1}(H^+)$ be the restriction of charts where $h_2(x, y) = h_1(x, c_\xi(x)y)$ for $x \in H^+ \cap H^-$. We define $g \colon (H^+, S^{n-1}) \to (E_F, F)$, where

$$g(x) = \begin{cases} h_2(f(x), y_0) & \text{for } f(x) \in H^+ \\ h_1(f(x), c_\xi(h(x))y_0) & \text{for } f(x) \in H^- \end{cases}$$

If $f(x) = h(x) = x'$, the two definitions of g agree; that is, $h_1(x', c_\xi(x')y_0) = h_2(x', y_0)$. Clearly, $pg = f$ and $g(S^{n-1}) \subset F$. For $x \in S^{n-1}$, we have $f(x) = -e_n$, $h(x) = x$, and $g(x) = h_1(-e_n, c_\xi(x)y_0)$, which equals $c_\xi(x)y_0$ viewed with values in F. Then $[f]$ generates $\pi_n(S^n)$ and $\partial([f]) = [c_F]$, which proves the theorem. Here we used an explicit form of ∂.

11. Homotopy Groups of Stiefel Varieties

Let $j: V_k(F^n) \to V_{k+1}(F^{n+1})$ be the map $j(v_1, \ldots, v_k) = (v_1, \ldots, v_k, e_{n+1})$. Then j can be viewed as the inclusion of the fibre over e_{n+1} of the map $p: V_{k+1}(F^{n+1}) \to V_1(F^{n+1})$, where $p(v_1, \ldots, v_{k+1}) = v_{k+1}$. From this fibre map, we have the homotopy sequence.

$$\pi_{i+1}(V_1(F^{n+1})) \to \pi_i(V_k(F^n)) \xrightarrow{j_*} \pi_i(V_{k+1}(F^n)) \to \pi_i(V_1(F^{n+1}))$$

This yields the next proposition immediately.

11.1 Proposition. *The homomorphism*

$$j_*: \pi_i(V_k(F^n)) \to \pi_i(V_{k+1}(F^{n+1}))$$

is an isomorphism for $i \leq c(n+1) - 3$.

For $c = 1$, the inequality has the form $i \leq n - 2$, and $\pi_i(V_1(\mathbf{R}^n)) = \pi_i(S^{n-1}) = 0$. This yields $\pi_i(V_k(\mathbf{R}^n)) = 0$ for $i \leq (n-k) - 1$, which is Theorem (5.1). The first nontrivial group is $\pi_{n-k}(V_k(\mathbf{R}^n))$. This we calculate in the next proposition.

11.2 Proposition. *The group $\pi_{n-k}(V_k(\mathbf{R}^n))$ is $\mathbf{Z}$ for $k = 1$ or $n - k$ even and Z_2 for $k \geq 2$ and $n - k$ odd.*

Proof. For the case $k = 1$, we have $\pi_{n-1}(V_1(\mathbf{R}^n)) = \pi_{n-1}(S^{n-1}) = \mathbf{Z}$. By (11.1) it suffices to prove the result for $k = 2$ because $n - 2 \leq (n+1) - 3$. For $k = 2$, we view $V_2(\mathbf{R}^n) \to S^{n-1}$ as the sphere bundle with fibre $F = S^{n-2}$ associated with $V_n(\mathbf{R}^n) \to S^{n-1}$. Then the characteristic map is $c_F(x) = c_\xi(x)y = \alpha(x)\alpha(e_{n-2})y$ by Theorem (10.4) and Corollary (9.9). By Theorem (10.4) there is the exact sequence

$$\mathbf{Z} = \pi_{n-1}(S^{n-1}) \xrightarrow{\partial} \pi_{n-2}(S^{n-2}) \to \pi_{n-2}(V_2(\mathbf{R}^n)) \to 0$$

where $\partial[1] = [c_F]$. By Theorem (10.1), we have $\deg c_F = (+1) + (-1)^{n-1}$. Then the above sequence becomes

$$\mathbf{Z} \xrightarrow{\partial} \mathbf{Z} \xrightarrow{\alpha} \pi_{n-2}(V_2(\mathbf{R}^n)) \to 0$$

For n and $n - 2$ even, $\partial(1) = 0$ and $\pi_{n-2}(V_2(\mathbf{R}^n)) = \mathbf{Z}$, and for n and $n - 2$ odd, $\partial(1) = 2$ and $\pi_{n-2}(V_2(\mathbf{R}^n)) = Z_2$. This proves the proposition.

For $c = 2$, the inequality of (4.1) becomes $i \leq 2n - 1$. Since $\pi_i(V_1(\mathbf{C}^n)) = \pi_i(S^{2n-1}) = 0$ for $i \leq 2n - 2$, we have $\pi_i(V_k(\mathbf{C}^n)) = 0$ for $i \leq 2(n-k)$ which is Theorem (5.1). Moreover, $\pi_{2(n-k)+1}(V_k(\mathbf{C}^n))$ can be calculated $n - k$ from the case $k = 1$ because $2(n-1) + 1 \leq 2n - 1$. This yields the next result.

11.3 Proposition. *The relation $\pi_{2(n-k)+1}(V_k(\mathbf{C}^n)) = \mathbf{Z}$ holds.*

Proof. Note that for $k = 1$ we have $\pi_{2n-1}(V_1(\mathbf{C}^n)) = \pi_{2n-1}(S^{2n-1}) = \mathbf{Z}$, and now we use (11.1).

For $c = 4$, the inequality of (11.1) becomes $i \leq 4n + 1$. Again, $\pi_i(V_k(\mathbf{H}^n)) = 0$ for $i \leq 4(n - k) + 2$. By the same argument as was used in (11.3), we have the next proposition.

11.4 Proposition. *The relation* $\pi_{4(n-k)+3}(V_k(\mathbf{H}^n)) = \mathbf{Z}$ *holds.*

12. Some of the Homotopy Groups of the Classical Groups

We consider some special calculations of low-dimensional homotopy groups.

12.1 π_0 of the classical groups. The inequality $0 \leq c(n + 1) - 3$ holds for all $n \geq 1$ if $c = 2, 4$ and for $n \geq 2$ if $c = 1$. Using the elementary calculations of $U_F(1)$, $SU_F(1)$, $SO(2)$, and $O(2)$, we have $\pi_0(O(n)) = Z_2$ for all $n \geq 1$ and $\pi_0(SO(n)) = \pi_0(U(n)) = \pi_0(SU(n)) = \pi_0(Sp(n)) = 0$ for all $n \geq 1$.

12.2 Relation Between $\pi_i(U_F(n))$ and $\pi_i(SU_F(n))$. From the fibre sequence $SO(n) \to O(n) \to Z_2$ we have $0 = \pi_{i+1}(Z_2) \to \pi_i(SO(n)) \to \pi_i(O(n)) \to \pi_i(Z_2)$. For $i \geq 1$, the inclusion $SO(n) \to O(n)$ induces isomorphisms of the homotopy groups:

$$\pi_i(SO(n)) \to \pi_i(O(n))$$

From the fibre sequence $SU(n) \to U(n) \to S^1$ we have

$$0 = \pi_{i+1}(S^1) \to \pi_i(SU(n)) \to \pi_i(U(n)) \to \pi_i(S^1)$$

For $i \geq 2$, the inclusion $SU(n) \to U(n)$ induces isomorphisms of the homotopy groups:

$$\pi_i(SU(n)) \to \pi_i(U(n))$$

For $i = 1$, there is the exact sequence

$$0 \to \pi_1(SU(n)) \to \pi_1(U(n)) \to \pi_1(S^1) = \mathbf{Z} \to 0$$

12.3 π_1 of the Classical Groups. Clearly $\pi_1(O(1)) = \pi_1(SO(1)) = 0$. Since $U(1) = SO(2) = S^1$, since $Sp(1) = S^3$, and since $1 \leq c(n + 1) - 2$ for $c = 2$ or 4, we have

$$\pi_1(O(2)) = \pi_1(SO(2)) = \mathbf{Z}$$

$$\mathbf{Z} = \pi_1(U(1)) = \pi_1(U(n)) \qquad \text{for } 1 \leq n \leq +\infty$$

$$0 = \pi_1(SU(1)) = \pi_1(SU(n)) \qquad \text{for } 1 \leq n \leq +\infty$$

$$0 = \pi_1(Sp(1)) = \pi_1(Sp(n)) \qquad \text{for } 1 \leq n \leq +\infty$$

Since $1 \leq (n + 1) - 3$ for $n \geq 3$, we must calculate $\pi_1(SO(3))$. For this we prove $SO(3)$ is homeomorphic to RP^3. This can be seen as follows: An element $u \in SO(3)$ leaves a point x_u of the sphere $\|x\| = 1$ in $\mathbf{R}^3$ invariant and rotates the plane orthogonal to x_u by an angle θ_u. We associate with u the point $\theta_u x_u$ in the ball $B(0, \pi)$ in $\mathbf{R}^3$ with antipodal points of $\partial B(0, \pi)$ identified. This is RP^2 with a cell attached in the correct way so that the resulting space is RP^3. Consequently, we have $Z_2 = \pi_1(RP^3) = \pi_1(SO(3)) = \pi_1(SO(n)) = \pi_1(O(n))$ for $3 \leq n \leq +\infty$.

12.4 π_2 of the Classical Groups. We prove that $\pi_2(U_F(n)) = \pi_2(SU_F(n)) = 0$ for all $n \geq 1$. Observe that $0 = \pi_2(S^1) = \pi_2(U(1)) = \pi_2(SU(1))$. Since $Sp(1) = SU(2) = S^3$ and $2 \leq c(n + 1) - 3$ for $c = 2$ or 4, we have $0 = \pi_2(Sp(n)) = \pi_2(U(n)) = \pi_2(SU(n))$ for $1 \leq n \leq +\infty$. From the above representations of $SO(n)$ and $O(n)$ we have $\pi_2(SO(n)) = \pi_2(O(n))$ for $n \leq 3$. To do the calculation in the stable range, we consider the following fibre sequence:

$$SO(3) \to SO(4) \to S^3$$

From the homotopy sequence, we have

$$0 = \pi_2(SO(3)) \to \pi_2(SO(4)) \to \pi_2(S^3) = 0$$

Therefore, we have $0 = \pi_2(SO(n)) = \pi_2(O(n))$ for $n \geq 4$ since $\pi_2(SO(3)) = \pi_2(RP^3) = \pi_2(S^3) = 0$.

This result concerning π_2 of the classical groups is true of π_2 of an arbitrary compact Lie group.

Using Theorems (10.1) and (10.4), we shall be able to calculate π_3 and π_4. First, we have the following simple results.

12.5 We have $\pi_i(O(2)) = \pi_i(SO(2)) = \pi_i(U(1)) = 0$ for $i > 1$ and equal to $\mathbf{Z}$ for $i = 1$.

12.6 π_3 of Unitary and Symplectic Groups. Again we use the relation $SU(2) = Sp(1) = S^3$, and we derive the result $\pi_3(U(2)) = \pi_3(SU(2)) = \pi_3(Sp(1)) = \mathbf{Z}$. Since $3 \leq 2n - 1$ when $n \geq 2$ for the complex case and $3 \leq 4n + 1$ for the quaternionic case [see (4.1)], we have

$$\mathbf{Z} = \pi_3(Sp(k)) \qquad \text{for } k \geq 1$$

$$\mathbf{Z} = \pi_3(U(k)) = \pi_3(SU(k)) \qquad \text{for } k \geq 2$$

For $k = 1$, $\pi_3(U(1)) = \pi_3(SU(1)) = 0$.

We view S^3 as the subspace of $x \in \mathbf{H}$ with $\|x\| = 1$, and we define two maps

$$r: S^3 \to SO(4) \qquad \text{and} \qquad s: S^3 \to SO(4)$$

by $r(x)y = xyx^{-1}$ and $s(x)y = xy$ for $x \in S^3 \subset \mathbf{H}$ and $y \in \mathbf{R}^4 = \mathbf{H}$. The properties of these two maps are summarized in the next two propositions. Note the relation with (12.3).

12.7 Proposition. *There is a commutative diagram where r' is a homeomorphism, f is the natural projection, and g is the natural inclusion, where $g(u)1 = 1 \in \mathbf{H} = \mathbf{R}^4$:*

$$
\begin{array}{ccc}
S^3 & \xrightarrow{\ r\ } & SO(4) \\
{\scriptstyle f}\downarrow & & \uparrow{\scriptstyle g} \\
RP^3 & \xrightarrow{\ r'\ } & SO(3)
\end{array}
$$

Moreover, the map r' is uniquely determined by this diagram.

Proof. Since $r(x)1 = 1$ and $r(x) = r(-x)$, the map r' exists. If $r(x)y = y$ for all y, then $xy = yx$ for all $y \in \mathbf{H}$. This happens if and only if x is real, and therefore $x = \pm 1$. This means that r' is injective. The formula $r(\cos\theta + i\sin\theta) = j\cos 2\theta + k\sin 2\theta$ and the formulas arising from cyclic permutations of i, j, k demonstrate that the image of r includes all rotations about the three axes i, j, and k. Therefore, r' is a bijection and a homeomorphism because RP^3 is compact.

12.8 Corollary. *The group $\pi_3(SO(3)) = \mathbf{Z}$ is an infinite cyclic group generated by $r: S^3 \to SO(3)$.*

Proof. The induced homomorphisms $(r')_*$ and $f_*: \pi_3(S^3) = \mathbf{Z} \to \pi_3(RP^3)$ are isomorphisms.

For $p: SO(4) \to S^3$, where $p(u) = u(1)$ in $\mathbf{H}$, we have $ps = 1$, and s is a cross section of the principal $SO(3)$-bundle $p: SO(4) \to S^3$.

12.9 Proposition. *The space $SO(4)$ is homeomorphic to $S^3 \times SO(3)$. Moreover, $\pi_3(SO(4)) = \mathbf{Z}[r] \oplus \mathbf{Z}[s]$.*

Proof. The first statement comes from 4(8.3). The cross section $[s]$ generates the image of $\pi_3(S^3) \to \pi_3(SO(4))$ and $[r]$ generates the image of $\pi_3(SO(3)) \to \pi_3(SO(4))$. Since $\pi_3(SO(4)) = \pi_3(S^3) \oplus \pi_3(SO(3))$, we have the result.

12.10 Proposition. *The characteristic map c_5 of the principal $SO(4)$-bundle $SO(5) \to S^4$ is given by $c_5(x) = r(x)^{-1}s(x)^2$.*

Proof. Consider $r(x)^{-1}s(x)^2 y = x^{-1}x^2yx = xyx$. If x and y are orthogonal, we have $xyx = y\bar{x}x = y$. If y lies on the circle from 1 to x, then $xy = yx$, and the map $y \to xyx = x^2y$ is a rotation in this circle of an angle equal to twice the angle from 1 to x. This can be seen immediately by parametrizing the $(1, x)$-plane. Now we use Theorem (9.5).

12.11 Proposition. $\pi_3(SO(k)) = \mathbf{Z}$ for $5 \leq k$.

Proof. By (12.10), $[c_5(x)] = -[r(x)] + 2[s(x)]$. By Theorem (10.4), there is the exact sequence

$$\pi_4(S^4) \xrightarrow{\partial} \pi_3(SO(4)) \to \pi_3(SO(5)) \to 0$$

Since the image of ∂ is generated by $-[r(x)] + 2[s(x)]$, we have $\pi_3(SO(5)) \cong$ coker $\partial = \mathbf{Z}$. Since $3 \leq k - 2$ for $k \geq 5$, we have $\pi_3(SO(k)) = \mathbf{Z}$ by (4.1).

Exercises

1. Verify that the formula in (9.1) has the desired properties.

2. Let $f: B_1 \to B$ be a map, and let ξ be a principal G-bundle over $S(B)$ with characteristic map $c_\xi: B \to G$. Then prove $c_\xi f$ is a characteristic map of $S(f)^*(\xi)$.

3. Let $\omega_0 = (E_0, p_0, B_0)$ be a universal G-bundle in dimensions $\leq n$. Let c denote the characteristic map for $f^*(\omega_0)$, where $f: S^n \to B_0$ is a map. If $\partial: \pi_n(B_0) \to \pi_{n-1}(G)$ is the boundary map, prove that $\partial[f] = [c]$.

4. Describe the characteristic maps of $U(n) \to S^{2n-1}$ and $Sp(n) \to S^{4n-1}$ as in Sec. 9. Carry through the discussion in Sec. 1 for these bundles and their characteristic maps.

5. By viewing $\mathbf{C}^2 = \mathbf{H}$, prove that $SU(2) = Sp(1)$. Then prove that $\pi_i(U(2)) = \pi_i(S^3)$ for $i \geq 2$.

6. Using the fact that $\pi_{k+1}(S^k) = Z_2$ for $k \geq 3$, calculate $\pi_4(Sp(k))$, $\pi_4(U(2))$, $\pi_4(SU(2))$, $\pi_4(SO(k))$ for $1 \leq k \leq 4$.

7. Calculate $\pi_{2(n-k)+2}(V_k(\mathbf{C}^n))$ and $\pi_{4(n-k)+4}(V_k(\mathbf{H}^n))$.

8. Prove that the inclusion $O(n) \to O(n + 1)$ yields the exact sequence $0 \to \mathbf{Z} \to \pi_{n-1}(O(n)) \to \pi_{n-1}(O(n+1)) \to 0$ for n even and $0 \to Z_2 \to \pi_{n-1}(O(n)) \to \pi_{n-1}(O(n+1)) \to 0$ for n odd.

9. Find all $U_F(n)$ principal bundles over S^m for $m = 1$, $m = 2$, $m = 3$, and $m = 4$.

PART II

ELEMENTS OF K-THEORY

CHAPTER 9

Stability Properties of Vector Bundles

Two vector bundles ξ and η are called s-equivalent provided $\xi \oplus \theta^n$ and $\eta \oplus \theta^m$ are isomorphic for some n and m where θ^m denotes the m dimensional trivial vector bundle. Stable equivalence, or s-equivalence, is an equivalence relation, and the stable classes form a ring (over finite-dimensional spaces), with $\oplus$ inducing the addition operation and $\otimes$ the multiplication operation. These are the K-rings of the space. We study the relation between isomorphism and stable equivalence. Also we consider elementary properties of the cofunctor $\tilde{K}$.

1. Trivial Summands of Vector Bundles

Let F denote $\mathbf{R}$, $\mathbf{C}$, or $\mathbf{H}$, and let c denote $\dim_{\mathbf{R}} F$. Using 2(7.1) and the fact that $F^n - 0$ is $(cn - 2)$-connected, we can decompose high-dimensional vector bundles into the Whitney sum of a trivial bundle and another bundle of lower dimension. For a vector bundle ξ, let ξ_0 denote the subbundle of nonzero vectors. Throughout this section, X denotes an n-dimensional CW-complex.

1.1 Proposition. *If ξ^k is a k-dimensional vector bundle with $n \leq ck - 1$, then ξ is isomorphic to $\eta \oplus \theta^1$ for some vector bundle η.*

Proof. The fibre of ξ_0 is $F^k - \{0\}$ and is $(ck - 2)$-connected. By Theorem 2(7.1), under hypothesis (H1), we have a cross section s of ξ_0. The map s can be viewed as an everywhere-nonzero cross section of ξ. This cross section determines a monomorphism $u: \theta^1 \to \xi$, where $u(b, a) = as(b)$ for $(b, a) \in$

$E(\theta^1)$. Let η be the coker u which is a vector bundle by 3(8.2), and since X is paracompact, by 3(9.6), there is an isomorphism between ζ and $\eta \oplus \theta^1$. This proves the proposition.

For $x \in \mathbf{R}$, let $\langle x \rangle$ denote the smallest integer n with $x \leq n$.

1.2 Theorem. *Let* $m = \langle ((n + 1)/c) - 1 \rangle$. *Then each* k-*dimensional vector bundle* ζ^k *is isomorphic to* $\eta^m \oplus \theta^{k-m}$ *for some* m-*dimensional vector bundle* η.

Proof. By induction on $k \geq m$, ζ^k is isomorphic to $\eta^m \oplus \theta^{k-m}$ for $n \leq c(m + 1) - 1$ in view of Proposition (1.1). This inequality also has the form $[(n + 1)/c] - 1 \leq m$.

Note: For $c = 1$ and real vector bundles, ζ^k is isomorphic to $\eta^n \oplus \theta^{k-n}$ for some η. For complex vector bundles, m is approximately $n/2$ and approximately $n/4$ for quaternionic bundles.

1.3 Remark. Theorem (1.2) says that a vector bundle over a point is trivial, or, in other words, it has a basis. This theorem can be regarded as the natural generalization to vector bundles of the basis theorem for vector spaces. In the case of vector spaces there is a uniqueness theorem which says that two bases have the same number of elements. Theorem (1.5) can be viewed as the proper generalization of this result to vector bundles.

1.4 Proposition. *If* u, $v: \theta^1 \to \zeta^k$ *are two monomorphisms of vector bundles with* $n \leq ck - 2$, *then coker* u *and coker* v *are isomorphic over* X.

Proof. As in (1.1), a monomorphism $\theta^1 \to \zeta$ is completely determined by a cross section of ζ_0, and a homotopy of monomorphism is determined by a cross section of $\zeta_0 \times I = (\zeta \times I)_0$ over $X \times \{0,1\}$, where $s|(X \times 0)$ corresponds to u and $s|(X \times 1)$ to v. Since $\dim(X \times I) = n + 1 \leq ck - 1$, we have a prolongation of s to $X \times I$ as a cross section of $\zeta \times I$. This cross section s^* determines a monomorphism $w: \theta^1 \to \zeta \times I$. Since coker $w|(X \times 0)$ is isomorphic to coker u and coker $w|(X \times 1)$ is isomorphic to coker v, there is an isomorphism between coker u and coker v by 3(4.6).

1.5 Theorem. *Let* $m = \langle ((n + 2)/c) - 1 \rangle$. *If* ζ_1^k *and* ζ_2^k *are two* k-*dimensional vector bundles such that* $m \leq k$ *and* $\zeta_1 \oplus \theta^l$ *and* $\zeta_2 \oplus \theta^l$ *are isomorphic for some* l, *then* ζ_1 *and* ζ_2 *are isomorphic.*

Proof. By induction on l, it follows that ζ_1 and ζ_2 are isomorphic for $n \leq c(k + 1) - 2$ in view of Proposition (1.4). This inequality also has the form $((n + 2)/c) - 1 \leq k$.

1.6 Remark. Every real vector bundle ξ^k with $k \geq n + 1$ is isomorphic to $\eta^{n+1} \oplus \theta^{k-n-1}$, where η is uniquely determined by ξ up to isomorphism. The inequality given in (1.5) is the best possible, in general, as can be seen from the isomorphism between $\tau(S^n) \oplus \theta^1$ and $\theta^n \oplus \theta^1 = \theta^{n+1}$ given in 2(4.7) and the fact that $\tau(S^n)$ is nontrivial for $n \neq 1, 3$, and 7.

Finally, observe that the methods in this section are elementary in the sense that the general theory of fibre bundles is not used here. These results can be derived from the refined homotopy classification theorem in 4(13.1).

2. Homotopy Classification and Whitney Sums

2.1 Definition. A classifying map of a vector bundle ξ over a space X is a map $f: X \to G_k(F^{k+m})$ such that ξ and $f^*(\gamma_k^{k+m})$ are isomorphic.

We can reformulate the homotopy classification theorem for vector bundles [Theorem 7(7.2)] as follows: Each k-dimensional vector bundle over a CW-complex X of dimension n with $n \leq c(m + 1) - 2$ has a classifying map $f: X \to G_k(F^{k+m})$ and f is unique up to homotopy equivalence. To calculate the classifying map of a Whitney sum, we define a morphism $(w, d): \gamma_{k^n} \times \gamma_l^m \to \gamma_{k+l}^{n+m}$, where $\gamma_{k^n} = (E_1, p_1, G_k(F^n))$, $\gamma_l^m = (E_2, p_2, G_l(F^m))$, and $\gamma_{k+l}^{n+m} = (E, p, G_{k+l}(F^{n+m}))$. We have the following commutative diagram, where $d(W_1, W_2) = W_1 \oplus W_2$ and

$$w((W_1, x_1), (W_2, x_2)) = (W_1 \oplus W_2, x_1 + x_2)$$

$$
\begin{array}{ccc}
E_1 \times E_2 & \xrightarrow{\;\;w\;\;} & E \\
\downarrow{\scriptstyle p_1 \times p_2} & & \downarrow{\scriptstyle p} \\
G_k(F^n) \times G_l(F^m) & \xrightarrow{\;\;d\;\;} & G_{k+l}(F^{n+m})
\end{array}
$$

2.2 Theorem. If $f: X \to G_k(F^n)$ is a classifying map for ξ, and if $g: X \to G_l(F^m)$ is a classifying map for η, then $d(f \times g)\Delta$ is a clasifying map for $\xi \oplus \eta$.

Proof. Consider the vector bundle morphism $h: f^*(\gamma_{kn}) \oplus g^*(\gamma_l^m) \to (d(f \times g)\Delta)^*(\gamma_{k+l}^{n+m})$ defined by the relation $h((b, W, y), (b, W', y')) = (b, W \oplus W', y + y')$. Since h is clearly a bijection, by 3(2.5), it is an isomorphism. This proves the proposition.

There are two important special cases of Theorem (2.2).

2.3 Corollary. If $f: X \to G_k(F^n)$ is a classifying map for a vector bundle ξ and if $i: G_k(F^n) \to G_k(F^{n+m})$ is the natural inclusion, then if is a classifying map for ξ.

Proof. The space $G_0(F^m)$ is just one point 0, and i is just $d: G_k(F^n) \times G_0(F^m) \to G_k(F^{n+m})$. For the unique map $g: X \to G_0(F^m)$, we have $g^*(\gamma_0^m) = 0$. Since $\xi \oplus 0$ is ξ, we have our result by (2.2).

2.4 Corollary. *If $f: X \to G_k(F^n)$ is a classifying map for a vector bundle ξ and if $j: G_k(F^n) \to G_{k+m}(F^{n+m})$ is the natural inclusion, where $j(W) = W \oplus F^m$, then jf is the classifying map for $\xi \oplus \theta^m$, where θ^m is the trivial m-dimensional vector bundle.*

Proof. The space $G_m(F^m)$ has only one point F^m, and j is just $d: G_k(F^n) \times G_m(F^m) \to G_{k+m}(F^{n+m})$. For the unique map $g: X \to G_m(F^m)$, we have $g^*(\gamma_m^m) \cong \theta^m$. We have our result by (2.2).

2.5 Remark. Let $\tau: G_k(F^n) \to G_{n-k}(F^n)$ be the map given by $\tau(W) = W^*$, the orthogonal complement of W. Then, using the notation of (2.3) and (2.4), we have the following commutative diagram for $k + l = n$:

$$
\begin{array}{ccc}
G_k(F^n) & \xrightarrow{\ \ i\ \ } & G_k(F^{n+m}) \\
\downarrow{\scriptstyle\tau} & & \downarrow{\scriptstyle\tau} \\
G_l(F^n) & \xrightarrow{\ \ j\ \ } & G_{l+m}(F^{n+m})
\end{array}
$$

2.6 Theorem. *For a CW-complex X of dimension n with $n \leqq c(m + 1) - 2$ and $n \leqq c(k + 1) - 2$, the functions $i_*: [X, G_k(F^{k+m})] \to [X, G_k(F^{k+m+1})]$ and $j_*: [X, G_k(F^{k+m})] \to [X, G_{k+1}(F^{k+m+1})]$ are bijections.*

Proof. The statement about i_* is Theorem 7(7.2), and the statement about j_* follows from the commutative diagram in (2.5) where τ is a homeomorphism.

2.7 Corollary. *With the hypothesis of (2.6), the function $j_* i_* [X, G_m(F^{2m})] \to [X, G_{m+1}(F^{2m+2})]$ is a bijection.*

Observe that Theorem (1.5) is a corollary of (2.6). Our proof of (1.5) was elementary in the sense that general fibre theory was not used.

3. The K Cofunctors

We begin by describing a general algebraic schema. First, we recall that a semiring is a triple (S, α, μ), where S is a set, $\alpha: S \times S \to S$ is the addition function usually denoted $\alpha(a, b) = a + b$, and $\mu: S \times S \to S$ is the multiplication function usually denoted $\mu(a, b) = ab$. A semiring is required to satisfy all the axioms of a ring except the existence of a negative or additive inverse. The natural numbers $\{0, 1, 2, 3, \ldots\}$ with the usual addition and multiplication is an example of a semiring. We speak of semirings with identity and which are commutative as with rings. Finally we recall that a semigroup is a pair (S, α) satisfying all the axioms for a group except the one on the existence of inverses.

3.1 Example. Let X be a topological space and let $\text{Vect}_F(X)$ be the set of isomorphism classes of F-vector bundles over X where $F = \mathbf{R}, \mathbf{C}$, or $\mathbf{H}$. These vector bundles need not have the same dimension on each component of X. For $F = \mathbf{R}$ or $\mathbf{C}$ the set $\text{Vect}_F(X)$ admits a natural commutative semiring structure where $(\xi, \eta) \mapsto \xi \oplus \eta$ is the addition function and $(\xi, \eta) \mapsto \xi \otimes \eta$ is the multiplication function. For $F = \mathbf{H}$, the set $\text{Vect}_\mathbf{H}(X)$ admits only a natural commutative semigroup structure, where $(\xi, \eta) \mapsto \xi \oplus \eta$ is the addition function. We used the same symbol for a vector bundle and for its isomorphism class.

Recall that a morphism from a semiring S to a semiring S' is a function $f: S \to S'$ of the underlying sets such that $f(a + b) = f(a) + f(b)$, $f(ab) = f(a)f(b)$, and $f(0) = 0$.

3.2 Example. For a pointed space X we have a semiring morphism rk: $\text{Vect}_F(X) \to \mathbf{Z}$, the rank or dimension of the vector bundle on the component of X containing the base point. For $\mathbf{H} = F$, rk is only a semigroup morphism.

Now we consider for general semirings the process of passing from a semiring to a ring in the most efficient way. When applied to the semiring of natural numbers, this process yields the ring of integers. A similar discussion applies to the group completion of commutative semigroups.

3.3 Definition. The ring completion of a semiring S is a pair (S^*, θ), where S^* is a ring and $\theta: S \to S^*$ is a morphism of semirings such that if $f: S \to R$ is any morphism into a ring there exists a ring morphism $g: S^* \to R$ such that $g\theta = f$. Moreover, g is required to be unique.

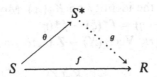

For the construction of S^* we consider pairs $(a, b) \in S \times S$ and put the following equivalence relation on these pairs; that is, (a, b) and (a', b') are equivalent provided there exists $c \in S$ with $a + b' + c = a' + b + c$. The reader can easily verify that this is an equivalence relation. Let $\langle a, b \rangle$ denote the equivalence class of (a, b) (thought of as $\langle a, b \rangle = a - b$). Let S^* denote the set of equivalence classes $\langle a, b \rangle$. Then we define $\langle a, b \rangle + \langle c, d \rangle = \langle a + c, b + d \rangle$ and $\langle a, b \rangle \langle c, d \rangle = \langle ac + bd, bc + ad \rangle$. The negative of $\langle a, b \rangle$ is $\langle b, a \rangle$ and $0 = \langle 0, 0 \rangle$. Finally, $\theta: S \to S^*$ is defined by $\theta(a) = \langle a, 0 \rangle$. We can also view S^* as the free abelian group generated by the set S modulo the subgroup generated by $(a + b) + (-1)a + (-1)b$, $a, b \in S$. We leave the details to the reader, for example, that multiplication prolongs linearly.

Finally, the uniqueness of (S^*, θ) follows from a commutative diagram involving a second completion (S_1^*, θ_1) of S, namely, the following:

Then gf and fg are identities. The situation with the group completion of a semigroup is completely analogous to the above.

3.4 Definition. The $K_F(X)$ ring (or group for $F = H$) of a space X is the ring (or group) completion of $\text{Vect}_F(X)$.

Observe that Vect_F is a cofunctor from the category of spaces and maps to the category of semirings (or semigroups). If $f: Y \to X$ is a map, then $\text{Vect}_F(f): \text{Vect}_F(X) \to \text{Vect}_F(Y)$ is defined by the relation $\text{Vect}_F(f)(\xi)$ equals the isomorphism class of $f^*(\xi)$ over Y. Here ξ denotes both a vector bundle and its isomorphism class.

Similarly, the functions K_F define a cofunctor. For a map $f: Y \to X$ there are the following morphisms:

$$
\begin{array}{ccc}
\text{Vect}_F(X) & \xrightarrow{\theta} & K_F(X) \\
{\scriptstyle \text{Vect}_F(f)} \downarrow & & \downarrow {\scriptstyle K_F(f)} \\
\text{Vect}_F(Y) & \xrightarrow{\theta} & K_F(Y)
\end{array}
$$

The requirement of commutativity defines the morphism $K_F(f)$. If $g: Z \to Y$ is a second map, we have $K_F(fg) = K_F(g)K_F(f)$ by the unicity statement in (3.3). Similarly, $K_F(1_X)$ is the identity on $K_F(X)$. More preciselyy, if $\xi - \eta \in K_F(X)$, we have $K_F(f)(\xi - \eta) = f^*(\xi) - f^*(\eta)$.

By (3.3), the morphism $rk: \text{Vect}_F(X) \to \mathbf{Z}$ factors as follows:

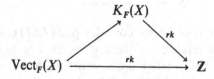

More precisely, we have $rk(\xi - \eta) = rk\xi - rk\eta$.

The multiplicative identity, denoted 1, in $K_F(X)$ is represented by the trivial line bundle and $rk(1) = 1$. Consequently, there is a morphism $\varepsilon: \mathbf{Z} \to K_F(X)$ such that $(rk)\varepsilon = 1_{\mathbf{Z}}$, where $\varepsilon(n)$ is the class of θ^n for $n \geq 0$. For a (pointed) map $f: Y \to X$, the following diagram is commutative.

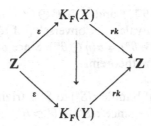

3.5 Definition. The reduced K_F cofunctor, denoted $\tilde{K}_F$, is the $\ker(rk\colon K_F \to \mathbf{Z})$.

From the above diagram, the cofunctor K_F splits $K_F(X) = \tilde{K}_F(X) \oplus \mathbf{Z}$. Moreover, $\tilde{K}_F(f)$ for a map $f\colon Y \to X$ is the restriction of $K_F(f)$, and $\tilde{K}_F$ is a cofunctor. The following description of $\tilde{K}_F(X)$ is very useful for future sections. First, we consider a definition.

3.6 Definition. Two vector bundles ξ and η over a space X are s-equivalent, denoted $\xi \sim \eta$, provided there exists q and n such that $\xi \oplus \theta^n$ and $\eta \oplus \theta^q$ are isomorphic over X. A bundle ξ that is s-equivalent to 0 is called s-trivial.

Stable equivalence, or s-equivalence, is clearly an equivalence relation, and isomorphic vector bundles are s-equivalent. Consequently, s-equivalence can be thought of as a relation on $\mathrm{Vect}_F(X)$.

3.7 Remark. Let X be an n-dimensional CW-complex. In the language of s-equivalence, Theorem (1.2) says that every vector bundle is s-equivalent to a k-dimensional bundle where $n \leq c(k+1) - 1$, and Theorem (1.5) says that two k-dimensional bundles ξ and η are isomorphic if and only if they are s-equivalent when $n \leq c(k+1) - 2$.

In the next theorem we determine the s-equivalence classes of vector bundles over a space.

3.8 Theorem. *Let X be a space with the following property (S): For each vector bundle ξ over X there exists a vector bundle η over X with $\xi \oplus \eta$ isomorphic to some θ^m. Then the function $\alpha\colon \mathrm{Vect}_F(X) \to \tilde{K}_F(X)$ defined by $\alpha(\xi) = \xi - rk(\xi)$ is a surjection, and $\alpha(\xi) = \alpha(\eta)$ if and only if ξ and η are s-equivalent. To form $\xi - rk(\xi)$, we view $\mathbf{Z} \subset K_F(X)$, using ε.*

With this theorem, we see that the s-equivalence classes can be identified with the elements of $\tilde{K}_F(X)$.

Proof. To prove that α is surjective, let $\xi - \eta \in \tilde{K}_F(X)$, where $rk\xi = rk\eta$. Let η' be a vector bundle such that $\eta \oplus \eta'$ is isomorphic to θ^m. Then in $\tilde{K}_F(X)$ we have $\xi - \eta = \xi \oplus \eta' - \eta \oplus \eta' = \xi \oplus \eta' - m = \xi \oplus \eta' - rk(\xi \oplus \eta') = \alpha(\xi \oplus \eta')$. Consequently, α is surjective.

Let ξ^n and η^m be two vector bundles such that $\xi - n = \alpha(\xi) = \alpha(\eta) = \eta - m$. Then there is a bundle ζ such that $\xi \oplus \theta^m \oplus \zeta$ and $\eta \oplus \theta^n \oplus \zeta$ are isomorphic. Let ζ' be a vector bundle such that $\zeta \oplus \zeta'$ and θ^q are isomorphic.

Then $\xi \oplus \theta^m \oplus \theta^q = \xi \oplus \theta^{m+q}$ and $\eta \oplus \theta^n \oplus \theta^q = \eta \oplus \theta^{n+q}$ are isomorphic and ξ and η are s-equivalent. Conversely, if $\xi \oplus \theta^n$ and $\eta \oplus \theta^m$ are isomorphic, we have $\alpha(\xi \oplus \theta^n) = \alpha(\eta \oplus \theta^m)$. Since $\alpha(\zeta \oplus \theta^q) = \alpha(\zeta)$, we have $\alpha(\xi) = \alpha(\eta)$. This proves the theorem.

3.9 Example. The tangent bundle $\tau(S^n)$ to S^n is trivial only for $n = 1, 3,$ or $7,$ and $\tau(S^n)$ is s-trivial for all n since $\tau(S^n) \oplus \theta^l \cong \theta^{n+1}$.

3.10 Remark. In 4(11.2) we classified all fibre bundles with given group G and fibre F, and in 3(7.2) we classified all vector bundles of given dimension. In the next secton we classify stability classes of all vector bundles; that is, we represent $\tilde{K}_F(X)$ as homotopy classes $[X, B_F]$, where B_F is an H-space. This will be proved for X, a finite CW-complex.

4. Corepresentations of $\tilde{K}_F$

There are two domains over which $\tilde{K}_F$ has a corepresentation: first, over the category of connected CW-complexes of dimension less than a fixed integer and, second, over the category of all finite CW-complexes. Neither category includes the other.

We begin by defining a morphism of functors $\phi_{X^n}: [X, G_n(F^{2n})] \to \tilde{K}_F(X)$ by the requirement that $\phi_{X^n}([g])$ equal the class of $g^*(\gamma_n^{2n}) - n$ in $\tilde{K}_F(X)$.

4.1 Proposition. *For an arbitrary category determined by paracompact spaces* X, *the family of functions* $\phi_{X^n}: [X, G_n(F^{2n})] \to \tilde{K}_F(X)$ *is a morphism of cofunctors.*

Proof. Let $f: Y \to X$ be a map. Then $(gf)^*(\gamma_n^{2n})$ and $f^*(g^*(\gamma_n^{2n}))$ are isomorphic. Therefore, the following diagram is commutative:

$$
\begin{array}{ccc}
[X, G_n(F^{2n})] & \xrightarrow{\phi_X} & \tilde{K}_F(X) \\
{\scriptstyle [f, G_n(F^{2n})]}\downarrow & & \downarrow{\scriptstyle \tilde{K}_F(f)} \\
[Y, G_n(F^{2n})] & \xrightarrow{\phi_Y} & \tilde{K}_F(Y)
\end{array}
$$

Therefore, ϕ is a morphism of cofunctors.

4.2 Theorem. *On the category of connected CW-complexes of dimension $\leq m$, the morphism* $\phi: [-, G_n(F^{2n})] \to \tilde{K}_F(-)$ *is an isomorphism for* $m \leq c(n + 1) - 2$.

Proof. By Theorem (3.8), the elements of $\tilde{K}_F(X)$ have the form $\xi - rk(\xi)$ and $\xi - rk(\xi) = \eta - rk(\eta)$ if and only if ξ and η are s-equivalent. By (3.8) the functions ϕ_{X^n} are surjective. Moreover, if $\phi_{X^n}([f]) = \phi_{X^n}([g])$, then $f^*(\gamma_n^{2n})$

and $g^*(\gamma_n^{2n})$ are s-equivalent and of the same dimension. By (3.8), this means that $f^*(\gamma_n^{2n})$ and $g^*(\gamma_n^{2n})$ are isomorphic over X, and, by 7(7.2), we have $[f] = [g]$. Therefore, ϕ_{X^n} is a bijection for each X, and ϕ^n is an isomorphism.

4.3 Definition. Let $B_{(F)}$ denote the $\bigcup\limits_{1 \leq n} G_n(F^{2n})$ with the inductive topology.

In the real case, we denote $B_{(F)}$ by B_0; in the complex case, by B_U; and in the quaternionic case, by B_{Sp}.

4.4 Proposition. *For each finite connected CW-complex X, there exists a k such that the natural inclusion $G_q(F^{2q}) \to B_{(F)}$ induces a bijection $[X, G_q(F^{2q})] \to [X, B_{(F)}]$ for $q \geq k$.*

Proof. By (2.7), for k such that $\dim X \leq c(k + 1) - 2$, the following bijections are induced by inclusions:

$$[X, G_k(F^{2k})] \to \cdots \to [X, G_q(F^{2q})] \to \cdots$$

Since X is compact, every map $f: X \to B_{(F)}$ has the image $f(X) \subset G_n(F^{2n})$ for some n for $k \leq n$, and the function $[X, G_q(F^{2q})] \to [X, B_{(F)}]$ is surjective. Since the image of a homotopy of maps $X \to B_{(F)}$ lies in some $G_n(F^{2n}) \subset B_{(F)}$ for some n for $k \leq n$, the function $[X, G_q(F^{2q})] \to [X, B_{(F)}]$ is injective.

On the category of finite CW-complexes we define for q with $c(q + 1) - 2 \geq \dim X$ the following sequence:

$$[X, B_{(F)}] \xrightarrow{\alpha_q} [X, G_q(F^{2q})] \xrightarrow{\phi^q} \tilde{K}_F(X)$$

Here α_q is the inverse of the bijection $[X, G_q(F^{2q})] \to [X, B_{(F)}]$. We denote the composition by $\theta: [-, B_{(F)}] \to \tilde{K}_F(-)$; it is independent of q for $k \leq q$. Since θ is an isomorphism of cofunctors on each subcategory generated by a finite number of finite CW-complexes, θ is an isomorphism. Consequently, we have the next theorem.

4.5 Theorem. *There is an isomorphism of cofunctors, defined on the category of finite connected CW-complexes, $\theta: [-, B_{(F)}] \to \tilde{K}_F(-)$. The cofunctors are viewed as having values in the category of sets. Moreover $B_{(F)}$ has an H-space structure such that θ is a morphism of cofunctors with values in the category of abelian groups.*

Proof. Only the last statement remains to be proved. For this, we use the following linear functions: $f_0: F^\infty \to F^\infty$ and $f_e: F^\infty \to F^\infty$ defined by $f_0(e_i) = e_{2i-1}$ and $f_e(e_i) = e_{2i}$ for $i \geq 1$. Note that f_0 and f_e are monomorphisms, and $f_0(F^n) \subset F^{2n}$ and $f_e(F^n) \subset F^{2n}$. Therefore, f_0 and f_e induce maps g_0, g_e: $G_k(F^{2n}) \to G_k(F^{4n})$, where $g_0(W) = f_0(W)$ and $g_e(W) = f_e(W)$. Consequently, there is a map $\psi_n: G_n(F^{2n}) \times G_n(F^{2n}) \to G_{2n}(F^{4n})$ where $\psi_n(W, W') = f_0(W) + f_e(W')$. Moreover, it is clear from the construction that $\psi_n^*(\gamma_{2n}^{4n}) = \gamma_n^{2n} \times \gamma_n^{2n}$. The maps ψ_n give rise to the following commutative diagram where the vertical maps are inclusions.

$$G_n(F^{2n}) \times G_n(F^{2n}) \xrightarrow{\psi_n} G_{2n}(F^{4n})$$

$$\downarrow \qquad\qquad\qquad\qquad \downarrow$$

$$G_{n+k}(F^{2(n+k)}) \times G_{n+k}(F^{2(n+k)}) \xrightarrow{\psi_{n+k}} G_{2(n+k)}F^{4(n+k)})$$

In this way we see that there exists a unique map $\psi: B_{(F)} \times B_{(F)} \to B_{(F)}$ such that the following diagram is commutative where the vertical maps are inclusions.

$$G_n(F^{2n}) \times G_n(F^{2n}) \xrightarrow{\psi_n} G_{2n}(F^{4n})$$

$$\downarrow \qquad\qquad\qquad\qquad \downarrow$$

$$B_{(F)} \times B_{(F)} \xrightarrow{\psi} B_{(F)}$$

For $[f], [g] \in [X, B_{(F)}]$, there exists an integer n such that we can view f, $g: X \to G_n(F^{2n})$. Then $\theta([f]) = f^*(\gamma_n^{2n}) - n$ and $\theta([g]) = g^*(\gamma_n^{2n}) - n$. Moreover, $(\psi_n(f \times g)\Delta)^*(\gamma_{2n}^{4n}) = \Delta^*(f^*(\gamma_n^{2n}) \times g^*(\gamma_n^{2n})) = f^*(\gamma_n^{2n}) \oplus g^*(\gamma_n^{2n})$. Therefore, in $\tilde{K}_F(X)$ we have $\theta([\psi_n(f \times g)\Delta]) = f^*(\gamma_n^{2n}) \oplus g^*(\gamma_n^{2n}) - 2n = (f^*(\gamma_n^{2n}) - n) + (g^*(\gamma_n^{2n}) - n) = \theta([f]) + \theta([g])$. This proves the theorem.

4.6 Remark. In view of the natural splitting $K_F(X) = \tilde{K}_F(X) \oplus Z$, there is an isomorphism $\theta: [-, B_{(F)} \times Z] \to K_F(-)$ for connected finite CW-complexes.

4.7 Remark. For a connected space X we can replace $[X, B_{(F)}]$ by $[X, B_{(F)}]_0$ in the isomorphism $\theta: [X, B_{(F)}] \to K_F(X)$. This amounts to working with s-equivalent bundles each with a fixed trivialization at the base point. The reader can easily verify that the same equivalence classes result. The H-space structure $\psi: B_{(F)} \times B_{(F)} \to B_{(F)}$ is with respect to the base point which is the image in $B_{(F)}$ of $\sum_{1 \le i \le n} Fe_i$ in $G_n(F^{2n})$.

4.8 Remark. For a contractible paracompact space X we have $\tilde{K}_F(X) = 0$, and for S^0, we have $\tilde{K}_F(S^0) = Z$.

5. Homotopy Groups of Classical Groups and $\tilde{K}_F(S^i)$

The following theorem is useful in calculating $\tilde{K}_F(S^i)$.

5.1 Theorem. *There is a group isomorphism $\tilde{K}_F(S^i) \to \pi_{i-1}(U_F)$ for $1 \le i$.*

Proof. By Theorem (4.5) and Remark (4.7) there is an isomorphism $\theta^{-1}: \tilde{K}_F(S^i) \to [S^i, B_{(F)}]_0$. By 1(4.1) the group structure on $[S^i, B_{(F)}]_0$ can be computed with the coH-space structure of S^i or the H-space structure of $B_{(F)}$. Using the isomorphism of (4.4), we have $\alpha: [S^i, G_n(F^{2n})]_0 \to [S^i, B_{(F)}]_0$ for some large n.

For the principal $U_F(n)$-bundle $V_n(F^{2n}) \to G_n(F^{2n})$, we have the homotopy exact sequence

$$0 = \pi_i(V_n(F^{2n})) \to \pi_i(G_n(F^{2n})) \xrightarrow{\partial} \pi_{i-1}(U_F(n)) \to \pi_{i-1}(V_n(F^{2n})) = 0$$

This holds by 7(5.1) for $i < (n+1)c - 2$. This yields the isomorphism $\partial: \pi_i(G_n(F^{2n})) \to \pi_{i-1}(U_F(n))$. By 7(4.1) for $i - 1 < c(n+1) - 3$, the inclusion $U_F(n) \to U_F$ induces an isomorphism $\pi_{i-1}(U_F(n)) \to \pi_{i-1}(U_F)$. We have the following sequence of isomorphisms:

$$\tilde{K}_F(S^i) \xrightarrow{\theta^{-1}} \pi_i(B_{(F)}) \xrightarrow{\alpha} \pi_i(G_n(F^{2n})) \xrightarrow{\partial} \pi_{i-1}(U_F(n)) \to \pi_{i-1}(U_F)$$

The composition is an isomorphism $\tilde{K}_F(S^i) \to \pi_{i-1}(U_F)$. This proves the theorem.

5.2 Corollary. *The calculations of $\tilde{K}_F(S^i)$ are as follows:*

$$\tilde{K}_R(S^0) = Z \qquad \tilde{K}_C(S^0) = Z \qquad \tilde{K}_H(S^0) = Z$$

$$\tilde{K}_R(S^1) = Z_2 \qquad \tilde{K}_C(S^1) = 0 \qquad \tilde{K}_H(S^1) = 0$$

$$\tilde{K}_R(S^2) = Z_2 \qquad \tilde{K}_C(S^2) = Z \qquad \tilde{K}_H(S^2) = 0$$

$$\tilde{K}_R(S^3) = 0 \qquad \tilde{K}_C(S^3) = 0 \qquad \tilde{K}_H(S^3) = 0$$

$$\tilde{K}_R(S^4) = Z \qquad \tilde{K}_C(S^4) = Z \qquad \tilde{K}_H(S^4) = Z$$

Proof. We apply Theorem (5.1), using (4.8) for $\tilde{K}_F(S^0)$, 7(12.1) for $\tilde{K}_F(S^1)$, 7(12.3) for $\tilde{K}_F(S^2)$, 7(12.4) for $\tilde{K}_F(S^3)$, and 7(12.6) and 7(12.11) for $\tilde{K}_F(S^4)$.

5.3 Notation. The groups $\tilde{K}_C$ are frequently denoted by $\tilde{K}$ or $\tilde{KU}$, the groups $\tilde{K}_R$ by $\tilde{KO}$, and the groups $\tilde{K}_H$ by $\tilde{KSp}$. We make use of both sets of notation.

5.4 Remarks. We will see later in Chapter 11 that $\tilde{K}_C(S^n) = \tilde{K}_C(S^{n+2})$, and also $\tilde{K}_R(S^n) = \tilde{K}_H(S^{n+4})$ and $\tilde{K}_H(S^n) = \tilde{K}_H(S^{m+4})$ holds. This is the Bott periodicity of K-theory. With this and 5.2 all the groups $\tilde{K}_F(S^i)$ are determined.

Exercises

1. Prove that $\theta: [-, B_{(F)} \times Z] \to K_F(-)$ is an isomorphism between cofunctors defined on the category of all finite CW-complexes.

2. How does the discussion in Secs. 3 and 4 change when we define $K(X)$ using vector bundles of contant dimension?

3. Describe the bundles that determine generators of the groups given in (5.2).

CHAPTER 10

Relative K-Theory

We define a collapsing or trivialization procedure for bundles over X which yields a bundle over X/A for a closed subset A of X. With this construction we are able to give alternative descriptions of $K(X, A) = \tilde{K}(X/A)$. For a finite CW-pair (X, A) we can define an exact sequence $K(A) \leftarrow K(X) \leftarrow K(X, A) \leftarrow K(S(A)) \leftarrow K(S(X))$, using an appropriate "coboundary operator." With this sequence we see that in some sense the K-cofunctor can be used to define a cohomology theory.

1. Collapsing of Trivialized Bundles

1.1 Definition. Let ξ^n be a vector bundle over a space X, and let A be a subset of X. A trivialization of ξ^n over A is a map $t: E(\xi|A) \to F^n$ which is a linear isomorphism $\xi_b \to F^n$ upon restriction to a fibre of $\xi|A$.

A trivialization is a means for collapsing all the fibres of ξ over A to a single fibre over $*$ in X/A.

1.2 Definition. Let t be a trivialization over A of a vector bundle ξ over X. The collapsing of ξ with respect to the trivialization t is a triple $(\xi/t, u, r)$, where ξ/t is a vector bundle over X/A, $(u, p_A): \xi \to \xi/t$ is a vector bundle morphism, and $r: (\xi/t)_* \to F^n$ is an isomorphism. It is assumed that the restriction of u to $\xi_x \to (\xi/t)_{p_A(x)}$ is a linear isomorphism and that $r: (\xi/t)_* \to F^n$ is an isomorphism such that t equals ru restricted to $E(\xi|A)$.

The following proposition states to what extent collapsed bundles exist.

1.3 Proposition. *Let ξ be a vector bundle over X with a trivialization t over a closed subset A of X. Then there exists a collapsing $(\xi/t, u, r)$ of ξ with respect* *to t if and only if there exists a trivialization t' of ξ over an open set U constaining A which prolongs t. If ((ξ/t)', u', r') is a second collapsing of ξ with respect to t, there is a morphism v: ξ/t → (ξ/t)' such that the following diagram is commutative.*

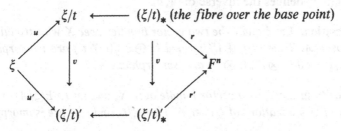

Moreover, v is unique with respect to this property, and v is an isomorphism. Finally, every vector bundle on X/A is isomorphic to ξ/t for some bundle ξ on X with trivialization t over A.

Note. The above theorem applies to each CW-pair (X, A); that is, X is a CW-complex with subcomplex A. So A is a neighborhood deformation retract.

Proof. If the quotient $(ξ/t, u, r)$ exists, there are a local coordinate chart (V, ϕ), where $* \in V$, and a map $\phi: V \times F^n \to (ξ/t)|V$ such that r is the restriction of ϕ^{-1} to the fibre of $ξ/t$ over $*$. Let $t' = \phi^{-1}u$ on $ξ$ over $p_A^{-1}(V) = U$. This construction demonstrates also the last statement of the proposition.

Conversely, we construct $E(ξ/t)$ as the quotient of $E(ξ)$ where x and x' are identified provided $t(x) = t(x')$. There are no further identifications, and $E(ξ/t) \to X/A$ with the induced projection is a bundle of vector spaces. Let $u: ξ \to ξ/t$ be the quotient map. It is a fibrewise linear isomorphism. The prolongation t' of t defines a trivialization of $ξ/t$ on $p_A(V)$, and $ξ/t$ is locally trivial at $*$. The coordinate charts of $ξ|(X - A)$ define coordinate charts of $ξ/t$ at other points of X/A.

Finally, for uniqueness the isomorphism $v: ξ/t \to (ξ/t)'$ is defined as $u'u^{-1}$ over points other than $*$ and by $r'^{-1}r$ near $*$, using charts that restrict to r and r'. This proves the proposition.

1.4 Proposition. *Let ξ and ξ' be two vector bundles over X and X' with trivializations t and t' over closed subspaces A and A', respectively. Let (v, f): ξ → ξ' be a vector bundle morphism such that $f(A) \subset A'$ and $t = t'v$. Then there exists a vector bundle morphism (w, g): ξ/t → ξ'/t' such that g is the quotient of f, and the following diagram is commutative:*

$$
\begin{array}{ccc}
ξ & \xrightarrow{(v, f)} & ξ' \\
\downarrow{u} & & \downarrow{u'} \\
ξ/t & \xrightarrow{(w, g)} & ξ'/t'
\end{array}
$$

Moreover, (w, g) is unique with respect to this property, and if (v, f) is an isomorphism with $f(A) = A'$, then (w, g) is an isomorphism.

Proof. The above diagram defines w uniquely. If (v, f) is an isomorphism, the inverse of (v, f) defines the inverse of (w, g).

1.5 Proposition. *Let ξ and ξ' be two vector bundles over X with trivializations t and t' over A. Then $(\xi/t) \oplus (\xi'/t')$ and $(\xi \oplus \xi')/(t \oplus t')$ are isomorphic, and $(\xi/t) \otimes (\xi'/t')$ and $(\xi \otimes \xi')/(t \otimes t')$ are isomorphic.*

1.6 Proposition. *Let ξ be a vector bundle over X, and let $t_s: E(\xi/A) \to F^n$ be a homotopy of trivializations of ξ over A. Then ξ/t_0 and ξ/t_1 are isomorphic over X/A.*

Proof. We view t_s as a trivializaton t of $\xi \times I$ over $A \times I$. Since $(X \times I)/(A \times I)$ and $(X/A) \times I$ are naturally isomorphic, it follows that ξ/t_0 is isomorphic to $((\xi \times I)/t)|(A \times 0)$ and ξ/t_1 is isomorphic to $((\xi \times I)/t)|(A \times 1)$. Since $(\xi \times I)/t$ is isomorphic to $\eta \times I$ for some η, there is an isomorphism between ξ/t_0 and ξ/t_1.

1.7 Remark. In subsequent discussions, it will be convenient occasionally to view a trivialization t of ξ over A as an isomorphism $t: \xi|A \to \theta^n$. Moreover, we shall be able to view a trivialization as a vector bundle morphism $t: \xi \to \theta^n$, which is an isomorphism when restricted to A, using the next proposition.

1.8 Proposition. *Let ξ_0 and ξ_1 be two vector bundles over X, and let $u: \xi_0|A \to \xi_1|A$ be a vector bundle morphism where (X, A) is a relative CW-complex. Then there exists a vector bundle morphism $v: \xi_0 \to \xi_1$ which prolongs u.*

Proof. We view u as a cross section of $\mathrm{Hom}(\xi_0, \xi_1)$ over A. Since the fibre of $\mathrm{Hom}(\xi_0, \xi_1)$ is a vector space, which is a contractible space, the cross section u prolongs to a cross section v of $\mathrm{Hom}(\xi_0, \xi_1)$ by 2(7.1).

It is easy to give a direct proof of (1.8) by constructing v over each skeleton cell by cell.

1.9 Convention. For a trivialization $t: \xi|A \to \theta^n$ of ξ over $A = \phi$, we define ξ/t to be ξ on X and θ^n on the discrete base point $*$ in X^+ outside X. Recall that X^+ equals $X \cup \{*\}$, that is, X/ϕ.

2. Exact Sequences in Relative K-Theory

All the exact sequences considered in this section arise from the following sequence using standard homotopy constructions. We use the characterization of $\widetilde{K}(X)$ given in 8(3.8).

2.1 Proposition. *Let A be a subcomplex of a finite CW-complex X. Then the maps $A \to X \to X/A$ induce an exact sequence $\tilde{K}(X/A) \overset{\beta}{\to} \tilde{K}(X) \overset{\alpha}{\to} \tilde{K}(A)$. Moreover, if A is contractible, $\beta: \tilde{K}(X/A) \to \tilde{K}(X)$ is an isomorphism.*

Proof. Since the composition $A \to X \to X/A$ is constant, $\alpha\beta$ is zero. Let ξ be a bundle over X such that ξ/A is s-trivial. Then there is an isomorphism $u: (\xi \oplus \theta^k)|A \to \theta^m$ which, when composed with the projection $\theta^m \to F^m$, defines a trivialization t of $\xi \oplus \theta^k$ over A. Then $(\xi \oplus \theta^k)/t$ is a vector bundle over X/A whose induced bundle over X by the projection $X \to X/A$ is s-equivalent to ξ.

For the last statement, if A is contractible, $\beta: \tilde{K}(X/A) \to \tilde{K}(X)$ is an epimorphism. Let ξ be a vector bundle over X/A such that $p_A^*(\xi)$ is s-trivial; that is, $p_A^*(\xi \oplus \theta^k) = p_A^*(\xi) \oplus \theta^k$ is trivial over X. Since $\xi \oplus \theta^k$ is isomorphic to $p_A^*(\xi \oplus \theta^k)/t$ for some trivialization t over A and since any two such trivializations are homotopic over A, it can be assumed that t is the restriction of a trivialization of $p_A^*(\xi \oplus \theta^k)$ over X. Consequently, $\xi \oplus \theta^k$ is trivial over X/A, and β is an isomorphism.

2.2 Corollary. *Since $K(X, A) = \tilde{K}(X/A)$ and since the inclusion $A \to X$ induces two morphisms $K(X) \to K(A)$ and $\tilde{K}(X) \to \tilde{K}(A)$ with the same kernel, the following sequence is exact:*

$$K(X, A) \to K(X) \to K(A)$$

The following homotopy constructions are introduced to help adapt Proposition (2.1) to the situation of a map $f: X \to Y$.

2.3 Definition. For a base point preserving map $f: X \to Y$ the mapping cylinder Z_f of f is the space $(X \times I) + Y$ modulo the relation $(x, 1) \in X \times I$ equals $f(x)$ in Y and $(*, t)$ equals $*$, and the mapping cone C_f of f is the space $C(X) + Y$ modulo the relation $\langle x, 1 \rangle \in C(X)$ equals $f(x)$ in Y.

If $f: X \to Y$ is a cellular map between CW-complexes, Z_f and C_f have the structure of a CW-complex. We assume for the remaining discussion that f is cellular and X and Y are finite CW-complexes.

2.4 We have the following diagram of maps and spaces:

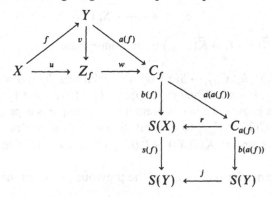

We define $u(x)$ to be the class of $(x, 0)$ and $v(y)$ the class of y in Z_f. The map w is the projection $Z_f \to Z_f/u(X) = C_f$, and $a(f)(y)$ is the class of y in C_f. We define $b(f)(y) = *$ and $b(f)(x, t) = \langle x, t \rangle$ in $S(X)$. We view $C_{a(f)}$ as $C(X) + C(Y)$ modulo the relation $\langle x, 1 \rangle$ equals $\langle f(x), 1 \rangle$. To define r, we require that $r(C(Y)) = *$ and that $r|C(X): C(X) \to S(X)$ be the natural projection. The homeomorphism j is defined by the relation $j(\langle x, t \rangle) = \langle x, 1 - t \rangle$.

2.5 Proposition. *The map v is a homotopy equivalence. The sequence $\tilde{K}(C_f) \to \tilde{K}(Y) \to \tilde{K}(X)$ derived from the diagram in (2.4) is exact.*

Proof. We define a map $v': Z_f \to Y$ by the requirement that $v'(y) = y$ and $v'(x, t) = f(x)$. Then clearly $v'v = 1$. We have a homotopy $k_s: Z_f \to Z_f$ defined by $k_s(y) = y$ and $k_s(x, t) = (x, 1 - s(1 - t))$. Then $vv' = k_0$ and $1 = k_1$, and this proves the first statement.

In the following diagram, which comes from (2.4), the bottom row is exact, by (2.1), and the vertical morphism is an isomorphism. From this we prove the proposition where $f^!$ denotes $\tilde{K}(f)$.

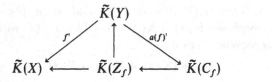

2.6 Proposition. *The sequence $\tilde{K}(S(X)) \to \tilde{K}(C_f) \to \tilde{K}(Y)$ derived from the diagram in (2.4) is exact.*

Proof. The map $a(f)$ is an inclusion map, and the projection $C_f \to C_f/a(f)(Y) = S(X)$ is just $b(f)$. The proposition now follows from (2.1).

2.7 Proposition. *The following diagram is commutative up to homotopy.*

$$
\begin{array}{ccc}
S(X) & \xleftarrow{\ r\ } & C_{a(f)} \\
{\scriptstyle s(f)}\downarrow & & \downarrow{\scriptstyle b(a(f))} \\
S(Y) & \xleftarrow{\ j\ } & S(Y)
\end{array}
$$

The morphism $\tilde{K}(S(X)) \to \tilde{K}(C_{a(f)})$ is an isomorphism.

Proof. We define $h_s: C_{a(f)} \to S(Y)$ by the relations $h_s(x, t) = (f(x), (1 - s)t)$ and $h_s(y, t) = (y, 1 - st)$. When $(x, 1)$ equals $(y, 1)$, we have $(f(x), (1 - s)1) = (y, 1 - 1s)$. This proves the first statement. The map r is a projection arising from pinching $C(Y) \subset C_{a(f)}$ to a point. Since $C(Y)$ is contractible, Proposition (2.1) applies, and $r^!: \tilde{K}(S(X)) \to \tilde{K}(C_{a(f)})$ is an isomorphism.

Now we summarize the results of the previous propositions.

2.8 Theorem. *Let $f: X \to Y$ be a cellular map between finite CW-complexes. The following sequence is exact:*

$$\tilde{K}(S(Y)) \xrightarrow{S(f)'} \tilde{K}(S(X)) \xrightarrow{b(f)'} \tilde{K}(C_f) \xrightarrow{a(f)'} \tilde{K}(Y) \xrightarrow{f'} \tilde{K}(X)$$

If A is a subcomplex of a finite CW-complex X with inclusion map j, there is a morphism $\delta = (q^!)^{-1}b(j)^!: \tilde{K}(S(X)) \to \tilde{K}(X/A)$ such that the following sequence is exact:

$$\tilde{K}(S(X)) \to \tilde{K}(S(A)) \xrightarrow{\delta} \tilde{K}(X/A) \to \tilde{K}(X) \to \tilde{K}(A)$$

Proof. The exactness of $\tilde{K}(S(X)) \to \tilde{K}(C_f) \to \tilde{K}(Y) \to \tilde{K}(X)$ follows from (2.5) and (2.6). From (2.7) we have the following commutative diagram where the top row is exact and the vertical morphism is an isomorphism.

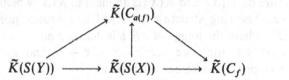

Consequently, the bottom row is exact, and the first sequence in the theorem is exact.

For the second statement, we have the following commutative diagram:

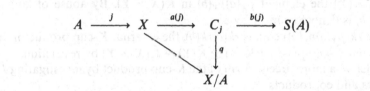

Since q is the projection arising from pinching $C(A)$ to a point, $q^!: \tilde{K}(X/A) \to \tilde{K}(C_j)$ is an isomorphism by (2.1). The top row is exact, and the vertical morphism is an isomorphism in the following diagram:

$$\tilde{K}(S(A)) \xrightarrow{b(j)'} \tilde{K}(C_j) \xrightarrow{a(j)'} \tilde{K}(X)$$

with δ and $q^!$ and $\tilde{K}(X/A)$.

This proves the last statement of the theorem.

2.9 Remark. If ξ is a vector bundle over $S(A)$, then $\xi|C(A)_+$ and $\xi|C(A)_-$ are trivial bundles, and their relation on A defines an automorphism $u: \theta^n \to \theta^n$ over A. This automorphism defines a trivialization $t: \theta^n|A \to F^n$. The image of the stability class of ξ under δ is the stability class of θ^n/t over X/A.

2.10 Remark. As in (2.8), the following exact sequence for a subcomplex A of a finite CW-complex X holds:

$$K(S(X)) \to K(S(A)) \to K(X, A) \to K(X) \to K(A)$$

If K were equal to H^n and KS were equal to $H^nS = H^{n-1}$, this would be a portion of the exact sequence for the cohomology theory $\{H^n\}$. This is the reasoning used by Atiyah and Hirzebruch [2] to define the K^*-cohomology theory.

3. Products in K-Theory

In this section we deal only with real or complex vector bundles over finite CW-complexes.

By the operation of the tensor product of vector bundles we have defined a ring structure on $K(X)$, and $\tilde{K}(X)$ is an ideal in $K(X)$ which is the kernel of $rk\colon K(X) \to \mathbf{Z}$. The ring structure on $K(X)$ is a group morphism $K(X) \otimes K(X) \to K(X)$, where the image of $a \otimes b$ is denoted ab.

In the next definition we use $p_X\colon X \times Y \to X$ and $p_Y\colon X \times Y \to Y$ to denote the two projections from the product.

3.1 Definition. For two spaces X and Y the external K-cup product is a group morphism $K(X) \otimes K(Y) \to K(X \times Y)$ which assigns to each $a \otimes b \in K(X) \otimes K(Y)$ the element $p_X^!(a)p_Y^!(b)$ in $K(X \times Y)$. By abuse of language, $p_X^!(a)p_Y^!(b)$ is denoted simply by ab.

Since $rk(p_X^!(a)p_Y^!(b))$ equals $rk(a)rk(b)$, the external K-cup product induces an external $\tilde{K}$-cup product $\tilde{K}(X) \otimes \tilde{K}(Y) \to \tilde{K}(X \times Y)$ by restriction.

We derive a more precise form of the $\tilde{K}$-cup product by investigating $\tilde{K}$ on products and coproducts.

3.2 Proposition. *For two pointed spaces X and Y let $q_X\colon X \to X \vee Y$ and $q_Y\colon Y \to X \vee Y$ be the two natural inclusions. Then the group morphism $(q_X^!, q_Y^!)\colon \tilde{K}(X \vee Y) \to \tilde{K}(X) + \tilde{K}(Y)$ is an isomorphism.*

Proof. Since a vector bundle and its stability class on $X \vee Y$ are uniquely determined by their restrictions to X and to Y, the morphism $(q_X^!, q_Y^!)$ is a monomorphism.

For a bundle ξ on X and η on Y with $\dim \xi = \dim \eta$, we can define a bundle ζ on $X \vee Y$ with $\zeta|X = \xi$ and $\zeta|Y = \eta$, and the morphism $(q_X^!, q_Y^!)$ is an epimorphism.

The reader can supply a purely functional proof of (3.2), using only the exact sequence in (2.1).

3.3 Corollary. *The group morphism*

$$(q_1^!, \ldots, q_n^!)\colon \tilde{K}(X_1 \vee \cdots \vee X_n) \to \tilde{K}(X_1) + \cdots + \tilde{K}(X_n)$$

is an isomorphism where $q_i\colon X_i \to X_1 \vee \cdots \vee X_n$ is the natural inclusion.

3.4 Proposition. *The inclusion $X \vee Y \subset X \times Y$ defines the following split exact sequence*:

$$0 \to \tilde{K}(X \wedge Y) \to \tilde{K}(X \times Y) \to \tilde{K}(X \vee Y) \to 0$$

Proof. By (2.1) the sequence $\tilde{K}(X \wedge Y) \to \tilde{K}(X \times Y) \to \tilde{K}(X \vee Y)$ is exact. If $p_X: X \times Y \to X$ and $p_Y: X \times Y \to Y$ are the usual projections, the following diagram is commutative.

$$\tilde{K}(X \wedge Y) \longrightarrow \tilde{K}(X \times Y) \longrightarrow \tilde{K}(X \vee Y)$$

$$p_X' \oplus p_Y' \Big\uparrow \qquad \swarrow {\scriptstyle (q_X', q_Y')}$$

$$\tilde{K}(X) + \tilde{K}(Y)$$

Therefore, the morphism $\tilde{K}(X \times Y) \to \tilde{K}(X \vee Y)$ is an epimorphism with right inverse $(p_X' \oplus p_Y')(q_X', q_Y')$.

Finally, the above discussion applies to $\tilde{K}$ replaced by $\tilde{K}S$, and $\tilde{K}(S(X \times Y)) \to \tilde{K}(S(X \vee Y))$ is an epimorphism in the following exact sequence of (2.8):

$$\tilde{K}(S(X \times Y)) \to \tilde{K}(S(X \vee Y)) \to \tilde{K}(X \wedge Y) \to \tilde{K}(X \times Y)$$

3.5 Proposition. *If the first morphism below is the $\tilde{K}$-cup product, the following composition is zero.*

$$\tilde{K}(X) \otimes \tilde{K}(Y) \to \tilde{K}(X \times Y) \to \tilde{K}(X \vee Y)$$

Proof. If ab is a cup product in $\tilde{K}(X \times Y)$, then $a = 0$ when projected into $\tilde{K}(Y)$ from $\tilde{K}(X \vee Y) = \tilde{K}(X) \oplus \tilde{K}(Y)$ and $b = 0$ when projected into $\tilde{K}(X)$. Therefore, the product is zero when projected into $\tilde{K}(X \vee Y)$.

3.6 Remark. In view of (3.5) and (3.4), there is a unique morphism $\tilde{K}(X) \otimes \tilde{K}(Y) \to \tilde{K}(X \wedge Y)$ which, composed with the monomorphism $\tilde{K}(X \wedge Y) \to \tilde{K}(X \times Y)$, is the $\tilde{K}$-cup product. Now we refer to this morphism $\tilde{K}(X) \otimes \tilde{K}(Y) \to \tilde{K}(X \wedge Y)$ as the $\tilde{K}$-cup product.

3.7 Definition. Let (X, A) and (Y, B) be two finite CW-pairs. Then the relative K-cup product is the morphism $K(X, A) \otimes K(Y, B) \to K(X \times Y, (X \times B) \cup (A \times Y))$ which is the $\tilde{K}$-cup product.

$$\tilde{K}(X/A) \otimes \tilde{K}(Y/B) \to \tilde{K}((X/A) \wedge (Y/B)) = \tilde{K}((X \times Y)/(X \times B) \cup (A \times Y))$$

Recall that $K(X, A) = \tilde{K}(X/A)$, etc., for the above definition.

4. The Cofunctor $L(X, A)$

To study $K(X, A)$, we define a new cofunctor $L(X, A)$ in a manner similar to that used in defining $K(X)$. We use equivalence classes of pairs of vector bundles ξ_0 and ξ_1 over X together with an isomorphism $\alpha: \xi_0|A_0 \to \xi_1|A$.

The following discussion applies to real, complex, or quaternionic vector bundles.

4.1 Definition. Let (X, A) be a pair of spaces. A difference isomorphism (over (X, A)) is a vector bundle morphism $\alpha: \xi_0 \to \xi_1$ such that the retriction $\alpha: \xi_0|A \to \xi_1|A$ is an isomorphism. Two difference isomorphisms $\alpha: \xi_0 \to \xi_1$ and $\beta: \eta_0 \to \eta_1$ are isomorphic provided there exist isomorphisms $u_i: \xi_i \to \eta_i$ (over X) for $i = 0$, 1 such that the following diagram of isomorphisms is commutative.

$$
\begin{array}{ccc}
\xi_0|A & \xrightarrow{\ \alpha\ } & \xi_1|A \\
\Big\downarrow{\scriptstyle u_0} & & \Big\downarrow{\scriptstyle u_1} \\
\eta_0|A & \xrightarrow{\ \beta\ } & \eta_1|A
\end{array}
$$

4.2 Notation. Let $S(X, A)$ [or, more precisely, $S_F(X, A)$] denote the semigroup of isomorphism classes of difference isomorphisms of F-vector bundles over (X, A). We define a commutative semigroup structure on $S(X, A)$, using the quotient function of the Whitney sum operation defined as usual by $\alpha \oplus \beta: \xi_0 \oplus \eta_0 \to \xi_1 \oplus \eta_1$ for $\alpha: \xi_0 \to \xi_1$ and $\beta: \eta_0 \to \eta_1$. If $f: (Y, B) \to (X, A)$ is a map, and if $\alpha: \xi_0 \to \xi_1$ is a difference isomorphism, then $f^*(\alpha): f^*(\xi_0) \to f^*(\xi_1)$ is a difference isomorphism over (Y, B). This operation is compatible with isomorphisms and Whitney sums, and therefore it defines a semigroup morphism $f^*: S(X, A) \to S(Y, B)$. With these definitions the following proposition is clear.

4.3 Proposition. *The semigroups $S(X, A)$ and induced morphisms f^* collect to define a cofunctor from pairs of spaces and maps to commutative semigroups and semigroup morphisms.*

The identities $1: \xi \to \xi$ define special elements in $S(X, A)$. In the next proposition we consider difference isomorphisms which are isomorphic to $1: \xi \to \xi$.

4.4 Proposition. *Let $\alpha: \xi_0 \to \xi_1$ be a difference isomorphism over (X, A). The restriction $\alpha: \xi_0|A \to \xi_1|A$ prolongs to an isomorphism if and only if $\alpha: \xi_0 \to \xi_1$ is isomorphic to $1: \xi \to \xi$. Observe that in this situation ξ is isomorphic to both ξ_0 and ξ_1.*

Proof. If $u: \xi_0 \to \xi_1$ is an isomorphism prolonging α, then $(u, 1)$ is an isomorphism from $\alpha: \xi_0 \to \xi_1$ to $1: \xi_1 \to \xi_1$. Conversely, if (u_0, u_1) is an isomorphism from $\alpha: \xi_0 \to \xi_1$ to $1: \xi \to \xi$, the $u_1^{-1} u_0$ is an isomorphism prolonging $\alpha: \xi_0|A \to \xi_1|A$.

4.5 Definition. Let $L(X, A)$ [or, more precisely, $L_F(X, A)$] denote the quotient semigroup of $S(X, A)$ defined by the equivalence relation where $\alpha: \xi_0 \to \xi_1$ and $\beta: \eta_0 \to \eta_1$ are related provided $\alpha \oplus 1: \xi_0 \oplus \zeta \to \xi_1 \oplus \zeta$ and

$\beta \oplus 1 : \eta_0 \oplus \zeta' \to \eta_1 \oplus \zeta'$ are equal in $S_F(X, A)$. Let $[\alpha : \xi_0 \to \xi_1]$ denote the element in $L(X, A)$ determined by $\alpha : \xi_0 \to \xi_1$.

Clearly, this is an equivalence relation compatible with the semigroup structure on $S(X, A)$, and $L(X, A)$ is a commuative semigroup. Moreover, this relation is preserved under induced morphisms $f^* : S(X, A) \to S(Y, B)$ defined by maps $f : (Y, B) \to (X, A)$. From this, there is a quotient semigroup morphism $f^* : L(X, A) \to L(Y, B)$. The next proposition is immediate.

4.6 Proposition. *The semigroups $L(X, A)$ and the induced morphisms f^* collect to define a cofunctor $L(X, A)$ from the category of pairs of spaces to the category of semigroups.*

4.7 Remark. The zero element of $L(X, A)$ is the class consisting of all isomorphisms $\alpha : \xi_0 \to \xi_1$ defined over X.

4.8 Special case. We calculate $L(X, \phi)$. In this case $S(X, \phi) = \text{Vect}_F(X) \times \text{Vect}_F(X)$, that is, pairs of isomorphism classes of vector bundles. Two pairs (ξ_0, ξ_1) and (η_0, η_1) determine the same element in $L(X, \phi)$ if and only if there exist ζ_1 and ζ_2 with $\xi_i \oplus \zeta_1$ isomorphic to $\eta_i \oplus \zeta_2$ for $i = 1, 2$. This implies that $\xi_0 \oplus \eta_1 \oplus (\zeta_1 \oplus \zeta_2)$ and $\xi_1 \oplus \eta_0 \oplus (\zeta_1 \oplus \zeta_2)$ are isomorphic, and the two pairs (ξ_0, ξ_1) and (η_0, η_1) determine the same element in $K(X)$. Conversely, if $\xi_0 \oplus \eta_1 \oplus \zeta$ and $\eta_0 \oplus \xi_1 \oplus \zeta$ are isomorphic, that is, $\xi_0 - \xi_1 = \eta_0 - \eta_1$ in $K(X)$, then for $\zeta_1 = \eta_1 \oplus \zeta$ and $\zeta_2 = \xi_1 \oplus \zeta$ the bundle $\xi_i \oplus \zeta_1$ is isomorphic to $\eta_i \oplus \zeta_2$ for $i = 1, 2$. As a quotient of the semigroups $\text{Vect}_F(X) \times \text{Vect}_F(X)$, we have $L(X, \phi) = K(X)$.

4.9 Remark. For a relative CW-complex (X, A) every $\alpha : \xi_0|A \to \xi_1|A$ (in particular, every isomorphism) prolongs to a morphism $\alpha' : \xi_0 \to \xi_1$ by (1.8). Difference isomorphisms could be defined as triples (ξ_0, α, ξ_1), where $\alpha : \xi_0|A \to \xi_1|A$ is an isomorphism.

5. The Difference Morphism

We wish to define an isomorphism $\Delta : L(X, A) \to K(X, A)$ of cofunctors which reduces to the identity on $L(X, \phi) = K(X, \phi) = K(X)$ for $A = \phi$. We demonstrated the equality $L(X, \phi) = K(X, \phi)$ in (4.8). All pairs in this section are finite CW-pairs.

5.1 Theorem. *For each pair (X, A) there is a function $\Delta : L(X, A) \to K(X, A)$ such that $\Delta([\alpha : \xi \to \theta^n]) = \{\xi/\alpha\} - n$, the class of ξ/α in $K(X, A) = \tilde{K}(X/A)$. This function Δ is unique with respect to this property, and for $A = \phi$ it is the identity. Moreover, the set of Δ for various (X, A) defined an isomorphism $L(X, A) \to K(X, A)$ of commutative group-valued cofunctors.*

Proof. For the first statement, each difference isomorphism $\alpha\colon \xi_0 \to \xi_1$ is equivalent in $L(X, A)$ to $\alpha \oplus 1\colon \xi_0 \oplus \zeta \to \xi_1 \oplus \zeta$. If ζ has the property that $\xi_1 \oplus \zeta$ and θ^m are isomorphic, the property $\Delta([\alpha\colon \xi_0 \to \xi_1]) = \{(\xi_0 \oplus \zeta)/ (\alpha \oplus 1)\} - m$ proves that Δ is uniquely defined. To prove that this relation can serve as a definition of Δ, we consider $[\alpha\colon \xi \to \theta^m] = [\beta\colon \eta \to \theta^n]$. There exist vector bundles ζ_1 and ζ_2 with $\alpha \oplus 1\colon \xi \oplus \zeta_1 \to \theta^m \oplus \zeta_1$ isomorphic to $\beta \oplus 1\colon \eta \oplus \zeta_2 \to \theta^n \oplus \zeta_2$. Since ζ_1 and ζ_2 are s-equivalent, there exists a vector bundle ζ with $\zeta_1 \oplus \zeta$ isomorphic to θ^p, $\zeta_2 \oplus \zeta$ isomorphic to θ^q, and $m + p = n + q = k$. Then $\alpha \oplus 1\colon \xi \oplus \theta^p \to \theta^k$ and $\beta \oplus 1\colon \eta \oplus \theta^q \to \theta^k$ are isomorphic, and in $K(X, A)$ we have $\{\xi/\alpha\} - m = \{(\xi \oplus \theta^p)/(\alpha \oplus 1)\} - k = \{(\eta \oplus \theta^q)/(\beta \oplus 1)\} - k = \{\eta/\beta\} - m$. Consequently, Δ is a well-defined function by the above relation.

If $\alpha\colon \xi \to \theta^m$ and $\beta\colon \eta \to \theta^n$ are difference isomorphisms, then $(\xi \oplus \eta)/ (\alpha \oplus \beta)$ and $(\xi/\alpha) \oplus (\eta/\beta)$ are isomorphic by (1.5). Consequently, we have $\Delta(a) + \Delta(b) = \Delta(a + b)$, for $a = [\alpha\colon \xi \to \theta^m]$, $b = [\beta\colon \eta \to \theta^n]$, and $a + b = [\alpha \oplus \beta\colon \xi \oplus \eta \to \theta^{n+m}]$.

If $f\colon (Y, B) \to (X, A)$ is a map, $f^*(\xi/\alpha)$ and $f^*(\xi)/f^*(\alpha)$ are isomorphic from (2.4), and the relation $f^!(\{\xi/\alpha\} - m) = \{f^*(\xi)/f^*(\alpha)\} - m$ or $f^!\Delta = \Delta f^*$ holds. Consequently, Δ is a morphism of group-valued cofunctors.

The function $\Delta\colon L(X, A) \to K(X, A)$ is surjective because each element of $K(X/A) = \tilde{K}(X/A)$ has the form $\{\eta\} - m$, where η is a vector bundle over X/A. By (1.3) the vector bundle η is isomorphic to ξ/α for some difference isomorphism $\alpha\colon \xi \to \theta^m$ over A. If $\{\xi/\alpha\} - m$ equals $\{\eta/\beta\} - n$ in $K(X, A)$, then $(\xi/\alpha) \oplus \theta^p \cong (\xi \oplus \theta^p)/(\alpha \oplus 1)$ and $(\eta/\beta) \oplus \theta^q \cong (\eta \oplus \theta^q)/(\beta \oplus 1)$ are isomorphic over X/A. By (1.3), the objects $\alpha \oplus 1\colon \xi \oplus \theta^p \to \theta^{p+m}$ and $\beta \oplus 1\colon \eta \oplus \theta^q \to \theta^{q+n}$ are isomorphic. In $L(X, A)$ we have $[\alpha\colon \xi \to \theta^m] = [\beta\colon \eta \to \theta^n]$, and Δ is injective. This proves the theorem.

5.2 Corollary. *Let $j\colon X \to (X, A)$ be the inclusion map, and let $[\alpha\colon \xi_0 \to \xi_1]$ be a member of $L(X, A)$. Then $j^!\Delta([\alpha\colon \xi_0 \to \xi_1]) = \{\xi_0\} - \{\xi_1\}$.*

5.3 Corollary. *Let $[\alpha\colon \xi_0 \to \xi_1]$ be a member of $L(X, *)$. Then $\Delta([\alpha\colon \xi_0 \to \xi_1]) = \{\xi_0\} - \{\xi_1\} \in K(X, *) = \tilde{K}(X)$.*

Proof. Let $j\colon X \to (X, *)$ be the natural inclusion inducing the inclusion $j^!\colon \tilde{K}(X) = K(X, *) \to K(X)$, and apply (5.2).

5.4 Corollary. *Let $\alpha_t\colon \xi_0 \to \xi_1$ be a homotopy of difference isomorphisms over A. Then $[\alpha_0\colon \xi_0 \to \xi_1] = [\alpha_1\colon \xi_0 \to \xi_1]$ in $L(X, A)$.*

Proof. We apply (1.6) to $\alpha_t \oplus 1\colon \xi_0 \oplus \zeta \to \xi_1 \oplus \zeta$, where $\xi_1 \oplus \zeta$ is trivial, and the fact that Δ is injective.

5.5 Corollary. *Let $\alpha\colon \xi_0 \to \xi_1$ and $\beta\colon \xi_1 \to \xi_0$ be two difference isomorphisms over A. If $\beta|A$ equals $(\alpha|A)^{-1}$, or if β equals α^*, the adjoint morphism for some*

riemannian metric on ξ_0 and ξ_1 defined fibrewise, then in $L(X, A)$ we have
$[\alpha: \xi_0 \to \xi_1] = -[\beta: \xi_1 \to \xi_0]$.

Proof. The result is true for $A = \phi$ or $A =$ point by a direct inspection. In general, we use the natural isomorphism $L(X/A, *) \to L(X, A)$ given by the commutative diagram of isomorphisms.

$$
\begin{array}{ccc}
L(X/A, *) & \xrightarrow{\;\Delta\;} & K(X/A, *) \\
\downarrow & & \downarrow \\
L(X, A) & \xrightarrow{\;\Delta\;} & K(X, A)
\end{array}
$$

6. Products in $L(X, A)$

We make use of the following lemma on vector spaces which is a special case of the Künneth formula.

6.1 Lemma. *Let $\alpha: U \to U'$ and $\beta: V \to V'$ be two linear transformations such that one is an isomorphism. The following sequence is exact.*

$$
0 \to U \otimes V \xrightarrow{(\alpha \otimes 1, 1 \otimes \beta)} (U' \otimes V) \oplus (U \otimes V') \xrightarrow{1 \otimes \beta - \alpha \otimes 1} U' \otimes V' \to 0
$$

Proof. Suppose that α is an isomorphism, and let $\{e_i\}$ be a basis of U for $i \in I$. Clearly, $(\alpha \otimes 1, 1 \otimes \beta)$ is a monomorphism, $1 \otimes \beta - \alpha \otimes 1$ is an epimorphism, and $(1 \otimes \beta - \alpha \otimes 1)(\alpha \otimes 1, 1 \otimes \beta) = 0$. If $\sum_i \alpha(e_i) \otimes y_i + \sum_i e_i \otimes y_i' \in$ ker$(1 \otimes \beta - \alpha \otimes 1)$, we have $\sum_i \alpha(e_i) \otimes (\beta(y_i) - y_i') = 0$ or $\beta(y_i) = y_i'$. Consequently, $(\alpha \otimes 1, 1 \otimes \beta)\left(\sum_i e_i \otimes y_i\right) = \sum_i \alpha(e_i) \otimes y_i + \sum_i e_i \otimes y_i'$. This proves the lemma.

Recall the notation $\xi \otimes \eta$ for $p_X^*(\xi) \otimes p_Y^*(\eta)$, where $p_X: X \times Y \to X$ and $p_Y: X \times Y \to Y$ are projections and ξ is a vector bundle over X and η over Y.

Let $\alpha: \xi_0 \to \xi_1$ be a difference isomorphism over (X, A) and $\beta: \eta_0 \to \eta_1$ over (Y, B). We form the following sequence:

$$
(\alpha, \beta): 0 \to \xi_0 \otimes \eta_0 \xrightarrow{(1 \otimes \beta, \alpha \otimes 1)} (\xi_0 \otimes \eta_1) \oplus (\xi_1 \otimes \eta_0) \xrightarrow{-\alpha \otimes 1 + 1 \otimes \beta} \xi_1 \otimes \eta_1 \to 0
$$

By Lemma (6.1) this sequence is exact on $(X \times B) \cup (A \times Y)$. For riemannian metrics on ξ_0, ξ_1, η_0, and η_1, we form the following difference isomorphism on $(X \times Y, (X \times B) \cup (A \times Y))$:

$$
\alpha\beta = \begin{bmatrix} 1 \otimes \beta & -\alpha^* \otimes 1 \\ \alpha \otimes 1 & 1 \otimes \beta^* \end{bmatrix} : (\xi_0 \otimes \eta_0) \oplus (\xi_1 \otimes \eta_1) \to (\xi_0 \otimes \eta_1) \oplus (\xi_1 \otimes \eta_0)
$$

6.2 Proposition. *The product $\alpha\beta$ is compatible with the equivalence relation yielding $L(X, A)$ and defines an L-cup product*

$$L(X, A) \otimes L(Y, B) \to L(X \times Y, (A \times Y) \cup (X \times B))$$

Proof. If either $\alpha: \xi_0 \to \xi_1$ is an isomorphism over X or $\beta: \eta_0 \to \eta_1$ over Y, the sequences (α, β), (α^*, β), and $(-\alpha, \beta^*)$ above are exact over $X \times Y$, and $\alpha\beta$ is an isomorphism over $X \times Y$. Consequently, if $\alpha = 0$ in $L(X, A)$ or $\beta = 0$ in $L(Y, B)$, then $a\beta = 0$ in $L(X \times Y, (X \times B) \cup (A \times Y))$. This proves the proposition.

6.3 Definition. The L-cup product of $[\alpha: \xi_0 \to \xi_1]$ and $[\beta: \eta_0 \to \eta_1]$ is defined to be the class of $\alpha\beta$ in $L(X \times Y, (A \times Y) \cup (X \times B))$.

Similarly there is the K-cup product

$$K(X, A) \otimes K(Y, B) \xrightarrow{K\text{-cup}} K(X \times Y, (A \times Y) \cup (X \times B))$$

[see (3.7)].

6.4 Theorem. *The following diagram is commutative.*

$$
\begin{array}{ccc}
L(X, A) \otimes L(Y, B) & \xrightarrow{\;L\text{-cup}\;} & L(X \times Y, (A \times Y) \cup (X \times B)) \\
\Big\downarrow{\scriptstyle \Delta \otimes \Delta} & & \Big\downarrow{\scriptstyle \Delta} \\
K(X, A) \otimes K(Y, B) & \xrightarrow{\;K\text{-cup}\;} & K(X \times Y, (A \times Y) \cup (X \times B))
\end{array}
$$

Proof. For $A = B = \phi$, the above diagram is commutative because in $K(X \times Y)$ we have $(\{\xi_0\} - \{\xi_1\})(\{\eta_0\} - \{\eta_1\}) = \{(\xi_0 \otimes \eta_0) \oplus (\xi_1 \otimes \eta_1)\} - \{(\xi_0 \otimes \eta_1) \oplus (\xi_1 \otimes \eta_0)\}$. For $A = *$ or ϕ and $B = *$ or ϕ, we have a monomorphism $K(X \times Y, (A \times Y) \cup (X \times B)) \to K(X \times Y)$, and commutativity holds in $K(X \times Y)$. Finally, in general, $K(X, A) = \tilde{K}(X/A)$, $K(Y, B) = \tilde{K}(Y/B)$, and $K(X \times Y, (A \times Y) \cup (X \times B)) = \tilde{K}((X/A) \times (Y/B))$, and the L-cup induced on $K(X, A) \otimes K(Y, B)$ by $\Delta \otimes \Delta$ and Δ is the K-cup product. This proves the theorem.

7. The Clutching Construction

A triad $(X; X_1, X_0)$ is referred to as a CW-triad provided X_1 and X_0 are subcomplexes of the CW-complex X with $X = X_0 \cup X_1$. With the result of the next proposition it is possible to "glue" a vector bundle (real or complex) ξ_1 over X_1 to a vector bundle ξ_0 over X_0 with an isomorphism $\xi_1|(X_0 \cap X_1) \to \xi_0|(X_0 \cap X_1)$.

7.1 Proposition. *Let $(X; X_1, X_0)$ be a CW-triad with $A = X_1 \cap X_0$, let ξ_i be a vector bundle over X_i for $i = 0, 1$, and let $\alpha: \xi_1|A \to \xi_0|A$ be a vector bundle* *isomorphism. Then there exists a triple (ξ, u_1, u_0) such that ξ is a vector bundle over X, $u_i\colon \xi_i \to \xi|X_i$ is an isomorphism for $i = 0$, 1, and $u_0\alpha = u_1$ over A. Moreover, if (η, v_1, v_0) is a second triple with the above properties, there is an isomorphism $w\colon \eta \to \xi$ with $u_i = wv_i$ for $i = 0$, 1.*

Proof. Let $E(\xi)$ be the space resulting from the identification of $x \in E(\xi_1|A) \subset E(\xi_1)$ with $\alpha(x) \in E(\xi_0|A) \subset E(\xi_0)$. There is a natural projection $E(\xi) \to X$ resulting from the projections of ξ_i, a natural vector space structure on each fibre of ξ, and natural bundle isomorphisms $u_i\colon \xi_i \to \xi|X_i$ for $i = 0$, 1.

To prove local triviality of ξ, we have only to find charts of ξ near points $x \in A$. For $x \notin A$ the existence of local charts is clear. Let U be an open neighborhood of x in X such that there is a retraction $r\colon U \cap X_0 \to U \cap A$ and such that there are maps $\phi_i\colon (U \cap X_i) \times F^n \to \xi_i|(U \cap X_i)$ which are local coordinate charts of ξ_i over $U \cap X_i$. Over the set $U \cap A$ we have $\alpha\phi_1(x, v) = \phi_0(x, f(x)v)$, where $f\colon U \cap A \to GL(F^n)$ is a map. Replacing $\phi_0(x, v)$ by $\phi_0(x, f(r(x))v)$ for $(x, v) \in (U \cap X_0) \times F^n$, we can assume that $\alpha\phi_1(x, v) = \phi_0(x, v)$ for $(x, v) \in (U \cap A) \times F^n$. Consequently, there is a local chart $\phi\colon U \times F^n \to \xi|U$ such that $u_i\phi_i = \phi|((U \cap X_i) \times F^n)$. This proves the existence statement.

For the uniqueness statement, let $w\colon \eta \to \xi$ be the isomorphism that is $u_iv_i^{-1}\colon \eta|X_i \to \xi|X_i$. Since $E(\eta)$ is the union of the closed subsets $E(\eta|X_1)$ and $E(\eta|X_0)$, w is a well-defined homeomorphism.

7.2 Notation. The bundle ξ in (7.1) is denoted by $\xi_1 \underset{\alpha}{\bigcup} \xi_0$. It is unique up to isomorphism in the sense of (7.1), and it is called the result of clutching ξ_1 and ξ_0 along A. The triple (ξ_1, α, ξ_0) is called clutching data over $(X; X_0, X_1)$, and α is called a clutching function.

From the uniqueness statment we have the following results immediately for a CW-triad $(X; X_0, X_1)$ and $A = X_0 \cap X_1$.

7.3 Proposition. *Let ξ be a bundle over X. Then $(\xi|X_1) \underset{1}{\bigcup} (\xi|X_0)$ and ξ are isomorphic.*

7.4 Proposition. *Let (ξ_1, α, ξ_0) and (η_1, β, η_0) be two sets of clutching data, and let $(w_i, f_i)\colon \xi_i \to \eta_i$ be vector bundle morphisms for $i = 0$, 1 such that $w_0\alpha = \beta w_1$ over A and $f_i = f|X_i$ for $f\colon (X, X_i) \to (Y, Y_i)$. Then there exists a vector bundle morphism $(w, f)\colon \xi_1 \underset{\alpha}{\bigcup} \xi_0 \to \eta_1 \underset{\beta}{\bigcup} \eta_0$ such that the following diagram is commutative.*

$$
\begin{array}{ccc}
\xi_i & \xrightarrow{\;(w_i, f_i)\;} & \eta_i \\[4pt]
{\scriptstyle (u_i, 1)}\Big\downarrow & & \Big\downarrow{\scriptstyle (v_i, 1)} \\[4pt]
\xi_1 \underset{\alpha}{\bigcup} \xi_0 & \xrightarrow[\;\beta\;]{(w, f)} & \eta_1 \underset{\beta}{\bigcup} \eta_0
\end{array}
$$

The morphism w is unique with respect to this property. If w_0 and w_1 are isomorphisms, w is an isomorphism.

Proof. The existence and uniqueness of w follow immediately from the definition of $E\left(\xi_1 \underset{\alpha}{\bigcup} \xi_0\right)$.

From the uniqueness statement in (7.1) we have the next proposition.

7.5 Proposition. *Let (ξ_1, α, ξ_0) and (η_1, β, η_0) be two sets of clutching data. Then $(\xi_1 \oplus \eta_1) \underset{\alpha \oplus \beta}{\bigcup} (\xi_0 \oplus \eta_0)$ and $\left(\xi_1 \underset{\alpha}{\bigcup} \xi_0\right) \oplus \left(\eta_1 \underset{\beta}{\bigcup} \eta_0\right)$ are isomorphic, and $(\xi_1 \otimes \eta_1) \underset{\alpha \otimes \beta}{\bigcup} (\xi_0 \otimes \eta_0)$ and $\left(\xi_1 \underset{\alpha}{\bigcup} \xi_0\right) \otimes \left(\eta_1 \underset{\beta}{\bigcup} \eta_0\right)$ are isomorphic. There is a similar formula for duality.*

7.6 Proposition. *Let ξ_i be a bundle over X_i, and let $\alpha_t: \xi_1|A \to \xi_0|A$ be a homotopy of clutching functions. Then $\xi_1 \underset{\alpha_0}{\bigcup} \xi_0$ and $\xi_1 \underset{\alpha_1}{\bigcup} \xi_0$ are isomorphic.*

Proof. We view a_t as a clutching function $\alpha: (\xi_1 \times I)|(A \times I) \to (\xi_0 \times I)|(A \times I)$. Then $\xi_1 \underset{\alpha_0}{\bigcup} \xi_0$ is isomorphic to $\left[(\xi_1 \times I) \underset{\alpha}{\bigcup} (\xi_0 \times I)\right]\bigg|(A \times 0)$ and $\xi_1 \underset{\alpha_1}{\bigcup} \xi_0$ to $\left[(\xi_1 \times I) \underset{\alpha}{\bigcup} (\xi_0 \times I)\right]\bigg|(A \times 1)$. Since $(\xi_1 \times I) \underset{\alpha}{\bigcup} (\xi_0 \times I)$ is isomorphic to $\eta \times I$ for some η, there is an isomorphism between $\xi_1 \underset{\alpha_0}{\bigcup} \xi_0$ and $\xi_1 \underset{\alpha_1}{\bigcup} \xi_0$.

In the next two sections, we outline the theory of $L_n(X, A)$ and the elementary properties of half-exact cofunctors. The details are left as exercises for the reader.

8. The Cofunctor $L_n(X, A)$

In this section we consider in outline form a generalization of the results of Secs. 4 to 6.

We consider complexes of length n consisting of vector bundles over X which are exact when restricted to A and denoted

$$\xi: 0 \to \xi_0 \overset{\alpha_1}{\to} \xi_1 \overset{\alpha_2}{\to} \xi_2 \to \cdots \to \xi_n \to 0$$

An isomorphism from a complex ξ to a complex η is a sequence of isomorphisms $u_i: \xi_i \to \eta_i$ for $0 \leq i \leq n$ such that the following diagram is commutative.

$$0 \longrightarrow \xi_0|A \xrightarrow{\alpha_1} \xi_1|A \xrightarrow{\alpha_2} \cdots \longrightarrow \xi^n|A \longrightarrow 0$$

$$\downarrow u_0 \qquad\qquad \downarrow u_1 \qquad\qquad\qquad \downarrow u_n$$

$$0 \longrightarrow \eta_0|A \xrightarrow{\beta_1} \eta_1|A \xrightarrow{\beta_2} \cdots \longrightarrow \eta^n|A \longrightarrow 0$$

8.1 Notations. Let $S_n(X, A)$ denote the set of isomorphism classes of complexes of length n defined over X and acyclic over A. We define a commutative semigroup structure on $S_n(X, A)$, using the quotient function induced by the Whitney sum of two complexes. The induced complex defines a semigroup morphism $f^*: S_n(X, A) \to S_n(Y, B)$ for each map $f: (Y, B) \to (X, A)$, from which it follows that $S_n(X, A)$ is a cofunctor.

An elementary complex is a complex $0 \to \xi_0 \xrightarrow{\alpha_1} \xi_1 \to \cdots \to \xi_n \to 0$ such that $\xi_i = 0$ for $i \neq k - 1$, k, and $\alpha_k: \xi_{k-1} \to \xi_k$ is an isomorphism over X. Two complexes ξ and η are said to be equivalent provided there are elementary complexes $\delta_1, \ldots, \delta_p$ and $\varepsilon_1, \ldots, \varepsilon_q$ such that $\xi \oplus \delta_1 \oplus \cdots \oplus \delta_p$ equals $\eta \oplus \varepsilon_1 \oplus \cdots \oplus \varepsilon_q$ in $S_n(X, A)$.

8.2 Definition. We denote by $L_n(X, A)$ the quotient semigroup cofunctor of $S_n(X, A)$ under the above relation on elements of $S_n(X, A)$.

Observe that $L_1(X, A)$ is just $L(X, A)$ of Sec. 4. One can prove that ξ defines the zero element in $L_n(X, A)$ if ξ is acyclic on X.

By viewing a complex of length n as a complex of length $n + 1$, we define a natural morphism $L_n(X, A) \to L_{n+1}(X, A)$. By composition there are natural morphisms $L_1(X, A) \to L_m(X, A)$.

For finite CW-pairs we have the following result which is left to the reader.

8.3 Theorem. *The natural morphism $L_1(X, A) \to L_m(X, A)$ is an isomorphism.*

This theorem is dependent upon the following result which follows from 2(7.1) or Lemma 7.2, Atiyah, Bott, and Shapiro [1].

8.4 Lemma. *Let ξ and η be vector bundles over X, and let $u: \xi|A \to \eta|A$ be a monomorphism. If $\dim \eta \to \dim \xi + \dim X$, there exists a prolongation of u to a monomorphism $\xi \to \eta$.*

As in Sec. 5, we consider morphisms $L_n(X, A) \to K(X, A)$.

8.5 Definition. An Euler characteristic for $L_n(X, A)$ is a morphism of cofunctors $\chi: L_n \to K$ such that for empty A there is the relation $\chi(\xi) = \sum_{0 \leq i \leq n} (-1)^i \xi_i$.

The following theorem is proved by using Theorems (8.3) and (5.1).

8.6 Theorem. *A unique Euler characteristic exists for L_n, and it is an isomorphism of cofunctors.*

As corollaries of this theorem there are the following results. Recall that two complexes ξ and η are homotopic on (X, A) provided there is a complex ζ on $(X \times I, A \times I)$ such that $\xi = \zeta|(X \times 0, A \times 0)$ and $\eta = \zeta|(X \times 1, A \times 1)$.

8.7 Corollary. *Two homotopic complexes determine the same element in* $L_n(X, A)$.

We construct an inverse to the isomorphism $L_1(X, A) \to L_n(X, A)$. For $\xi \in S_n(X, A)$ we can put riemannian metrics on ξ_i and define $\alpha_i^*: \xi_i \to \xi_{i-1}$. We consider $\eta = (0 \to \eta_0 \xrightarrow{\beta} \eta_1 \to 0)$, where $\eta_0 = \sum_i \xi_{2i}$ and $\eta_1 = \sum_i \xi_{2i+1}$ and $\beta(x_0, x_2, x_4, \dots) = (\alpha_1(x_0), \alpha_2^*(x_2) + \alpha_3(x_2), \alpha_4^*(x_4) + \alpha_5(x_4), \dots)$.   Since $\xi_{2i+1} = \alpha_{2i}(\xi_{2i}) \oplus \alpha_{2i+1}^*(\xi_{2i+2})$ over A, the element η defines a class in $S_1(X, A)$. If $\xi \in S_1(X, A)$, then ξ equals η. This is easily seen to define an inverse $L_n(X, A) \to L_1(X, A)$, by (8.6) and (8.7).

8.8 Products. If $\xi \in S_n(X, A)$ and if $\eta \in S_m(Y, B)$, then the tensor product $\xi \otimes \eta \in S_{n+m}(X \times Y, (X \times B) \cup (A \times Y))$ by the Künneth formula.

This defines a biadditive pairing $L_n(X, A) \otimes L_m(T, B) \to L_{n+m}(X \times Y, (X \times B) \cup (A \times Y))$ such that $\chi(ab) = \chi(a)\chi(b)$, where χ is the Euler characteristic.

9. Half-Exact Cofunctors

The following notion is due to Dold [5]. Again we give only an outline of the results.

9.1 Definition. Let $\mathbf{W}$ denote the category of finite CW-complexes with base points and homotopy classes of maps preserving base points. A half-exact cofunctor U is a cofunctor defined on $\mathbf{W}$ with values in the category of abelian groups such that for each subcomplex A of X the following sequence is exact:

$$U(A) \leftarrow U(X) \leftarrow U(X/A)$$

9.2 Example. Ordinary reduced singular cohomology $\tilde{H}^n(-; G)$ is a half-exact cofunctor. The natural quotient map $X \to X/A$ defines an isomorphism $\tilde{H}^n(X/A; G) \to H^n(X, A; G)$. The half-exactness property is derived from this and the exact sequence for cohomology.

9.3 Example. Real, complex, and quaternionic $\tilde{K}$-group cofunctors are half exact by (2.1).

9.4 Example. Let H be a homotopy associative and commutative H-space. Then $U_H = [-, H]_0$ is a half-exact cofunctor.

9.5 Example. Let U be an arbitrary half-exact cofunctor, and let Z be a (finite) cell complex. Then the functions that assign to X the group $U(X \wedge Z)$ and to $f: Y \to X$ the group morphism $U(f \wedge 1_Z)$ collect to define a half-exact cofunctor.

9.6 Proposition. *Let U be a half-exact cofunctor. If X is a point, $U(X)$ equals 0. If f is null homotopic, then $U(f)$ equals 0.*

The discussion in Secs. 2 and 3 applies to half-exact functors, and we have the following theorems.

9.7 Theorem. *Let $f: X \to Y$ be a cellular map of finite CW-complexes. Then, with the notations of Sec. 2, there is the following exact sequence:*

$$U(X) \xleftarrow{U(f)} U(Y) \xleftarrow{U(a(f))} U(C_f) \xleftarrow{U(b(f))} U(S(X)) \xleftarrow{U(S(f))} U(S(Y))$$

This theorem corresponds to (2.8), and corresponding to (2.10) is the next remark.

9.8 Remark. For a subcomplex A of X there is a coboundary morphism $U(S(A)) \to U(X/A)$ such that the following sequence is exact.

$$U(A) \leftarrow U(X) \leftarrow U(X/A) \leftarrow U(S(A)) \leftarrow U(S(X))$$

We leave it to the reader to state and verify (3.2) to (3.4) for half-exact cofunctors.

The next theorem is an important uniqueness theorem which can be used in giving a general K-theory setting for the periodicity theorem of Bott. It is useful in comparing rational K-theory and rational cohomology theory.

9.9 Theorem. *Let $\phi: U \to V$ be a morphism of half-exact cofunctors. If $\phi(S^m)$ is an isomorphism for each $m \geq 0$, then ϕ is an isomorphism.*

This theorem is proved by induction on the number of cells of X for $\phi(X)$.

Finally, (2.6) and (9.7) are related to the following basic result known as the Puppe mapping sequence (see Puppe [1]). We use the notations of (2.6) and (9.7). Everything is for base point preserving maps.

9.10 Theorem. *Let $f: X \to Y$ be a map, and let Z be a space. Then the following sequence is exact as a sequence of sets with base points.*

$$[X,Z]_0 \xleftarrow{f^*} [Y,Z]_0 \xleftarrow{a(f)^*} [C_f,Z]_0 \xleftarrow{b(f)^*} [SX,Z]_0 \xleftarrow{Sf^*} [SY,Z]_0$$

Exercises

1. Develop the results of this chapter for vector bundles over a pair (X, A) where X is a compact space and A is a closed subspace.

2. Fill in the details in Secs. 8 and 9.

CHAPTER 11

Bott Periodicity in the Complex Case

Using elementary methods in analysis and vector bundle theory, we present the Atiyah-Bott proof of the complex periodicity theorem. In this setting we see the role of the Hopf bundle in defining the periodicity isomorphism.

1. K-Theory Interpretation of the Periodicity Result

Ket $k = 2$ for complex K-theory K and $k = 8$ for real K-theory KO. From the sequence of spaces

$$X \vee S^k \to X \times S^k \to X \wedge S^k$$

there is the following exact sequence in either real or complex K-theory:

$$0 \to \tilde{K}(X \wedge S^k) \to \tilde{K}(X \times S^k) \to \tilde{K}(X \vee S^k) \to 0$$

The external cup product or tensor product yields a morphism

$$K(X) \otimes K(S^k) \to K(X \times S^k)$$

From this we have the following commutative diagram:

$$K(X) \otimes K(S^k) = (\tilde{K}(X) \otimes \tilde{K}(S^k)) \oplus \tilde{K}(X) \oplus \tilde{K}(S^k) \oplus \mathbf{Z}$$
$$\downarrow \text{cup}$$
$$K(X \times S^k) = \tilde{K}(X \wedge S^k) \oplus \tilde{K}(X) \oplus \tilde{K}(S^k) \oplus \mathbf{Z}$$

The morphism cup restricted to $\tilde{K}(X) \oplus \tilde{K}(S^k) \oplus \mathbf{Z}$ is the identity when restricted to the last three summands. This leads to the following result.

1.1 Proposition. *The external cup product $K(X) \otimes K(S^k) \to K(X \times S^k)$ is an isomorphism if and only if the external cup product $\tilde{K}(X) \otimes \tilde{K}(S^k) \to \tilde{K}(X \wedge S^k)$ is an isomorphism.*

Now we state the K-theory formulation of Bott periodicity.

1.2 Theorem. *Let X be a compact space. The external cup products $K(X) \otimes K(S^2) \to K(X \times S^2)$ in complex K-theory and $KO(X) \otimes KO(S^8) \to KO(X \times S^8)$ in real K-theory are isomorphisms. In addition, $K(S^2)$ is a free abelian group on two generators 1 and η class of the complex Hopf bundle, and $KO(S^8)$ is the free abelian group on two generators 1 and η_8, where η_8 is the class of the real eight-dimensional Hopf bundle.*

In view of Proposition (1.1), Theorem (1.2) is equivalent to:

1.3 Theorem. *Let X be a compact space. The function that assigns to a $\in \tilde{K}(X)$ the element $a \otimes (\eta - 1)$ in $\tilde{K}(X \wedge S^2)$ is a cofunctor isomorphism, and the function that assigns to a $\in \tilde{KO}(X)$ the element $a \otimes (\eta_8 - 8)$ in $\tilde{KO}(X \wedge S^8)$ is a cofunctor isomorphism.*

We shall prove Theorem (1.2) for complex vector bundles.

2. Complex Vector Bundles Over $X \times S^2$

2.1 Notations. We view S^2 as the Riemann sphere of complex numbers and the point at infinity. Let D_0 denote the disk of z with $|z| \leq 1$, and let D_∞ denote the disk of z with $|z| \geq 1$. Then $D_0 \cap D_\infty$ equals S^1, the unit circle.

2.2 Notations. For a compact space X we denote the natural projections on X by $\pi_0: X \times D_0 \to X$, $\pi_\infty: X \times D_\infty \to X$, and $\pi: X \times S^1 \to X$. Observe that we have $(X \times D_0) \cup (X \times D_\infty) = X \times S^2$ and $(X \times D_0) \cap (X \times D_\infty) = X \times S^1$. Let $s: X \to X \times S^2$ be the map defined by the relation $s(x) = (x, 1)$. We use these notations throughout this chapter.

Using the clutching construction of Chap. 10, Sec. 7, we are able to give a description of complex vector bundles over $X \times S^2$.

2.3 Proposition. *Let ξ be a complex vector bundle over $X \times S^2$, and let ζ equal $s^*(\xi)$. Then there is an automorphism $u: \pi^*(\zeta) \to \pi^*(\zeta)$ such that ξ and $\pi_0^*(\zeta) \bigcup_u \pi_1^*(\zeta)$ are isomorphic and $u: \zeta|X \times 1 \to \zeta|X \times 1$ is homotopic to the identity in the space of automorphisms. Moreover, u is unique with respect to the above properties.*

Proof. We consider $s: X \to X \times D_0$ or $X \times D_\infty$. The compositions $s\pi_0: X \times D_0 \to X \times D_0$ and $s\pi_\infty: X \times D_\infty \to X \times D_\infty$ are homotopy equivalences. Consequently, the natural isomorphism $\xi|(X \times 1) \to \pi^*(\zeta)|(X \times 1)$ prolongs to two isomorphisms $u_0: \xi|(X \times D_0) \to \pi_0^*(\zeta)$ and $u_\infty: \xi|(X \times D_\infty) \to \pi_\infty^*(\zeta)$.

Two such extensions over $X \times D_0$ or over $X \times D_\infty$ differ by an auto-morphism over $X \times D_0$ or over $X \times D_\infty$, which is the identity on $X \times 1$. Then $u = u_\infty^{-1} u_0$ is the desired automorphism, for by Proposition 10(7.1) there is an isomorphism between ξ and $\pi_0^*(\zeta) \underset{u}{\bigcup} \pi_\infty^*(\zeta)$. This proves the proposition.

The task in the following sections is to consider clutching automorphisms $u: \pi^*(\zeta) \to \pi^*(\zeta)$ for vector bundles ζ over X and the projection $\pi: X \times S^1 \to X$. Such a clutching map u is given by a continuous family of automorphisms $u(x,z): \zeta_x \to \zeta_x$ for $x \in X$ and $|z| = 1$.

2.4 Definition. A Laurent polynomial clutching map u for ζ is a clutching map of the form

$$u(x,z) = \sum_{|k| \leq n} a_k(x) z^k$$

where $a_k: \zeta \to \zeta$ is a morphism of ζ. A Laurent polynomial clutching map u is a polynomial clutching map provided $u(x,z) = a_0(x) + a_1(x)z + \cdots + a_n(x)z^n$ and is a linear clutching map provided $u(x,z) = a(x) + b(x)z$.

In the next proposition we see that any complex vector bundle ξ over $X \times S^2$ is isomorphic to $\pi_0^*(\zeta) \underset{u}{\bigcup} \pi_\infty^*(\zeta)$, where u is a Laurent polynomial clutching map.

2.5 Proposition. *Let ξ be a complex vector bundle over $X \times S^2$, and let ζ equal $s^*(\xi)$. Then ξ is isomorphic to $\pi_0^*(\zeta) \underset{v}{\bigcup} \pi_\infty^*(\zeta)$, where v is a Laurent polynomial clutching map.*

Proof. Let ξ be isomorphic to $\pi_0^*(\zeta) \underset{u}{\bigcup} \pi_\infty^*(\zeta)$ as in Proposition (2.3), where u is an arbitrary clutching map. Let $a_k: \zeta \to \zeta$ be the morphism defined by the integral

$$a_k(x) = \frac{1}{2\pi i} \int_{S^1} z^{-k} u(x,z) \frac{dz}{z}$$

We define $s_k(x,z)$ to be $\sum_{|j| \leq k} a_j(x) z^j$ and $u_n(x,z)$ to be $1/(n+1) \sum_{0 \leq k \leq n} s_k(x,z)$. Then $u_n(x,z)$ is the nth partial Cesaro sum of a Fourier series, and by an extension of Fejer's theorem u_n converges uniformly in x and z to u.

Let $v = u_n$ for large n, where u_n is so close to u that it is a clutching map homotopic to u. This is possible because homotopy classes of clutching maps are open sets in the uniform topology. Then there is an isomorphism between ξ and $\pi_0^*(\zeta) \underset{v}{\bigcup} \pi_\infty^*(\zeta)$. This proves the proposition.

2.6 Example. Let γ denote the canonical line bundle over $S^2 = CP^1$, and let η denote the dual line bundle of γ. Then γ is isomorphic to the bundle obtained by clutching two trivial line bundles over D_0 and D_∞ with the clutching map

$u(z) = z$, and η is isomorphic to the bundle obtained with the clutching map $u(z) = z^{-1}$. To see this, we recall that $E(\gamma)$ is the subspace of $(\langle z_0, z_1 \rangle, (w_0, w_1)) \in CP^1 \times \mathbf{C}^2$. If V_i denotes the open set of $\langle z_0, z_1 \rangle$ with $z_i \neq 0$ for $i = 0$, 1, we have charts $h_i: V_i \times \mathbf{C} \to E(\gamma|V_i)$, where $h_0(\langle z_0, z_1 \rangle, a) = (\langle z_0, z_1 \rangle, (a, z_1 a/z_0))$ and $h_1(\langle z_0, z_1 \rangle, a) = (\langle z_0, z_1 \rangle, (z_0 a/z_1, a))$. Since the clutching map $u(z)$ for $|z| = 1$ has the property that $h_0(z, a) = h_1(z, u(z)a)$ for $z = z_1/z_0$, we have $u(z) = z$. Similarly for the dual bundle the clutching map is $u(z) = z^{-1}$ from the tensor product property of clutching maps, see 10(7.4), and of dual bundles.

2.7 Notation. Let ζ be a complex vector bundle over X, and let u be a clutching map $\pi^*(\zeta) \to \pi^*(\zeta)$. Then $\pi_0^*(\zeta) \underset{u}{\bigcup} \pi_\infty^*(\zeta)$ is denoted by $[\zeta, u]$. Denote $\gamma = [\theta^1, z]$ and $\eta = [\theta^1, z^{-1}]$.

2.8 Proposition. *Let ξ be isomorphic to $[\zeta, u]$ over $X \times S^2$. Then $[\zeta, uz^n]$ is isomorphic to $\xi \hat{\otimes} \gamma^n$ or $\xi \hat{\otimes} \eta^{-n}$, where γ^n denotes the n-fold tensor product of γ and $\hat{\otimes}$ is the external tensor product.*

Proof. This follows from tensor product properties of clutching maps 10(7.4); that is, $\pi_0^*(\zeta) \underset{uz^n}{\bigcup} \pi_\infty^*(\zeta)$ is isomorphic to $(\pi_0^*(\zeta) \otimes \theta^1) \underset{u \otimes z^n}{\bigcup} (\pi_\infty^*(\zeta) \otimes \theta^1)$ which in turn is isomorphic to $\zeta \hat{\otimes} \gamma^n$.

By (2.8) we have a complete picture of vector bundles over $X \times S^2$ whose clutching map is a monomial; that is, $[\zeta, az^n]$ is isomorphic to $\zeta \hat{\otimes} \gamma^n$.

3. Analysis of Polynomial Clutching Maps

By (2.5) every complex vector bundle ξ over $X \times S^2$ is isomorphic to $[\zeta, z^{-n}p]$, where p is a polynomial clutching function. By (2.8), $[\zeta, z^{-n}p]$ is isomorphic to $[\zeta, p] \hat{\otimes} \eta^n$ over $X \times S^2$. In this section we linearize polynomial clutching functions in a manner analogous to the transformation used in reducing nth-order differential equations to linear differential equations.

3.1 Notation. Let $p = \underset{0 \leq k \leq n}{\sum} p_k z^k$ be a polynomial clutching function for a complex vector bundle ζ over X. Let $L^n(p)$ denote the linear polynomial clutching function for the complex vector bundle $L^n(\zeta) = \zeta \oplus \overset{(n+1)}{\cdots} \oplus \zeta$ given by the following matrix.

$$L^n(p) = \begin{bmatrix} p_0 & p_1 & p_2 & \cdots & p_{n-1} & p_n \\ -z & 1 & 0 & \cdots & 0 & 0 \\ 0 & -z & 1 & \cdots & 0 & 0 \\ \hdotsfor{6} \\ 0 & 0 & 0 & \cdots & 1 & 0 \\ 0 & 0 & 0 & \cdots & -z & 1 \end{bmatrix}$$

Observe that $L^n(p)$ is the product of three matrices where $p_r^*(z) = \sum\limits_{r \leq k \leq n} p_k z^{k-r}$
and $p_r^*(z) - zp_{r+1}^*(z) = p_r$.

$$\begin{bmatrix} 1 & p_1^*(z) & \cdots & p_n^*(z) \\ 0 & 1 & \cdots & 0 \\ \hdotsfor{4} \\ 0 & 0 & \cdots & 0 \\ 0 & 0 & \cdots & 1 \end{bmatrix} \begin{bmatrix} p(z) & 0 & \cdots & 0 & 0 \\ 0 & 1 & \cdots & 0 & 0 \\ \hdotsfor{5} \\ 0 & 0 & \cdots & 1 & 0 \\ 0 & 0 & \cdots & 0 & 1 \end{bmatrix} \begin{bmatrix} 1 & 0 & \cdots & 0 & 0 \\ -z & 1 & \cdots & 0 & 0 \\ \hdotsfor{5} \\ 0 & 0 & \cdots & 1 & 0 \\ 0 & 0 & \cdots & -z & 1 \end{bmatrix}$$

Consequently, $L^n(p) = (1 + N_1)(p \oplus 1_n)(1 + N_2)$, where N_1 and N_2 are nilpotent. Then $L_t^n(p) = (1 + tN_1)(p \oplus 1_n)(1 + tN_2)$ is a homotopy of clutching functions of $L^n(\zeta)$. This yields the following result.

3.2 Proposition. *For a polynomial clutching map*

$$p(z) = \sum_{0 \leq k \leq n} p_k z^k \text{ for } \zeta \text{ over } X$$

$[L^n(\zeta), L^n(p)]$ *and* $[L^n(\zeta), p \oplus 1_n]$ *are isomorphic vector bundles over* $X \times S^2$.

Using these notations, we have the following two propositions for a polynomial clutching map $p(z) = \sum\limits_{0 \leq k \leq n} p_k z^k$ for ζ over X. We may view $p(z)$ as a polynomial of degree $n + 1$.

3.3 Proposition. *There is an isomorphism between* $[L^{n+1}(\zeta), L^{n+1}(p)]$ *and* $[L^n(\zeta), L^n(p)] \oplus [\zeta, 1]$.

Proof. We use the following homotopy of clutching maps:

$$\begin{bmatrix} p_0 & p_1 & \cdots & p_{n-1} & p_n & 0 \\ -z & 1 & \cdots & 0 & 0 & 0 \\ \hdotsfor{6} \\ 0 & 0 & \cdots & 1 & 0 & 0 \\ 0 & 0 & \cdots & -z & 1 & 0 \\ 0 & 0 & \cdots & 0 & -(1-t)z & 1 \end{bmatrix}$$

For $t = 0$ we have $[L^{n+1}(\zeta), L^{n+1}(p)]$ and for $t = 1$, $[L^n(\zeta), L^n(p)] \oplus [\zeta, 1]$.

3.4 Proposition. *There is an isomorphism between* $[L^{n+1}(\zeta), L^{n+1}(zp)]$ *and* $[L^n(\zeta), L^n(p)] \oplus [\zeta, z]$.

Proof. We use the following homotopy of clutching maps.

$$\begin{bmatrix} 0 & p_0 & p_1 & \cdots & p_{n-1} & p_n \\ -z & 1-t & 0 & \cdots & 0 & 0 \\ 0 & -z & 1 & \cdots & 0 & 0 \\ \cdots\cdots\cdots\cdots\cdots\cdots\cdots\cdots\cdots\cdots\cdots \\ 0 & 0 & 0 & \cdots & 1 & 0 \\ 0 & 0 & 0 & \cdots & -z & 1 \end{bmatrix}$$

For $t = 0$ we have $[L^{n+1}(\zeta), L^{n+1}(zp)]$, and for $t = 1$ we have $[L^n(\zeta), L^n(p)] \oplus [\zeta, z]$.

As a corollary we get two relations involving each of the Hopf bundles γ an dη.

3.5 Corollary. *There is an isomorphism between $\gamma^2 \oplus \theta^1$ and $\gamma \oplus \gamma$, or between $\eta^{-2} \oplus 1$ and $\eta^{-1} \oplus \eta^{-1}$.*

Proof. By (3.4) there is an isomorphism between $[\theta^2, z^2] = [L^2(\theta^1), z^2]$ and $[\theta^1, z] \oplus [\theta^1, z] = \gamma \oplus \gamma$, and by (3.2) there is an isomorphism between $[\theta^2, z^2]$ and $[\theta^1, z^2] \oplus [\theta^1, 1] = \gamma^2 \oplus 1$.

The proof of Corollary (3.5) follows directly from the following homotopy of clutching maps.

$$\begin{bmatrix} z & 0 \\ 0 & 1 \end{bmatrix} \begin{bmatrix} \cos(\pi t/2) & -\sin(\pi t/2) \\ \sin(\pi t/2) & \cos(\pi t/2) \end{bmatrix} \begin{bmatrix} z & 0 \\ 0 & 1 \end{bmatrix} \begin{bmatrix} \cos(\pi t/2) & \sin(\pi t/2) \\ -\sin(\pi t/2) & \cos(\pi t/2) \end{bmatrix}$$

This map is $\begin{bmatrix} z^2 & 0 \\ 0 & 1 \end{bmatrix}$ for $t = 0$ and $\begin{bmatrix} z & 0 \\ 0 & z \end{bmatrix}$ for $t = 1$.

4. Analysis of Linear Clutching Maps

Let $p(x, z) = a(x)z + b(x)$ be a linear clutching map for a bundle ζ over X. We wish to prove that ζ decomposes into a Whitney sum $\zeta_+^0 \oplus \zeta_-^0$ such that $p|\zeta_+^0: \zeta_+^0 \to \zeta_+^\infty$ is nonsingular for all z with $|z| \geq 1$ and $p|\zeta_-^0: \zeta_-^0 \to \zeta_-^\infty$ is nonsingular for all z with $|z| \leq 1$. Then the bundle $[\zeta, a(x)z + b(x)]$ is isomorphic to $[\zeta_+^0, z] \oplus [\zeta_-^0, 1]$.

If $p(x, z)$ has the form $z + b(x)$, the fibre of ζ_+ at x will be the sum of the eigenspaces of $b(x)$ for values λ with $|\lambda| < 1$, and the fibre of ζ_- at x will be the sum of the eigenspaces of $b(x)$ for value λ with $|\lambda| > 1$.

The following treatment of this case was pointed out by J. F. Adams. The author is very grateful for his permission to use it here.

4.1 Notations. The linear clutching map $a(x)z + b(x) = p(x, z): \zeta \to \zeta$ is nonsingular for $|z| = 1$ and all x. By the compactness of the base space, the map $p(x, z)$ is nonsingular for all z with $1 - \varepsilon \leq |z| \leq 1 + \varepsilon$ and all $x \in X$. Form

$$p_0(x) = \frac{1}{2\pi i} \int_{|z|=1} (a(x)z + b(x))^{-1} a(x)\, dz$$

and

$$p_\infty(x) = \frac{1}{2\pi i} \int_{|z|=1} a(x)(a(x)z + b(x))^{-1}\, dz$$

where $p_0, p_\infty: \zeta \to \zeta$ are bundle morphisms independent of z.

The main properties of p_0 and p_∞ are contained in the next proposition.

4.2 Proposition. *We have $p(x,z)p_0(x) = p_\infty(x)p(x,z)$ for all $x \in X$ and $|z| = 1$, and $p_0(x)$ and $p_\infty(z)$ are each projections; that is, $p_0(x)p_0(x) = p_0(x)$ and $p_\infty(x)p_\infty(x) = p_\infty(x)$.*

Proof. First, there is the following relation for $z \neq w$.

$$(aw + b)^{-1} a(az + b)^{-1} = (az + b)^{-1} a(aw + b)^{-1}$$

$$= \frac{(az + b)^{-1}}{w - z} + \frac{(aw + b)^{-1}}{z - w} \qquad (R)$$

To see this, we make the following calculation:

$$\frac{(az + b)^{-1}}{w - z} + \frac{(aw + b)^{-1}}{z - w} = (aw + b)^{-1} \frac{aw + b}{w - z}(az + b)^{-1}$$

$$+ (aw + b)^{-1} \frac{az + b}{z - w}(az + b)^{-1}$$

$$= (aw + b)^{-1} a(az + b)^{-1}$$

Now we use the symmetry between z and w. It should be noted that the first equality in relation (R) holds also for $z = w$.

To derive the relation $p(x,z)p_0(x) = p_\infty(x)p(x,z)$, we multiply relation (R) by $az + b$ on the left and right and then integrate. We get

$$(az + b)p_0 = \frac{1}{2\pi i} \int_{|z|=1} (az + b)(aw + b)^{-1} a\, dw$$

$$= \frac{1}{2\pi i} \int_{|z|=1} a(aw + b)^{-1}(az + b)\, dw$$

$$= p_\infty(az + b)$$

To prove that $p_0 p_0 = p_0$, we choose r_1 and r_2 with $1 - \varepsilon < r_2 < r_1 < 1 + \varepsilon$. Then we calculate the following expression, using relation (R).

$$p_0 p_0 = \frac{1}{(2\pi i)^2} \int_{|z|=r_1} \int_{|w|=r_2} (az + b)^{-1} a(aw + b)^{-1} a\, dz\, dw$$

$$= \frac{1}{(2\pi i)^2} \int_{|z|=r_1} \int_{|w|=r_2} \left[\frac{1}{w - z}(az + b)^{-1} a + \frac{1}{z - w}(aw + b)^{-1} a \right] dz\, dw$$

$$= \frac{1}{2\pi i} \int_{|w|=r_2} (aw + b)^{-1} a \, dw$$

$$= p_0$$

Observe that

$$\int_{|w|=r_2} \frac{dw}{w - z} = 0 \qquad \text{for } |z| = r_1 > r_2$$

A similar calculation yields the relation $p_\infty p_\infty = p_\infty$. This proves the proposition.

4.3 Remark. The vector bundle projections $p_0: \zeta \to \zeta$ and $p_\infty: \zeta \to \zeta$ are of constant rank. This is a general property of projections and is easily seen by considering local cross sections $s_1, \ldots, s_n$ of ζ such that $s_1(x), \ldots, s_r(x)$ is a basis of $\ker(p_0)_x$ and $s_{r+1}(x), \ldots, s_n(x)$ is a basis of $\text{im}(p_0)_x$. Then $(1 - p_0)s_1(y)$, $\ldots, (1 - p_0)s_r(y)$ is a basis of $\ker(p_0)_y$ near x, and $p_0 s_{r+1}(y), \ldots, p_0 s_n(y)$ is a basis of $\text{im}(p_0)_y$ for y near x in the base space of ζ.

4.4 Notations. We denote the vector bundle $\text{im } p_0$ by ζ_+^0, $\text{im } p_\infty$ by ζ_+^∞, $\ker p_0$ by ζ_-^0, and $\ker p_\infty$ by ζ_-^∞. The relation $p(x, z)p_0(x) = p_\infty(x)p(x, z)$ implies that the following restrictions of $p(x, z)$ are defined:

$$p_+(-, z): \zeta_+^0 \to \zeta_+^\infty$$

$$p_-(-, z): \zeta_-^0 \to \zeta_-^\infty$$

4.5 Proposition. *The restriction* $p_+(-, z): \zeta_+^1 \to \zeta_+^\infty$ *is an isomorphism for* $|z| \geq 1$, *and the restriction* $p_-(-, z): \zeta_-^0 \to \zeta_-^\infty$ *is an isomorphism for* $|z| \leq 1$.

Proof. Let v be in the fibre of ζ over x such that $(a(x)w + b(x))v = 0$ for $|w| \neq 1$. Then $(a(x)z + b(x))v = (z - w)a(x)v$ and $(a(x)z + b(x))^{-1}a(x)v = (z - w)^{-1}v$ for $|z| = 1$. If we integrate over the circle $|z| = 1$, we get the relation

$$p_0(x)v = \begin{cases} v & \text{for } |w| < 1 \\ 0 & \text{for } |w| > 1 \end{cases}$$

If $v \in \ker(a(x)w + b(x))$ and $|w| < 1$, then $v \in \zeta_+^0$ and $p_-(x, z)$ is a monomorphism for $|z| \leq 1$. If $v \in \ker(a(x)w + b(x))$ and $|w| > 1$, then $v \in \zeta_-^0$ and $p_+(x, z)$ is a monomorphism for $|z| \geq 1$. For reasons of dimension, p_+ and p_- are isomorphisms for $|z| \geq 1$ or $|z| \leq 1$, respectively. This proves the proposition.

4.6 Proposition. *Let* $p_+ = a_+z + b_+$ *and* $p_- = a_-z + b_-$ *where* p_+ *and* p_- *are defined as in* (4.4), *and let* $p' = p_+^t + p_-^t$, *where* $p_+^t = a_+z + tb_+$ *and* $p_-^t = ta_-z + b_-$ *for* $0 \leq t \leq 1$. *Then* p' *is a homotopy of linear clutching maps from*

$a_+z + b_-$ to p. Moreover, the bundles $[\zeta, p]$ and $[\zeta^0_+, z] \oplus [\zeta^0_-, 1]$ are isomorphic.

Proof. By (4.5), p^t_+ and p^t_- are isomorphisms onto their images for all t, $0 \leq t \leq 1$. Then $[\zeta, p]$ and $\left(\zeta^0_+ \underset{a_+z}{\bigcup} \zeta^\infty_+ \right) \oplus \left(\zeta^0_- \underset{b}{\bigcup} \zeta^\infty_- \right)$ are isomorphic bundles over $X \times S^2$. Since $a_+ : \zeta^0_+ \to \zeta^\infty_+$ and $b_- : \zeta^0_- \to \zeta^\infty_-$ are isomorphisms, there are isomorphisms between $[\zeta^0_+, z]$ and $\zeta^0_+ \underset{a_+z}{\bigcup} \zeta^\infty_+$ and between $[\zeta^0_-, 1]$ and $\zeta^0_- \underset{b}{\bigcup} \zeta^\infty_-$. This proves the proposition.

4.7 Notation. Let p be a polynomial clutching map for ζ of degree $\leq n$. Then the bundle $L^n(\zeta) = (n+1)\zeta$ decomposes with respect to the linear clutching map $L^n(p)$, as in (4.6). We denote this as follows:

$$L^n(\zeta) = L^n(\zeta, p)_+ \oplus L^n(\zeta, p)_-$$

Then (4.6) says that the bundles $[L^n(\zeta), L^n(p)]$ and $[L^n(\zeta, p)_+, z] \oplus [L^n(\zeta, p)_-, 1]$ are isomorphic.

From the results of (3.3) and (3.4) and the above analysis, we have the following result.

4.8 Proposition. *Let $p(x, z)$ be a polynomial clutching map of degree less than n for a vector bundle ζ. For $L^{n+1}(p)$, there are the following isomorphisms:*

$$L^{n+1}(\zeta, p)_+ \cong L^n(\zeta, p)_+ \quad \text{and} \quad L^{n+1}(\zeta, p)_- \cong L^{n+1}(\zeta, p)_- \oplus \zeta$$

For $L^{n+1}(zp)$, there are the following isomorphisms:

$$L^{n+1}(\zeta, zp)_+ \cong L^n(\zeta, p)_+ \oplus \zeta \quad \text{and} \quad L^{n+1}(\zeta, zp)_- \cong L^n(\zeta, p)_-$$

5. The Inverse to the Periodicity Isomorphism

The periodicity morphism $\mu : K(X) \otimes K(S^2) \to K(X \times S^2)$ is just the external K-cup product. We wish to define an inverse morphism $v : K(X \times S^2) \to K(X) \otimes K(S^2)$. For a vector bundle ξ over X, the element of $K(X)$ determined by ξ is also denoted by ξ.

5.1 Notations. Let u be a clutching map for a vector bundle ζ over X. By (2.4), there is a Laurent polynomial clutching map $z^{-n}p_n(x, z)$ arbitrarily close and, therefore, homotopic to u for large n. Here $p_n(x, z)$ is a polynomial clutching map where $\deg p_n(x, z) \leq 2n$. In $K(X) \otimes K(S^2)$, we consider the following element:

$$v_n(\zeta, u) = L^{2n}(\zeta, p_n)_+ \otimes (\eta^{n-1} - \eta^n) + \zeta \otimes \eta^n$$

Since $(1 - \eta)\eta = 1 - \eta$ in $K(S^2)$, by (3.5), we have

$$v_n(\zeta, u) = L^{2n}(\zeta, p_n)_+ \otimes (1 - \eta) + \zeta \otimes \eta^n$$

5.2 Proposition. *With the above notation, we have* $v_n(\zeta, u) = v_{n+1}(\zeta, u)$.

Proof. Using (4.8), we have the following relations in $K(X)$:

$$L^{2n+2}(\zeta, p_{n+1})_+ = L^{2n+2}(\zeta, zp_{n+1})_+ = L^{2n+1}(\zeta, zp_n)_+ = L^{2n}(\zeta, p_n)_+ \oplus \zeta$$

From this we have

$$
\begin{aligned}
v_{n+1}(\zeta, u) &= L^{2n+2}(\zeta, p_{n+1})_+ \otimes (\eta^n - \eta^{n+1}) + \zeta \otimes \eta^{n+1} \\
&= L^{2n}(\zeta, p_n)_+ \otimes (1 - \eta) + \zeta \otimes (\eta^n - \eta^{n+1}) + \zeta \otimes \eta^{n+1} \\
&= L^{2n}(\zeta, p_n)_+ \otimes (\eta^{n-1} - \eta^n) + \zeta \otimes \eta^n \\
&= v_n(\zeta, u)
\end{aligned}
$$

This proves the proposition.

5.3 Remarks. In view of (5.2), $v(\zeta, u)$ may be written for $v_n(\zeta, u)$, since v is independent of n. If u' is a second clutching map near u, the line segment joining $p_n = z^n f_n$ to $q_n = z^n g_n$ is a homotopy of clutching maps, where f_n approximates u and g_n approximates u'. Consequently, we have $v(\zeta, u) = v(\zeta, u')$ and $v(\xi)$ equal to $v(\zeta, u)$, where $\xi = [\zeta, u]$ is well defined. Observe that $v(\xi \oplus \xi') = v(\xi) + v(\xi')$ for vector bundles on $X \times S^2$. Consequently, a morphism $v \colon K(X \times S^2) \to K(X) \otimes K(S^2)$ is defined.

Now we are in a position to prove the complex periodicity theorem.

5.4 Theorem. *The external K-cup product* $\mu \colon K(X) \otimes K(S^2) \to K(X \times S^2)$ *is an isomorphism with v as its inverse.*

Proof. First, we show that $v\mu = 1$. For this it suffices to prove that $v\mu(\zeta \otimes \eta^n) = \zeta \otimes \eta^n$, where ζ is a vector bundle. Since $\mu(\zeta \otimes \eta^n) = [\zeta, z^{-n}]$, we calculate

$$v_n([\zeta, z^{-n}]) = L^{2n}(\zeta, 1)_+ \otimes (\eta^{n-1} - \eta^n) + \zeta \otimes \eta^n = \zeta \otimes \eta^n$$

This follows from the fact that $L^{2n}(\zeta, 1)_+ = L^0(\zeta, 1)_+ = 0$. We have $v\mu = 1$.

To prove $\mu v = 1$, we make use of two relations. The first comes from (4.6).

$$
\begin{aligned}
[L^{2n}(\zeta, p_n)_+, z] &= [L^{2n}(\zeta), L^{2n}(p_n)] - L^{2n}(\zeta, p_n)_-, 1] \\
[L^{2n}(\zeta, p_n)_-, z^{1-n}] &= [L^{2n}(\zeta), L^{2n}(p_n)] \otimes \eta^n - [L^{2n}(\zeta, p_n)_-, 1] \otimes \eta^n
\end{aligned}
\tag{a}
$$

The second comes from (3.2).

$$[L^{2n}(\zeta), L^{2n}(p_n)] = [\zeta, p_n(x, z)] + 2n\zeta \tag{b}$$

For $v(\xi) = v([\zeta, u]) = v([\zeta, z^{-n}p_n]) = L^{2n}(\zeta, p_n)_+ \otimes (\eta^{n-1} - \eta^n) + \zeta \otimes \eta^n$, we make the following calculation:

$$\mu v(\xi) = [L^{2n}(\zeta, p_n)_+, z^{1-n}] - [L^{2n}(\zeta, p_n)_+, z^{-n}] + [\zeta, z^{-n}]$$

$$= [L^{2n}(\zeta), L^{2n}(p_n)] \otimes \eta^n - [L^{2n}(\zeta, p_n)_-, 1] \otimes \eta^n - [L^{2n}(\zeta, p_n)_+, 1] \otimes \eta^n$$

$$+ [\zeta, z^{-n}] \qquad \text{by } (a)$$

$$= 2n(\zeta \otimes \eta^n) + [\zeta, z^{-n} p_n] - [L^{2n}(\zeta), 1] \otimes \eta^n + \zeta \otimes \eta^n \qquad \text{by } (b)$$

$$= [\zeta, z^{-n} p_n(x, z)]$$

$$= \xi$$

This proves the complex periodicity theorem.

As a corollary of the periodicity theorem, we can calculate $\tilde{K}(S^m)$.

5.5 Theorem. *There is a relation $\tilde{K}(S^{2n+1}) = 0$, and $\tilde{K}(S^{2n})$ is infinite cyclic with generator β_{2n}, where the ring structure is given by $\beta_{2n}^2 = 0$. Moreover, the natural map $S^2 \times \overset{(n)}{\cdots} \times S^2 \to S^{2n}$ induces a monomorphism $\tilde{K}(S^{2n}) \to \tilde{K}(S^2 \times \overset{(n)}{\cdots} \times S^2)$ and the image of β_{2n} in $\tilde{K}(S^2 \times \overset{(n)}{\cdots} \times S^2)$ equals a product $a_1 \cdots a_n$, where $a_i = \{\zeta_i\} - 1$ such that ζ_i is a line bundle on $S^2 \times \overset{(n)}{\cdots} \times S^2$.*

Proof. Since the external cup product with β_2 yields an isomorphism $\tilde{K}(S^i) \to \tilde{K}(S^i \wedge S^2)$, we have the result on the additive structure from the relations $\tilde{K}(S^0) = \mathbf{Z}$ and $\tilde{K}(S^1) = 0$ of 9(5.2).

The induced morphism $\tilde{K}(S^{2n}) \to \tilde{K}(S^2 \times \overset{(n)}{\cdots} \times S^2)$ is a monomorphism by 9(3.4). The image β_{2n} in $\tilde{K}(S^2 \times \overset{(n)}{\cdots} \times S^2)$ equals the image of the $\tilde{K}$-cup product of β_{2n-2} and β_2 under the monomorphism $\tilde{K}(S^{2n-2} \times S^2) \to \tilde{K}(S^2 \times \overset{(n)}{\cdots} \times S^2)$. Since β_2 equals $\{\zeta_i\} - 1$, where ζ_i is a line bundle induced from S^2 to $S^2 \times \overset{(n)}{\cdots} \times S^2$, and since $\beta_2^2 = 0$ from the relation $\beta_2^2 = (\{\zeta_i\} - 1)^2 = \{\zeta_i\}^2 + 1 - 2\{\zeta_i\} = 0$, we have the last statement of the theorem.

The relation $\beta_{2n}^2 = 0$ follows from the fact that the image of β_{2n} in $K(S^2 \times \cdots \times S^2)$ has a square that is zero.

5.6 Remark. The above proof of the periodicity theorem was given by Atiyah and Bott in *Acta Mathematica*, vol. 112, 1964. This proof was modivated by their work on complex elliptic boundary value problems. In the spring of 1966, motivated by the study of real elliptic operators, Atiyah gave a proof of the real periodicity theorem similar in character to the above proof of the complex periodicity theorem.

The first proof of the periodicity theorem was given by Bott in [3] using geometric arguments. Soon after, J. C. Moore proved the periodicity theorem using homology theory; see the H. Cartan Seminaire, 1959–1960.

CHAPTER 12
Clifford Algebras

Using methods from the theory of quadratic forms, one is able to construct vector bundles over spheres and projective spaces. We develop some general properties of Clifford algebras and completely calculate the Clifford algebras that arise in topology. Apart from constructing vector fields on a sphere, the topological applications are left to later chapters. Using Clifford algebras, we can give a concrete description of Spin(n).

1. Unit Tangent Vector Fields on Spheres: I

In this section we consider the very classical problem of determining when a sphere S^n has a single unit tangent vector field on it. Our first means of constructing vector fields is contained in the next proposition.

1.1 Proposition. *If S^{n-1} has k orthonormal tangent vector fields $v_1, \ldots, v_k$, then S^{nq-1} has k orthonormal tangent vector fields $v_1^*, \ldots, v_k^*$.*

Proof. We can view $v_i \colon S^{n-1} \to \mathbf{R}^n$ such that $(x|v_i(x)) = 0$ and $(v_i(x)|v_j(x)) = \delta_{i,j}$ for all $x \in S^{n-1}$ and $1 \le i, j \le k$. Next, the sphere S^{nq-1} can be considered as the join of q spheres S^{n-1}; that is, for $x \in S^{nq-1}$ we can write $x = (\alpha(1)x(1), \ldots, \alpha(q)x(q))$, where $x(i) \in S^{n-1}$ and $\sum_i \alpha(i)^2 = 1$, $\alpha(i) \ge 0$. We define $v_i^* \colon S^{nq-1} \to \mathbf{R}^{nq}$ by the relation $v_i^*(\alpha(1)x(1), \ldots, \alpha(q)x(q)) = \alpha(1)v_i(x(1)) + \cdots + \alpha(q)v_i(x(q))$. Then we have $(x|v_i^*(x)) = 0$ and $(v_i^*(x)|v_j^*(x)) = \delta_{i,j}$ for $1 \le i, j \le k$ by a direct calculation, using $(x(i)|x(j)) = 0$ for $i \ne j$.

1.2 Corollary. *Every odd-dimensional sphere S^{m-1} has a nonzero vector field on it.*

Proof. Since $m = 2q$, it suffices, in view of (1.1), to prove that S^1 has a unit vector field defined on it. Let $v(x_1, x_2) = (-x_2, x_1)$. Then $(x|v(x)) = 0$ and $\|v(x)\| = 1$ for each $x = (x_1, x_2) \in S^1$. The map v is a 90° rotation.

See Exercise 1 for a similar application.
In the next proposition we derive a property of S^n with unit vector fields.

1.3 Proposition. *Let S^n be a sphere with a unit tangent vector field $v(x)$. Then the antipodal map $x \mapsto -x$ is homotopic to the identity.*

Proof. Let $h_t(x) = (\cos \pi t)x + (\sin \pi t)v(x)$. Then $h_t: S^n \to S^n$ is a homotopy with $h_0(x) = x$ and $h_1(x) = -x$. This deformation is on the great circle from x to $-x$ in the direction of $v(x)$.

The degree of $x \to -x$ as a map $S^n \to S^n$ is $(-1)^{n+1}$. From (1.2) and (1.3) and this remark we get the following equivalences.

1.4 Theorem. *For a sphere S^n the following are equivalent.*

(1) *n is odd.*
(2) *$x \mapsto -x$ is of degree 1.*
(3) *$x \mapsto -x$ is homotopic to the identity.*
(4) *S^n has a unit tangent vector field.*

By negation, the following statements are also equivalent properties of S^n.

(1) *n is even.*
(2) *$x \mapsto -x$ is of degree -1.*
(3) *$x \mapsto -x$ is not homotopic to the identity.*
(4) *S^n has no unit tangent vector field.*

2. Orthogonal Multiplications

In this section we see that orthogonal multiplications give rise to ortho-normal vector fields on a sphere. The problem of the existence of orthogonal multiplications is a purely algebraic problem.

2.1 Definition. A function $\mu: \mathbf{R}^k \times \mathbf{R}^n \to \mathbf{R}^n$ is called an orthogonal multiplication provided μ is bilinear and $\|\mu(y, x)\| = \|y\| \|x\|$ for each $y \in \mathbf{R}^k$ and $x \in \mathbf{R}^n$.

For $y \in S^{k-1} \subset \mathbf{R}^k$, the function $x \to \mu(y, x)$ is an orthogonal transformation, and for $x \in S^{n-1} \subset \mathbf{R}^n$ the function $y \to \mu(y, x)$ is an isometry; i.e., it is inner product preserving since $(x|y) = (1/2)(\|x + y\| - \|x\| - \|y\|)$.

An orthogonal multiplication μ is normalized provided $\mu(e_k, x) = x$ for each $x \in \mathbf{R}^n$, where $e_k = (0, \ldots, 0, 1)$. If $\mu(e_k, x) = u(x)$, then $\mu(y, u^{-1}(x))$ is a normalized orthonormal multiplication.

In the next theorem we can see how orthogonal multiplications can be used to define vector fields on spheres.

2.2 Theorem. *If there exists an orthogonal multiplication* $\mu: \mathbf{R}^k \times \mathbf{R}^n \to \mathbf{R}^n$, *there exist* $k - 1$ *orthonormal vector fields on* S^{n-1}.

Proof. By the above remark, we can assume that μ is normalized. Then, for each $x \in S^{n-1}$, the vectors $\mu(e_1, x), \ldots, \mu(e_{k-1}, x), x = \mu(e_k, x)$ are orthonormal. Then the $v_1(x) = \mu(e_i, x)$, where $1 \leq i \leq k - 1$ orthonormal vector fields on S^{n-1}.

2.3 Remarks. The existence of vector fields is closely related to the existence of orthogonal multiplications. Observe that $v_i(-x) = -v_i(x)$ in Theorem (2.2), and therefore $v_1(x), \ldots, v_{k-1}(x)$ define vector fields on RP^{n-1} which are orthonormal. Scalar multiplication $\mathbf{C} \times \mathbf{C}^n \to \mathbf{C}^n$ defines an orthogonal multiplication $\mathbf{R}^2 \times \mathbf{R}^{2n} \to \mathbf{R}^{2n}$. Corollary (1.2) follows also from (2.2).

In the next theorem we derive an algebraic version of the notion of orthogonal multiplication. This leads to the study of Clifford algebras in the following sections.

2.4 Theorem. *The set of normalized orthogonal multiplications* $\mu: \mathbf{R}^k \times \mathbf{R}^n \to \mathbf{R}^n$ *are in bijective correspondence with sets of* $u_1, \ldots, u_{k-1} \in O(n)$ *such that* $u_i^2 = -1$ *and* $u_i u_j + u_j u_i = 0$ *for* $i \neq j$. *The correspondence is achieved by defining* $u_i(x) = \mu(e_i, x)$ *for given* μ.

Proof. Clearly, normalized bilinear μ are in bijective correspondence with linear transformations $u_1, \ldots, u_{k-1}$. To check the orthogonality condition it is necessary to prove only that $\|\mu(y, x)\| = 1$ if $\|x\| = \|y\| = 1$. Let $u_k = 1$. The condition of orthogonality reduces to showing that $\sum_i a_i u_i \in O(n)$ if and only if $(a_1, \ldots, a_k) \in S^{k-1}$.

Recall for a linear $v: \mathbf{R}^n \to \mathbf{R}^n$ that $v \in O(n)$ if and only if $vv^* = 1$, where v^* is the transpose of v. The orthogonality property of μ is equivalent to

$$1 = \left(\sum_i a_i u_i \right) \left(\sum_j a_j u_j^* \right) = \sum_i a_i^2 u_i u_i^* + \sum_{i < j} a_i a_j (u_i u_j^* + u_j u_i^*)$$

Since $u_i u_i^* = 1$, since $\sum_i a_i^2 = 1$, and since (a_i) are arbitrary on S^{k-1}, the orthogonality property of μ is equivalent to the relation $u_i u_j^* + u_j u_i^* = 0$ for $i < j$. For $j = k$ we have $u_k = 1$ and $u_i = -u_i^* = -u_i^{-1}$. Therefore, we have $u_i^2 = -1$. For $i < j < k$, we have $u_i(-u_j) + u_j(-u_i) = 0$ or $u_i u_j + u_j u_i = 0$ for $i \neq j$. Conversely, these two conditions on the u_i's imply $u_i u_j^* + u_j u_i^* = 0$ for $i < j$. This proves the theorem.

In the general theory of Clifford algebras we shall derive properties of algebras which have a set of generators $e_1, \ldots, e_k$ and satisfy the relations $e_i^2 = -1$ and $e_i e_j + e_j e_i = 0$ for $i \neq j$.

If $u_1, \ldots, u_{k-1}$ are linear transformations of a real vector space M onto itself such that $u_i^2 = -1$ and $u_i u_j + u_j u_i = 0$ for $i \neq j$, there is an inner product $(x|y)$ on M such that $(u_i(x)|u_i(y)) = (x|y)$ for each $x, y \in M$.

In effect, let Γ_{k-1} be the group generated by $\pm u_i$ for $1 \leq i \leq k - 1$. An element of Γ_{k-1} is of the form $\pm u_{i(1)}, \ldots, u_{i(r)}$, where $i(1) < \cdots < i(r)$, and Γ_{k-1} is a group with 2^k elements. Let $\langle x|y \rangle$ be any inner product on M, and form $(x|y) = 2^{-k} \sum_{\sigma \in \Gamma_{k1}} \langle \sigma(x)|\sigma(y) \rangle$. Since $u_i \in \Gamma_{k-1}$, we have $(u_i(x)|u_i(y)) = (x|y)$ for each i.

2.5 Remark. The above discussion demonstrates that every module over a Clifford algebra arises from an orthogonal multiplication.

3. Generalities on Quadratic Forms

Let R denote a general commutative ring with 1. All modules M are unitary, that is, $1x = x$ for each $x \in M$. We use the following definition.

3.1 Definition. A quadratic form is a pair (M, f), where M is an R-module and $f: M \times M \to R$ is a symmetric bilinear form; i.e., the following relations hold.

(1) $f(ax + a'x', y) = af(x, y) + a'f(x', y)$ for $a, a' \in R$ and $x, x', y \in M$.
(2) $f(x, by + b'y') = bf(x, y) + b'f(x, y')$ for $b, b' \in R$ and $x, y, y' \in M$.
(3) $f(x, y) = f(y, x)$ for $x, y \in M$.

Much of the following discussion can be carried through for antisymmetric forms where axiom (3) is replaced by $f(x, y) = -f(y, x)$ for $x, y \in M$ and for sesquilinear forms over $\mathbf{C}$ where axiom (3) is replaced by $f(x, y) = \overline{f(y, x)}$ for all $x, y \in M$ and axiom (2) is replaced by $f(x, by) = \overline{b}f(x, y)$ for $x, y \in M$ and $b \in \mathbf{C}$.

3.2 Remark. There is a second definition of a quadratic form (M, Q), where $Q: M \to R$ is a function such that $Q(ax) = a^2 Q(x)$ and $f_Q(x, y) = Q(x + y) - Q(x) - Q(y)$ is a symmetric bilinear form. If $1/2 \in R$, these two notions are equivalent, and $f(x, y) = (1/2)(f(x + y, x + y) - f(x, x) - f(y, y))$.

Let $\langle x, u \rangle = u(x)$ be the canonical pairing $M \times M^+ \to R$, where M^+ is the dual of M. For each quadratic form (M, f) its correlation is the unique homomorphism $c_f: M \to M^+$ such that $\langle x, c_f(y) \rangle = f(x, y)$ for each $x, y \in M$. For sesquilinear forms, M^+ denotes the conjugate dual of conjugate linear functionals.

3.3 Definition. A quadratic form (M, f) is nondegenerate provided its correlation c_f is a monomorphism and is nonsingular provided c_f is an isomorphism.

A quadratic form (M, f) is nondegenerate if and only if $f(x, y) = 0$ for all $y \in M$ implies that $x = 0$. A quadratic form (M, f) is nonsingular if and only if for each linear form $u: M \to R$ there exists a unique $y \in M$ such that $u(x) = f(x, y)$ for all $x \in M$. The two notions are equivalent if R is a field.

If $e_1, \ldots, e_n$ is a basis of M, the elements $f(e_i, e_j)$ completely determine f. For $x = \sum_i x_i e_i$ and $y = \sum_j y_j e_j$, we have $f(x, y) = f(\sum x_i e_i, \sum y_j e_j) = \sum_{i,j} f(e_i e_j) x_i y_j$. Moreover, the symmetry of f is equivalent to $f(e_i, e_j) = f(e_j, e_i)$ for all i, j. When R is an integral domain, (M, f) is nondegenerate if and only if $\det[f(e_i, e_j)] \neq 0$, and it is nonsingular if and only if $\det[f(e_i, e_j)]$ is a unit.

3.4 Examples. On $\mathbf{R}^n$, n-dimensional euclidean space, $(x|y)$ and $-(x|y)$ are two very useful quadratic forms.

3.5 Example. If (M, f) is a quadratic form and if E is a subspace of M, then (E, f_E), where $f_E = f|E \times E$, is a quadratic form.

3.6 Definition. An orthogonal splitting of a quadratic form (M, f), denoted $M = E_1 \perp \cdots \perp E_r$, is a direct sum decomposition of $M = E_1 \oplus \cdots \oplus E_r$ such that $f(x, y) = 0$ for $x \in E_i$ and $y \in E_j$, where $i \neq j$, $1 \leq i, j \leq r$.

The following splitting result is very general and useful.

3.7 Proposition. *Let (M, f) be a quadratic form and E a subspace of M such that (E, f_E) is nonsingular. If E^* is the subspace of all $y \in M$ with $f(x, y) = 0$ for each $x \in E$, then $M = E \perp E^*$ is an orthogonal splitting. If M is nonsingular, then E^* is nonsingular.*

Proof. It suffices to show $M = E \oplus E^*$. Since E is nonsingular, $f(E, y) = 0$ for $y \in E$ implies $y = 0$, and consequently we have $E \cap E^* = 0$. If $x \in M$, then $y \mapsto f(x, y)$ is a linear functional on E. By nonsingularity it is of the form $f(x_1, y)$, where $x_1 \in E$. Then $f(x, y) = f(x_1, y)$ or $f(x - x_1, y) = 0$ for each $y \in E$. Therefore, $x - x_1 = x_2 \in E^*$ and $M = E \oplus E^*$.

For the second statement, let $u: E^* \to R$ be a linear functional. We prolong u to M by $u(E) = 0$. There exists a unique $y \in M$ with $f(x, y) = \langle x, u \rangle$. Since $f(x, y) = 0$ for $x \in E$, we have $y \in E^*$. This proves the proposition. □

3.8 Example. Let (M, f) be a quadratic form, where $M = Re$ has a single basic element. We have $f(xe, ye) = f(e, e)xy$. Then f is nondegenerate if and only if $f(e, e)$ is not a zero divisor and nonsingular if and only if $f(e, e)$ is a unit. In the next theorem, we consider a case where a quadratic form is an orthogonal sum of one-dimensional forms.

3.9 Theorem. *Let (M, f) be a quadratic form over a field R with $2 \neq 0$. Then $M = E_1 \perp \cdots \perp E_r$, where E_i is one-dimensional for $1 \leq i \leq r$.*

Proof. If $f(x,y) = 0$ for all $y \in M$, where $x \neq 0$, let L be a supplementary subspace to x in M. Then $M = L \perp Rx$. If no such x exists, M is nonsingular and $f(x,x) \neq 0$ for some x; otherwise $f(x+y, x+y) = f(x,x) + 2f(x,y) + f(y,y) = 0$ would hold for all $x, y \in M$, and f would be zero. If $f(x,x) \neq 0$, then Rx is nonsingular and $M = L \perp Rx$, where $L = Rx^*$ by (3.7). Now the inductive hypothesis is applied to L.

In terms of bases, Theorem (3.9) says that there is a basis $e_1, \ldots, e_r$ of M such that $f(e_i, e_j) = 0$ for $i \neq j$. If $R = \mathbf{C}$, the complex numbers, we replace e_i by e_i divided by $f(e_i, e_i)^{1/2}$ when $f(e_i, e_i) \neq 0$. If $R = \mathbf{R}$, the field of real numbers, we replace e_i by e_i divided by $f(e_i, e_i)^{1/2}$ for $f(e_i, e_i) > 0$ and by e_i divided by $(-f(e_i, e_i))^{1/2}$ for $f(e_i, e_i) < 0$. In summary, we have the following result.

3.10 Theorem. *Every quadratic form (M,f) over $\mathbf{C}$ has a basis $e_1, \ldots, e_n \in M$ such that with respect to this basis $f(x,y) = x_1 y_1 + \cdots + x_r y_r$, where $x = x_1 e_1 + \cdots + x_n e_n$ and $y = y_1 e_1 + \cdots + y_n e_n$. Every quadratic form (M,f) over $\mathbf{R}$ has a basis $e_1, \ldots, e_n \in M$ such that $f(x,y) = -x_1 y_1 - \cdots - x_k y_k + x_{k+1} y_{k+1} + \cdots + x_r y_r$, where k is the dimension of the largest subspace E such that $f(x,x) < 0$ for $x \in E$, $x \neq 0$.*

The integer k is called the index of f. The index is independent of the basis $e_1, \ldots, e_n$ that is used. This is an exercise.

4. Clifford Algebra of a Quadratic Form

For a quadratic form (M,f), we consider linear functions $u: M \to A$, where A is an R-algebra such that $u(x)^2 = f(x,x)1$. The Clifford algebra is universal with respect to such linear functions.

4.1 Definition. The Clifford algebra of a quadratic form (M,f) is a pair $(C(f), \theta)$, where $C(f)$ is an R-algebra, and $\theta: M \to C(f)$ is a linear function such that $\theta(x)^2 = f(x,x)1$ for each $x \in M$. We assume the following universal property: For all linear functions $u: M \to A$ with $u(x)^2 = f(x,x)1$ there exists an algebra morphism $u': C(f) \to A$ such that $u'\theta = u$ and u' is unique with respect to this property.

4.2 Theorem. *A Clifford algebra $(C(f), \theta)$ exists for each quadratic form (M,f). If $(C(f), \theta)$ and $(C(f)', \theta')$ are two Clifford algebras, there is an algebra morphism $u: C(f) \to C(f)'$ such that $\theta' = u\theta$. Moreover, u is an isomorphism and is unique.*

Proof. For the first part, let $C(f)$ be the tensor algebra $T(M) = \sum_{0 \leq k} T^k(M)$ modulo the ideal generated by $x \otimes x - f(x,x)1$. Here $T^k(M)$ is the k-fold tensor product of M. Let θ be the composition of the injection $M \to T^1(M) \subset T(M)$ and the projection $T(M) \to C(f)$. If $u: M \to A$ is a linear function with $u(x)^2 = f(x,x)1$, then u factors as $M \to T(M) \overset{u''}{\to} A$ and as $M \overset{\theta}{\to} C(f) \overset{u'}{\to} A$. Moreover, u' is unique because im θ generates $C(f)$.

Finally, the uniqueness of $(C(f), \theta)$ follows from the existence of two algebra morphisms $u: C(f) \to C(f)'$ and $v: C(f)' \to C(f)$ with $\theta' = u\theta$ and $\theta = v\theta'$. This implies that $\theta = vu\theta$ and $\theta' = uv\theta'$. From the uniqueness property of factorizations, $1 = vu$ and $1 = uv$. This proves the theorem.

4.3 The Z_2-Grading of $C(f)$. The algebra morphism $-\theta: M \to C(f)$ determines an involution $\beta: C(f) \to C(f)$ with $\beta\theta(x) = -\theta(x)$. Usually, we write $\beta(x) = \bar{x}$. Then we denote the subalgebra of $x \in C(f)$ with $\bar{x} = x$ by $C(f)^0$ and the submodule of $x \in C(f)$ with $\bar{x} = -x$ by $C(f)^1$. Then $C(f)^0$ is the image of $\sum_{0 \leq k} T^{2k}(M)$ in $C(f)$, and $C(f)^1$ is the image of $\sum_{0 \leq k} T^{2k+1}(M)$ in $C(f)$. Finally, we have $C(f) = C(f)^0 \oplus C(f)^1$, which is a Z_2-grading.

4.4 Z_2-Graded Algebras. An algebra A is Z_2-graded provided $A = A^0 \oplus A^1$, where $Z_2 = \{0, 1\}$, and $A^i A^j \subset A^{i+j}$. A morphism $f: A \to B$ between two Z_2-graded algebras is an algebra morphism such that $f(A^i) \subset B^i$. The tensor product of two Z_2-graded algebras A and B, denoted $A \hat{\otimes} B$, is the tensor product of the underlying modules with $(A^0 \otimes B^0) \oplus (A^1 \otimes B^1) = (A \hat{\otimes} B)^0$ and $(A^1 \otimes B^0) \oplus (A^0 \otimes B^1) = (A \hat{\otimes} B)^1$ and multiplication given by $(x \hat{\otimes} y)(x' \hat{\otimes} y') = (-1)^{ij}(xx') \hat{\otimes} (yy')$ for $x' \in A^i$ and $y \in B^j$. Finally, two elements $x \in A^i$ and $y \in A^j$ commute (in the Z_2-graded sense) provided $xy = (-1)^{ij}yx$, and, in general, two elements $x, y \in A$ commute provided their homogeneous components commute. Note that $C(f)$ is not graded commutative.

4.5 Example. Let $u: M \to N$ be a morphism where (M, f) and (N, g) are quadratic forms such that $f(x, x') = g(u(x), u(x'))$ for each pair $x, x' \in M$. Then u defines morphism $C(u): C(f) \to C(g)$ of Z_2-graded algebras.

4.6 Proposition. *Let $f: A \to C$ and $g: B \to C$ be two Z_2-algebra morphisms of Z_2-graded algebras such that $f(x)$ commutes with $g(y)$ for each $x \in A, y \in B$. Then the module homomorphism $h: A \otimes B \to C$ defined by $h(x \otimes y) = f(x)g(y)$ is a morphism of Z_2-graded algebras.*

Proof. Since $(x, y) \to f(x)g(y)$ is a bilinear map $A \times B \to C$, h is defined as a Z_2-graded module homomorphism. Let $x \otimes y_i \in A \otimes B^i$ and $x_j \otimes y \in A^j \otimes B$, and compute $h((x \otimes y_i)(x_j \otimes y)) = (-1)^{ij}h(((xx_j) \otimes (y_iy))) = (-1)^{ij}f(x)f(x_j)g(y_i)g(y) = f(x)g(y_i)f(x_j)g(y) = h(x \otimes y_i)h(x_j \otimes y)$. This proves the proposition.

Let (M, f) be a quadratic form, and let $M = E_1 \perp E_2$. We denote the restriction $f|E_i$ by f_i for $i = 1$ or 2. Let $(C(f_i), \theta_i)$ be the Clifford algebra for (E_i, f_i) and define $\phi: M \to C(f_1) \hat{\otimes} C(f_2)$ by the relation $\phi(x_1, x_2) = (\theta_1(x_1) \hat{\otimes} 1) + (1 \hat{\otimes} \theta_2(x_2))$. Now we calculate $\phi(x_1, x_2)^2 = \theta_1(x_1)^2 \hat{\otimes} 1 + \theta_1(x_1) \hat{\otimes} \theta_2(x_2) - \theta_1(x_1) \hat{\otimes} \theta_2(x_2) + 1 \hat{\otimes} \theta_2(x_2)^2 = \theta_1(x_1)^2 \hat{\otimes} 1 + 1 \hat{\otimes} \theta_2(x_2)^2 = f_1(x_1, x_1) + f_2(x_2, x_2) = f((x_1, x_2), (x_1, x_2))$. Therefore, there exists an algebra morphism $u: C(f) \to C(f_1) \hat{\otimes} C(f_2)$ such that the following diagram is commutative.

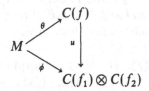

4.7 Theorem. *The above algebra morphism u is an isomorphism.*

Proof. We construct $v: C(f_1) \hat{\otimes} C(f_2) \to C(f)$, an inverse of u. Let $q_1: E_i \to M$ be the inclusion morphisms $q_1(x_1) = (x_1, 0)$ and $q_2(x_2) = (0, x_2)$ which define $\mathbb{Z}_2$-graded algebra morphisms $C(q_i): C(f_i) \to C(f)$ for $i = 1, 2$. If $f(z, z') = 0$ in M, we have $\theta(z)\theta(z') + \theta(z')\theta(z) = 0$, since $(\theta(z) + \theta(z'))^2 = f(z + z', z + z') = f(z, z) + f(z', z') = \theta(z)^2 + \theta(z')^2$. Therefore, we have $C(q_1)(x_1)C(q_2)(x_2) = -C(q_2)(x_2)C(q_1)(x_1)$ for $x_i = \theta_i(y_i)$ with $y_i \in E_i$ and $i = 1, 2$, and the hypothesis of Proposition (4.6) is satisfied on the generators im θ_1 of $C(f_1)$ and im θ_2 of $C(f_2)$. Therefore, there is a $\mathbb{Z}_2$-graded algebra morphism of $v: C(f_1) \hat{\otimes} C(f_2) \to C(f)$ such that $v(x_1 \hat{\otimes} x_2) = C(f_1)(x_1)C(f_2)(x_2)$.

The elements of the form $\theta(x_1, 0)$ and $\theta(0, x_2)$ generate $C(f)$. For these elements, we have $vu\theta(x_1, 0) = v\phi(x_1, 0) = v(\theta_1(x_1)) = \theta(x_1, 0)$ and similarly $vu\theta(0, x_2) = \theta(0, x_2)$. Consequently, the relation $vu = 1$ holds. The elements of the form $\theta_1(x_1) \hat{\otimes} 1$ and $1 \hat{\otimes} \theta_2(x_2)$ generate the tensor product $C(f_1) \hat{\otimes} C(f_2)$. We compute $uv(\theta_1(x_1) \hat{\otimes} 1) = u\theta(x_1, 0) = \phi(x_1, 0) = \theta_1(x_1) \hat{\otimes} 1$, and similarly we have $uv(1 \hat{\otimes} \theta_2(x_2)) = 1 \hat{\otimes} \theta_2(x_2)$. Therefore, v is the inverse of u, and u is an isomorphism.

Let (M, f) be a quadratic form, and let $M = E_1 \perp \cdots \perp E_r$.

4.8. Corollary. *There is an isomorphism $u: C(f) \to C(f_1) \hat{\otimes} \cdots \hat{\otimes} C(f_r)$ such that $u(\theta(0, \ldots, 0, x_j, 0, \ldots, 0)) = 1 \hat{\otimes} \cdots \hat{\otimes} \theta_j(x_j) \hat{\otimes} \cdots \hat{\otimes} 1$ for $1 < j < r$.*

5. Calculations of Clifford Algebras

5.1 Proposition. *Let (M, f) be a quadratic form, where M has one basis element e and $a = f(e, e)$. Then $C(f) = R1 \oplus Re$ with the relation $e^2 = a$.*

Proof. The tensor algebra is freely generated by e, and the relation $e^2 = a$ holds in the Clifford algebra.

The reader can easily verify the universal property for $C(f) = R1 \oplus Re$. Note that $\dim_R C(f) = 2$. Moreover, $C(f)^0 = R1$ and $C(f)^1 = Re$.

5.2 Theorem. *Let (M, f) be a quadratic form where M has a basis $e_1, \ldots, e_r$ with $f(e_i, e_j) = 0$ for $i \neq j$ and $a_i = f(e_i, e_i)$. Then $C(f)$ is generated by $e_1, \ldots, e_r$ with relations $e_i^2 = a_i$ and $e_i e_j + e_j e_i = 0$ for $i \neq j$. The elements $e_{i(1)} \cdots e_{i(s)}$, where $i(1) < \cdots < i(s)$ and $1 \leq s \leq r$, together with 1 form a base of $C(f)$. The dimension of $C(f)$ is 2^r.*

Proof. The first statement is immediate, and the relation $e_i^2 = a_i$ clearly holds. As in the proof of (4.7), we have $e_i e_j + e_j e_i = 0$ since $f(e_i, e_j) = 0$. From the isomorphism $u: C(f) \to C(f_1) \hat{\otimes} \cdots \hat{\otimes} C(f_r)$, we have $u(e_{i(1)} \cdots e_{i(s)}) = x_1 \hat{\otimes} \cdots \hat{\otimes} x_r$, where $x_i = e_{i(j)}$ for $i = i(j)$ and $x_i = 1$ for $i \neq i(j)$, all j with $1 \leq j \leq s$. The elements of the form $u(e_{i(1)} \cdots e_{i(s)})$ and 1 form a base of $C(f_1) \hat{\otimes} \cdots \hat{\otimes} C(f_r)$. This proves the theorem.

5.3 Notation. Let C_k denote $C(-(x|y))$, where $-(x|y)$ is a form on $\mathbf{R}^k$, and let C_k' denote $C((x|y))$, where $(x|y)$ is a form on $\mathbf{R}^k$. There are two real algebras of dimension 2^k. Let C_k^c be the complexification $C_k \otimes \mathbf{C} = C_k' \otimes \mathbf{C}$. Then $C_k^c = C(-(z|w)) = C((z|w))$, where $-(z|w)$ is a form on $\mathbf{C}^k$.

We wish to calculate the algebras C_k, C_k', and C_k^c. The next proposition is the first step. We use the notation $F(n)$ for the algebra of $n \times n$ matrices with coefficients in F.

5.4 Proposition. *As algebras over $\mathbf{R}$, there are the relations $C_1 \cong \mathbf{C}$, $C_2 \cong \mathbf{H}$, $C_1' \cong \mathbf{R} \oplus \mathbf{R}$, and $C_2' \cong \mathbf{R}(2)$, and as algebras over $\mathbf{C}$, there are $C_1^c \cong \mathbf{C} \oplus \mathbf{C}$ and $C_2^c \cong \mathbf{C}(2)$.*

Proof. First, C_1 is two-dimensional over $\mathbf{R}$ with basis elements 1 and e, where $e^2 = -1$, and this is $\mathbf{C}$. Also, C_2 is four-dimensional over $\mathbf{R}$ with basis elements 1, e_1, e_2 and $e_1 e_2$, where $e_1^2 = e_2^2 = -1$ and $e_1 e_2 = -e_2 e_1$. If we map 1 to 1, e_1 to i, e_2 to j, and $e_1 e_2$ to k and prolong to $C_2 \to \mathbf{H}$, we get an algebra isomorphism.

For C_1' there are two basis elements 1 and e, where $e^2 = 1$. If we map $1 \to (1, 1)$ and $e \to (1, -1)$ and prolong to $C_1' \to \mathbf{R} \oplus \mathbf{R}$, we get an algebra isomorphism. For C_2' there are four basis elements 1, e_1, e_2 and $e_1 e_2$, where $e_1^2 = e_2^2 = 1$ and $e_1 e_2 = -e_2 e_1$. If we map

$$e_1 \to \begin{bmatrix} 1 & 0 \\ 0 & -1 \end{bmatrix} \quad \text{and} \quad e_2 \to \begin{bmatrix} 0 & 1 \\ 1 & 0 \end{bmatrix}$$

and prolong by linearity, we get an algebra isomorphism $C_2' \to \mathbf{R}(2)$.

For the last statement we use the relations $F \otimes \mathbf{R}(n) = F(n)$ for $F = \mathbf{C}$ or $\mathbf{H}$ and $F \otimes (\mathbf{R} \oplus \mathbf{R}) = F \oplus F$.

The next calculations will be useful.

5.5 Proposition. *There are isomorphisms* $\mathbf{C} \underset{\mathbf{R}}{\bigotimes} \mathbf{C} \cong \mathbf{C} \oplus \mathbf{C}$, $\mathbf{C} \underset{\mathbf{R}}{\bigotimes} \mathbf{H} \cong \mathbf{C}(2)$, *and* $\mathbf{H} \underset{\mathbf{R}}{\bigotimes} \mathbf{H} \cong \mathbf{R}(4)$.

Proof. The first two follow from $\mathbf{C} \otimes C_k \cong \mathbf{C} \otimes C_k'$ for $k = 1$, 2 in Proposition (5.4). For the third, we define an isomorphism $w: \mathbf{H} \underset{\mathbf{R}}{\bigotimes} \mathbf{H} \to \mathrm{Hom}_\mathbf{R}(\mathbf{H}, \mathbf{H}) = \mathbf{R}(4)$ by the relation $w(x_1 \otimes x_2)x = x_1 x \bar{x}_2$. Then w is an algebra morphism, and it suffices to show w is epimorphic since $\dim \mathbf{H} \underset{\mathbf{R}}{\bigotimes} \mathbf{H} = 16 = \dim \mathbf{R}(4)$. First, note that $w(1 \otimes 1) = 1$ and $w(i \otimes i)1 = 1$, $w(i \otimes i)i = i$, $w(i \otimes i)j = -j$, and $w(i \otimes i)k = -k$. Similar relations hold for $w(j \otimes j)$ and $w(k \otimes k)$. Consequently, $w((1 \otimes 1 + i \otimes i + j \otimes j + k \otimes k)/4)$ projects 1 on 1 and i, j, k onto 0. Moreover, we calculate $w(i \otimes j)\, 1 = k$, $w(i \otimes j)i = j$, $w(i \otimes j)j = i$, $w(i \otimes j)k = -1$, and a similar calculation holds for $w(j \otimes k)$ and $w(i \otimes k)$. Consequently, every matrix with only one nonzero entry is in im w. These generate $\mathbf{R}(4)$, and therefore w is epimorphic.

The following periodicity result is basic for the calculation of C_k and C_k'.

5.6 Theorem. *There exist isomorphisms* $u: C_{k+2} \to C_k' \otimes C_2$ *and* $v: C_{k+2}' \to C_k \otimes C_2'$.

Proof. Let $e_1, \ldots, e_k$ be the basic generators of C_k and $e_1', \ldots, e_k'$ for C_k'. We define $u': \mathbf{R}^{k+2} \to C_k' \otimes C_2$ by $u'(e_i) = 1 \otimes e_i$ for $1 \leq i \leq 2$, and $u'(e_i) = e_{i-2}' \otimes e_1 e_2$ for $3 \leq i \leq k + 2$. We calculate $u'(e_i)^2 = (1 \otimes e_i)(1 \otimes e_i) = 1 \otimes e_i^2 = -1$ for $i \leq 2$ and $u'(e_i)^2 = e_{i-2}'^2 \otimes e_1 e_2 e_1 e_2 = 1 \otimes 1(-1) = -1$. Also note that $u'(e_i)u'(e_j) + u'(e_j)u'(e_i) = 0$ for $i \neq j$. Therefore, u' prolongs to $u: C_{k+2} \to C_k' \otimes C_2$. Since u carries distinct basis elements into distinct basis elements, u is injective. For reasons of dimension it is an isomorphism.

For v we require $v(e_i') = 1 \otimes e_i'$ for $1 \leq i \leq 2$ and $v(e_2') = e_{i-2} \otimes e_1' e_2'$ for $3 \leq i \leq k + 2$. Then, as in the previous case, v is an isomorphism. This proves the theorem. $\square$

5.7 Corollary. *There are isomorphisms* $C_{k+4} \to C_k \underset{\mathbf{R}}{\bigotimes} C_4$ *and* $C_{k+4}' \to C_k' \underset{\mathbf{R}}{\bigotimes} C_4'$, *where* $C_4 \cong C_4' \cong C_2 \otimes C_2' \cong \mathbf{H}(2)$.

Proof. Let $k = 2$ in (5.6); then there are isomorphisms $C_4 \cong C_2 \otimes C_2' \cong C_4'$ and $C_2 \otimes C_2' = \mathbf{H} \otimes \mathbf{R}(2) = \mathbf{H}(2)$. For the first part, we have isomorphisms $C_{k+4} \to C_{k+2}' \otimes C_2 \to C_k \otimes (C_2' \otimes C_2) = C_k \otimes C_4$ and similarly $C_{k+4}' \to C_{k+2} \otimes C_2' \to C_k' \otimes C_4'$.

5.8 Corollary. *There are isomorphisms* $C_{k+8} \to C_k \otimes \mathbf{R}(16)$ *and* $C'_{k+8} \to C'_k \otimes$ $\mathbf{R}(16)$.

Proof. We iterate (5.7) and use $\mathbf{H}(2) \otimes \mathbf{H}(2) = \mathbf{H} \otimes \mathbf{H} \otimes \mathbf{R}(2) \otimes \mathbf{R}(2) \cong$ $\mathbf{R}(4) \otimes \mathbf{R}(4) \cong \mathbf{R}(16)$.

5.9 Corollary. *There is an isomorphism* $C^c_{k+2} \to C^c_k \underset{C}{\otimes} \mathbf{C}(2)$.

Proof. From (5.6) there is an isomorphism $C_{k+2} \underset{R}{\otimes} \mathbf{C} = C^c_{k+2} \to$

$$C'_k \underset{R}{\otimes} \mathbf{C} \underset{R}{\otimes} C_2 = \left(C'_k \underset{R}{\otimes} \mathbf{C} \right) \underset{C}{\otimes} \left(C_2 \underset{R}{\otimes} \mathbf{C} \right) = C^c_k \underset{C}{\otimes} \mathbf{C}(2). \qquad \square$$

5.10. Table of Clifford algebras.

k	C_k	C'_k	C^c_k
1	$\mathbf{C}$	$\mathbf{R} \oplus \mathbf{R}$	$\mathbf{C} \oplus \mathbf{C}$
2	$\mathbf{H}$	$\mathbf{R}(2)$	$\mathbf{C}(2)$
3	$\mathbf{H} \oplus \mathbf{H}$	$\mathbf{C}(2)$	$\mathbf{C}(2) \oplus \mathbf{C}(2)$
4	$\mathbf{H}(2)$	$\mathbf{H}(2)$	$\mathbf{C}(4)$
5	$\mathbf{C}(4)$	$\mathbf{H}(2) \oplus \mathbf{H}(2)$	$\mathbf{C}(4) \oplus \mathbf{C}(4)$
6	$\mathbf{R}(8)$	$\mathbf{H}(4)$	$\mathbf{C}(8)$
7	$\mathbf{R}(8) \oplus \mathbf{R}(8)$	$\mathbf{C}(8)$	$\mathbf{C}(8) \oplus \mathbf{C}(8)$
8	$\mathbf{R}(16)$	$\mathbf{R}(16)$	$\mathbf{C}(16)$

Our success in calculating the algebras C_k, C'_k, and C^c_k depended on the fact that in certain cases the graded tensor product could be replaced by the ordinary tensor product, for example, $C_k \oplus C_4 = C_{k+4} = C_k \oplus C_4$ and $C^c_k \otimes C^c_2 = C^c_{k+2}$.

6. Clifford Modules

6.1 Definition. A Z_2-graded module M over a Z_2-graded algebra A is an A-module M with $M = M^0 \oplus M^1$ such that $A^i M^j \subset M^{i+j}$ for $i, j \in Z_2$. A Clifford module is a Z_2-graded module M over a Clifford algebra $C(f)$.

We are particularly interested in Clifford modules over the algebras C_k and C^c_k. In the next two propositions we reduce the study of Clifford modules to ordinary (ungraded) modules.

6.2 Proposition. *The function* $\phi(x) = e_0 \otimes x$ *for* $x \in C(f)$ *prolongs to an (un-graded) algebra isomorphism* $\phi: C(f) \to C^0(-xy \oplus f)$, *where* $C^0(-xy \oplus f) \subset$

$C(-xy \oplus f) = C(-xy) \hat{\otimes} C(f)$ and e_0 is the generator of $C(-xy)$ such that $e_0^2 = -1$.

Proof. Let $e_1, \ldots, e_n$ be a basis of M for the form (M, f) with $f(e_i, e_j) = 0$ for $i \neq j$. We calculate $\phi(e_i)^2 = (e_0 \otimes e_i)(e_0 \otimes e_i) = -e_0^2 \otimes e_i^2 = e_i^2$. Therefore, the prolongation exists and is a monomorphism since distinct basis elements are carried into distinct basis elements. Since dim $C(f) = $ dim $C^0(-xy + f)$, ϕ is an isomorphism.

A direct picture of the isomorphism $\phi: C_{k-1} \to C_k^0$ is given by the formula $\phi(x_0 + x_1) = x_0 + e_k x_1$ for $x_0 + x_1 \in C_{k-1}^0 \oplus C_{k-1}^1$. Here ϕ is a vector space isomorphism, and the multiplicative character follows from the relation $\phi(x_0 + x_1)\phi(y_0 + y_1) = (x_0 + e_k x_1)(y_0 + e_k y_1) = x_0 y_0 + e_k x_1 e_k y_1 + e_k(x_1 y_0 + x_0 y_1) = (x_0 y_0 + x_1 y_1) + e_k(x_1 y_0 + x_0 y_1) = \phi((x_0 + x_1)(y_0 + y_1))$.

Let A be a Z_2-graded algebra. Let $M(A)$ denote the free abelian group with irreducible Z_2-graded A-modules (e.g., modules with no submodules) as free generators, and let $N(A)$ denote the free abelian group with irreducible A-modules as free generators.

6.3 Proposition. *Let (M, f) be a quadratic form. The functor R, which assigns to each graded $C(f)$ module $M = M^0 + M^1$ the $C(f)^0$ module M^0, induces a group isomorphism $M(C(f)) \to N(C(f)^0)$.*

Proof. Let S be the functor that assigns to each $C(f)$ module L the graded $C(f)^0$-module $C(f) \underset{C(f)^0}{\bigotimes} L$.

An isomorphism $SR \to 1$ of functors is given by the scalar multiplication $C(f) \underset{C(f)^0}{\bigotimes} M^0 \to M$ in the module, and an isomorphism $1 \to RS$ is given by $L \to 1 \otimes L$. This proves the proposition.

Let M_k denote $M(C_k)$, M_k^c denote $M(C_k^c)$, N_k denote $N(C_k)$, and N_k^c denote $N(C_k^c)$. Using (6.1) and (6.3), we have group isomorphisms $M(C_k) \to N(C_k^0) \to N(C_{k-1})$ and $M(C_k^c) \to N(C_k^{c0}) \to N(C_{k-1}^c)$. In summary we have the next proposition.

6.4 Proposition. *There are group isomorphisms $M_k \to N_{k-1}$ and $M_k^c \to N_{k-1}^c$.*

From general properties of matrix algebras we know the irreducible modules over $F(n)$ and $F(n) \oplus F(n)$. In the case of $F(n)$, there is only one, namely, the action of $F(n)$ on F^n, and its dimension (over F) is n. In the case of $F(n) \oplus F(n)$, there are two, namely, the two projections $F(n) \oplus F(n) \to F(n)$ followed by the action of $F(n)$ on F^n, and both have dimension n (over F).

Finally, if a_k denotes the dimension of M^0 over $\mathbf{R}$, where M is an irreducible Z_2-graded C_k-module, and a_k^c of M^0 over $\mathbf{C}$, where M is an irreducible Z_2-graded C_k^c-module, then we have Table 6.5.

Table 6.5. Table of Clifford modules.

k	C_k	N_k	M_k	a_k	C_k^c	N_k^c	M_k^c	a_k^c
1	**C**	**Z**	**Z**	1	**C**⊕**C**	**Z**⊕**Z**	**Z**	1
2	**H**	**Z**	**Z**	2	**C**(2)	**Z**	**Z**⊕**Z**	1
3	**H**⊕**H**	**Z**⊕**Z**	**Z**	4	**C**(2)⊕**C**(2)	**Z**⊕**Z**	**Z**	2
4	**H**(2)	**Z**	**Z**⊕**Z**	4	**C**(4)	**Z**	**Z**⊕**Z**	2
5	**C**(4)	**Z**	**Z**	8	**C**(4)⊕**C**(4)	**Z**⊕**Z**	**Z**	4
6	**R**(8)	**Z**	**Z**	8	**C**(8)	**Z**	**Z**⊕**Z**	4
7	**R**(8)⊕**R**(8)	**Z**⊕**Z**	**Z**	8	**C**(8)⊕**C**(8)	**Z**⊕**Z**	**Z**	8
8	**R**(16)	**Z**	**Z**⊕**Z**	8	**C**(16)	**Z**	**Z**⊕**Z**	8

Then $N_{k+8} \cong N_k$, $M_{k+8} \cong M_k$, $a_{k+8} = 16a_k$, $N_{k+2}^c \cong N_k^c$, $M_{k+8}^c \cong M_k^c$, and $a_{k+2}^c = 2a_k^c$.

For the next calculation we need the center of C_k, that is, the set of elements commuting with every element of C_k.

6.6 Proposition. *The center of C_k is* $\mathbf{R}1$ *for* $k = 2r$ *and* $\mathbf{R}1 + \mathbf{R}\, e_1 \cdots e_k$ *for* $k = 2r + 1$. *The center of C_k^0 is* $\mathbf{R}1$ *for* $k = 2r - 1$ *and* $\mathbf{R}1 + \mathbf{R}\, e_1 \cdots e_k$ *for* $k = 2r$.

Proof. An element $x \in C_k$ is in the center of C_k if and only if $e_i x = x e_i$ for each i, $1 \leqq i \leqq k$. Since $e_i^{-1} = -e_i$, we must have $e_i x e_i = -x$. Let

$$x = \sum a_{i(1),\ldots,i(k)} e_1^{i(1)} \cdots e_k^{i(k)}$$

Then we have

$$e_s x e_s^{-1} = \sum (-1)^{i(s)} a_{i(1),\ldots,i(k)} e_1^{i(1)} \cdots e_k^{i(k)} + \sum (-1)^{i(s)+1} a_{i(1),\ldots,i(k)} e_1^{i(1)} \cdots e_k^{i(k)}$$

where the first sum is for indices with $i(1) + \cdots + i(k)$ even and the second sum is for $i(1) + \cdots + i(k)$ odd. Therefore, $e_s x e_s^{-1} = x$ for each s, $1 \leqq s \leqq k$, if and only if $a_{i(1),\ldots,i(k)} = 0$ for $(i(1), \ldots, i(k)) \neq (0,\ldots,0)$ and k even and $a_{i(1),\ldots,i(k)} = 0$ for $(i(1),\ldots,i(k)) \neq (0,\ldots,0)$ and $(1,\ldots,1)$ and k odd. For the statement about C_k^0 we use in addition the isomorphism $C_k \to C_{k+1}^0$ of (6.2). This proves the proposition.

6.7 Confunctor Properties of M and N. Let $u\colon A \to B$ be an algebra morphism of semisimple algebras; i.e., all modules over these algebras are direct sums of irreducible modules. Then we define $u^*\colon N(B) \to N(A)$ by requiring, for each B-module L, the scalar multiplication of the A-module $u^*(L)$ to be given by the relation that ax in $u^*(L)$ is $u(a)x$ in L. The rest of the structure of L is unchanged. If, in addition, $u\colon A \to B$ is a Z_2-graded morphism, then $u^*\colon M(B) \to M(A)$ is defined as above.

6.8 Examples. We consider the following isomorphisms for a quadratic form (M, f).

(1) For $b \in M - \{0\} \subset C(f)$, $\alpha_b \colon C(f) \to C(f)$ is the inner automorphism $\alpha_b(a) = bab^{-1}$, and its restriction to $C(f)^0$, $\alpha_b | C(f)^0$, is denoted α_b^0.

(2) We have $\beta \colon C(f) \to C(f)$, where $\beta(x_0 + x_1) = x_0 - x_1$ for $x_0 \in C(f)^0$ and $x_1 \in C(f)^1$.

(3) By (6.2) there is an isomorphism $\phi \colon C_k \to C_{k+1}^0$ given by $\phi(x_0 + x_1) = x_0 + e_{k+1} x_1$ for $x_0 \in C_k^0$ and $x_1 \in C_k^1$.

Finally, we have an involution c of $M(C(f))$, where $c(M)^0 = M^1$ and $c(M)^1 = M^0$.

6.9 Proposition. *The following diagrams of isomorphisms are commutative, using the above notations.*

$$
\begin{array}{ccccc}
M_{k+1} \xleftrightarrow{\ 1\ } M_{k+1} \xrightarrow{\ R\ } N(C_{k+1}^0) & & N(C_{k+1}^0) \xrightarrow{\ \phi^*\ } N_k \\
c\downarrow \quad\quad \alpha_b^*\downarrow \quad\quad \alpha_b^{0*}\downarrow & & \alpha_{e_{k+1}}^{0*}\downarrow \quad\quad \beta^*\downarrow \\
M_{k+1} \xleftrightarrow{\ 1\ } M_{k+1} \xrightarrow{\ R\ } N(C_{k+1}^0) & & N(C_{k+1}^0) \xrightarrow{\ \phi^*\ } N_k
\end{array}
$$

Proof. For the first square, observe that multiplication by b is an isomorphism $M^0 \to M^1$ and $M^1 \to M^0$. The second square is commutative by the definition of α_b^0 from α_b. For the last square we compute $\alpha_{e_{k+1}} \phi(x_0 + x_1) = e_{k+1}(x_0 + e_{k+1} x_1)(-e_{k+1}) = -e_{k+1}^2 x_0 + e_{k+1}^3 x_1 = x_0 - e_{k+1} x_1 = \beta\phi(x_0 + x_1)$. Therefore, $\phi^* \beta^* = \alpha_{e_{k+1}}^* \phi^*$. This proves the proposition.

6.10 Proposition. *Let m_1 and m_2 correspond to the two distinct irreducible graded modules in M_{4m} or M_{2m}^c. Then $c(m_1) = m_2$ and $c(m_2) = m_1$.*

Proof. By (6.9), the action of c on M_{4m} corresponds to the action of β^* on N_{4m-1}. By (6.6), the center of C_{4m-1} has a basis 1 and $\omega = e_1 \cdots e_{4m-1}$, where $\omega^2 = 1$. Then multiplication by $(1 + \omega)/2$ and by $(1 - \omega)2$ is a projection of C_{4m-1} on two ideals which are direct summands. Since $\beta(\omega) = \omega$, the automorphism interchanges these two ideals, and β^* interchanges the two irreducible C_{4m-1} modules. The above argument applies to C_{2m-1}^c with $\omega = i^m e_1 \cdots e_{2m-1}$.

The results in (6.6) to (6.10) are preliminary to the calculation of $L_k = \operatorname{coker}(i^* \colon M_{k+1} \to M_k)$, where $i^* \colon M_{k+1} \to M_k$ is induced by the inclusion $i \colon C_k \to C_{k+1}$. Similarly, we denote by L_k^c the coker $(i^* \colon M_{k+1}^c \to M_k^c)$.

6.11 Proposition. *The following diagrams are commutative where $\pi \colon C_{k+1} \to C_{k+1}$ is the automorphism with $\pi(e_i) = e_i$ for $i \leq k - 1$, $\pi(e_k) = e_{k+1}$, and $\pi(e_{k+1}) = e_k$.*

$$
\begin{array}{ccc}
M_{k+1} & \xrightarrow{\ i^*\ } & M_k \\
\phi^* R\pi^* \downarrow & & \quad\downarrow \phi^* R \searrow L_k \longrightarrow 0 \\
N_k & \xrightarrow{\ i^*\ } & N_{k-1}
\end{array}
$$

The similar diagram in the complex case is also commutative.

Proof. Using the following commutative diagram, we have the result by applying the cofunctor N.

$$
\begin{array}{ccc}
C_{k-1} & \xrightarrow{\ i\ } & C_k \\
\phi \downarrow & & \downarrow \phi \\
C_k^0 & \xrightarrow{\ j\ } & C_{k+1}^0
\end{array}
$$

Here $j(e_i) = e_i$ *for* $i \leq k - 1$ *and* $j(e_k) = e_{k+1}$. Moreover, we use the following commutative diagram:

$$
\begin{array}{ccc}
M_{k+1} & \xrightarrow{\ i^*\ } & M_k \\
R\pi^* \downarrow & & \downarrow R \\
N(C_{k+1}^0) & \xrightarrow{\ j^*\ } & N(C_k^0)
\end{array}
$$

The proposition is proved by putting these two diagrams together.

6.12. Table of L_k and L_k^c

k	C_k	N_k	M_k	L_k	a_k
1	$\mathbf{C}(1)$	$\mathbf{Z}$	$\mathbf{Z}$	$\mathbf{Z}_2$	1
2	$\mathbf{H}(1)$	$\mathbf{Z}$	$\mathbf{Z}$	$\mathbf{Z}_2$	2
3	$\mathbf{H}(1) \oplus \mathbf{H}(1)$	$\mathbf{Z} \oplus \mathbf{Z}$	$\mathbf{Z}$	0	4
4	$\mathbf{H}(2)$	$\mathbf{Z}$	$\mathbf{Z} \oplus \mathbf{Z}$	$\mathbf{Z}$	4
5	$\mathbf{C}(4)$	$\mathbf{Z}$	$\mathbf{Z}$	0	8
6	$\mathbf{R}(8)$	$\mathbf{Z}$	$\mathbf{Z}$	0	8
7	$\mathbf{R}(8) \oplus \mathbf{R}(8)$	$\mathbf{Z} \oplus \mathbf{Z}$	$\mathbf{Z}$	0	8
8	$\mathbf{R}(16)$	$\mathbf{Z}$	$\mathbf{Z} \oplus \mathbf{Z}$	$\mathbf{Z}$	8

Moreover, we have $C_{k+8} = C_k \otimes C_8$, $N_{k+8} = N_k$, $M_{k+8} = M_k$, and $a_{k+8} = 16a_k$.

k	C_k^c	N_k^c	M_k^c	L_k^c	a_k^c
1	$\mathbf{C}(1) \oplus \mathbf{C}(1)$	$\mathbf{Z} \oplus \mathbf{Z}$	$\mathbf{Z}$	0	1
2	$\mathbf{C}(2)$	$\mathbf{Z}$	$\mathbf{Z} \oplus \mathbf{Z}$	$\mathbf{Z}$	1

Moreover, we have $C_{k+2}^c = C_k^c \bigotimes_C C(2)$, $N_{k+2}^c = N_k^c$, $M_{k+2}^c = M_k^c$, $L_{k+2}^c = K_k^c$, and $a_{k+2}^a = 2a_k^c$.

To prove the above statements, for $k = 1, 2$ we have $L_1 \cong \text{coker}(N_1 \to N_0)$ and $L_2 \cong \text{coker}(N_2 \to N_1)$, which are defined by the restriction to $\mathbf{R}$ of $\mathbf{C}$ acting on $\mathbf{C}$ and to $\mathbf{C}$ of $\mathbf{H}$ acting on $\mathbf{H}$, respectively. Two components result, and the cokernel is Z_2.

For $k = 5, 6$, we have $L_5 \cong \text{coker}(N_5 \to N_4)$ and $L_6 \cong \text{coker}(N_6 \to N_5)$, which are defined by restricting $\mathbf{C}(4)$ to $\mathbf{H}(2)$ acting on $\mathbf{H}^2$ and $\mathbf{R}(8)$ to $\mathbf{C}(4)$ acting on $\mathbf{C}^4$. Irreducible modules go into irreducible modules, and $L_5 = L_6 = 0$.

For $k = 3, 7$, we have $L_3 \cong \text{coker}(N_3 \to N_2)$ and $L_7 \cong \text{coker}(N_7 \to N_6)$ which are defined by restriction of $\mathbf{H}(1) \oplus \mathbf{H}(1)$ and of $\mathbf{R}(8) \oplus \mathbf{R}(8)$ to the first factor. Then $N_3 \to N_2$ and $N_7 \to N_6$ are epimorphisms, and $L_3 = L_7 = 0$.

For $k = 4, 8$, we have $L_4 \cong \text{coker}(M_5 \to M_4)$ and $L_8 \cong \text{coker}(M_9 \to M_8)$. Since there is only one generator z of M_{4m+1} we have $c(z) = z$. Then its image in M_{4m} also has this property. Consequently, z projects to $m_1 + m_2$ which is invariant under c by (6.10) for reasons of dimension. Then coker $(M_{4m+1} \to M_{4m}) = \mathbf{Z}$.

The reasoning used in the last two paragraphs applies to the complex case. This verifies the tables.

7. Tensor Products of Clifford Modules

An isomorphism $\phi_{k,l}: C_{k+l} \to C_k \hat{\otimes} C_l$ is defined by the relations $\phi_{k,l}(e_i) = e_i \hat{\otimes} 1$ for $1 \leq i \leq k$ and $\phi_{k,l}(e_i) = 1 \hat{\otimes} e_{i-k}$ for $k + 1 \leq i \leq k + l$.

7.1 Definition. Let M be a Z_2-graded A-module, and let N be a Z_2-graded B-module. The Z_2-graded tensor product, denoted $M \hat{\otimes} N$, is the ordinary tensor product with the following grading $(M \hat{\otimes} N)^0 = (M^0 \otimes N^0) \oplus (M^1 \otimes N^1)$ and $(M \hat{\otimes} N)^1 = (M^1 \otimes N^0) \oplus (M^0 \otimes N^1)$ and with the following scalar multiplication by $A \hat{\otimes} B$.

$$(a \hat{\otimes} b)(x \hat{\otimes} y) = (-1)^{ij}(ax \hat{\otimes} by) \qquad \text{for } b \in B^i \text{ and } x \in M^j$$

The operation $(M, N) \mapsto \phi_{k,l}^*(M \hat{\otimes} N)$ is a bilinear map and defines a group morphism

$$M_k \bigotimes_Z M_l \to M_{k+l}$$

Similarly, there is a morphism $M_k^c \bigotimes_Z M_l^c \to M_{k+l}^c$. Therefore, $M_* = \sum_{0 \leq k} M_k$ becomes a graded ring, and similarly $M_*^c = \sum_{0 \leq k} M_k^c$ is a graded ring.

7.2 Proposition. *For $u \in M_k$ (or M_k^c), $v \in M_l$ (or M_l^c), and $w \in M_m$ (or M_m^c), there are the following relations:*

(1) $u(vw) = (uv)w$.

(2) $c(uv) = uc(v)$.

(3) For the inclusions $i: C_{k-1} \to C_k$ and $i^*: M_k \to M_{k-1}$, there is the relation $(ui^*(v)) = i^*(uv)$.

(4) $uv = vu$ for kl even and $uv = c(vu)$ for kl odd.

Proof. Statements (1) to (3) follow easily from the definitions. For the last statements, we have the following diagram:

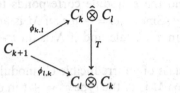

Here $T(x_p \,\hat{\otimes}\, y_q) = (-1)^{pq} y_q \,\hat{\otimes}\, x_p$ for $x_p \in C_k^p$ and $y_q \in C_l^q$. Let $\sigma = \phi_{l,k}^{-1} T \phi_{k,l}$ which is an automorphism of C_{k+l}. Observe that $\sigma(e_i) = e_{l+i}$ for $1 \le i \le k$ and $\sigma(e_i) = e_{i-k}$ for $k+1 \le i \le k+l$. Then σ is a composition of kl inner automorphisms α_b, where $b \in R^k - 0$. For example, α_b where $b = (e_1 + e_2)/\sqrt{2}$ interchanges e_1 and e_2. By (6.9) this is kl applications of c when applied to M_{k+l}. Since $T^*(N \,\hat{\otimes}\, M) = M \,\hat{\otimes}\, N$ and since $\sigma^* \phi_{l,k}^* = \phi_{k,l}^* T^*$, we have $\phi_{k,l}^*(N \,\hat{\otimes}\, M) = c^{pq} \phi_{l,k}^*(M \,\hat{\otimes}\, N)$. This proves (4).

As an application of (7.2), we have the following proposition and theorem.

7.3 Proposition. *Let λ denote the class of an irreducible graded module over C_8. Then multiplication by λ defines an isomorphism $M_k \to M_{k+8}$. If μ denotes the class of an irreducible graded module over C_2^c, multiplication by μ defines an isomorphism $M_k^c \to M_{k+2}^c$.*

Proof. By the table for a_k, the result follows from reasons of dimension for $k \ne 4m$. For $k = 4m$ there are two generators m_1 and m_2 of C_{4m} with $c(m_1) = m_2$ and $c(m_2) = m_1$, as in (6.10).

By (2) in (7.2) we have $\lambda m_2 = \lambda c(m_1) = c(\lambda n_1)$. For reasons of dimension and by (6.10), λm_1 and λm_2 are the two generators of M_{4m+8}. This argument applies to the complex case. This proves the proposition.

By (3) in (7.2), the image $i^*: M_* \to M_*$ is an ideal, and the quotient graded ring $L_* = \sum_{0 \le k} L_k$ is defined as in (6.11). Let the image of λ in L_8 be denoted by λ_8. Similarly for L_*^c we have an induced ring structure, and μ generates L_2^c.

7.4 Theorem. *One can choose generators $1 \in L_0$, $\lambda_1 \in L_1$, $\lambda_4 \in L_4$, and $\lambda_8 \in L_8$ of the graded ring L_* satisfying the relations $2\lambda_1 = 0$, $\lambda_1^3 = 0$, $\lambda_1 \lambda_4 = 0$, $\lambda_4^2 = 4\lambda_8$, and 1 is the unit of L_*. One can choose generators $1 \in L_0^c$ and $\mu \in L_2^c$ of the graded ring L_*^c where 1 is the unit. The ring homomorphism $c: L_* \to L_*^c$*

given by complexification is defined by $c(\lambda_1) = 0$, $c(\lambda_4) = 2\mu^1$, and $c(\lambda_8) = \mu^4$, and the group homomorphism $r: L^c_* \to L_*$ given by restriction from C^c_k to C_k is defined by $r(\mu) = r(\mu^3) = 0$, $r(\mu^2) = \lambda_4$, and $r(\mu^4) = 2\lambda_8$.

Proof. Since $L_1 = Z/2Z$, it is generated by λ_1 with $2\lambda_1 = 0$. Since $a_1 = 1$, $a_2 = 2$, it follows that λ_1^2 generates $L_2 = Z/2Z$.

Let $\omega = e_1 \cdots e_k$ in C_k. For $k = 2q$ we have $\omega^2 = (-1)^q$ and $k = 4m$ we have $\omega^2 = 1$. If M is an irreducible graded C_k-module, ω acts on M^0 as the scalar $\varepsilon = \pm 1$, and the ε-module corresponds to the action of ω equal to multiplication by ε. Since $e_i\omega = -\omega e_i$, if M is an ε-module, $c(M)$ is a $(-\varepsilon)$-module. If M is an ε-module and if N is an ε'-module, then $M \hat{\otimes} N$ is an $\varepsilon\varepsilon'$-module.

Let λ_4 be the class of an irreducible C_4-module M in L_4 which is a (-1)-module. Then $M \hat{\otimes} M$ is of the type $\varepsilon^2 = +1$ in cases over C_8. Let λ_8 be the class of the $(+1)$-module W of C_8. Then we have $M \hat{\otimes} M = W^4$; for reasons of dimension, $8^2 = 4 \cdot 16$, we have $\mu^2 = 4\lambda$. This proves the statement for the real case.

For $k = 2q$ we have $\omega^2 = (-1)^q$. If M is an irreducible graded C^c_k-module, ω acts like $\pm i^q$ on M^0. Again M is called an ε-module if ω acts by scalar multiplication by ε on M^0. Let $\mu^c_q \in M^c_{2q}$ denote the generator corresponding to an irreducible i^q-module. Then we have $\mu^c_q = (\mu^c_1)^8$ by considering dimensions of modules in the class. This proves the statement for the complex case.

For the relation between the real and complex cases, let M be a real ε-module for C_{4m}. Then $M \underset{R}{\otimes} C$ is a complex $(-1)^m$ ε-module for C^c_{4m}. The complexification $L_k \to L^c_k$ is given by $\lambda_4 \to 2\mu^2_\mu$ for reasons of dimension.

For the statement about r we use the relations $rc(y) = 2y$ and $cr(x) = x + \bar{x}$, where $\bar{x}$ is the conjugate of x. Observe that $z = -\xi$. This proves the theorem.

7.5 Remark. In the paper of Atiyah, Bott, and Shapiro [1], group morphisms

$$L_k \to \widetilde{KO}(S^k) \quad \text{and} \quad L^c_k \to \widetilde{K}(S^k)$$

are defined. These morphisms are proved to be isomorphisms using the graded ring structure of L_* and L^c_* described in (7.4) and the Bott periodicity.

8. Unit Tangent Vector Fields on Spheres: II

Let $\rho(n)$ be the number such that there exist $(\rho(n) - 1)$ orthonormal tangent vector fields on S^{n-1} by Clifford algebra constructions, see Theorem (2.2) and Remarks (2.3). There exists an orthogonal multiplication $R^k \times R^n \to R^n$ if and only if R^n is a C_{k-1}-module, by (2.5).

8.1 Table of Irreducible Modules Over C_{k-1}. Let b_k be the minimum dimension n such that $\mathbf{R}^n$ has the structure of a C_{k-1}-module.

8.1 Table of Irreducible Modules Over C_{k-1}. Let b_k be the minimum dimension n such that $\mathbf{R}^n$ has the structure of an irreducible C_{k-1}-module.

k	1	2	3	4	5	6	7	8	9
C_{k-1}	$\mathbf{R}$	$\mathbf{C}$	$\mathbf{H}$	$\mathbf{H} \oplus \mathbf{H}$	$\mathbf{H}(2)$	$\mathbf{C}(4)$	$\mathbf{R}(8)$	$\mathbf{R}(8) \oplus \mathbf{R}(8)$	$\mathbf{R}(16)$
b_k	1	2	4	4	8	8	8	8	16

Moreover, $b_{k+8} = 16b_k$. Note that if $\mathbf{R}^n$ admits the structure of an irreducible C_k-module, then S^{n-1} has k orthonormal tangent vector fields.

8.2 Theorem. *If* $n = (\text{odd})2^{c(n)}16^{d(n)}$, *where* $0 \leq c(n) \leq 3$, *then* $\rho(n) = 2^{c(n)} + 8d(n)$.

Proof. The formula holds by inspection for $1 \leq n \leq 8$. For $n \geq 8$, we use $b_{k+8} = 16b_k$ which follows from $C_{k+8} = C_k \otimes \mathbf{R}(16)$.

It is a theorem of Adams that there are not $\rho(n)$ orthonormal tangent vector fields on S^{n-1}; see Chap. 16.

9. The Group Spin(k)

Let $(x_1 \otimes \cdots \otimes x_p)^* = x_p \otimes \cdots \otimes x_1$ define an operation on $T(M) = \sum_{0 \leq k} T^k(M)$. Since $(x \otimes x - f(x,x)1)^* = x \otimes x - f(x,x)$, the linear involution $*\colon C_k \to C_k$ is defined where

$$(e_{i(1)} \cdots e_{i(r)})^* = e_{i(r)} \cdots e_{i(1)} = (-1)^{r(r-1)/2} e_{i(1)} \cdots e_{i(r)}$$

Moreover, $(xy)^* = y^*x^*$.

9.1 Definition. Let pin(k) denote the subgroup of the multiplicative group of units in C_k generated by S^{k-1} where $S^{k-1} \subset \mathbf{R}^k \subset C_k$, and let Spin$(k)$ denote the subgroup pin$(k) \cap C_k^0$ of pin(k).

For $u \in C_k$, we have $u \in$ pin(k) if and only if $u = x_1 \cdots x_m$, where $x_i \in S^{k-1}$ for $1 \leq i \leq m$. Moreover, we have $uu^* = +1$ if and only if m is even, that is, $u \in C_k^0$, and $uu^* = -1$ if and only if m is odd, that is, $u \in C_k^1$. Consequently, $uu^* = +1$ for $u \in$ Spin(k).

For $u \in$ pin(k), we define $\phi(u) \in 0(k)$ by the relation $\phi(u)x = uxu^*$. Clearly, $\phi(u)$ is linear and satisfies the relation

$$\|\phi(u)x\|^2 = (uxu^*)(uxu^*) = uxxu^* = -xx = \|x\|^2$$

The properties of Spin(k) are contained in the next theorem.

9.2 Theorem. *The map* ϕ: pin(k) $\to O(k)$ *is a continuous group epimorphism with* $\phi^{-1}(SO(k)) = $ Spin(k) *and* ker $\phi = \{+1, -1\}$ *for the restriction* ϕ: Spin(k) $\to SO(k)$. *Moreover, for* $k \geq 3$, Spin(k) *is the universal covering group of* $SO(n)$ *and* $\pi_0($Spin$(k)) = \pi_1($Spin$(k)) = 0$.

Proof. Clearly, ϕ is continuous, and the relation $\phi(uv)x = uvx(uv)^* = uvxv^*u^* = u(\phi(v)x)u^* = \phi(u)\phi(v)x$ yields the group morphism property of ϕ.

To prove ϕ is surjective, we begin by proving that $\phi(u)$ for $u \in S^{n-1}$ is a reflection through the hyperplane perpendicular to u. Let $x = tu + u'$, where $(u|u') = 0$. Then $\phi(u)x = u(tu + u')u^* = tuuu^* + uu'u^* = -tu - u'uu^* = -tu + u'$. This proves the statement about $\phi(u)$ for $u \in S^{k-1}$. Since these reflections generate $O(k)$, the map ϕ is surjective.

For $u \in S^{n-1}$, det $\phi(u) = -1$ and det $\phi(u_1 \cdots u_r) = (-1)^r$ for $u_i \in S^{n-1}$ and $1 \leq i \leq r$. Therefore, $u \in$ Spin(k) if and only if $\phi(u) \in SO(k)$.

For $u \in$ ker ϕ we have $ue_iu^* = e_i$, or $ue_i = e_iu$ since $uu^* = 1$, and, conversely, these conditions imply that $u \in$ ker ϕ. This happens if and only if u is in the center of C_k intersected with C_k^0, that is, $u \in \mathbf{R}1$. This is equivalent to $u = \pm 1$ since $uu^* = u^2 = 1$.

For the last statement, it suffices to prove that $+1$ and -1 in Spin(k) are connected since ϕ: Spin(k) $\to SO(k)$ is locally trivial. Since the element $(e_1 \cos t + e_2 \sin t)(e_1 \cos t - e_2 \sin t) = -\cos 2t - e_1e_2 \sin 2t$ is a member of Spin (n), we get a path in Spin(n) from -1 to $+1$ by varying t from 0 to $\pi/2$. This construction can be used to prove that Spin(k) is connected directly.

Exercises

1. Using the fact that S^3 has three orthonormal tangent vector fields and that S^7 has seven orthonormal vector fields, prove that S^{n-1} has $2^c - 1$ orthonormal tangent vector fields where $n = (\text{odd})2^c16^d$ and $c \leq 3$.

2. A morphism $u: (M, f) \to (N, g)$ between two quadratic forms is a module morphism $u: M \to N$ such that $g(u(x), u(x')) = f(x, x')$ for all $x, x' \in M$.

3. Describe explicitly $i: C_{k-1} \to C_k$ for the table of C_k in (5.10).

4. For $x, y \in S^{k-1} \subset C_k$, verify the formula $xy + yx = 2(x|y)$ directly.

5. Prove that $\pi_i(SO(k)) = \pi_i($Spin$(k))$ for $i \geq 2$.

6. Calculate the kernel of ϕ: pin(k) $\to O(k)$.

7. Referring to (4.5), in what sense is $(M, f) \to C(f)$ a functor?

8. Let 2 be a unit in R. Prove that a module morphism $u: M \to N$ prolongs to an algebra morphism $v: C(M, f) \to C(N, g)$ if and only if u is a morphism of quadratic forms.

CHAPTER 13

The Adams Operations and Representations

Every representation M of a topological group G and every principal bundle α over a space X determine a fibre bundle $\alpha[M]$ over X that admits the structure of a vector bundle. For a given α the function that assigns $\alpha[M]$ to M prolongs to a group morphism $R(G) \to K(X)$, where $R(G)$ is the representation ring of G. We study $K(X)$ using this morphism; in particular, properties of operations in $K(X)$ can be derived from properties of operations in $R(G)$.

1. λ-Rings

1.1 Definition. A λ-semiring is a (commutative) semiring R together with functions $\lambda^i \colon R \to R$ for $i \geq 0$ satisfying the following properties:

(1) $\lambda^0(x) = 1$ and $\lambda^1(x) = x$ for each $x \in R$.
(2) For each $x, y \in R$, $\lambda^k(x + y) = \sum_{i+j=k} \lambda^i(x)\lambda^j(y)$.

A morphism $u \colon (R, \lambda^i) \to (R', \lambda^i)$ between two λ-semirings is a semiring morphism $u \colon R \to R'$ such that $\lambda^i(u(x)) = u(\lambda^i(x))$ for each $x \in R$ and $i \geq 0$. A λ-ring is a λ-semiring whose underlying semiring is a ring.

1.2 Example. The semiring of isomorphism classes of vector bundles $\mathrm{Vect}_F(X)$ is a λ-semiring for $F = \mathbf{R}$ and $\mathbf{C}$, where we define $\lambda^i[\xi] = [\Lambda^i \xi]$, using the ith exterior power which exists by 5(6.9). Then axioms (1) and (2) follow from corresponding properties of exterior powers of vector spaces and Theorem 5(6.3).

If $f: Y \to X$ is a map and if ξ is a vector bundle, by $5(6.9)f^*(\Lambda^i \xi)$ and $\Lambda^i f^*(\xi)$ are Y-isomorphic. Consequently, $\mathrm{Vect}_F(f): \mathrm{Vect}_F(X) \to \mathrm{Vect}_F(Y)$ is a morphism of λ-semirings.

To define a λ-ring structure on $K_F(X)$, we use the following construction. For a λ-semiring R with operations λ^i, we define $\lambda_t: R \to 1 + R[[t]]^+$ to be $\lambda_t(x) = \sum_{i \geq 0} \lambda^i(x)t^i$. By axiom (2) we have $\lambda_t(x + y) = \lambda_t(x)\lambda_t(y)$. Here $1 + R[[t]]^+$ denotes the semigroup of power series $1 + a_1 t + a_2 t^2 + \cdots$. If R is a ring, then $1 + R[[t]]^+$ is a commutative group.

1.3 Proposition. *Let $\tilde{R}$ denote the ring completion of the underlying semiring R of the λ-semiring (R, λ^i). There exists a λ-ring structure on $\tilde{R}$ given by operations $\tilde{\lambda}^i$ such that the natural morphism $\theta: R \to \tilde{R}$ is a λ-semiring morphism. Moreover, the operations $\tilde{\lambda}^i$ are unique with respect to this property.*

Proof. The underlying additive group structure of $\tilde{R}$ is the group completion of the underlying semigroup structure of R. We consider $\lambda_t: R \to 1 + \tilde{R}[[t]]^+$, which is the multiplicative group of power series with 1 as a constant term. This defines to a group morphism $\tilde{\lambda}_t: \tilde{R} \to 1 + \tilde{R}[[t]]^+$ such that $\tilde{\lambda}_t \theta = \lambda_t$. Moreover, $\tilde{\lambda}_t$ is unique with respect to this property. Let $\tilde{\lambda}^i(x)$ be defined by the relation $\tilde{\lambda}_t(x) = \sum_{i \geq 0} \tilde{\lambda}^i(x)t^i$. Then $\tilde{\lambda}_i$ are operations on $\tilde{R}$ satisfying (1) and (2). From the relation $\tilde{\lambda}_t \theta = \lambda_t$, the function θ is a morphism of λ-semirings. Since $\tilde{\lambda}_t$ is unique, $\tilde{\lambda}^i$ is uniquely defined.

1.4 Application. The rings $K(X)$ and $KO(X)$ admit a λ-ring structure such that $\lambda^i[\xi] = [\Lambda^i \xi]$ for a vector bundle class $[\xi]$. If $f: Y \to X$ is a map, then $f^!: K(X) \to K(Y)$ and $f^!: KO(X) \to KO(Y)$ are λ-ring morphisms.

2. The Adams ψ-Operations in λ-Ring

Associated with the operations λ^i in a λ-ring are the Adams operations ψ^k.

2.1 Definition. Let $\psi_t(x) = \sum_{i \geq 1} \psi^k(x)t^k$ be given by the relation $\psi_{-t}(x) = -t((d/dt)\lambda_t(x))/\lambda_t(x)$ for a λ-ring R. The functions $\psi^k: R \to R$ are the Adams ψ-operations in R.

The ψ-operations on a λ-ring R have the following properties.

2.2 Proposition. *The function $\psi^k: R \to R$ is additive, and if $u: R \to R'$ is a λ-ring morphism, then $u(\psi^k(x)) = \psi^k(u(x))$ for $x \in R$.*

Proof. We have

$$\psi_{-t}(x + y) = -t(d(dt)\lambda_t(x + y))/\lambda_t(x + y)$$

$$= -t[((d/dt)\lambda_t(x))\lambda_t(y) + \lambda_t(x)((d/dt)\lambda_t(y))]/\lambda_t(x)\lambda_t(y)$$

$$= -t((d/dt)\lambda_t(x))/\lambda_t(x) - t((d/dt)\lambda_t(y))/\lambda_t(y)$$

$$= \psi_{-t}(x) + \psi_{-t}(y)$$

and therefore, by comparing coefficients, we have $\psi^k(x + y) = \psi^k(x) + \psi^k(y)$. For the last statement, we apply u to the coefficients of $\psi_t(x)$ and $\lambda_t(x)$.

2.3 Proposition. *If $\lambda^i(x) = 0$ for $i > 1$, then $\psi^k(x) = x^k$.*

Proof. We have $\lambda_t(x) = 1 + tx$ and $(d/dt)\lambda_t(x) = x$. Then $\psi_{-t}(x) = -tx/(1 + tx)$ or $\psi_t(x) = tx/(1 - tx) = \sum_{k \geq 1} x^k t^k$. Since $\psi_t(x) = \sum_{k \geq 1} \psi^k(x) t^k$, we have $\psi^k(x) = x^k$.

2.4 Remark. From a universal formula relating symmetric functions we have $\psi^k(x) = s_k^n(\lambda_1(x), \ldots, \lambda_n(x))$ for $n \geq k$, where $x_1^k + \cdots + x_n^k = s_k^n(\sigma_1, \ldots, \sigma_n)$ and σ_i are the elementary symmetric functions; see Proposition (1.8) of the next chapter.

2.5 Proposition. *There is the following relation between λ^i and ψ^k for $x \in R$:*

$$\psi^k(x) - \lambda^1(x)\psi^{k-1}(x) + \cdots + (-1)^{k-1}\lambda^{k-1}(x)\psi^1(x) + (-1)^k k\lambda^k(x) = 0$$

Proof. From the definition $\lambda_t(x)\psi_{-t}(x) + t(d/dt)\lambda_t(x) = 0$, we have

$$\left(\sum_{i \geq 0} \lambda^i(x)t^i \right)\left(\sum_{i \geq 1} (-1)^i \psi^i(x)t^i \right) + t \sum_{k \geq 0} (k + 1)\lambda^{k+1}(x)t^k = 0. \text{ Therefore,}$$

$$\sum_{k \geq 1} \left(\sum_{i+j=k} (-1)^i \lambda^i(x)\psi^j(x) + k\lambda^k(x) \right) t^k = 0$$

This leads to the stated formula.

2.6 Special cases. For $k = 1$, we have $\psi^1(x) + (-1)\lambda^1(x) = 0$, or

$$\psi^1(x) = x$$

For $k = 2$, we have $\psi^2(x) - \lambda^1(x)\psi^1(x) + 2\lambda^2(x) = 0$, or

$$\psi^2(x) = x^2 - 2\lambda^2(x)$$

and for $k = 3$, we have $\psi^3(x) - \lambda^1(x)\psi^2(x) + \lambda^2(x)\psi^1(x) + (-1)^3 3\lambda^3(x) = 0$, or

$$\psi^3(x) = x^3 + 3\lambda^3(x) - 3x\lambda^2(x)$$

2.7 Definition. A λ-semiring with line elements is a triple (R, λ^i, L), where (R, λ^i) is a λ-semiring and L is a multiplicatively closed subset of R consisting of $x \in R$ with $\lambda^i(x) = 0$ for $i > 1$.

In the case of $\text{Vect}_F(X)$ or $K_F(X)$ for $F = \mathbf{R}$ or $\mathbf{C}$, the set L consists of the classes containing line bundles.

2.8 Definition. A λ-semiring R splits, provided for each finite set $x_1, \ldots,$ $x_r \in R$ there exists a λ-semiring monomorphism $u: R \to R'$, where (R', λ^i, L') is a λ-semiring with line elements L' such that each $u(x_i)$ is a sum of line elements in L'.

Later in this chapter we shall prove that certain representation rings are split λ-rings. This will allow us to prove that the result of Theorem (2.9), which follows, holds for the λ-rings $K(X)$ and $KO(X)$. These additional properties are very important.

2.9 Theorem. *Let R be a split λ-semiring. Then $\psi^k: R \to R$ is a semiring morphism for each $k \geq 0$, and $\psi^k \psi^l = \psi^{kl}$.*

Proof. Let $x, y \in R$, and let $u: R \to R'$ be a semiring monomorphism, where $u(x) = x_1 + \cdots + x_r$, $u(y) = y_1 + \cdots + y_s$, and $x_1, \ldots, x_r, y_1, \ldots, y_s$ are line elements of R'. Now we compute

$$u(\psi^k(xy)) = \psi^k(u(x)u(y)) = \sum_{i,j} \psi^k(x_i y_j) = \sum_{i,j} x_i^k y_j^k$$

$$= \left(\sum_i x_i^k\right)\left(\sum_j y_j^k\right) = \left(\sum_j \psi^k(x_i)\right)\left(\sum_j \psi^k(y_j)\right)$$

$$= \psi^k(u(x))\psi^k(u(y)) = u(\psi^k(x)\psi^k(y))$$

Since u is a monomorphism, $\psi^k(xy) = \psi^k(x)\psi^k(y)$.

With the same notations, we have $u(\psi^k\psi^l(x)) = \psi^k\psi^l(u(x)) = \psi^k\psi^l(x_1 + \cdots + x_r) = \psi^k(x_1^l + \cdots + x_r^l) = x_1^{kl} + \cdots + x_r^{kl} = \psi^{kl}(x_1 + \cdots + x_r) = u(\psi^{kl}(x))$. Again, since u is a monomorphism, $\psi^k\psi^l = \psi^{kl}$.

2.10 Remark. Let (R, λ) be a λ-ring where the ψ^k operations satisfy the following:

(1) The functions ψ^k are semiring morphisms.
(2) $\psi^k\psi^l = \psi^{kl}$.

If $u: R' \to R$ is a monomorphism of λ-semirings, then R' satisfies (1) and (2).

2.11 Definition. A λ-semiring with line elements and conjugation is a 4-tuple $(R, \lambda^i, L, *)$, where (R, λ^i, L) is a λ-semiring with line elements and $*: R \to R$ is a λ-semiring involution such that $xx^* = 1$ for $x \in L$.

2.12 Remark. Examples of the above concept are $\text{Vect}_C(X)$ and $K_C(X)$, where x^* is the complex conjugate bundle. For a λ-semiring R with line elements and conjugation, $\psi^{-k}(x) = \psi^k(x^*) = \psi^k(x)^*$ is defined and (1) and (2) hold.

3. The γ^i Operations

In a λ-ring R, the ψ^k are polynomial combinations of the λ^i. We consider here other operations which are combinations of the λ^i. These operations are useful in studying the representation rings of Spin(n) and the immersion theory of manifolds.

3.1 Definition. The γ-operations in a λ-ring R, denoted $\gamma^i \colon R \to R$, are defined by the requirement that $\gamma_t(x) = \lambda_{t/(1-t)}(x)$, where $\gamma_t(x) = \sum_{0 \le i} \gamma^i(x)t^i$.

We have the relation $\lambda_s(x) = \gamma_{s/(1+s)}(x)$ and the relation

$$\sum_{0 \le i} \gamma^i(x)t^i = \sum_{0 \le i} \lambda^i(x)t^i(1-t)^{-i} = \sum_{0 \le i} \lambda^i(x)t^i(1 + it + \cdots)$$

3.2 Proposition. *The γ-operations in a λ-ring R have the following properties*:

(1) $\gamma^0(x) = 1$ *and* $\gamma^1(x) = x$ *for each* $x \in R$.
(2) *For each* $x, y \in R$, $\gamma^k(x + y) = \sum_{i+j=k} \gamma^i(x)\gamma^j(y)$.
(3) $\gamma^k(x) = \lambda^k(x) + \sum_{i<k} a_{i,k}\lambda^i(x)$ *and* $\lambda^k(x) = \gamma^k(x) + \sum_{i<k} b_{i,k}\gamma^i(x)$,

where $a_{i,k}$ and $b_{i,k}$ are integers.

Proof. The first two statements follow from the relations $\gamma^0(x) = \lambda^0(x)$, $\gamma^1(x) = \lambda^1(x)$, and $\gamma_t(x + y) = \gamma_t(x)\gamma_t(y)$. For the last relation we consider the kth coefficient of $\lambda_{t/(1-t)}(x)$ and of $\gamma_{s/1+s}(x)$.

The operations γ^i are used in applications of K-theory to immersion theory. We use the γ^i operations in Chap. 14, Sec. 10.

3.3 Examples. From the definition

$$\gamma^0(x) + \gamma^1(x)t + \gamma^2(x)t^2 + \gamma^3(x)t^3 + \cdots$$
$$= \lambda^0(x) + \lambda^1(x)t(1 + t + t^2 + \cdots) + \lambda^2(x)t^2(1 + 2t + 3t^2 \cdots)$$
$$+ \lambda^3(x)t^3(1 + 3t + \cdots) + \cdots$$

we get the following relations:

$$\gamma^0(x) = \lambda^0(x) = 1$$
$$\gamma^1(x) = \lambda^1(x) = x$$
$$\gamma^2(x) = \lambda^2(x) + \lambda^1(x)$$
$$\gamma^3(x) = \lambda^3(x) + 2\lambda^2(x) + \lambda^1(x)$$
$$\gamma^4(x) = \lambda^4(x) + 3\lambda^3(x) + 3\lambda^2(x) + \lambda^1(x)$$

4. Generalities on G-Modules

Let F denote $\mathbf{R}, \mathbf{C}$, or $\mathbf{H}$. Then each finite-dimensional F-vector space M has a unique topology such that any vector space isomorphism $F^n \to M$ is a homeomorphism. Moreover, this is a norm topology. Frequently, in the next three sections, special considerations are required for $F = \mathbf{H}$ because of its noncommutative character. These considerations will be left to the reader.

4.1 Definition. Let G be a topological group. A G-module M is a G-space such that the action of $s \in G$ on M is linear.

For $s \in G$ and a G-module M, let s_M denote the linear automorphism of M such that $s_M(x) = sx$. Then the relations $1_M = 1$, $(st)_M = s_M t_M$, and $(s^{-1})_M = (s_M)^{-1}$ hold. The function $(s, x) \mapsto sx$ of $G \times M \to M$ is continuous.

4.2 Example. The space F^n is a $U_F(n)$-module and $SU_F(n)$-module with the G-module action given by a linear transformation acting on F^n.

Other examples will arise in the next paragraphs.

4.3 Definition. A function $f: M \to N$ is a G-morphism between G-modules (both over F) provided f is F-linear and $f(sx) = sf(x)$ for $x \in M$ and $s \in G$.

The identity on M is a G-morphism, and the composition $vu: M \to L$ of G-morphisms $u: M \to N$ and $v: N \to L$ is a G-morphism. The kernel, image, and cokernel of a G-morphism are defined and admit the structure of a G-module in a natural way. Let $\operatorname{Hom}_G(M, N)$ denote the set of all G-morphisms $M \to N$. Then $\operatorname{Hom}_G(M, N)$ is a subspace (or subgroup for $F = \mathbf{H}$) of $\operatorname{Hom}_F(M, N)$.

4.4 Definition. Let M and N be two G-modules. Then the direct sum of M and N, denoted $M \oplus N$, is a G-module, where $s(x, y) = (sx, sy)$ for $s \in G$ and $(x, y) \in M \oplus N$. The vector space structure on $M \oplus N$ is the direct sum structure.

The direct sum plays the role of the product and coproduct in the category of G-modules. A similar definition applies to the direct sum of n modules.

4.5 Definition. The tensor product $M \otimes N$ and exterior product $\Lambda^r(M)$ of G-modules M and N are defined by the relations $s(x \otimes y) = sx \otimes sy$ and $s(x_1 \wedge \cdots \wedge x_r) = sx_1 \wedge \cdots \wedge sx_r$.

The operations of direct sum, tensor product, and exterior product are related in the next proposition.

4.6 Proposition. Let $M_1, \ldots, M_n$ be one-dimensional G-modules. Then $\Lambda^r(M_1 \oplus \cdots \oplus M_n)$ and $\displaystyle\sum_{i(1) < \cdots < i(r)} M_{i(1)} \otimes \cdots \otimes M_{i(r)}$ are isomorphic as G-modules.

Proof. The isomorphism given by $x_1 \wedge \cdots \wedge x_r \to x_1 \otimes \cdots \otimes x_r$ preserves the action of G.

4.7 Definition. Let M and N be two G-modules. Then the relation $sf = s_N f s_M^{-1}$ defines a G-module structure on $\operatorname{Hom}_F(M, N)$.

Observe that for $f \in \operatorname{Hom}_F(M, N)$ we have $f \in \operatorname{Hom}_G(M, N)$ if and only if $sf = f$ for each $s \in G$. For the action $sa = a$ on F, we have the structure of a G-module on the dual module $M^+ = \operatorname{Hom}_F(M, \bar{F})$, the module of conjugate linear functionals.

4.8 Definition. A G-module M is simple provided M has no G-submodules, i.e., no subspaces N with $sN = N$ for all $s \in G$.

4.9 Proposition. *Let M be a simple G-module. Then every G-morphism $f: M \to N$ is either zero or a monomorphism, and every G-morphism $g: L \to M$ is either zero or an epimorphism.*

Proof. Either $\ker f$ equals M or 0, and either $\operatorname{coker} g$ equals M or 0 in the second case.

As a corollary, we have the next theorem.

4.10 Theorem. (Schur's lemma). *If $f: M \to N$ is a G-morphism between two simple G-modules, f is either zero or an isomorphism. If $M = N$ and if F is algebraically closed, f is multiplication by a scalar.*

Proof. The first statement follows from (4.9). For the second statement, let λ be an eigenvalue of f, and let $\ker (f - \lambda) = L$ be a G-submodule of M. Since $L \neq 0$, we have $L = M$ and $f(x) = \lambda x$. This proves the theorem.

The proof in the next proposition is straightforward and is left to the reader; see Cartan and Eilenberg [1, chap. 1].

4.11 Proposition. *For a G-module M, the following statements are equivalent.*

(1) *The module M is a sum of simple G-submodules.*
(2) *The module M is a direct sum of simple G-submodules.*
(3) *For each G-submodule N of M, there exists a G-submodule N' with $M = N \oplus N'$.*

4.12 Definition. A G-module M is semisimple provided it satisfies the three equivalent properties in (4.11).

In Sec. 6 we prove that for a compact group every G-module is semisimple.

5. The Representation Ring of a Group G and Vector Bundles

For a topological group G, let $\mathbf{M}_F(G)$ denote the set of isomorphism classes $[M]$ of G-modules M over F. The operations $[M] + [N] = [M \oplus N]$ and $[M][N] = [M \otimes N]$ make $\mathbf{M}_F(G)$ into a semiring (only a semigroup for

$F = \mathbf{H}$). For $F = \mathbf{R}$ or $\mathbf{C}$, the functions $\lambda_i[M] = [\Lambda^i M]$ define a λ-semiring structure on $\mathbf{M}_F(G)$.

5.1 Definition. The representation ring $R_F(G)$ of a topological group G is the ring associated with the semiring $\mathbf{M}_F(G)$; see 9(3.3).

The elements of $R_F(G)$ are of the form $[M] - [N]$, where M and N are G-modules, and there is a natural morphism $\mathbf{M}_F(G) \to R_F(G)$. Moreover by (1.3), $R_F(G)$ admits a natural λ-ring structure.

5.2 Definition. Let $u: G \to H$ be a morphism of topological groups; that is, u is continuous and preserves the group structure. Let M be an H-module. We define $u^*(M)$ to be the underlying vector space of M with G-module structure given by $sx = u(s)_M(x)$ for $s \in G$ and $x \in M$.

Then for a topological group morphism $u: G \to H$ and H-modules M and N we have $u^*(M \oplus N) = u^*(M) \oplus u^*(N)$, $u^*(M \otimes N) = u^*(M) \otimes u^*(N)$, and $u^*(\Lambda^i M) = \Lambda^i u^*(M)$. Consequently, u^* defines a λ-semiring morphism $\mathbf{M}_F(u): \mathbf{M}_F(H) \to \mathbf{M}_F(G)$ by $\mathbf{M}_F(u)[M] = [u^*(M)]$, and this defines a unique λ-ring morphism $R_F(u): R_F(H) \to R_F(G)$ such that the following diagram is commutative.

$$
\begin{array}{ccc}
\mathbf{M}_F(H) & \xrightarrow{\;\mathbf{M}_F(u)\;} & \mathbf{M}_F(G) \\
\downarrow & & \downarrow \\
R_F(H) & \xrightarrow{\;R_F(u)\;} & R_F(G)
\end{array}
$$

5.3 Proposition. *With the above notations, R_F is a cofunctor from the category of topological groups to the category of λ-rings (only groups for $F = \mathbf{H}$).*

As with vector bundles, we denote $R_\mathbf{C}(G)$ by $R(G)$, $R_\mathbf{R}(G)$ by $RO(G)$, and $R_\mathbf{H}(G)$ by $RSp(G)$.

5.4 Remark. Vector bundles and G-modules are related by the following mixing construction. For each locally trivial principal G-bundle α over a space X and each G-module M, the fibre bundle $\alpha[M]$ is formed. Since the action of G preserves the vector space operations on M, there is a natural vector bundle structure on $\alpha[M]$. Since $\alpha[M \oplus N]$ and $\alpha[M] \oplus \alpha[N]$ are isomorphic, since $\alpha[M \otimes N]$ and $\alpha[M] \otimes \alpha[N]$ are isomorphic, and since $\Lambda^i \alpha[M]$ and $\alpha[\Lambda^i M]$ are isomorphic, there is a morphism of λ-semiring $\tilde{\alpha}$: $\mathbf{M}_F(G) \to \mathrm{Vect}_F(X)$ defined by the relation $\tilde{\alpha}([M])$ equals the isomorphism class of $\alpha[M]$. The morphism $\tilde{\alpha}$ defines a λ-ring morphism, also denoted $\tilde{\alpha}$, $R_F(G) \to K_F(X)$. We shall discuss this morphism further in a later section.

5.5 Example. View Z_2 as $\{+1, -1\}$, a subgroup of S^1, the circle group in the complex plane. Then $S^1 \to S^1$ mod Z_2 is a principal Z_2-bundle. If $M = \mathbf{R}$ is the standard representation of $O(1) = Z_2$, then $\alpha[M]$ is the canonical line bundle on $S^1 = RP^1$.

6. Semisimplicity of *G*-Modules over Compact Groups

All topological groups are compact in this section.

5.1 Haar Measure. We outline the properties of the Haar measure that will be used in this section. A rapid proof of the existence and uniqueness can be found in Pontrjagin [1, chap. 4, sec. 25].

Let V be a normed vector space over F, and let G be a compact topological group. Let $C_V(G)$ denote the normed space of all continuous $f: G \to V$, where $\|f\| = \sup\|f(s)\|$ for $s \in G$. Let $C(G)$ denote $C_R(G)$.

The Haar measure is the linear function $\mu: C(G) \to R$ with the following properties:

(1) If $f \geq 0$, then $\mu(f) \geq 0$, and if, in addition, $f(s) > 0$ for some $s \in G$, then $\mu(f) > 0$.
(2) For the function 1, $\mu(1) = 1$.
(3) $\mu(L_a f) = \mu(f)$, where $L_a f(s) = f(as)$ for $a, s \in G$.

It is a theorem that μ exists and is unique with respect to properties (1) to (3). See Pontrjagin [1].

The Haar measure defines a unique Haar measure $\mu_V: C_V(G) \to V$ for each normed linear space V from the requirement that $u\mu_V(f) = \mu(uf)$ for each R-linear $u: V \to R$. If $e_1, \ldots, e_n$ is a basis of V, and if $f = f_1 e_1 + \cdots + f_n e_n \in C_V(G)$, then $\mu_V(f) = \mu(f_1)e_1 + \cdots + \mu(f_n)e_n$.

6.2 Definition. A hermitian form on an F-module is a function $\beta: M \times M \to F$ such that $x \mapsto \beta(x, y)$ is linear for each $y \in M$, $\beta(x, y) = \overline{\beta(y, x)}$ for each $x, y \in M$, and $\beta(x, x) > 0$ for each $x \in M$, $x \neq 0$. A hermitian form β is G-invariant provided $\beta(sx, sy) = \beta(x, y)$ for each $x, y \in M$ and $s \in G$. The correlation associated with β is the morphism $c_\beta: M \to M^+$ defined by the requirement that $c_\beta(x)(y) = \beta(x, y)$ for $x, y \in M$.

Observe that β is G-invariant if and only if c_β is a G-morphism.

6.3 Proposition. *Every G-module M has a G-invariant hermitian metric β.*

Proof. Let $\beta'(x, y)$ be any hermitian metric on M. The function $s \to \beta'(sx, sy)$ is continuous and has a Haar integral $\beta(x, y)$ for each $x, y \in M$. From the linearity of the Haar integral, the form $\beta(x, y)$ is linear in x and conjugate linear in y. Moreover, we have $\beta(x, y) = \overline{\beta(y, x)}$ and $\beta(x, x) > 0$ for each $x \in M$ with $x \neq 0$.

6.4 Corollary. *If M is a simple G-module, $c_\beta: M \to M^+$ is an isomorphism of G-modules for each G-invariant β, and M and M^+ are G-isomorphic.*

6.5 Theorem. *Every G-module M is semisimple.*

Proof. Let L be a G-submodule, and let β be a G-invariant hermitian form on M. Let L' denote the subset of $y \in M$ with $\beta(x, y) = 0$ for all $x \in L$. Clearly, L' is a G-submodule of M and $L \cap L'$ equals 0. For each $x \in M$ the function $y \to \beta(x, y)$ is an element of L^+, and since $c_\beta : L \to L^+$ is an isomorphism, we have $z \in L$ with $\beta(x, y) = \beta(z, y)$ for all $y \in L$. Then $x - z = z'$ is a member of L', and x equals $z + z'$ with $z \in L$, $z' \in L'$. Therefore, we have $M = L \oplus L'$. By (4.11) and (4.12), M is semisimple.

6.6 Corollary. *The set of G-module classes $[M]$, where M is simple, generate the group $R_F(G)$.*

7. Characters and the Structure of the Group $R_F(G)$

We wish to prove that the simple module classes freely generate $R_F(G)$ as an abelian group. For this we introduce characters.

7.1 Definition. The character of a representation M, denoted χ_M, is the element of $C_F(G)$ given by $s \mapsto \operatorname{Tr} s_M$. Let $\operatorname{ch}_F G$ denote the subring of $C_F(G)$ generated by the characters χ_M of representations.

The characters have the following properties all of which follow from elementary properties of Tr.

7.2 Proposition. *If M and N are isomorphic G-modules, then χ_M equals χ_N. For two modules M and N, we have $\chi_{M \oplus N} = \chi_M + \chi_N$ and $\chi_{M \otimes N} = \chi_M \chi_N$. For a character χ, we have $\chi(tst^{-1}) = \chi(s)$ for all $s, t \in G$.*

There is a natural ring morphism $R_F(G) \to \operatorname{ch}_F G$ defined by the function $[M] \mapsto \chi_M$. This is clearly an epimorphism.

7.3 Notation. For each $u \in \operatorname{Hom}_F (M, N)$ one can form $s_N u s_M^{-1}$, which is a function of s, and integrate it over G to get an element of $\operatorname{Hom}_F(M, N)$, denoted $\tilde{u}$. By the invariance of the integration process, $\tilde{u}$ is a member of $\operatorname{Hom}_G (M, N)$. Moreover, $\tilde{u} = u$ if and only if $u \in \operatorname{Hom}_G(M, N)$. For two G-modules M and N, we define $\langle \chi_M, \chi_N \rangle$ equal to the integral of $\chi_M(s)\chi_N(s^{-1})$ over G.

Next we consider the Schur orthogonality relations.

7.4 Theorem. *Let M and N be two simple G-modules. Then $\langle \chi_M, \chi_N \rangle = 0$ if M and N are nonisomorphic and $\langle \chi_M, \chi_N \rangle > 0$ if M and N are isomorphic. Moreover, $\langle \chi_M, \chi_M \rangle = 1$ if $F = \mathbf{C}$.*

Proof. If $f : M \to F$ is F-linear and $y \in N$, then for $u(x) = f(x)y$ we have $\tilde{u} = 0$ when M and N are nonisomorphic. If $a(s)$ is a matrix element of s_M and $b(s)$ of s_N, the integral of $a(s)b(s^{-1})$ over G is zero. Since $\langle \chi_M, \chi_N \rangle$ equal the integral of $\chi_M(s)\chi_N(s^{-1})$ over G is a sum of integrals of the form $a(s)b(s^{-1})$ over G, we have $\langle \chi_M, \chi_N \rangle = 0$ when M and N are nonisomorphic.

For the second statement, let $F = C$ and $e_1, \ldots, e_n$ be an F-basis of M. Let $E'_{i,j}$ denote the integral of $sE_{i,j}s^{-1}$ and $E''_{i,j}$ of $s^{-1}E_{i,j}s$ over G, where $E_{i,j}e_k = \delta_{i,k}e_j$. By Theorem (4.10) $E'_{i,j}$ is multiplication by $\lambda'_{i,j}$, and $E''_{i,j}$ is multiplication by $\lambda''_{i,j}$. Looking at the matrix elements of $E'_{i,j}$ and $E''_{i,j}$, we have $\lambda'_{i,j}\delta_{k,l} = \lambda''_{k,l}\delta_{i,j}$. This means that $\lambda'_{i,i} = \lambda''_{j,j} = \lambda$ and $\lambda'_{i,j} = \lambda''_{i,j} = 0$ for $i \neq j$. Since $1 = E_{1,1} + \cdots + E_{n,n}$, we have 1 equal to n times the integral of $sE_{i,i}s^{-1}$ or $n\lambda$. Therefore, λ equals $(\dim M)^{-1}$.

Finally, if $a(s)$ and $b(s)$ are two matrix elements of s_M, the integral of $a(s)b(s^{-1})$ is zero if $a \neq b$ and is $\lambda = 1/n$ if $a = b$ is a diagonal matrix element. Therefore, $\langle \chi_M, \chi_M \rangle$ equal $n(1/n)$ or 1. If M is real, we form $N = M \otimes C$, and $\langle \chi_N, \chi_N \rangle = \langle \chi_M, \chi_M \rangle$ is strictly positive. This proves the theorem.

7.5 Corollary. *The set of isomorphism classes of simple G-modules freely generates the abelian group $R_F(G)$, and the ring morphism $[M] \mapsto \chi_M$ of $R_G(G) \to \mathrm{ch}_F\, G$ is an isomorphism.*

Proof. By (6.6) the set of isomorphism classes of simple G-modules generate $R_F(G)$. If a sum $\sum_L a_L[L] = 0$ in $R_F(G)$, we have $\sum_L a_L\chi_L = 0$ in $\mathrm{ch}_F\, G$, and by taking the inner product with χ_M, we have $a_M\langle \chi_M, \chi_M \rangle = 0$ for each simple module M. Since $\langle \chi_M, \chi_M \rangle \neq 0$, a_M equals 0 for each distinct class of simple modules. This proves the corollary.

7.6 Corollary. *If M and N are two G-modules with $\chi_M = \chi_N$, then M and N are isomorphic.*

Proof. By (7.5), M and N are each the direct sum of simple modules in the same isomorphism classes.

7.7 Corollary. *Let $u: G \to H$ be a morphism of topological groups such that for each $t \in H$ there exists $s \in H$ with $sts^{-1} \in u(G)$. Then $R_F(u): R_F(H) \to R_F(G)$ is a monomorphism.*

Proof. For an H-module M we have $\chi_{u^*(M)} = \chi_M u$ on G. If $\chi_M u = \chi_N u$, we have $\chi_M = \chi_N$ since $u(G)$ intersects each conjugate class of elements in H and since characters are constant on conjugate classes, by (7.2). By (7.6) M and N are isomorphic. Therefore, if $R_F(u)[M] = R_F(u)[N]$, we have $[M] = [N]$, and this verifies the corollary.

7.8 Corollary. *Let $u: G \to G$ be an inner automorphism. Then $R_F(u)$ is the identity on $R_F(G)$.*

Proof. By the last properties of characters in (7.2), u induces the identity on $\mathrm{ch}_F\, G$ and, therefore, on $R_F(G)$.

This corollary can be deduced from the definition of $R_F(G)$ directly.

8. Maximal Tori

An important class of compact groups are the tori.

8.1 Definition. The n-dimensional torus, denoted $\mathbf{T}^n$, is the quotient topological group $\mathbf{R}^n$ mod $\mathbf{Z}^n$. A torus is any topological group isomorphic to an n-dimensional torus.

There follow the well-known results about Lie groups which we assume for the discussion in this section.

8.2 Background Results. See Chevalley [1] for proofs.

(1) For $n \neq m$ the tori T^n and T^m are nonisomorphic. Consequently, the dimension of a torus is well defined.
(2) A topological group is a torus if and only if it is a compact, abelian, connected Lie group. The natural quotient map $\mathbf{R}^n \rightarrow \mathbf{T}^n$ is a Lie group morphism.
(3) Every element in a connected, compact Lie group G is a member of a subgroup T of G, where T is a torus.
(4) A closed subgroup of a Lie group is a Lie group.

We use the following definition of maximal torus in a topological group.

8.3 Definition. A subgroup T is a maximal torus of a compact group G provided T is a torus with $G = \bigcup_{s \in G} sTs^{-1}$.

The condition $G = \bigcup_{s \in G} sTs^{-1}$ says that each conjugate class of G intersects T. Since T is connected, G is also connected. By examples [see (8.4)], we see that a wide class of groups have maximal tori. Their importance lies in the fact that Corollary (7.7) applies to the inclusion $T \rightarrow G$, where T is a maximal torus of G. In general, every compact, connected Lie group has a maximal torus.

8.4 Examples. A maximal torus of $U(n)$ and of $SU(n)$ consists of all diagonal matrices. Let $D(\theta)$ denote the matrix

$$\begin{bmatrix} \cos \theta & -\sin \theta \\ \sin \theta & \cos \theta \end{bmatrix}$$

let $D(\theta_1, \ldots, \theta_r)$ denote the $2r \times 2r$ matrix with $D(\theta_1)$, ..., $D(\theta_r)$ on the diagonal, and let $D(\theta_1, \ldots, \theta_r, *)$ denote the $(2r + 1) \times (2r + 1)$ matrix with $D(\theta_1), \ldots, D(\theta_r)$, 1 on the diagonal. Then the subgroup of all $D(\theta_1, \ldots, \theta_r)$ is a maximal torus of $SO(2r)$, and the subgroup of all $D(\theta_1, \ldots, \theta_r, *)$ is a maximal torus of $SO(2r + 1)$.

We use the following interpretation of the well-known theorem of Kronecker in number theory (see Hardy and Wright [1, theorem 442, p. 380]).

8.5 Theorem. *Let $a_1, \ldots, a_n \in \mathbf{R}$ such that $1, a_1, \ldots, a_n$ are $\mathbf{Q}$-linearly independent. Let a denote the class of $(a_1, \ldots, a_n)$ in $\mathbf{T}^n$. Then the set of all a^k, where $k \geq 0$, is dense in $\mathbf{T}^n$.*

8.6 Definition. An element a of a topological group G is a generator provided the set of all a^k, where $k \geq 0$, is dense in G. A topological group is monic provided it has a generator.

Theorem (8.5) says that each torus is monic.

8.7 Theorem. *Let G be a topological group with a maximal torus T. If T' is any torus subgroup of G, then $T' \subset sTs^{-1}$ for some $s \in G$. Moreover, T' is a maximal torus if and only if $T' = sTs^{-1}$.*

Proof. If a is a generator of T', we have $a \in sTs^{-1}$ for some $s \in G$. This means that $a^k \in sTs^{-1}$ for all $k \geq 0$ and $T' \subset sTs^{-1}$ for this $s \in G$. If $T' = sTs^{-1}$, we have $G = \bigcup_{s \in G} sT's^{-1}$, and T' is a maximal torus. If T' is a maximal torus, we have $T' \subset sTs^{-1}$ and $T \subset tT't^{-1}$ or $T \subset tsT(ts)^{-1}$. But for reasons of dimension [see (1) in (8.2)], $T = tsT(ts)^{-1}$ and $T' = sTs^{-1}$. This proves the theorem.

8.8 Remarks. By Theorem (8.7) a maximal torus of G is a maximal element in the ordered set (by inclusion) of torus subgroups of G. If G has a maximal torus in the sense of Definition (8.3), the maximal tori of G are precisely the maximal elements of the ordered set of torus subgroups of G. It is a theorem of E. Cartan, for which A. Weil and G. Hunt (see Hunt [1]) have given proofs, that each maximal element in the ordered set of torus subgroups of a connected Lie group G is a maximal torus in the sense of Definition (8.3).

8.9 Definition. Let G be a topological group with maximal torus T. Then the rank of G is the dimension of T.

Observe that the rank of a group is independent of the maximal torus T by (8.7) and (1) in (8.2).

We can improve on the result of (8.7) as interpreted in (8.8). We can prove that a maximal torus is actually a maximal element in the set of abelian closed subgroups of G for a compact Lie group G. For this we use the next lemma.

8.10 Lemma. *Let G be an abelian Lie group whose connected component is a torus T and G/T is a finite cyclic group with m elements. Then G is a monic group.*

Proof. Let a be a generator of T, and let $b \in G$ such that the image of b generates G/T. Then b^m is in T, and there exists $c \in T$ with $b^m c^m = a$. Then bc is the desired generator.

8.11 Theorem. *Let T be a torus subgroup of a connected, compact Lie group G. Let s commute with all elements of T. Then there exists a torus subgroup T' of G with $T \subset T'$ and $s \in T'$.*

Proof. Let A be the closed subgroup of G generated by T and s, and let T_0 be the connected component of the identity in A. Then T_0 is connected, compact, and abelian, and, therefore, a closed subgroup of a Lie group. By (4) and (2) of (8.2), T_0 is a torus subgroup. Then A/T_0 is a finite group since it is compact and discrete. Let A' be the subgroup of A generated by T_0 and s. By (8.10), A' has a generator x. Therefore, there is a torus T' with $A' \subset T'$; moreover, we have $T \subset T'$ and $s \in T'$.

8.12 Remark. If T is a maximal torus, then we have $s \in T$ for all $s \in G$ commuting with each element T. Consequently, maximal tori are maximal elements of the ordered set of closed abelian subgroups of a compact, connected Lie group. On the contrary, it is not true that all maximal abelian closed subgroups are tori.

For a torus subgroup T of G, let N_T denote the normalizer of T in G, that is, all $s \in G$ with $sTs^{-1} = T$. Then T is a normal subgroup of N_T.

8.13 Definition. The Weyl group $W(G)$ of a compact group G is N_T/T, where T is a maximal torus of G.

The function $t \mapsto sts^{-1}$ is an automorphism of T for $s \in N_T$, which is the identity for $s \in T$. Therefore, $W(G)$ acts as an automorphism group of T. We shall compute the Weyl groups for the classical groups $U(n)$, $SU(n)$, $SO(n)$, $Sp(n)$, and Spin(n) in the next chapter.

8.14 Theorem. *Let T be a maximal torus for a compact Lie group G, and let $W(G)$ be the Weyl group of G. Then as a transformation group of T only the identity of $W(G)$ acts as the identity on T, and $W(G)$ is a finite group.*

Proof. The first statement follows immediately from (8.12). For the second, the groups N_T and $W(G)$ are compact. The action of an element u of $W(G)$ on T^n determines a linear transformation $\theta_u : \mathbf{R}^n \to \mathbf{R}^n$, where $\theta_u(\mathbf{Z}^n) \subset \mathbf{Z}^n$. This construct defines a continuous injection of $W(G)$ into a discrete space. This is possible only if $W(G)$ is finite.

8.15 Remark. If T is a maximal torus in G and if W is the Weyl group, then by (7.7) the morphism $R(G) \to R(T)$ induced by the inclusion is a monomorphism. The Weyl group W acts on T, and therefore on $R(T)$. The subring $R(T)^W$ of elements left elementwise fixed by W contains the image of $R(G)$ since an inner automorphism of G induces the identity on $R(G)$ by (7.8).

9. The Representation Ring of a Torus

We begin with a general result on the complex representation rings of abelian groups.

9.1 Theorem. *Let G be an abelian group, and let M be a simple G-module over $\mathbf{C}$. Then M is one-dimensional over $\mathbf{C}$, and s_M is multiplication by λ_s, where $s \mapsto \lambda_s$ is a group morphism $G \to \mathbf{C}^\times$, with $\mathbf{C}^\times$ the multiplicative group of nonzero elements in $\mathbf{C}$.*

Proof. For $s \in G$, the fact that G is abelian means that $s_M \colon M \to M$ is a G-morphism. By Theorem (4.10) s_M is multiplication by a complex number. Since each one-dimensional subspace of M is a G-submodule and since M is simple, M is one-dimensional. The last statement is immediate from the relation $\lambda_s \lambda_t x = \lambda_{s+t} x$.

9.2 Example. Let $M(k_1, \ldots, k_n)$ denote the one-dimensional representation of T^n, where the action of T^n is given by the relation $(\theta_1, \ldots, \theta_n)z = \exp[2\pi i(k_1\theta_1 + \cdots + k_n\theta_n)]z$ for $(k_1, \ldots, k_n) \in \mathbf{Z}^n$. Since $s(x \otimes y) = sx \otimes sy$ is the relation defining the action of G on a tensor product, $M(k_1, \ldots, k_n) \otimes M(l_1, \ldots, l_n) = M(k_1 + l_1, \ldots, k_n + l_n)$.

9.3 Theorem. *The simple $T(n)$-modules are the one-dimensional $T(n)$-modules, and each one is isomorphic to precisely one module of the form $M(k_1, \ldots, k_n)$. The ring $RT(n)$ is the polynomial ring $\mathbf{Z}[\alpha_1, \alpha_1^{-1}, \ldots, \alpha_n, \alpha_n^{-1}]$, where α_i is the class of $M(0, \ldots, 0, \overset{(i)}{1}, 0, \ldots, 0)$.*

Proof. The first statement follows from (9.1) and the fact that all group morphisms $T(n) \to \mathbf{C}^\times$ are of the form $(\theta_1, \ldots, \theta_n) \to \exp[2\pi i(k_1\theta_1 + \cdots + k_n\theta_n)]$. For the second statement, observe that $\alpha_1^{k_1} \cdots \alpha_n^{k_n}$ equals the class of $M(k_1, \ldots, k_n)$. These monomials form a $\mathbf{Z}$-base of $RT(n)$ and the multiplicative structure follows from (9.2).

9.4 Remark. In view of (7.8) every ring $R(G)$ is a subring of $\mathbf{Z}[\alpha_1, \alpha_1^{-1}, \ldots, \alpha_n, \alpha_n^{-1}]$, where G is of rank n. This gives some information about the nature of $R(G)$.

9.5 Theorem. *For a topological group G with maximal torus T, the complex representation ring $R(G)$ is a split λ-ring with involution. For the real representation ring $RO(G)$ there is a λ-ring monomorphism $\varepsilon_U \colon RO(G) \to R(G)$ given by complexification; that is, $\varepsilon_U([L]) = [L \otimes \mathbf{C}]$.*

Proof. A monomorphism $R(G) \to R(T)$ is induced by inclusion. The line elements of $R(T)$ are the classes $\alpha_1^{k(1)} \cdots \alpha_n^{k(n)} = [M(k(1), \ldots, k(n))]$. The involution

$[M]^*$ is $[\overline{M}]$, where $\overline{M}$ is the conjugate module of M and ax in $\overline{M}$ is $\bar{a}x$ in M. We have $(\alpha_1^{k(1)} \cdots \alpha_n^{k(n)})^* = \alpha_1^{-k(1)} \cdots \alpha_n^{-k(n)}$.

Finally, if $\varepsilon_0 : R(G) \to RO(G)$ is the group homomorphism given by restriction of scalars from $\mathbf{C}$ to $\mathbf{R}$, we have $\varepsilon_0(\varepsilon_U(x)) = 2x$. Since $R(G)$ is a free abelian group by (7.5), ε_U is a monomorphism. This proves the theorem. The results of (2.9) and (2.12) apply to $R(G)$ and $RO(G)$.

10. The ψ-Operations on $K(X)$ and $KO(X)$

We wish to verify the formula $\psi^k \psi^l = \psi^{kl}$ and prove that ψ^k is a ring morphism as defined on $K(X)$ and $KO(X)$. To do this, we shall use the result that these statements hold for $R(G)$ and $RO(G)$, where G is a compact Lie group with a maximal torus. An oriented vector bundle is one with structure group $SO(n)$.

We begin with some preliminary propositions.

10.1 Proposition. *Let ξ be an n-dimensional vector bundle, and let ζ be a line bundle over X. Then the bundles $\Lambda^{n+1}(\xi \oplus \zeta)$ and $\Lambda^n(\xi) \otimes \zeta$ are isomorphic over X.*

Proof. This is true for vector spaces with a functorial isomorphism. By 5(6.4) it is true for vector bundles.

10.2 Corollary. *Every vector bundle class $\{\xi\}$ in $KO(X)$ is of the form $a_1 - a_2$, where a_1 is the class of an oriented vector bundle and a_2 is the class of a line bundle.*

Proof. In (10.1), let ζ equal $\Lambda^n\xi$, where n is the dimension of ξ. Then $\xi \oplus \Lambda^n\xi$ is orientable, and in $KO(X)$ we have $\{\xi\} = \{\xi \oplus \Lambda^n\xi\} - \{\Lambda^n\xi\}$.

Let ζ^k denote the k-fold tensor product of a vector bundle ζ.

10.3 Proposition. *Let ξ be an n-dimensional vector bundle, and let ζ be a line bundle over X. Then $\Lambda^k(\xi \otimes \zeta)$ and $\Lambda^k(\xi) \otimes \zeta^k$ are X-isomorphic, and in $K(X)$ or $KO(X)$ there is the relation $\psi^k(\{\xi\}\{\zeta\}) = \psi^k(\{\xi\})\psi^k(\{\zeta\})$.*

Proof. For vector spaces there is a functorial isomorphism between $\Lambda^k(\xi \otimes \zeta)$ and $\Lambda^k(\xi) \otimes \zeta^k$. By 5(6.4) there is a natural isomorphism between $\Lambda^k(\xi \otimes \zeta)$ and $\Lambda^k(\xi) \otimes \zeta^k$.

For the formula involving ψ^k we use induction on k and the relation given in (2.5).

10.4 Notations. Let ρ_m denote the regular complex representation $U(m)$ and its class in $RU(m)$ or the regular real representation of $SO(m)$ and its class in $RO(SO(m))$. Let ξ^m and η^n be two vector bundles over X with principal bundles α or β. Using the constructions of (5.4), we have a morphism $\widehat{\alpha \oplus \beta}: R(U(m) \oplus U(n)) \to K(X)$ such that $(\widehat{\alpha \oplus \beta})(\rho_m \oplus 0) = \tilde{\alpha}(\rho_m) = \{\xi\}$ and $(\widehat{\alpha \oplus \beta})(0 \oplus \rho_n) = \tilde{\beta}(\rho_n) = \{\eta\}$ in $K(X)$ for complex vector bundles. For real vector bundles the morphism $\widehat{\alpha \oplus \beta}$ is defined $RO(SO(m) \oplus SO(n)) \to KO(X)$.

With these notations we are able to state and prove the main result of this section.

10.5 Theorem. *In $K(X)$ and $KO(X)$ the Adams operations ψ^k are ring morphisms, and the relation $\psi^k \psi^l = \psi^{kl}$ holds for the action of ψ^k on these groups.*

Proof. Observe, if T is a maximal torus of G and T' of G', that $T \oplus T'$ is a maximal torus of $G \oplus G'$.

Then the theorem holds for $K(X)$ and for the subgroup of $KO(X)$ generated by classes of oriented real vector bundles from the relation $\psi^k(\widehat{\alpha \oplus \beta}) = (\widehat{\alpha \oplus \beta})\psi^k$ and from the fact that the above relations hold in $R(U(m) \oplus U(n))$ and in $RO(SO(m) \oplus SO(n))$ by Theorem (9.5). Finally, the theorem holds for all of $KO(X)$, by (10.2) and (10.3).

10.6 Remark. We leave it to the reader to define the action of ψ^{-k} on $K(X)$ and $KO(X)$ such that ψ^{-1} is the complex conjugate of ξ for a line bundle ξ.

11. The ψ-Operations on $\tilde{K}(S^n)$

The following is a general calculation of ψ^k for elements defined by line bundles.

11.1 Proposition. *Let ζ be a line bundle over X. Then we have $\psi^k(\{\zeta\}) = \{\zeta\}^k$, and if $(\{\zeta\} - 1)^2 = 0$, then we have $\psi^k(\{\zeta\} - 1) = k(\{\zeta\} - 1)$.*

Proof. The first statement follows from (2.3), and the second statement from the following calculation:

$$\psi^k(\{\zeta\} - 1) = \{\zeta\}^k - 1 = (\{\zeta\} - 1 + 1)^k - 1 = k(\{\zeta\} - 1)$$

We recall from 11(5.5) that $\tilde{K}(S^{2m+1}) = 0$ and $\tilde{K}(S^{2m}) = \mathbf{Z}\beta_{2m}$, where β_{2m} is the generator of a cyclic group and $\beta_{2m}^2 = 0$. Moreover, the image of β_{2m} in $\tilde{K}(S^2 \times \overset{(m)}{\cdots} \times S^2)$ under the natural monomorphism $\tilde{K}(S^{2m}) \to \tilde{K}(S^2 \times \overset{(m)}{\cdots} \times S^2)$ equals a product $a_1 \cdots a_m$, where $a_i^2 = 0$ and $a_i = \{\zeta_i\} - 1$ such that ζ_i is a line bundle on $S^2 \times \cdots \times S^2$. The image of $\psi^k(\beta_{2m})$ is

$\psi^k(a_1) \cdots \psi^k(a_m) = k^m a_1 \cdots a_m$, which is k^m times the image of β_{2m}. Since $\tilde{K}(S^{2m}) \to \tilde{K}(S^2 \times \cdots \times S^2)$, we have the following theorem.

11.2 Theorem. *The group $\tilde{K}(S^{2m})$ is infinite cyclic with generator β_{2m} such that $\beta_{2m}^2 = 0$. Moreover, $\psi^k(\beta_{2m}) = k^m \beta_{2m}$.*

CHAPTER 14

Representation Rings of Classical Groups

In the previous chapter we saw the importance of the relation between the representation rings $R(G)$ and $K(X)$. In this chapter we give a systematic calculation of $R(G)$ for G equal to $U(n)$, $SU(n)$, $Sp(n)$, $SU(n)$, and Spin(n). Finally, we consider real representations of Spin(n), which were successfully used by Bott for a solution of the vector field problem.

1. Symmetric Functions

Let R denote an arbitrary commutative ring with 1, and let $R[x_1,\ldots,x_n]$ denote the ring of polynomials in n variables. Let S_n denote the group under composition of all bijections $\{1,2,\ldots,n\} \to \{1,2,\ldots,n\}$, that is, the permutation group on n letters.

For each polynomial $P \in R[x_1,\ldots,x_n]$ and $\tau \in S_n$ we define $^\tau P$ by the relation $^\tau P(x_1,\ldots,x_n) = P(x_{\tau(1)},\ldots,x_{\tau(n)})$.

1.1 Definition. A polynomial $P(x_1,\ldots,x_n)$ is symmetric provided $^\tau P = P$ for each $\tau \in S_n$.

1.2 Example. In the ring $R[x_1,\ldots,x_n,z]$ we form the product $\prod_{1 \leq i \leq n} (z + x_i)$ and write it as a polynomial in z with coefficients in $R[x_1,\ldots,x_n]$, that is,

$$\prod_{1 \leq i \leq n} (z + x_i) = \sum_{0 \leq i \leq n} \sigma_i(x_1,\ldots,x_n)z^{n-i}.$$ Since the product of linear polynomials is invariant under the action of S_n, the polynomials $\sigma_i(x_1,\ldots,x_n)$ are symmetric. We call $\sigma_i(x_1,\ldots,x_n)$ the ith elementary symmetric function. Occasionally we write σ_i^n to denote the number of variables n in σ_i.

1.3 Remarks. Observe that $\sigma_0^n = 1$, $\sigma_1^n(x_1,\ldots,x_n) = x_1 + \cdots + x_n$, $\sigma_n^n(x_1,\ldots,x_n) = x_1 \cdots x_n$, and $\sigma_k^n(x_1,\ldots,x_n) = \sum x_{i(1)} \cdots x_{i(k)}$, where $1 \leq i(1) < \cdots < i(k) \leq n$. By convention we define $\sigma_i^n = 0$ for $i > n$. Finally, we have $\sigma_i^{n-1}(x_1,\ldots,x_{n-1}) = \sigma_i^n(x_1,\ldots,x_{n-1},0)$, and σ_i^n is homogeneous of degree i in n variables.

1.4 Example. If $g(y_1,\ldots,y_m)$ is a polynomial, $g(\sigma_1^n,\ldots,\sigma_m^n)$ is a symmetric polynomial in n variables.

1.5 Definition. In the ring $R[y_1,\ldots,y_m]$, the weight of a monomial $y_1^{a(1)} \cdots y_m^{a(m)}$ is defined to be $a(1) + 2a(2) + \cdots + ma(m)$. The weight of an arbitrary polynomial is the maximum weight of the nonzero monomials of which it is a sum.

If $g(y_1,\ldots,y_m)$ is a polynomial of weight k, then $g(\sigma_1(x_1,\ldots,x_n),\ldots,\sigma_m(x_1,\ldots,x_n))$ is a polynomial of degree k in $x_1,\ldots,x_n$.

The next theorem is the fundamental theorem on elementary symmetric functions.

1.6 Theorem. *The subring $R[\sigma_1,\ldots,\sigma_n]$ of $R[x_1,\ldots,x_n]$ contains all the symmetric functions, and the elementary symmetric functions are algebraically independent.*

Proof. We prove this by induction on n. For $n = 1$, $\sigma_1(x_1) = x_1$, and the result holds. We let f be a symmetric polynomial of degree k and assume the result for all polynomials of degree $\leq k - 1$. Then $f(x_1,\ldots,x_{n-1},0) = g(\sigma_1',\ldots,\sigma_{n-1}')$, where $\sigma_i'(x_1,\ldots,x_{n-1}) = \sigma_i(x_1,\ldots,x_{n-1},0)$. We form the symmetric polynomial $f_1(x_1,\ldots,x_n) = f(x_1,\ldots,x_n) - g(\sigma_1,\ldots,\sigma_{n-1})$. Then we have $\deg f_1 \leq k$ and $f_1(x_1,\ldots,x_{n-1},0) = 0$. Since f_1 is symmetric, x_n and $\sigma_n = x_1 \cdots x_n$ divide $f_1(x_1,\ldots,x_n)$. Therefore, $f(x_1,\ldots,x_n) = \sigma_n f_2(x_1,\ldots,x_n) + g(\sigma_1,\ldots,\sigma_{n-1})$, where f_2 is symmetric and $\deg f_2 < \deg f$. By applying the inductive hypothesis to f_2, we have $f(x_1,\ldots,x_n) = h(\sigma_1,\ldots,\sigma_n)$.

To show that the $\sigma_1, \ldots, \sigma_n$ are algebraically independent, we use induction on n again. For $n = 1$, we have $\sigma_1 = x_1$. Let $f(\sigma_1,\ldots,\sigma_n) = 0$, where $f(\sigma_1,\ldots,\sigma_n) = f_m(\sigma_1,\ldots,\sigma_{n-1})(\sigma_n)^m + \cdots + f_0(\sigma_1,\ldots,\sigma_{n-1}) = 0$ and m is minimal. Let $x_n = 0$ in σ_i, and the result is $\sigma_n = 0$ and $f_0(\sigma_1',\ldots,\sigma_{n-1}') = 0$, where $\sigma_i'(x_1,\ldots,x_{n-1}) = \sigma_i(x_1,\ldots,x_{n-1},0)$. By inductive hypothesis, we have $f_0 = 0$, and this contradicts the minimal character of m. Therefore, $m = 0$ and $f = 0$. This proves the theorem.

1.7 Application. There exist "universal" polynomials in n variables s_k^n such that

$$x_1^k + \cdots + x_n^k = s_k^n(\sigma_1,\ldots,\sigma_n)$$

The polynomials s_k^n arise from another construction. Recall that a formal series $a_0 + a_1 t + \cdots$ in $R[[t]]$ is a unit if and only if a_0 is a unit in R. Let $1 + R[[t]]^+$ denote the multiplicative group of formal series $1 + a_1 t + \cdots$.

1.8 Proposition. *Let* $f(t) = 1 + y_1 t + \cdots + y_n t^n$ *in* $1 + R[[t]]^+$. *Then*

$$\sum_{k \geq 0} s_k^n(y_1, \ldots, y_n)(-t)^k = -t[(d/dt)f(t)]/f(t).$$

Proof. We prove the result for the polynomial ring $R[\sigma_1, \ldots, \sigma_n] \subset R[x_1, \ldots, x_n]$ and then substitute y_i for σ_i. Here σ_i is the ith elementary symmetric function in the $x_1, \ldots, x_n$. For $f(t) = 1 + \sigma_1 t + \cdots + \sigma_n t^n$ we have

$$f(t) = (1 + x_1 t) \cdots (1 + x_n t)$$

and

$$
\begin{aligned}
-t[(d/dt)f(t)]/f(t) &= -t(d/dt)\log f(t) \\
&= -tx_1/(1 + x_1 t) - \cdots - tx_n/(1 + x_n t) \\
&= \sum_{k \geq 1} (x_1^k + \cdots + x_n^k)(-t)^k \\
&= \sum_{k \geq 1} s_k^n(\sigma_1, \ldots, \sigma_n)(-t)^k
\end{aligned}
$$

This proves the proposition.

2. Maximal Tori in $SU(n)$ and $U(n)$

Let $x_1, \ldots, x_n$ be an orthonormal base of $\mathbf{C}^n$, and let $T(x_1, \ldots, x_n)$ denote the subgroup of $u \in U(n)$ with $u(x_j) = a_j x_j$. Let $ST(x_1, \ldots, x_n)$ denote the subgroup $T(x_1, \ldots, x_n) \cap SU(n)$ of $SU(n)$. As usual, let $e_1, \ldots, e_n$ denote the canonical base of $\mathbf{C}^n$. Every orthonormal base $x_1, \ldots, x_n$ of $\mathbf{C}^n$ is of the form $w(e_1), \ldots, w(e_n)$, where $w \in U(n)$. The element w is uniquely determined by the base $x_1, \ldots, x_n$. The base is called special provided $w \in SU(n)$.

2.1 Theorem. *With the above notations, the subgroups* $T(x_1, \ldots, x_n)$ *are the maximal tori of* $U(n)$, *and the subgroups* $ST(x_1, \ldots, x_n)$ *for special bases* $x_1, \ldots, x_n$ *are the maximal tori of* $SU(n)$. *The rank of* $U(n)$ *is* n *and of* $SU(n)$ *is* $n - 1$. *The Weyl group of* $U(n)$ *and of* $SU(n)$ *is the symmetric group of all permutations of the indices of the coordinates* $(a_1, \ldots, a_n)$.

Proof. First, we observe that $u \in T(x_1, \ldots, x_n)$ is determined by $u(x_j) = a_j x_j$, where $a_j \in \mathbf{C}$ and $|a_j| = 1$ for $1 \leq j \leq n$. For $ST(x_1, \ldots, x_n)$ there is the additional condition that $a_1 \cdots a_n = 1$. Consequently, $T(x_1, \ldots, x_r)$ is an n-dimensional torus, and $ST(x_1, \ldots, x_n)$ is $(n - 1)$-dimensional. Next, we observe that $wT(e_1, \ldots, e_n)w^{-1} = T(w(e_1), \ldots, w(e_n))$ for $w \in U(n)$ and $wST(e_1, \ldots, e_n)w^{-1} = ST(w(e_1), \ldots, w(e_n))$ for $w \in SU(n)$. For each $u \in U(n)$ there is a base $x_1, \ldots, x_n$ of $\mathbf{C}^n$ such that $u(x_j) = a_j x_j$, where the x_j are eigenvectors of u. There exist $w \in U(n)$ with $x_j = w(e_j)$ and $u \in wT(e_1, \ldots, e_n)w^{-1}$. Since the x_j can be changed by a scalar multiple a, where $|a| = 1$, we can

assume $w \in SU(n)$ and $a_1 \cdots a_n = 1$ for $u \in SU(n)$. We have proved that
$U(n) = \bigcup_{w \in U(n)} wT(e_1, \ldots, e_n)w^{-1}$ and $SU(n) = \bigcup_{w \in SU(n)} wST(e_1, \ldots, e_n)w^{-1}$.

Finally, to compute the Weyl groups, we let $u \in ST(e_1, \ldots, e_n)$ with $u(e_j) = a_j e_j$ and $a_j \neq a_k$ for $j \neq k$. For $wuw^{-1} \in ST(e_1, \ldots, e_n)$ or $T(e_1, \ldots, e_n)$ we have $wuw^{-1}(e_j) = b_j e_j$ and $uw^{-1}(e_j) = b_j w^{-1}(e_j)$. Therefore, $w^{-1}(e_j) = ae_k$, with $|a| = 1$ and $j = k$. The Weyl group W of $U(n)$ and of $SU(n)$ is the full permutation group on the set of n elements $\{e_1, \ldots, e_n\}$ since conjugation by w, where $w(e_1) = e_2$, $w(e_2) = -e_1$, and $w(e_i) = e_i$ for $i \geq 3$, permutes the first two coordinates of $(a_1, \ldots, a_n) \in T(e_1, \ldots, e_n)$ or $ST(e_1, \ldots, e_n)$. Similarly all permutations are in the Weyl group. This proves the theorem.

3. The Representation Rings of $SU(n)$ and $U(n)$

In the next theorem we determine $R(U(n))$ and $R(SU(n))$, using properties of elementary symmetric functions. We use the notation λ_i for the class in $RU(n)$ or $RSU(n)$ of the ith exterior power $\Lambda^i \mathbf{C}^n$, where $U(n)$ or $SU(n)$ acts on $\mathbf{C}^n$ in the natural manner by substitution. If ρ_n denotes the class of $\mathbf{C}^n$ in the λ-ring $RU(n)$ or $RSU(n)$, then λ_i equals $\lambda^i(\rho_n)$.

3.1 Theorem. *The ring* $RU(n)$ *equals* $\mathbf{Z}[\lambda_1, \ldots, \lambda_n, \lambda_n^{-1}]$, *where there are no polynomial relations between* $\lambda_1, \ldots, \lambda_n$. *As a subring of*

$$RT(n) = \mathbf{Z}[\alpha_1, \alpha_1^{-1}, \ldots, \alpha_n, \alpha_n^{-1}]$$

the relation $\lambda_k = \sum_{i(1) < \cdots < i(k)} \alpha_{i(1)} \cdots \alpha_{i(k)}$ *holds. The* $RSU(n)$ *equals the polynomial ring* $\mathbf{Z}[\lambda_1, \ldots, \lambda_{n-1}]$.

Proof. As a $T(e_1, \ldots, e_n)$-module, $\mathbf{C}^n$ is a direct sum of one-dimensional modules corresponding to $\alpha_1, \ldots, \alpha_n$. The representation as an elementary symmetric function $\lambda_k = \sum_{i(1) < \cdots < i(k)} \alpha_{i(1)} \cdots \alpha_{i(k)}$ follows from Proposition 13(4.6). Observe that $\lambda_n = \alpha_1 \cdots \alpha_n$ is a one-dimensional class where $uy = (\det u)y$ for $u \in U(n)$. Then the one-dimensional class λ_n^{-1} is defined by $uy = (\det u)^{-1}y$ for $y \in \mathbf{C}$. Therefore, we have $\mathbf{Z}[\lambda_1, \ldots, \lambda_n, \lambda_n^{-1}] \subset RU(n) \subset RT(n)^W \subset RT(n)$. From properties of elementary symmetric functions Theorem (1.6), we know there are no polynomial relations between $\lambda_1, \ldots, \lambda_n$.

Finally, we prove that $\mathbf{Z}[\lambda_1, \ldots, \lambda_n, \lambda_n^{-1}] = RT(n)^W$, which also implies it equals $RU(n)$. Let $f(\alpha_1, \ldots, \alpha_n)$ be a polynomial in $\alpha_1, \alpha_1^{-1}, \ldots, \alpha_n, \alpha_n^{-1}$ which is invariant under permutations of $\alpha_1, \ldots, \alpha_n$. Then $f(\alpha_1, \ldots, \alpha_n) = (\lambda_n)^{-k}g(\alpha_1, \ldots, \alpha_n)$, where g is a polynomial in $\alpha_1, \ldots, \alpha_n$ and invariant under all permutations of $\alpha_1, \ldots, \alpha_n$. By Theorem (1.6), we have $g(\alpha_1, \ldots, \alpha_n) = h(\lambda_1, \ldots, \lambda_n)$ for some polynomial h. This proves the above statement concerning $RU(n)$.

For the ring $RSU(n)$, observe that the ring $R(ST(e_1,\ldots,e_n))$ is the quotient of $RT(e_1,\ldots,e_n)$ by the ideal generated by $\lambda_n - 1 = \alpha_1 \cdots \alpha_n - 1$. Therefore, $RSU(n)$ is the polynomial ring $\mathbf{Z}[\lambda_1,\ldots,\lambda_{n-1}]$. This proves the theorem.

4. Maximal Tori in $Sp(n)$

4.1 Description of $Sp(n)$. We view $\mathbf{H}^n$ as $\mathbf{C}^{2n}$. Then multiplication by j is a conjugate linear $J\colon \mathbf{C}^{2n} \to \mathbf{C}^{2n}$ defined by $J(e_i) = e_{i+n}$ for $1 \leq i \leq n$ and $J(e_i) = -e_{i-n}$ for $n + 1 \leq i \leq 2n$. We have $J^2 = -1$, and J determines the quaternionic structure on the complex vector space $\mathbf{C}^{2n}$. The identification $\mathbf{H}^n = \mathbf{C}^{2n}$ allows us to view $Sp(n)$ as a subgroup of $U(2n)$. For $u \in U(2n)$ we have $u \in Sp(n)$ if and only if $uJ = Ju$ or, in other words, $u \in Sp(n)$ if and only if $\beta(u(x), u(y)) = \beta(x, y)$, where $\beta(x, y) = (x|J(y))$. The form β is an antiher- mitian form, that is, $\beta(y, x) = -(J^2 y|Jx) = -(Jy|x) = -\overline{\beta(x, y)}$. In terms of $(2n)$-tuples $x, y \in \mathbf{C}^{2n}$, we have $\beta(x, y) = \displaystyle\sum_{1 \leq i \leq n} (x_i \bar{y}_{i+n} - x_{i+n} \bar{y}_i)$. For a diago- nal element $u = \mathrm{diag}(a_1,\ldots,a_{2n}) \in U(2n)$ we have $u \in Sp(n)$ if and only if $a_i = \bar{a}_{i+n}$ for $1 \leq i \leq n$ from the relation $uJ = Ju$.

As in Sec. 2, let $T(x_1,\ldots,x_n)$ denote the subgroup of $u \in Sp(n)$ with $u(x_k) = \exp(2\pi i \theta_k)x_k$ for an orthonormal base of $\mathbf{H}^n$. As usual, let $e_1,\ldots,e_n$ denote the canonical base of $\mathbf{H}^n$. Every orthonormal base $x_1,\ldots,x_n$ of $\mathbf{H}^n$ is of the form $w(e_1),\ldots,w(e_n)$, where $w \in Sp(n)$.

4.2 Theorem. *With the above notations, the subgroups $T(x_1,\ldots,x_n)$ are the maximal tori of $Sp(n)$. The rank of $Sp(n)$ is n, and the Weyl group of $Sp(n)$ is the symmetric group of all permutations of the indices of the coordinates $(\theta_1,\ldots,\theta_n)$ together with maps of the form $(\theta_1,\ldots,\theta_n) \mapsto (\pm\theta_1,\ldots,\pm\theta_n)$. It has $2^n n!$ elements.*

Proof. As with $U(n)$, we have $wT(x_1,\ldots,x_n)w^{-1} = T(w(x_1),\ldots,w(x_n))$, and it suffices to prove that $T(e_1,\ldots,e_n)$ is an n-dimensional torus. This is clear since it consists of all $\mathrm{diag}(a_1,\ldots,a_{2n})$ in $U(2n)$ with $|a_1| = \cdots = |a_n| = 1$ and $a_i = \bar{a}_{i+n}$ for $1 \leq i \leq n$. To prove that each $T(x_1,\ldots,x_n)$ is a maximal torus, it suffices to show that every $w \in Sp(n)$ is a member of some $T(x_1,\ldots,x_n)$. For this, we let x_1 be an eigenvector of w over $\mathbf{C}$; that is, we have $w(x_1) = a_1 x_1$, and $w(Jx_1) = Jw(x_1) = J(a_1 x_1) = \bar{a}_1 Jx_1$. Continuing this process on the orthogonal complement of x_1 and Jx_1 in $\mathbf{C}^{2n}$, we get the desired base $x_1,\ldots,x_n$ of $\mathbf{H}^n$ with $w \in T(x_1,\ldots,x_n)$.

Finally, to compute the Weyl group, we let $u \in T(e_1,\ldots,e_n)$ with $u(e_k) = a_k e_k$ and $a_j \neq a_k$ for $j \neq k$. For $wuw^{-1} \in T(e_1,\ldots,e_n)$, we have $wuw^{-1}(e_k) = b_k e_k$ and $uw^{-1}(e_k) = b_k w^{-1}(e_k)$. Therefore, $w^{-1}(e_k)$ equals some e_l and b_k equals a_l and $\bar{a}_l$, and the action of $u \mapsto wuw^{-1}$ permutes the coordinates and conju- gates some of the coordinates. All such permutations and conjugations are possible by inner automorphisms. This proves the theorem.

5. Formal Identities in Polynomial Rings

To compute the ring $RSp(n)$, we need some formal identities in the ring $Z[\alpha_1, \alpha_1^{-1}, \ldots, \alpha_n, \alpha_n^{-1}]$.

5.1 Proposition. *There exists a polynomial $f_m(y) \in Z[y]$ such that $x^m + x^{-m} = f_m(x + x^{-1})$. This polynomial satisfies the recursion formula $f_{m+1}(y) = yf_m(y) - f_{m-1}(y)$ with $f_0(y) = 1$ and $f_1(y) = y$.*

Proof. It suffices to establish the recursion formula. For this, we calculate $(x^m + x^{-m})(x + x^{-1}) = (x^{m+1} + x^{-(m+1)}) + (x^{m-1} + x^{-(m-1)})$ or $f_m(y)y = f_{m+1}(y) + f_{m-1}(y)$.

5.2 Proposition. *Let $f \in R[x, x^{-1}]$ such that $f(x) = f(1/x)$, where R is an arbitrary commutative ring with 1. Then f is a member of $R[x + x^{-1}] \subset R[x, x^{-1}]$.*

Proof. We have $f(x) = \sum_m a_m x^m$ with $a_m = a_{-m}$ since $f(x) = f(1/x)$. Therefore, f has the form $\sum_{m \geq 0} a_m(x^m + x^{-m}) = \sum_{m \geq 0} a_m j_m(x + x^{-1})$ by (5.1). This proves the proposition.

5.3 Corollary. *Let $f \in R[\alpha_1, \alpha_1^{-1}, \ldots, \alpha_r, \alpha_r^{-1}]$ with $f(\alpha_1, \ldots, \alpha_i, \ldots, \alpha_r) = f(\alpha_1, \ldots, \alpha_i^{-1}, \ldots, \alpha_r)$ for each i. Then we have*

$$f \in R[\alpha_1 + \alpha_1^{-1}, \ldots, \alpha_r + \alpha_r^{-1}]$$

Let $\sigma_1, \ldots, \sigma_r$ denote the elementary symmetric functions in the r variables $\alpha_1 + \alpha_1^{-1}, \ldots, \alpha_r + \alpha_r^{-1}$ and $\lambda_1, \ldots, \lambda_r$ denote the elementary symmetric function in the $2r$ variables $\alpha_1, \ldots, \alpha_r, \alpha_1^{-1}, \ldots, \alpha_r^{-1}$ (case 1) or the $2r + 1$ variables $\alpha_1, \ldots, \alpha_r, \alpha_r^{-1}, \ldots, \alpha_r^{-1}, 1$ (case 2).

5.4 Proposition. *With the above notations we have*

$$\lambda_1, \ldots, \lambda_r \in Z[\sigma_1, \ldots, \sigma_r] \qquad and \qquad \lambda_k = \sigma_k + \sum_{l < k} a_l \sigma_l$$

for $a_l \in Z$ and $k \leq r$. In addition, $\lambda_1, \ldots, \lambda_r$ are algebraically independent, and $Z[\sigma_1, \ldots, \sigma_r] = Z[\lambda_1, \ldots, \lambda_r]$.

Proof. The other statements follow easily from the relation $\lambda_k = \sigma_k + \sum_{l < k} a_l \sigma_l$.
For each sequence $1 \leq i(1) < \cdots < i(k) \leq r$ and numbers $\varepsilon(j) = \pm 1$, the monomial $\alpha_{i(1)}^{\varepsilon(1)} \cdots \alpha_{i(k)}^{\varepsilon(k)}$ appears in σ_k exactly once. The polynomial λ_k is a sum of monomials $\alpha_{i(1)}^{\varepsilon(1)} \cdots \alpha_{i(l)}^{\varepsilon(l)}$ with $i(1) \leq \cdots \leq i(l)$, where $i(p) = i(p + 1)$ implies $\varepsilon(p) = -\varepsilon(p + 1)$ and $l = k$ or $k - 1$ (with $k - 1$ only in case 2). For each $l < k$, if the monomial $\alpha_{i(1)}^{\varepsilon(1)} \cdots \alpha_{i(l)}^{\varepsilon(l)}$ appears in $\lambda_k - \sigma_k$ with coefficient a_l, the result of any permutation of the indices $i(1), \ldots, i(l)$ and any substitution

$\varepsilon(i) \mapsto -\varepsilon(i)$ appears with coefficient a_l. Therefore, we have $\lambda_k - \sigma_k = \sum_{l<k} a_l \sigma_l$. This proves the proposition.

6. The Representation Ring of $Sp(n)$

To compute the ring $RSp(n)$, we denote the class of the exterior power $\Lambda^i \mathbf{C}^{2n}$ in $RSp(n)$ by λ_i, where $Sp(n)$ acts on $\mathbf{C}^{2n} = \mathbf{H}^n$ by substitution.

6.1 Theorem. *The ring $RSp(n)$ equals the polynomial ring $\mathbf{Z}[\lambda_1, \ldots, \lambda_n]$, where as a subring of $\mathbf{Z}[\alpha_1, \alpha_1^{-1}, \ldots, \alpha_n, \alpha_n^{-1}]$ the element λ_k is the kth elementary symmetric function in the $2n$ variables $\alpha_1, \alpha_1^{-1}, \ldots, \alpha_n, \alpha_n^{-1}$.*

Proof. As a $T(e_1, \ldots, e_n)$-module, $\mathbf{C}^{2n} = \mathbf{H}^n$ is a direct sum of $2n$ one-dimensional modules corresponding to $\alpha_1, \alpha_1^{-1}, \ldots, \alpha_n, \alpha_n^{-1}$, and the representation of λ_k as the elementary symmetric function in $\alpha_1, \alpha_1^{-1}, \ldots, \alpha_n, \alpha_n^{-1}$ follows from Proposition 13(4.6). Since the Weyl group W consists of all permutations of $\{1, \ldots, n\}$ composed with substitutions $\alpha_i \mapsto \alpha_i^{\pm 1}$, we have by (5.2) the inclusions $\mathbf{Z}[\lambda_1, \ldots, \lambda_n] \subset RSp(n) \subset \mathbf{Z}[\sigma_1, \ldots, \sigma_n] = \mathbf{Z}[\alpha_1, \alpha_1^{-1}, \ldots, \alpha_n, \alpha_n^{-1}]^W$, where σ_k is the kth elementary symmetric function in the n variables $\alpha_1 + \alpha_1^{-1}, \ldots, \alpha_n + \alpha_n^{-1}$. By (5.3) and (5.4) we have $RSp(n) = \mathbf{Z}[\lambda_1, \ldots, \lambda_n]$ as a polynomial ring. This proves the theorem.

7. Maximal Tori and the Weyl Group of $SO(n)$

Let $x_1, \ldots, x_n$ denote an orthonormal base of $\mathbf{R}^n$, and let $T(x_1, \ldots, x_n)$ denote the subgroup of $u \in SO(n)$ with $u(\mathbf{R}e_{2i-1} + \mathbf{R}e_{2i}) \subset \mathbf{R}e_{2i-1} + \mathbf{R}e_{2i}$ for $1 \leq i \leq n/2$. We have $u(\mathbf{R}e_{2i-1} + \mathbf{R}e_{2i}) \in SO(2)$ and $u(e_n) = e_n$ for n odd. As usual, let $e_1, \ldots, e_n$ denote the canonical base of $\mathbf{R}^n$. Each orthonormal base $x_1, \ldots, x_n$ of $\mathbf{R}^n$ with $x_1 \wedge \cdots \wedge x_n = ae_1 \wedge \cdots \wedge e_n$ and $a > 0$ is of the form $w(e_1), \ldots, w(e_n)$, where $w \in SO(n)$. Such a base $w(e_1), \ldots, w(e_n)$ with $w \in SO(n)$ is called special. For $w \in SO(n)$ we have $wT(x_1, \ldots, x_n)w^{-1} = T(w(x_1), \ldots, w(x_n))$.

Let $D(\theta)$ denote the rotation

$$\begin{bmatrix} \cos 2\pi\theta & -\sin 2\pi\theta \\ \sin 2\pi\theta & \cos 2\pi\theta \end{bmatrix}$$

in $SO(2)$, and let $D(\theta_1, \ldots, \theta_r)$ equal $\mathrm{diag}(D(\theta_1), \ldots, D(\theta_r))$ in $SO(2r)$ and equal $\mathrm{diag}(D(\theta_1), \ldots, D(\theta_r), 1)$ in $SO(2r + 1)$. Then $T(e_1, \ldots, e_n)$ is the group of all $D(\theta_1, \ldots, \theta_r)$ for $n = 2r$ or $2r + 1$ and for $0 \leq \theta_i \leq 1$.

7.1 Theorem. *With the above notations, the subgroups $T(x_1, \ldots, x_n)$ for some special basis $x_1, \ldots, x_n$ of $\mathbf{R}^n$ are the maximal tori of $SO(n)$. The rank of $SO(n)$*

is r, where $n = 2r$ or $2r + 1$. *The Weyl group W of $SO(2r + 1)$ consists of the $2^r r!$ permutations of the indexes of $(\theta_1, \ldots, \theta_r)$ composed with substitutions $(\theta_1, \ldots, \theta_r) \mapsto (\pm\theta_1, \ldots, \pm\theta_r)$, and Weyl group W of $SO(2r)$ consists of the $2^{r-1} r!$ permutations of the indexes of $(\theta_1, \ldots, \theta_r)$ composed with substitutions $(\theta_1, \ldots, \theta_r) \mapsto (\varepsilon_1\theta_1, \ldots, \varepsilon_r\theta_r)$ with $\varepsilon_i = \pm 1$ and $\varepsilon_1 \cdots \varepsilon_r = 1$.*

Proof. Clearly, $T(e_1, \ldots, e_n)$ is an r-dimensional torus $n = 2r$ or $2r + 1$. Since $wT(e_1, \ldots, e_n)w^{-1} = T(w(e_1), \ldots, w(e_n))$, we have only to show that each $u \in SO(n)$ is a member of some $T(x_1, \ldots, x_n)$ to prove the first two statements. For this, we let $c = a + ib$ be a complex eigenvalue of u with eigenvector $x_1 + ix_2$. Then we have $u(x_1 + ix_2) = (a + ib)(x_1 + ix_2) = (ax_1 - bx_2) + i(bx_1 + ax_2)$. Since $|a + ib| = 1$, we can represent $u|(\mathbf{R}x_1 + \mathbf{R}x_2)$ by $D(\theta_1)$ for some θ_1. If all eigenvalues are real, then $\theta_1 = 0$ or $1/2$ for $c = 1$ or -1, respectively. This procedure is applied to u restricted to the orthogonal complement of $\mathbf{R}x_1 + \mathbf{R}x_2$ to get $u \in T(x_1, \ldots, x_n)$.

As with $SU(n)$, all permutations of the coordinates $(\theta_1, \ldots, \theta_r)$ are realized by conjugation by elements of $SO(n)$. Since $\operatorname{diag}(-1, 1)D(\theta)\operatorname{diag}(-1, 1) = D(-\theta)$, conjugation of $D(\theta_1, \ldots, \theta_r)$ by $\operatorname{diag}(-1, 1, \ldots, 1, -1)$ yields $D(-\theta_1, \theta_2, \ldots, \theta_r)$ for $n = 2r + 1$ and $D(-\theta_1, \ldots, \theta_{r-1}, -\theta_r)$ for $n = 2r$. Composing with permutations, we see that the Weyl group is at least as large as was stated in the theorem. To prove the Weyl group is no larger than this, we let $u \in T(e_1, \ldots, e_n)$, where $u = D(\theta_1, \ldots, \theta_r)$ and $\theta_i \neq \theta_j$ for $i \neq j$. If we view $u \in U(n)$ where $SO(n) \subset U(n)$, then u has eigenvectors $e_{2j-1} \pm ie_{2j}$ with eigenvalues $\exp(\pm 2\pi i\theta_j)$ for $1 \leq j \leq r$, where $n = 2r$ or $2r + 1$. If $w \in SO(n)$ with $wuw^{-1} \in T(e_1, \ldots, e_n)$, then from the proof of (2.1) we know that $u \to wuw^{-1}$ is a permutation of the indices of e_i, $1 \leq i \leq n$. This is a permutation of the indices of $\theta_1, \ldots, \theta_r$ together with a coordinate substitution $(\theta_1, \ldots, \theta_r) \mapsto (\pm\theta_1, \ldots, \pm\theta_r)$ as described above. This proves the theorem.

8. Maximal Tori and the Weyl Group of Spin(n)

8.1 Notations. As in Chap. 12, let $\phi \colon \operatorname{Spin}(n) \to SO(n)$ be the natural covering morphism $\phi(u)x = uxu^*$ for $u \in \operatorname{Spin}(n)$ and $x \in \mathbf{R}^n$. Let $\omega_j \colon S^1 = \mathbf{R}/\mathbf{Z} \to \operatorname{Spin}(n)$ be the homomorphism given by the relation $\omega_j(\theta) = \cos 2\pi\theta + e_{2j-1}e_{2j}\sin 2\pi\theta$ for $1 \leq j \leq n/2$. First, observe that $\omega_j(\theta + 1/2) = -\omega_j(\theta)$ and that

$$[\phi\omega_j(\theta)]e_k = \begin{cases} e_k & \text{for } k \neq 2j - 1, 2j \\ e_{2j-1}\cos 4\pi\theta + e_{2j}\sin 4\pi\theta & \text{for } k = 2j - 1 \\ -e_{2j-1}\sin 4\pi\theta + c_{2j}\cos 4\pi\theta & \text{for } k = 2j \end{cases}$$

In other words, $\phi\omega_j(\theta) = D(0, \ldots, 0, 2\theta, 0, \ldots, 0)$ for all $\theta \in \mathbf{R}/\mathbf{Z}$. Let $\omega \colon \mathbf{T}^r \to \operatorname{Spin}(n)$ be defined by $\omega(\theta_1, \ldots, \theta_r) = \omega_1(\theta_1) \cdots \omega_r(\theta_r)$ for $(\theta_1, \ldots, \theta_r) \in \mathbf{T}^r$ and $n = 2r$ or $2r + 1$.

We have the following diagram where $T' = \omega(T(r))$.

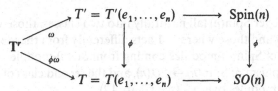

Then $\ker \omega$ consists of $(\theta_1, \ldots, \theta_r)$ with $\theta_i = 0$ or $1/2$ (mod 1) and $\theta_1 + \cdots + \theta_r = 0$ (mod 1), and there are 2^{r-1} elements in $\ker \omega$. Then $\ker \phi$ consists of two elements, 1 and -1. Note that we have $1 = \omega(\theta_1, \ldots, \theta_r)$ for $\theta_i = 0$ or $1/2$ (mod 1) and $\theta_1 + \cdots + \theta_r = 0$ (mod 1) and $-1 = \omega(\theta_1, \ldots, \theta_r)$ for $\theta_i = 0$ or $1/2$ (mod 1) and $\theta_1 + \cdots + \phi_r = 1/2$ (mod 1). Finally, we have $\phi\omega(\theta_1, \ldots, \theta_r) = (2\theta_1, \ldots, 2\theta_r)$ in $T(e_1, \ldots, e_n)$.

8.2 Proposition. *The torus $T'(e_1, \ldots, e_n)$ is a maximal torus of* Spin(n), *and the Weyl group of* Spin(n) *is the Weyl group of* SO(n).

Proof. If $u \in$ Spin(n) such that $uT'(e_1, \ldots, e_n)u^{-1} = T'(e_1, \ldots, e_n)$, we have

$$\phi(u)\phi(T'(e_1, \ldots, e_n))\phi(u)^{-1} = \phi(T'(e_1, \ldots, e_n))$$

or

$$\phi(u)T(e_1, \ldots, e_n)\phi(u)^{-1} = T(e_1, \ldots, e_n)$$

Then we have $N_{T'} = \phi^{-1}(N_T)$ and $T' = \phi^{-1}(T)$. This means that $N_{T'}/T'$ is isomorphic to N_T/T and that $\bigcup_{u \in \text{Spin}(n)} uT'u^{-1} = $ Spin(n). This proves the proposition.

The homomorphisms $T' \xrightarrow{\omega} T' \xrightarrow{\phi} T$ induce the following ring morphisms:

$$RT \to RT' \to RT^r$$

where $\phi\omega$ induces the morphism

$$\mathbf{Z}[\alpha_1, \alpha_1^{-1}, \ldots, \alpha_r, \alpha_r^{-1}] \to \mathbf{Z}[\alpha_1^{1/2}, \alpha_1^{-1/2}, \ldots, \alpha_r^{1/2}, \alpha_r^{-1/2}]$$

On $T(e_1, \ldots, e_n)$ we have $\alpha_i(\theta_1, \ldots, \theta_r) = \theta_i$, and on T^r we have

$$\alpha_i^{1/2}(\theta_1, \ldots, \theta_r) = \theta_i, \qquad \alpha_i(\theta_1, \ldots, \theta_r) = 2\theta_i$$

and $(\alpha_1 \cdots \alpha_r)^{1/2}(\theta_1, \ldots, \theta_r) = \theta_1 + \cdots + \theta_r$. Observe that $\alpha_1, \ldots, \alpha_r$, and $(\alpha_1 \cdots \alpha_r)^{1/2}$ is 0 on $\ker \omega$. Therefore, we have $\alpha_1, \ldots, \alpha_r$, and $(\alpha_1 \cdots \alpha_r)^{1/2} \in RT'$. This leads to the following result stated with the above notations.

8.3 Proposition. *The morphism $\phi \colon T' \to T$ induces the inclusion $RT = \mathbf{Z}[\alpha_1, \alpha_1^{-1}, \ldots, \alpha_r, \alpha_r^{-1}] \to \mathbf{Z}[\alpha_1, \alpha_1^{-1}, \ldots, \alpha_r, \alpha_r^{-1}, (\alpha_1 \cdots \alpha_r)^{1/2}] = RT'$.*

Proof. Since $T' \to T$ is a twofold covering, the rank of RT' as a RT is at most 2, and therefore RT' is $\mathbf{Z}[\alpha_1, \alpha_1^{-1}, \ldots, \alpha_r, \alpha_r^{-1}, (\alpha_1 \cdots \alpha_r)^{1/2}]$.

9. Special Representations of $SO(n)$ and $\mathrm{Spin}(n)$

The modules over $\mathrm{Spin}(n)$ fall naturally into two classes: those where -1 acts as the identity and those where -1 acts differently from the identity. The first class consists of $\mathrm{Spin}(n)$-modules coming from $SO(n)$-modules with the natural homomorphism $\phi\colon \mathrm{Spin}(n) \to SO(n)$, and the second class of $\mathrm{Spin}(n)$ comes from Clifford modules over C_k (see Chap. 12).

9.1 Proposition. *A* $\mathrm{Spin}(n)$*-module* M *decomposes into a direct sum of submodules* $M_1 \oplus M_2$ *where* -1 *acts as the identity on* M_1 *and* M_1 *is an* $SO(n)$*-module and where* -1 *acts as multiplication by* -1 *on* M_2 *and* M_2 *is a* C_{n-1}*-module with* $\mathrm{Spin}(n)$ *coming from the inclusion* $\mathrm{Spin}(n) \subset C_n^0$ *and the isomorphism* $C_n^0 \cong C_{n-1}$.

Proof. Observe that $-1 \in \mathrm{Spin}(n)$ and $(-1)^2 = +1$. We decompose a $\mathrm{Spin}(n)$-module M into $M = M_1 \oplus M_2$ where -1 acts as the identity on M_1 and as multiplication by -1 on M_2. The action of $e_i e_n$ in $\mathrm{Spin}(n)$ for $1 \leq i \leq n - 1$ on M_2 satisfies the properties of u_i in Theorem 12(2.4), and we have an orthogonal multiplication on M_2. This in turn defines a C_{n-1}-module structure on M_2 compatible with the action of $\mathrm{Spin}(n)$ on M_2 given by the inclusion $\mathrm{Spin}(n) \subset C_n^0$ and the isomorphism of $C_n^0 \cong C_{n-1}$ [see 12(2.5) and 12(6.2)]. This proves the proposition.

9.2 Some $SO(n)$-Modules. The canonical $SO(n)$-module structure on $\mathbf{R}^n$ is defined by requiring the scalar product to be substitution. The ith exterior product $\Lambda^i \mathbf{R}^n$ when tensored with $\mathbf{C}$ yields an element λ_i of $RSO(n)$ and $R\,\mathrm{Spin}(n)$. Viewing $RSO(n) \subset \mathbf{Z}[\alpha_1, \alpha_1^{-1}, \ldots, \alpha_r, \alpha_r^{-1}]$ and $R\,\mathrm{Spin}(n) \subset \mathbf{Z}[\alpha_1, \alpha_1^{-1}, \ldots, \alpha_r, \alpha_r^{-1}, (\alpha_1 \cdots \alpha_r)^{1/2}]$, where $n = 2r$ or $2r + 1$, we have λ_i equal to the ith elementary symmetric function in $\alpha_1, \alpha_1^{-1}, \ldots, \alpha_r, \alpha_r^{-1}$ for $n = 2r$ and in $\alpha_1, \alpha_1^{-1}, \ldots, \alpha_r, \alpha_r^{-1}, 1$ for $n = 2r + 1$. This follows from the fact that the action of $D(\theta_j)$ on $e_{2j-1} - ie_{2j}$ is multiplication by $\exp(2\pi i\theta_j)$ and on $e_{2j-1} + ie_{2j}$ is multiplication by $\exp(-2\pi i\theta_j)$.

We have a linear isomorphism $f\colon \Lambda^k \mathbf{R}^n \to \Lambda^{n-k} \mathbf{R}^n$, where $f(e_{i(1)} \wedge \cdots \wedge e_{i(k)}) = \mathrm{sgn}(\sigma) e_{j(1)} \wedge \cdots \wedge e_{j(n-k)}$ and σ is the permutation of $\{1, \ldots, n\}$ with $\sigma(p) = i(p)$ for $p \leq k$ and $\sigma(p) = j(p - k)$ for $k < p \leq n$. Since we have $e_1 \wedge \cdots \wedge e_n = u(e_1) \wedge \cdots \wedge u(e_n)$, for each $u \in SO(n)$, f is an isomorphism of $SO(n)$-modules. Moreover, we have $ff = (-1)^{k(n-k)}$, and therefore, in $RSO(n)$ and $R\,\mathrm{Spin}(n)$ we have $\lambda_i = \lambda_{n-i}$ for $i < n/2$.

For $n = 2r$, we have $f\colon \Lambda^r \mathbf{R}^{2r} \to \Lambda^r \mathbf{R}^{2r}$ with $ff = (-1)^{rr} = (-1)^r$. For $r \equiv 0$ (mod 2), f has two eigenvalues ± 1 and $\lambda_r = \lambda_r^+$ where $\lambda_r^{\pm}$ is the eigenspace corresponding to ± 1; for $r \equiv 1$ (mod 2), f has two eigenvalues $\pm i$ and $\lambda_r = \lambda_r^+ + \lambda_r^-$, where $\lambda_r^{\pm}$ is the eigenspace corresponding to $\pm i$ in $\Lambda^r \mathbf{R}^{2r} \otimes \mathbf{C}$.

9.3 The Spin Modules. Since $\mathrm{Spin}(n + 1)$ is a subgroup of the group of units in $C_{n+1}^0 \cong C_n$, each C_n-module determines a real $\mathrm{Spin}\,(n + 1)$-module, and each C_n^c-module determines a complex Spin$(n + 1)$-module. Let Δ denote the Spin$(2r + 1)$-module corresponding to the simple module over $C_{2r}^c = \mathbf{C}(2^r)$ of dimension 2^r, and let Δ^+, Δ^- denote the Spin$(2r)$-module corresponding to the two simple modules over $C_{2s-1}^c = \mathbf{C}(2^{r-1}) \oplus \mathbf{C}(2^{r-1})$, each of dimension 2^{r-1}. To be specific, let $\Delta^\pm$ correspond to the module where $i^r e_1 \cdots e_{2r}$ acts as a multiplication by ± 1, respectively. The references for the above remarks are 12(5.9) and 12(6.5).

9.4 Proposition. *In* $R\,\text{Spin}(2r + 1)$,

$$\Delta = \prod_{1 \leq j \leq r} (\alpha_j^{1/2} + \alpha_j^{-1/2})$$

$$= \sum_{\varepsilon(j) = \pm 1} \alpha_1^{\varepsilon(1)/2} \cdots \alpha_r^{\varepsilon(r)/2}$$

In $R\,\text{Spin}(2r)$,

$$\Delta^+ = \sum_{\varepsilon(1) \cdots \varepsilon(r) = 1} \alpha_1^{\varepsilon(1)/2} \cdots \alpha_r^{\varepsilon(r)/2}$$

$$\Delta^- = \sum_{\varepsilon(1) \cdots \varepsilon(r) = -1} \alpha_1^{\varepsilon(1)/2} \cdots \alpha_r^{\varepsilon(r)/2}$$

In $RSO(2r)$,

$$\lambda_r^\pm = \sum_{i(1) \leq \cdots \leq i(r),\, \varepsilon(1) \cdots \varepsilon(r) = \pm 1} \alpha_{i(1)}^{\varepsilon(1)} \cdots \alpha_{i(r)}^{\varepsilon(r)}$$

Proof. Let T', Spin(n), and $C_n^+ \otimes \mathbf{C} \cong C_{n-1}^c$ act on the left of C_{n-1}^c by ring multiplication. Then we get $2^{2r}/2^r = 2^r$ factors of Δ for $n = 2r + 1$ and $2^{2(r-1)}/2^{r-1} = 2^{r-1}$ factors of each Δ^+, Δ^-, and $\Delta = \Delta^+ + \Delta^-$ for $n = 2r$. Using the homomorphism $\omega: T(r) \to T'$ in (8.1) we compute the character of Δ on $T(r)$. The diagonal entries of the action T' on C_{n-1}^c as functions on $T(r)$ are all of the form $\prod_{1 \leq j \leq r} \cos 2\pi\theta_j$. For $n = 2r + 1$ there are 2^{2r} such entries, and for $n = 2r$ there are 2^{2r-1} such entries. For $n = 2r + 1$, the character of $2^r\Delta$ equals $2^{2r} \prod_{1 \leq j \leq r} \cos 2\pi\theta_j$; for $n = 2r$, the character of $2^{r-1}\Delta$ equals $2^{2r-1} \prod_{1 \leq j \leq r} \cos 2\pi\theta_j$. In both cases, the character of Δ equals $2^r \prod_{1 \leq j \leq r} \cos 2\pi\theta_j = \prod_{1 \leq j \leq r} (e^{2\pi i \theta j} + e^{-2\pi i \theta j})$.

In the ring $R\,\text{Spin}(n)$, we conclude that

$$\Delta = \prod_{1 \leq j \leq r} (\alpha_j^{1/2} + \alpha_j^{-1/2})$$

$$= \sum_{\varepsilon(j) = \pm 1} \alpha_1^{\varepsilon(1)/2} \cdots \alpha_r^{\varepsilon(r)/2}$$

For the case $n = 2r$, Δ splits as $\Delta^+ + \Delta^-$. Since the elements Δ^+ and Δ^- are both invariant under the Weyl group, we have Δ^+ equal to one of the following expressions and Δ^- equal to the other.

$$\sum_{\varepsilon(1) \cdots \varepsilon(r) = \pm 1} \alpha_1^{\varepsilon(1)/2} \cdots \alpha_r^{\varepsilon(r)/2}$$

To distinguish these two cases, observe that $\omega(1/4,\ldots,1/4) = e_1 \cdots e_{2r}$. In Δ^+, $\omega(1/4,\ldots,1/4)$ is multiplication by i^r, and therefore the character evaluated at $(1/4,\ldots,1/4)$ must be $2^{r-1}i^r$ and we must have $\Delta^+ =$

$$\sum_{\varepsilon(1)\cdots\varepsilon(r)=1} \alpha_1^{\varepsilon(1)/2} \cdots \alpha_r^{\varepsilon(r)/2}.$$

The above reasoning applies to $\lambda_r^\pm$. This proves the proposition.

Now we are in a position to calculate $RSO(n)$ and $R\,\mathrm{Spin}(n)$, knowing enough about certain elements in these rings.

10. Calculation of $RSO(n)$ and $R\,\mathrm{Spin}(n)$

We begin by considering what happens to the elements $\lambda^k(\rho_n)$ under restriction in $R\,\mathrm{Spin}(n) \to R\,\mathrm{Spin}(n-1)$ and $RSO(n) \to RSO(n-1)$. For the study of these restriction properties it is more convenient to work with the operations γ^k. Let n denote the class of the trivial n-dimensional G-module viewed as an element of RG. Then we have $\lambda_t(n) = (1+t)^n$ and $\gamma_t(n) = (1-t)^{-n}$.

10.1 Proposition. *The elements $\gamma^i(\rho_n - n)$ in $R\,\mathrm{Spin}(n)$ or $RSO(n)$ restrict to $\gamma^i(\rho_{n-1} - (n-1))$ in $R\,\mathrm{Spin}(n-1)$ or $RSO(n-1)$, respectively.*

Proof. For $n = 2r + 1$ we have

$$\lambda_t(\rho_n - n) = \lambda_t(\rho_n)\lambda_t(n)^{-1}$$

$$= (1 + \alpha_1 t)(1 + \alpha_1^{-1}t)\cdots(1 + \alpha_r t)(1 + \alpha_r^{-1}t)(1+t)(1+t)^{-2r-1}$$

Also, $\lambda_t(\rho_n - n)$ equals this expression for $n = 2r$. Then we have

$$\gamma_t(\rho_n - n) = \lambda_{t/1-t}(\rho_n - n)$$

$$= \left(1 + \frac{t}{1-t}\alpha_1\right)\cdots\left(1 + \frac{t}{1-t}\alpha_r^{-1}\right)(1-t)^{2r}$$

$$= (1 - t + \alpha_1 t)\cdots(1 - t + \alpha_r^{-1}t)$$

Observe that $\gamma_t(\rho_{2r+1} - (2r+1)) = \gamma_t(\rho_{2r} - 2r)$, and since the image of α_r in $R\,\mathrm{Spin}(2r-1)$ or $RSO(2r-1)$ is 1, we have $\gamma_t(\rho_{2r} - 2r) = \gamma_t(\rho_{2r-1} - (2r-1))$. This proves the proposition.

10.2 Remark. Observe that by examining coefficients of t as in 12(3.2) we have $\gamma^k(\rho_n - n) = \lambda^k(\rho_n) + \sum_{i<k} a_{i,k}\lambda^i(\rho_n)$ and $\lambda^k(\rho_n) = \gamma^k(\rho_n - n) + \sum_{i<k} b_{i,k}\gamma^i(\rho_n - n)$ for integers $a_{i,k}$ and $b_{i,k}$.

The object of this section is to prove the following theorem.

10.3 Theorem. *In the case $n = 2r + 1$, $R\,\mathrm{Spin}(2r+1)$ equals the polynomial ring $\mathbf{Z}[\lambda^1(\rho_{2r+1}),\ldots,\lambda^{r-1}(\rho_{2r+1}),\Delta_{2r+1}]$, and $RSO(2r+1)$ equals the polyno-*

mial ring $\mathbf{Z}[\lambda^1(\rho_{2r+1}'), \ldots, \lambda^{r-1}(\rho_{2r+1})]$. The following relation holds: $\Delta_{2r+1}^2 = \lambda^r(\rho_{2r+1}) + \cdots + \lambda^1(\rho_{2r+1}) + 1$.

In the case $n = 2r$, $R \operatorname{Spin}(2r)$ equals the polynomial ring $\mathbf{Z}[\lambda^1(\rho_{2r}), \ldots, \lambda^{r-2}(\rho_{2r}), \Delta_{2r}^+, \Delta_{2r}^-]$, and $RSO(2r)$ equals the ring generated by $\lambda^1(\rho_{2r}), \ldots, \lambda^{r-1}(\rho_{2r}), \lambda_+^r(\rho_{2r}), \lambda^r(\rho_{2r})$ with one relation

$$(\lambda_+^r + \lambda^{r-2} + \cdots)(\lambda_-^r + \lambda^{r-2} + \cdots) = (\lambda^{r-1} + \lambda^{r-3} + \cdots)^2$$

In $R \operatorname{Spin}(2r)$, the following relations hold:

$$\Delta_{2r}^+ \Delta_{2r}^+ = \lambda_+^r(\rho_{2r}) + \lambda^{r-2}(\rho_{2r}) + \cdots$$

$$\Delta_{2r}^+ \Delta_{2r}^- = \lambda^{r-1}(\rho_{2r}) + \lambda^{r-3}(\rho_{2r}) + \cdots$$

$$\Delta_{2r}^- \Delta_{2r}^- = \lambda_-^r(\rho_{2r}) + \lambda^{r-2}(\rho_{2r}) + \cdots$$

Proof. We begin with the case of $n = 2r + 1$. First we prove the character formula $\Delta_{2r+1} \Delta_{2r+1} = 1 + \lambda^1(\rho_{2r+1}) + \cdots + \lambda^r(\rho_{2r+1})$. We consider the following polynomial:

$$\lambda_t(\rho_{2r+1}) = (1 + \alpha_1 t)(1 + \alpha_1^{-1} t) \cdots (1 + \alpha_r t)(1 + \alpha_r^{-1} t)(1 + t)$$

Since the coefficient of t^j is $\lambda^j(\rho_{2r+1})$ in $\lambda_t(\rho_{2r+1})$, we have $\lambda_t(\rho_{2r+1}) = 1 + \lambda^1 + \cdots + \lambda^{2r+1} = 2(1 + \lambda^1 + \cdots + \lambda^r)$ because λ^j equals λ^{2r+1-j}. Since $(1 + \alpha_j)(1 + \alpha_j^{-1}) = \alpha_j + 2 + \alpha_j^{-1} = (\alpha_j^{1/2} + \alpha_j^{-1/2})^2$, we have $1 + \lambda^1 + \cdots + \lambda^r = (1/2)\lambda_t(\rho_{2r+1}) = \left(\prod_{1 \le j \le r} (\alpha_j^{1/2} + \alpha_j^{-1/2}) \right)^2 = \Delta_{2r+1}^2$. This is the desired formula. Before considering the corresponding results for $n = 2r$, we finish the case $n = 2r + 1$.

To prove the statement about $RSO(2r + 1)$, we proceed as in (6.1). Since the Weyl group consists of all permutations of $\{1, \ldots, n\}$ composed with substitutions between α_j and α_j^{-1}, we have by (5.2) the inclusions $\mathbf{Z}[\lambda^1(\rho_{2r+1}), \ldots, \lambda^r(\rho_{2r+1})] \subset RSO(2r + 1) \subset \mathbf{Z}[\sigma_1, \ldots, \sigma_r] = \mathbf{Z}[\alpha_1, \alpha_r^{-1}, \ldots, \alpha_r, \alpha_r^{-1}]^W$. We have denoted by σ_j the jth elementary symmetric function in the r variables $\alpha_1 + \alpha_1^{-1}, \ldots, \alpha_r + \alpha_r^{-1}$. By (5.4), $RSO(2r + 1)$ is the polynomial ring $\mathbf{Z}[\lambda^1(\rho_{2r+1}), \ldots, \lambda^r(\rho_{2r+1})]$.

To prove the statement about $R \operatorname{Spin}(2r + 1)$, we view $RSO(2r + 1) \subset R \operatorname{Spin}(2r + 1)$, using the twofold covering morphism $\phi \colon \operatorname{Spin}(2r + 1) \to SO(2r + 1)$. We let T be the covering transfromation, where $T(\pm 1) = \mp 1$. Then T carries Δ_{2r+1} into $-\Delta_{2r+1}$ and leaves $RSO(2r + 1)$ elementwise fixed. From the discussion in (9.1) and (9.3) we see that every element of $R \operatorname{Spin}(2r + 1)$ has a unique representation of the form $a\Delta_{2r+1} + b$, where $a \in \mathbf{Z}$ and $b \in RSO(2r + 1)$. A polynomial relation

$$0 = f(\lambda^1, \ldots, \lambda^{r-1}, \Delta_{2r+1}) = 0$$

decomposes as $f_1(\lambda^1, \ldots, \lambda^r) + \Delta_{2r+1} f_2(\lambda^1, \ldots, \lambda^r) = 0$. This implies that as polynomials $f_1 = f_2 = 0$. Therefore, $R \operatorname{Spin}(2r + 1)$ is the polynomial ring

$\mathbf{Z}[\lambda^1(\rho_{2r+1}),\dots,\lambda^{r-1}(\rho_{2r+1}),\Delta_{2r+1}]$. This proves the theorem for the case $n = 2r + 1$.

For the case $n = 2r$, by the argument in the first paragraph of the proof, there is the following relation:

$$\Delta_{2r}^+\Delta_{2r}^+ + 2\Delta_{2r}^+\Delta_{2r}^- + \Delta_{2r}^-\Delta_{2r}^-$$

$$= \Delta_{2r}\Delta_{2r}$$

$$= 2 + 2\lambda^1(\rho_{2r}) + \cdots + 2\lambda^{r-1}(\rho_{2r}) + \lambda_+^r(\rho_{2r}) + \lambda_-^r(\rho_{2r})$$

To examine this relation further, we write

$$\Delta_{2r}^2 = \sum_{\varepsilon(j),\delta(j)=\pm 1} \alpha_1^{(\varepsilon(1)+\delta(1))/2} \cdots \alpha_r^{(\varepsilon(r)+\delta(r))/2}$$

For a given term, let u equal the number of j's with $\varepsilon(j) = \delta(j) = 1$, a the number of j's with $\varepsilon(j) = 1$ and $\delta(j) = -1$, b the number of j's with $\varepsilon(j) = -1$ and $\delta(j) = 1$, and v the number of j's with $\varepsilon(j) = \delta(j) = -1$. Then we have $u + a + b + v = r$. Such a term belongs to $\Delta_{2r}^+\Delta_{2r}^+$ if and only if $b + v$ and $a + v$ are each even; such a term belongs to $\Delta_{2r}^+\Delta_{2r}^-$ if and only if exactly one of the two numbers $b + v$ and $a + v$ is even; and such a term belongs to $\Delta_{2r}^-\Delta_{2r}^-$ if and only if $b + v$ and $a + v$ are each odd. A monomial in the above sum representation of Δ_{2r}^2 has the form $\alpha_1^{\eta(1)}\cdots\alpha_r^{\eta(r)}$, where $\eta(j)$ equals 0 or ± 1. Then u is the number of $\eta(j) = +1$, $a + b$ of $\eta(j) = 0$, and v of $\eta(j) = -1$.

First, we prove the following splitting of the relation for $\Delta_{2r}\Delta_{2r}$.

$$\Delta_{2r}^+\Delta_{2r}^- = \lambda^{r-1}(\rho_{2r}) + \lambda^{r-3}(\rho_{2r}) + \cdots$$

$$\Delta_{2r}^+\Delta_{2r}^+ + \Delta_{2r}^-\Delta_{2r}^- = \lambda_+^r(\rho_{2r}) + \lambda_-^r(\rho_{2r}) + 2(\lambda^{r-2}(\rho_{2r}) + \cdots)$$

To see this, we consider an element $\lambda^{r-i} = \sum \alpha_{j(1)}^{\pm 1}\cdots\alpha_{j(r-1)}^{\pm 1}$ whose terms cancel in pairs if at all. For $\alpha_1^{\eta(1)}\cdots\alpha_r^{\eta(r)}$ to be a monomial in the above sum for λ^{r-i}, we must have $u + v = r - i \pmod 2$. If $O(\varepsilon)$ denotes the number of $\varepsilon(j) = -1$ and $O(\delta)$ of $\delta(j) = -1$, then we have $u + v = r - a - b = r + 2v - O(\varepsilon) - O(\delta) = r - O(\varepsilon) - O(\delta) \pmod 2$. Consequently, if the monomial $\alpha_1^{\eta(1)}\cdots\alpha_r^{\eta(r)}$ appears in λ^{r-i}, we have $i = O(\varepsilon) + O(\delta) \pmod 2$. Since $O(\varepsilon) = 0$ or $1 \pmod 2$ implies $\alpha_1^{\varepsilon(1)/2}\cdots\alpha_r^{\varepsilon(r)/2}$, the monomial appears in Δ_{2r}^+ or Δ_{2r}^-, respectively. This yields the splitting $\Delta_{2r}^2 = (\Delta_{2r}^+\Delta_{2r}^+ + \Delta_{2r}^-\Delta_{2r}^-) + 2\Delta_{2r}^+\Delta_{2r}^-$.

Second, we prove the following splitting of the relation for $\Delta_{2r}^+\Delta_{2r}^+ + \Delta_{2r}^-\Delta_{2r}^-$. Because of the symmetry under the substitution of $\alpha_j^\mp$ for $\alpha_j^\pm$, $\Delta_{2r}^+\Delta_{2r}^+ - \Delta_{2r}^-\Delta_{2r}^-$ is a sum of monomials occurring in λ^r. Consequently, $\Delta_{2r}^\pm\Delta_{2r}^\pm$ equals $\lambda_\pm^r + \lambda^{r-2} + \cdots$ or $\lambda_\mp^r + \lambda^{r-2} + \cdots$. To distinguish between the two cases, we use the notation of (5.2), where σ_j is the jth elementary symmetric function in $\alpha_1 + \alpha_1^{-1},\dots,\alpha_r + \alpha_r^{-1}$. In this case σ_r splits into σ_r^+ and σ_r^-, where

$$\sigma_r^\pm = \sum_{\eta(1)\cdots\eta(r)=\pm 1} \alpha_1^{\eta(1)}\cdots\alpha_r^{\eta(r)}$$

As in (5.4), $\lambda_{\pm}^r$ equals $\sigma_r^{\pm}$ plus terms linear in $\sigma_1, \ldots, \sigma_{r-2}$. Consequently,

$$\Delta_{2r}^{\pm} \Delta_{2r}^{\pm} = \lambda_{\pm}^r(\rho_{2r}) + \lambda^{r-2}(\rho_{2r}) + \cdots$$

This proves the relations between the elements $\Delta_{2r}^{\pm}$ and $\lambda^i(\rho_{2r})$. The relation stated for $RSO(2r)$ follows from $\Delta_{2r}^+ \Delta_{2r}^+ \Delta_{2r}^- \Delta_{2r}^- = \Delta_{2r}^+ \Delta_{2r}^- \Delta_{2r}^+ \Delta_{2r}^-$.

For $R\,\mathrm{Spin}(2r)$, $\mathbf{Z}[\lambda^1(\rho_{2r}), \ldots, \lambda^{r-2}(\rho_{2r}), \Delta_{2r}^+, \Delta_{2r}^-]$ is a polynomial ring because these elements generate $R\,\mathrm{Spin}(2r)$ from the discussion in (9.1) and (9.3), and the following inclusions are given by restriction: $R\,\mathrm{Spin}(2r + 1) \subset R\,\mathrm{Spin}(2r)$ and $RSO(2r + 1) \subset RSO(2r)$. The rings $R\,\mathrm{Spin}(2r + 1)$ and $RSO(2r + 1)$ are made up of the elements left fixed by an involution T of $R\,\mathrm{Spin}(2r)$ and $RSO(2r)$, respectively. By (10.1) and (10.2), the elements $\lambda^i(\rho_n)$ can be replaced by $\gamma^i(\rho_n - n)$, and the polynomial generators $\gamma^1(\rho_{2r+1} - (2r + 1)), \ldots, \gamma^{r-1}(\rho_{2r+1} - (2r + 1)), \Delta_{2r+1}$ of $R\,\mathrm{Spin}(2r + 1)$ restrict to $\gamma^1(\rho_{2r} - 2r), \ldots, \gamma^{r-1}(\rho_{2r} - 2r)$ and $\Delta_{2r}^+ + \Delta_{2r}^-$. The involution T of $R\,\mathrm{Spin}(2r)$ can be thought of as an element of the Weyl group of $\mathrm{Spin}(2r + 1)$, where each ring is viewed as a subring of $RT'(r)$, and we have $T(\Delta_{2r}^{\pm}) = \Delta_{2r}^{\mp}$. Therefore, over $\mathbf{Q}$, the rational numbers $\gamma^1(\rho_{2r} - 2r), \ldots, \gamma^{r-2}(\rho_{2r} - 2r), \Delta_{2r}^+$ and Δ_{2r}^- are algebraically independent. Consequently, the ring $R\,\mathrm{Spin}(2r)$ contains the stated polynomial ring. It remains to check that $\lambda^1, \ldots, \lambda^{r-2}, \Delta^+$, Δ^- generate $R\,\mathrm{Spin}(2r)$. We sketch this as follows. For $u \in R\,\mathrm{Spin}(2r)$ note that $u + Tu$ is divisible by $\Delta^+ - \Delta^-$. Hence $2u$ is a polynomial P in λ_j, Δ^+, and Δ^-. Moreover this polynomial must be divisible by 2 otherwise there would be a relation mod 2 between λ_i, Δ^+, and Δ^-.

As for $RSO(2r)$, this ring is clearly generated by $\lambda^1(\rho_{2r}), \ldots, \lambda^{r-1}(\rho_{2r}), \lambda_+^r(\rho_{2r})$, and $\lambda_-^r(\rho_{2r})$ from the inclusion $RSO(2r + 1) \subset RSO(2r)$. In this inclusion $RSO(2r + 1)$ is the subring of elements left fixed by an involution of $RSO(2r)$ which interchanges $\lambda_+^r(\rho_{2r})$ and $\lambda_-^r(\rho_{2r})$. The elements $\lambda^1(\rho_{2r}), \ldots, \lambda^{r-1}(\rho_{2r}), \lambda_+^r(\rho_{2r}) + \lambda_-^r(\rho_{2r})$ are algebraically independent, and $\lambda_+^r(\rho_{2r}) - \lambda_-^r(\rho_{2r})$ satisfies the quadratic relation in the statement of the theorem. This proves the theorem.

11. Relation Between Real and Complex Representation Rings

In the previous sections we have calculated the complex representations of various groups. In this section we consider the relation of $R(G)$ to $RO(G)$ and $RSp(G)$.

11.1 Notation. We have the following morphisms of group-valued confunctors:

$$\varepsilon_U \colon RO \to R \qquad \varepsilon_0 \colon R \to RO \qquad \varepsilon_{Sp} \colon R \to RSp \qquad \varepsilon_U \colon RSp \to R$$

The morphism ε_U is defined by tensoring with $\mathbf{C}$, and ε_{Sp} is defined by tensoring with $\mathbf{H}$. The morphism ε_0 is defined by the restricting scalars from

C to **R**, and ε_U by restricting the scalars from **H** to **C**. The morphism ε_U preserves the λ-ring structures on RO and R.

11.2 Remark. If α is a principal G-bundle over a space X, we have the following commutative diagrams where $\tilde{\alpha}$ is defined in 13(5.4).

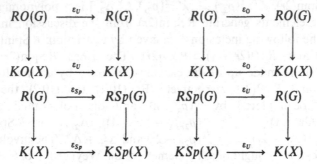

As with vector bundles, there are immediately the following relations involving the operations ε.

11.3 Proposition. *Between* $R(G)$ *and* $RO(G)$ *the relations* $\varepsilon_0\varepsilon_U = 2$ *and* $\varepsilon_U\varepsilon_0 = 1 + \psi_{-1}$ *hold, where* ψ_{-1} *is the* ψ *operation in the* λ-*ring* $R(G)$ *and refers to the conjugate representation of bundles. Between* $R(G)$ *and* $RSp(G)$ *the relations* $\varepsilon_{Sp}\varepsilon_U = 2$ *and* $\varepsilon_U\varepsilon_{Sp} = 1 + \psi_{-1}$ *hold, where* $\psi_{-1}x$ *denotes the conjugate class of* x.

In the next theorem we derive a criterion for a complex representation to be the complexification of a real representation or the restriction of a quaternionic representation. In the remainder of this section, G is a compact, connected Lie group; that is, G has a maximal torus.

11.4 Theorem. *Let* M *be a complex* G-*module with a* G-*invariant hermitian form* η. *Then* M *is the complexification of a real* G-*module if and only if* M *admits a* G-*invariant nondegenerate symmetric form* β, *and* M *is the restriction of a quaternionic* G-*module if and only if* M *admits a* G-*invariant nondegenerate antisymmetric form* γ.

Proof. If M is the complexification of L, we get a form β by complexifying the symmetric nondegenerate G-invariant form $\mathbf{Re}\{\eta|L \times L\}$. If $j \in \mathbf{H}$ acts on M, we define $\gamma(x,y) = \mathbf{Re}\{\eta(x,jy) - \eta(y,jx)\}$. Since $\gamma(x,jx) \neq 0$ for $x \neq 0$ and since $\gamma(x,y) = -\gamma(y,x)$, the form γ is a nondegenerate, antisymmetric form. This proves the direct implications in the theorem.

For the converse, let $\alpha(x,y)$ be a nondegenerate bilinear form with $\alpha(y,x) = \varepsilon\alpha(x,y)$ for all x, $y \in M$ and $\varepsilon = \pm 1$. Let $u: M \to M$ be the real G-automorphism with $\alpha(x,y)$ equal to the complex conjugate of $\overline{\eta(u(x),y)}$. Now we calculate $\eta(u(x),y) = \overline{\alpha(x,y)} = \varepsilon\overline{\alpha(y,x)} = \varepsilon\overline{\eta(u(y),x)} = \varepsilon\overline{\eta(x,u(y))}$. Since $\eta(u^2(x),y) = \eta(x,u^2(y))$, the automorphism u^2 is self-adjoint. Thus we can decompose M as the direct sum $M_1 \oplus \cdots \oplus M_r$ where u^2 has the eigenvalue λ_i on M_i. The λ_i are real, and we can suppose that $\lambda_i \neq \lambda_j$ for $i \neq j$. This implies that $\eta(M_i, M_j) = 0$ for $i \neq j$. The relation $u^2(x) = \lambda_i x$ yields the relation $u^2(u(x)) = \lambda_i u(x)$, and we have $u(M_i) \subset M_i$. Since $\eta(u(x), u(x)) = \varepsilon \eta(u^2(x), x) = \varepsilon \lambda_i \eta(x, x) > 0$ for $x \in M_i$, we have sgn ε = Sgn λ_i for all i. We replace η on M by η^*, where $\eta^*|(M_i \times M_i)$ equals $\sqrt{|\lambda_i|}(\eta|(M_i \times M_i))$, and we replace u by u^*, where $u^*|M_i$ equals $(1/\sqrt{|\lambda_i|})(u|M_i)$. Therefore, we have $(u^*)^2 = \varepsilon$ on M.

For $\varepsilon = +1$, let M_+ be the eigenspace of $+1$ and M_- of -1 for u^*. Since $u^*(ix) = -iu^*(x)$, we have $iM_\pm = M_\mp$. Then the scalar multiplication function $\mathbf{C} \bigotimes_{\mathbf{R}} M_+ \to M$ defines a G-isomorphism.

For $\varepsilon = -1$, let $j(x) = u^*(x)$. Then M admits $\mathbf{H}$ as a field of scalars extending the action of $\mathbf{C}$ so that M is a G-module over $\mathbf{H}$. This proves the theorem.

11.5 Corollary. *Let M be a complex G-module which is either the complexification of a real G-module or the result of restricting the scalars to $\mathbf{C}$ from a quaternionic G-module. Then M and its dual M^+ are isomorphic complex G-modules.*

Proof. The nondegenerate G-invariant symmetric or antisymmetric form on M has associated with it a correlation isomorphism $M \to M^+$ of G-modules.

11.6 Corollary. *Let M be a simple complex G-module with $M = M^+$. Then M is either the complexification of a (simple) real G-module, in which case $\lambda^2 M$ does not contain the trivial one-dimensional G-module, or the restriction of the scalars to $\mathbf{C}$ of a (simple) quaternionic G-module, in which case $S^2 M$, the symmetric product, does not contain the trivial one-dimensional G-module.*

Proof. The natural duality pairing $M \otimes M^+ \to \mathbf{C}$ defines the trivial one-dimensional G-submodule of $(M \otimes M^+)^+$, where $(M \otimes M^+)^+$ and $M \otimes M$ are isomorphic. But $M \otimes M$ is isomorphic to $\lambda^2 M \oplus S^2 M$, and one or the other contains the trivial one-dimensional G-module. If $S^2 M$ contains the trivial one-dimensional G-module, there is a symmetric nondegenerate form on M. By (11.4), M is the complexification of a real G-module. If $\lambda^2 M$ contains the trivial one-dimensional G-module, there is an antisymmetric nondegenerate form on M. By (11.4), M is the restriction of the scalars to $\mathbf{C}$ from a quaternionic G-module. This proves the corollary.

11.7 Corollary. *Let M be a complex G-module. Then the number of simple components of M coming from real G-modules minus the number of simple components of M coming from quaternionic G-modules equals the number of times that 1 occurs in $\psi^2 M$ in RG.*

Proof. If M is simple, $\psi^2 M = M \otimes M - 2\lambda^2 M$ will contain 1 with scalar multiple $+1$ if M is the complexification of a real G-module, with scalar multiple -1 if M is the restriction of a quaternionic module, and with scalar multiple 0 if M and M^+ are nonisomorphic. This follows immediately from (11.6).

12. Examples of Real and Quaternionic Representations

12.1 Examples. The usual action of $SO(n)$ on $\mathbf{R}^n$ and $\Lambda^i \mathbf{R}^n$ defines elements of $RO(SO(n))$ whose complexifications are ρ_n and $\lambda^i \rho_n$, respectively. Since the $\lambda^i \rho_n$ generate $RSO(2r + 1)$, the morphism $\varepsilon_U: RO(SO(2r + 1)) \to RSO(2r + 1)$ is an isomorphism. Recall that the relation $\varepsilon_0 \varepsilon_U = 2$ and the fact that $R(G)$ is a free abelian group imply that ε_U is a monomorphism. For $SO(2r)$, the module $\Lambda^r R^{2r}$ splits over $\mathbf{R}$ and $\mathbf{C}$ if r is even and splits over $\mathbf{C}$ for r odd. Then $\varepsilon_U: RO(SO(2r)) \to RSO(2r)$ is an isomorphism for r even and has the cokernel of order 2 for r odd.

12.2 Examples. The action of $Sp(n)$ on $\mathbf{H}^n$ defines an element of $RSp(Sp(n))$. After restriction of scalars to $\mathbf{C}$, there is an element of $RSp(n)$ whose exterior powers generate $RSp(n)$ as a ring.

12.3 Examples. Recall that $\mathrm{Spin}(n) \subset C_n^0 \cong C_{n-1}$ and that the simple C_{n-1}-modules determine precisely the simple $\mathrm{Spin}(n)$-modules From Table 11(5.10) and the periodicity relation $C_{+8} = C_k \otimes \mathbf{R}(16)$ we have the following real Spin modules: Δ_{8k-1}, Δ_{8k}^+, Δ_{8k}^-, Δ_{8k+1}. The Spin modules Δ_{8k+3}, Δ_{8k+4}^+, Δ_{8k+4}^-, and Δ_{8k+5} are the restriction of quaternionic Spin modules.

12.4. Table of Real Spin Representations.

Spin(n)	Real Spin module	Dimension
$8m + 1$	Δ_{8m+11}	16^m
$8m + 2$	$\Delta_{8m+2}^+ + \Delta_{8m+2}^-$	$2 \cdot 16^m$
$8m + 3$	$2\Delta_{8m+3}$	$4 \cdot 16^m$
$8m + 4$	$2\Delta_{8m+4}^+, 2\Delta_{8m+4}^-$	$4 \cdot 16^m$
$8m + 5$	$2\Delta_{8m+5}$	$8 \cdot 16^m$
$8m + 6$	$\Delta_{8m+6}^+ + \Delta_{8m+6}^-$	$8 \cdot 16^m$
$8m + 7$	Δ_{8m+7}	$8 \cdot 16^m$
$8m + 8$	$\Delta_{8m+8}^+, \Delta_{8m+8}^-$	$8 \cdot 16^m$

This table follows from 12(5.10) by tensoring the simple C_{n-1}-modules with $\mathbf{C}$ and comparing them with the simple C_{n-1}^c-modules described by Δ_n or Δ_n^+ and Δ_n^-.

Now we consider a different approach to determining whether or not a Spin representation is real or not; that is, we use the criterion given in Corollary (11.7). First, we must be able to calculate $\psi^k(a)$ in RG; for this, we use the next proposition.

12.5 Proposition. *Let T^r be a maximal torus of a compact group G, and consider RG as a subring of $RT^r = \mathbf{Z}[\beta_1, \beta_1^{-1}, \ldots, \beta_r, \beta_r^{-1}]$. If an element $a \in RG$ is of the form $a = P(\beta_1, \beta_1^{-1}, \ldots, \beta_r, \beta_r^{-1})$, where P is a polynomial, then $\psi^k(a) = P(\beta_1^k, \beta_1^{-k}, \ldots, \beta_r^k, \beta_r^{-k})$.*

Proof. First, ψ^k is compatible with the subring inclusion of RG in RT^r. For one-dimensional elements β we have $\psi^k(\beta) = \beta^k$ by 13(2.3). Then we have

$$\psi^k(a) = \psi^k P(\beta_1, \beta_1^{-1}, \ldots, \beta_r, \beta_r^{-1})$$
$$= P(\psi^k(\beta_1), \psi^k(\beta_1^{-1}), \ldots, \psi^k(\beta_r), \psi^k(\beta_r^{-1}))$$
$$= P(\beta_1^k, \beta_1^{-k}, \ldots, \beta_r^k, \beta_r^{-k})$$

This proves the proposition.

12.6 Corollary. *For $\Delta_m \in R\,\mathrm{Spin}(m)$, where $R\,\mathrm{Spin}(m) \subset \mathbf{Z}[\alpha_1^{1/2}, \alpha_1^{-1/2}, \ldots, \alpha_r^{1/2}, \alpha_r^{-1/2}]$ with $m = 2r$ or $2r + 1$,*

$$\psi^k(\Delta_m) = \prod_{1 \leq j \leq r} (\alpha_j^{+k/2} + \alpha_j^{-k/2})$$

We are particularly interested in the case where $k = 2$ and $\psi^2(\Delta_m) = \prod_{1 \leq j \leq r} (\alpha_j + \alpha_j^{-1})$.

12.7 Remark. We wish to write $\psi^k(\Delta_m)$ as a polynomial $P(\sigma_1, \ldots, \sigma_r)$, where σ_k is the kth elementary symmetric function in $\alpha_1, \alpha_1^{-1}, \ldots, \alpha_r, \alpha_r^{-1}$. Then $P(0, \ldots, 0)$ is the number of times that the trivial one-dimensional representation appears in $\psi^2(\Delta_m)$.

Consider the roots $z_1, z_1^{-1}, \ldots, z_r, z_r^{-1}$ of the equation $x^{2r} + 1 = 0$. Since

$$(x - z_1)(x - z_1^{-1}) \cdots (z - z_r)(z - z_r^{-1})$$
$$= \sum_{0 \leq k \leq 2r} (-1)^k \sigma_k(z_1, z_1^{-1}, \ldots, z_m, z_m^{-1}) x^{2r-k}$$
$$= 1 + x^{2r}$$

we have $\sigma_k(z_1, z_1^{-1}, \ldots, z_r, z_r^{-1}) = 0$ for $0 < k < 2r$. Therefore, we have $P(0, \ldots, 0) = \prod_{1 \leq j \leq r} (z_j + z_j^{-1})$.

Following a method suggested by Hirzebruch after a lecture by Adams at the Seattle Summer Institute in 1963, we look for a polynomial $f_m(x)$ with roots $z_1 + z_1^{-1}, \ldots, z_r + z_r^{-1}$. Then its constant term is $(-1)^r P(0,\ldots,0) =$ $(-1)^r \prod_{1 \le j \le r} (z_j + z_j^{-1})$. By (5.1) there is a polynomial f_r of degree r with $x^r + x^{-r} = f_r(x + x^{-1})$. Since $0 = z_j^r + z_j^{-r} = f_r(z_j + z_j^{-1})$, the r roots of f_r are $z_1 + z_1^{-1}, \ldots, z_r + z_r^{-1}$. To calculate the constant term, we substitute $x = i$. Then $x + x^{-1} = i + i^{-1} = 0$, and $f_r(0) = f_r(i + i^{-1}) = i^r + i^{-r} = i^r(1 + (-1)^r)$. In summary, we have the following proposition.

12.8 Proposition. *The trivial one-dimensional representation appears in* Δ_{2r} *or* Δ_{2r+1}

$$2 \text{ times for } r \equiv 0 \ (\mathrm{mod}\ 4)$$

$$-2 \text{ times for } r \equiv 2 \ (\mathrm{mod}\ 4)$$

$$0 \text{ times for } r \equiv 1, 3 \ (\mathrm{mod}\ 4)$$

Using (12.8), (11.7), and the relation $\Delta_{2r} = \Delta_{2r}^+ + \Delta_{2r}^-$, we have a second proof that Δ_{8r}^+ and Δ_{8r}^- are real and that Δ_{8r+4}^+ and Δ_{8r+4}^- are quaternionic. About the pair Δ_{8r+2}^+ and Δ_{8r+2}^- and the pair Δ_{8r+6}^+ and Δ_{8r+6}^- we can say nothing from this calculation. For example, it is possible that both Δ_n^+ and Δ_n^- are complex or that one is real and the other is quaternionic for $n = 8r + 2$ or $8r + 6$.

13. Spinor Representations and the K-Groups of Spheres

The following proposition follows easily from the previous analysis of $R\,\mathrm{Spin}(k)$ and $RO\,\mathrm{Spin}(k)$.

13.1 Proposition. (1) *The inclusion* $\mathrm{Spin}(2n) \to \mathrm{Spin}(2n + 1)$ *defines a monomorphism* $R\,\mathrm{Spin}(2n + 1) \to R\,\mathrm{Spin}(2n)$, *and* $R\,\mathrm{Spin}(2n)$ *is a free* $R\,\mathrm{Spin}(2n + 1)$*-module with two free generators* 1 *and* Δ_{2n}^+.

(2) *The inclusion* $\mathrm{Spin}(8n) \to \mathrm{Spin}(8n + 1)$ *defines a monomorphism* $RO\,\mathrm{Spin}(8n + 1) \to RO\,\mathrm{Spin}(8n)$, *and* $RO\,\mathrm{Spin}(8n)$ *is a free* $RO\,\mathrm{Spin}(8n + 1)$*-module with two free generators* 1 *and* Δ_{8n}^+.

(3) *In addition, there is an involution of* $R\,\mathrm{Spin}(2n)$, *leaving* $R\,\mathrm{Spin}(2n + 1)$ *elementwise fixed with* $\Delta_{2n}^{\pm} \mapsto \Delta_{2n}^{\mp}$, *and an involution of* $RO\,\mathrm{Spin}(8n)$, *leaving* $RO\,\mathrm{Spin}(8n + 1)$ *elementwise fixed with* $\Delta_{8n}^{\pm} \mapsto \Delta_{8n}^{\mp}$.

13.2 Notation. Let α_k denote the principal $\mathrm{Spin}(k)$-bundle over S^k.

$$\mathrm{Spin}(k) \to \mathrm{Spin}(k + 1) \to \mathrm{Spin}(k + 1)/\mathrm{Spin}(k) = S^k$$

The following result we state without proof (see Atiyah, Bott, and Shapiro [1]). Also, the relation with the periodicity theorem and the multiplicative properties of $\tilde{\alpha}_k$ are outlined in this article.

13.3 Theorem. *The principal bundles α_k define group isomorphisms*

$$\tilde{\alpha}_{2n}: R\,\mathrm{Spin}(2n)/R\,\mathrm{Spin}(2n+1) \to K(S^{2n})$$

$$\tilde{\alpha}_{8n}: RO\,\mathrm{Spin}(8n)/RO\,\mathrm{Spin}(8n+1) \to KO(S^{8n})$$

CHAPTER 15

The Hopf Invariant

We give a K-theory definition of the Hopf invariant of maps $S^{2n-1} \to S^n$ and then derive its elementary properties. In particular, we present the elementary Atiyah proof of the nonexistence of elements of Hopf invariant 1. We relate the K-theory definition to the cohomology definition using the Chern character in Chap. 20, Sec. 10.

1. K-Theory Definition of the Hopf Invariant

Recall that $\tilde{K}(S^{2n})$ is infinite cyclic as a group with a generator β_{2n}. The external K-cup product $\beta_{2n}\beta_{2m}$ is $\beta_{2(n+m)}$. We require β_2 to be the class determined by the canonical line bundle on $CP^1 = S^2$. Recall that $\tilde{K}(S^{2n+1}) = 0$.

1.1 Notations. Let $f: S^{2n-1} \to S^n$ be a map, and form the following mapping sequence:

$$S^{2n-1} \to S^n \to C_f \to SS^{2n-1} \overset{Sf}{\to} SS^n$$

Then there is the following exact sequence of K-groups:

$$0 \leftarrow \tilde{K}(S^n) \overset{\phi}{\leftarrow} \tilde{K}(C_f) \overset{\psi}{\leftarrow} \tilde{K}(S^{2n})$$

Let b_f or $b(f)$ denote $\psi(\beta_{2n})$ in $\tilde{K}(C_f)$, and let a_f or $a(f)$ denote any element of $\tilde{K}(C_f)$ such that $\phi(a^f) = \beta_n$ for n even and $a_f = 0$ for n odd. Two choices of a_f differ by a multiple of b_f. Since ϕ and ψ are ring homomorphisms, $b_f^2 = 0$, $a_f b_f = 0$, and a_f^2 is equal to a multiple of b_f. The relation $a_f b_f = 0$ holds since a_f (resp. b_f) is zero or the $n-1$ (resp. $2n-1$) skeleton.

1.2 Definition. With the above notations, the Hopf invariant of a map f: $S^{2n-1} \to S^n$ is the integer h_f such that $a_f^2 = h_f b_f$ in $\tilde{K}(C_f)$.

Since $a_f = 0$ for n odd, $h_f = 0$ in this case. We use the notation $h(f)$ for h_f occasionally. Since $a_f^2 = (a_f + mb_f)^2$, the Hopf invariant h_f is independent of the choice for a_f. Henceforth, we shall assume that n is even. Then the sequence of K-groups becomes the following sequence:

$$0 \leftarrow \tilde{K}(S^n) \leftarrow \tilde{K}(C_f) \leftarrow \tilde{K}(S^{2n}) \leftarrow 0$$

To prove the homotopy invariance of h_f, we use the notation j_t: $X \to (X \times I)_* = (X \times I)/(* \times I)$ for $j_t(x)$ equal to the class of (x, t). Then j_t is a homotopy inverse of the projection $(x, t) \mapsto x$.

1.3 Proposition. *If $f_t: S^{2n-1} \to S^n$ is a base point preserving homotopy, then $h(f_0) = h(f_1)$.*

Proof. We view f_t as a map $f: (S^{2n-1} \times I)_* \to S^n$, and then we have the following commutative diagram of spaces and maps for each $t \in I$:

$$\begin{array}{ccccccccc}
(S^{2n-1} \times I)_* & \xrightarrow{f} & S^n & \longrightarrow & C_f & \longrightarrow & S(S^{2n-1} \times I)_* & \longrightarrow & SS^n \\
\uparrow{\scriptstyle j_t} & & \uparrow{\scriptstyle 1} & & \uparrow{\scriptstyle k_t} & & \uparrow{\scriptstyle Sj_t} & & \uparrow{\scriptstyle 1} \\
S^{2n-1} & \xrightarrow{f_t} & S^n & \longrightarrow & C_{f_t} & \longrightarrow & SS^{2n-1} & \longrightarrow & SS^n
\end{array}$$

Next we have the following commutative diagram of K-groups:

The morphism $\beta = (Sj_t)^!$ is a ring isomorphism with inverse $(Sp)^!$, where $p: (S^{2n-1} \times I)_* \to S^{2n-1}$ is the projection. Therefore, the morphism α, which equals $(k_t)^!$, is a ring isomorphism with $\alpha(b(f)) = b(f_t)$ and $\alpha(a(f)) = a(f_t)$, where a and b are defined in (1.1). This means that $h(f_t)$ is independent of t.

2. Algebraic Properties of the Hopf Invariant

From Proposition (1.3) we see that the Hopf invariant can be viewed as a function $h: \pi_{2n-1}(S^n) \to \mathbf{Z}$.

2.1 Proposition. *The function $h: \pi_{2n-1}(S^n) \to \mathbf{Z}$ is a group morphism.*

Proof. Let $f_i: S^{2n-1} \to S^n$ be two base point preserving maps for $i = 1, 2$. In $\pi_{2n-1}(S^n)$ we have $[f_1] + [f_2] = [(f_1 \vee f_2)\theta]$, where $\theta: S^{2n-1} \to S^{2n-1} \vee S^{2n-1}$

is the map pinching the equator to a point. Let $q_i: S^{2n-1} \to S^{2n-1} \vee S^{2n-1}$ denote the inclusion maps into the coproduct, let g denote the map $(f_1 \vee f_2)\theta$, and let $r_i: C_{f_i} \to C_g$ denote the inclusion map induced by q_i. We have the following commutative diagrams of K-groups for $i = 1$ and 2:

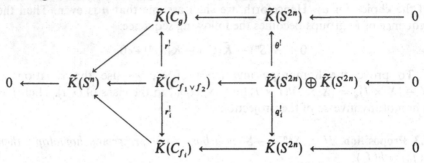

As a group, $\tilde{K}(C_{f_1 \vee f_2})$ has three free generators a, b_1, and b_2, where $r_1{}^!(a) = a(f_1)$, $r_2{}^!(a) = a(f_2)$, $r_1{}^!(b_1) = b(f_1)$, $r_2{}^!(b_2) = b(f_2)$, and $r_1{}^!(b_2) = r_2{}^!(b_1) = 0$. The ring structure is given by $ab_1 = ab_2 = b_1^2 = b_2^2 = 0$ and $a^2 = h_1 b_1 + h_2 b_2$, where h_1, $h_2 \in \mathbf{Z}$. By commutativity of the lower square, $a(f_i)^2 = r_i{}^!(a)^2 = h_i r_i{}^!(b_i) = h(f_i)b(f_i)$ or $h_i = h(f_i)$ for $i = 1$, 2. Finally, from the additivity property of $\theta^!$, we have $r^!(b_i) = b(g)$ for $i = 1$, 2 and $a(g)^2 = (h_1 + h_2)b(g)$. Therefore, we have $h(g) = h(f_1) + h(f_2)$, and h is a group homomorphism.

2.2 Proposition. *Let* $u: S^{2n-1} \to S^{2n-1}$, $f: S^{2n-1} \to S^n$, *and* $v: S^n \to S^n$ *be three maps. Then* $h_{vfu} = (\deg v)^2 h_f (\deg u)$.

Proof. It suffices to prove that $h_{vf} = (\deg v)^2 h_f$ and $h_{fu} = h_f(\deg u)$. We prove the first relation and leave the second to the reader. We have the following commutative diagram of spaces:

$$
\begin{array}{ccccccccc}
S^{2n-1} & \longrightarrow & S^n & \longrightarrow & C_f & \longrightarrow & S^{2n} & \longrightarrow & S^{n+1} \\
\downarrow{\scriptstyle 1} & & \downarrow{\scriptstyle v} & & \downarrow{\scriptstyle w} & & \downarrow & & \downarrow \\
S^{2n-1} & \xrightarrow{\ f\ } & S^n & \longrightarrow & C_{vf} & \longrightarrow & S^{2n} & \longrightarrow & S^{n+1}
\end{array}
$$

This diagram yields the following commutative diagram of K-groups:

$$
\begin{array}{ccccc}
0 \longleftarrow & \tilde{K}(S^n) & \longleftarrow & \tilde{K}(C_f) & \\
 & \uparrow{\scriptstyle v^!} & & \uparrow{\scriptstyle w^!} & \searrow \tilde{K}(S^{2n}) \longleftarrow 0 \\
0 \longleftarrow & \tilde{K}(S^n) & \longleftarrow & \tilde{K}(C_{vf}) & \nearrow
\end{array}
$$

We have $w^!(b_{vf}) = b_f$ and $w^!(a_{vf}) - (\deg v)a_f = mb_f$ for some $m \in \mathbf{Z}$. Therefore, we have $(\deg v)^2 a_f{}^2 = w^!(a_{vf}{}^2) = h_{vf} w^!(b_{vf}) = h_{vf} b_f$. Since $a_f{}^2 = h_f b_f$, we have $h_{vf} = (\deg v)^2 h_f$.

3. Hopf Invariant and Bidegree

3.1 Notation. Recall that the join $X * Y$ of two spaces X and Y is the quotient of $X \times [0,1] \times Y$ with cosets denoted $\langle x, t, y \rangle$, where $\langle x, 0, y \rangle = \langle x', 0, y \rangle$ and $\langle x, 1, y \rangle = \langle x, 1, y' \rangle$ for $x, x' \in X$ and $y, y' \in Y$. If X and Y have base points x_0 and y_0, then $\langle x_0, t, y \rangle = \langle x, t', y_0 \rangle$ is the base point of $X * Y$ after further identification. Recall that the suspension SZ of a space Z is the quotient of $Z \times [0,1]$ with cosets denoted $\langle z, t \rangle$, where $\langle z, 0 \rangle = \langle z', 0 \rangle$ and $\langle z, 1 \rangle = \langle z', 1 \rangle$ for $z, z' \in Z$. If Z has a base point z_0, then $\langle z_0, t \rangle = \langle z_0, t' \rangle$ is a base point of SZ after further identification. Observe that SZ is $Z * S^0$.

3.2 Proposition. *A homeomorphism* $u: S^{n-1} * S^{m-1} \to (B^n \times S^{m-1}) \cup (S^{n-1} \times B^m) = \partial(B^n \times B^m)$ *is given by* $u\langle x, t, y \rangle = (2\min(t, 1/2)x, 2\min(1 - t, 1/2)y)$, *and a homeomorphism* $v: S^{n-1} * S^{m-1} \to S^{n+m-1}$ *is given by* $v\langle x, t, y \rangle = (\sin(\pi t/2)x, \cos(\pi t/2)y)$.

Proof. By inspection, we see that the maps u and v are continuous bijections, and therefore, since u and v are defined on compact spaces, they are homeomorphisms.

3.3 Definition. The Hopf construction H assigns to each map $f: X \times Y \to Z$ a map $H(f): X * Y \to SZ$ defined by the relation $H(f)\langle x, t, y \rangle = \langle f(x, y), t \rangle$. If f preserves base points, then $H(f)$ preserves base points.

3.4 Proposition. *If* $f_s: X \times Y \to Z$ *is a homotopy,* $H(f_s): X * Y \to SZ$ *is a homotopy, and if* f_s *preserves base points, then* $H(f_s)$ *preserves base points.*

Proof. The map $f: X \times Y \times I \to Z$ defines a map $X \times I \times Y \times I \to Z \times I$ by $(x, t, y, s) \to (f_s(x, y), t)$. Since I is compact, the space $(X * Y) \times I$ has the quotient topology, and the function that assigns $H(f_s)\langle x, y, t \rangle$ to $(\langle x, t, y \rangle, s)$ is continuous.

The next theorem relates the bidegree of $f: S^{n-1} \times S^{n-1} \to S^{n-1}$ to the Hopf invariant of $H(f): S^{2n-1} \to S^n$ for n even. This is the main tool for constructing maps of given Hopf invariant.

3.5 Theorem. *If* $f: S^{n-1} \times S^{n-1} \to S^{n-1}$ *is a map of didegree* (d_1, d_2), *the Hopf invariant of* $H(f)$ *is* $d_1 d_2$.

Proof. Let $B_1 = B_2 = B^n$ and $S_1 = S_2 = S^{n-1}$ in $\mathbf{R}^n$. Let H_+ and H_- denote the upper and lower hemispheres of $SS^{n-1} = S^n$. We view the Hopf construction of f as a map $H(f): (B_1 \times S_2) \cup (S_1 \times B_2) \to S^n = H_+ \cup H_-$, and we denote $C_{H(f)} = (B_1 \times B_2) \cup S^n$ by X. The attaching map of $B_1 \times B_2$ to S^n in X is denoted $g: (B_1 \times B_2, B_1 \times S_2, S_1 \times B_2) \to (X, H_+, H_-)$.

We have the cup product and the following isomorphisms:

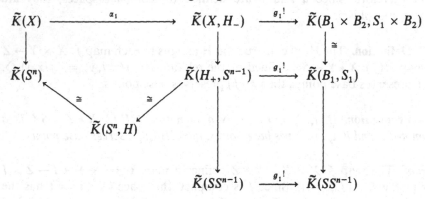

The image of $\beta_n \otimes \beta_n$ under this composition is β_{2n}. Let a_1 denote the image of β_n in $\tilde{K}(B_1 \times B_2, S_1 \times B_2)$, and a_2 of β_n in $\tilde{K}(B_1 \times B_2, B_1 \times S_2)$. Then the cup product $a_1 a_2$ is the generator of $\tilde{K}(B_1 \times B_2, B_1 \times S_2 \cup S_1 \times B_2)$ which projects to β_{2n}.

Next, we consider the following diagram:

The group homomorphism $g_1^!$ is multiplication by d_1. Moreover, we have $g_1^!(\alpha_1(a)) = d_1 a_1$. From a similar diagram we have $g_2^!(\alpha_2(a)) = d_2 a_2$. Finally, the element a^2, which is in the image of $\tilde{K}(S^{2n}) \to \tilde{K}(X)$, corresponds to the product $\alpha_1(a)\alpha_2(a)$ in $\tilde{K}(X, H_+ \cup H_1) = \tilde{K}(X, S^n)$ under the isomorphism $\tilde{K}(S^{2n}) \to \tilde{K}(X, H_+ \cup H_-)$. In $\tilde{K}(B_1 \times B_2, S_1 \times B_2 \cup B_1 \times S_2)$ the element a^2 corresponds to $d_1 d_2 a_1 a_2$, which is $d_1 d_2$ times the image of the canonical generator of $\tilde{K}(S^{2n})$ in $\tilde{K}(B_1 \times B_2, S_1 \times B_2 \cup B_1 \times S_2)$. Therefore, $a_{H(f)}^2 = d_1 d_2 b_{H(f)}$.

3.6 Corollary. *For n even, there is always a map $S^{2n-1} \to S^n$ with the Hopf invariant equal to any even number.*

Proof. By (3.5) and 8(10.3) we have a map of the Hopf invariant -2. By (2.1) we have a map of the Hopf invariant equal to any even number.

3.7 Corollary. *For $n = 2$, 4, and 8 there are maps $S^{2n-1} \to S^n$ of the Hopf invariant 1 and, in fact, with the Hopf invariant equal to any integer.*

Proof. The Clifford algebras C_1, C_3, and C_7 for the quadratic form $-(x|x)$ have irreducible modules of dimension 2, 4, and 8, respectively, by 12(6.5). By 12(2.3) there is an orthogonal multiplication $\mathbf{R}^m \times \mathbf{R}^m \to \mathbf{R}^n$ for $m = 2$, 4, and 8. From this multiplication with a two-sided unit, there is a map $S^{m-1} \times S^{m-1} \to S^{m-1}$ of bidegree $(1, 1)$. Now the corollary follows from (3.5).

3.8 Remark. The orthogonal multiplications $\mathbf{R}^m \times \mathbf{R}^m \to \mathbf{R}^m$ for $m = 2$, 4, or 8 are derived from the complex, quaternionic, and Cayley numbers, respectively.

4. Nonexistence of Elements of Hopf Invariant 1

In (3.7) we saw that there were maps $S^{2n-1} \to S^n$ with the Hopf invariant 1 for $n = 2, 4$, and 8. By a method due to Atiyah, we shall prove that these are the only dimensions in which there are elements of the Hopf invariant 1.

4.1 Let $f: S^{4n-1} \to S^{2n}$ be a map of the Hopf invariant ± 1. Then there is an exact sequence

$$0 \leftarrow \tilde{K}(S^{2n}) \leftarrow \tilde{K}(C_f) \leftarrow \tilde{K}(S^{4n}) \leftarrow 0$$

The group $\tilde{K}(C_f)$ has two free generators a_f and b_f, where $a_f^2 = \pm b_f$ as in (1.1). From the nature of the Adams operations in $\tilde{K}(S^m)$, the Adams operations have the following form: $\psi^k(a_f) = k^n a_f + q(k)b_f$ and $\psi^k(b_f) = k^{2n}b_f$.

4.2 Lemma. *If $a_f^2 = \pm b_f$, then $q(2)$ is odd.*

Proof. In terms of the exterior power operations λ_1 and λ_2, we have $\psi^2 = \lambda_1^2 - 2\lambda_2$ with $\lambda_1 = 1$. Then we calculate $\psi^2(a_f) = a_f^2 - 2\lambda_2(a_f) = \pm b_f - 2(-2^{n-1}a_f + mb_f)$, and we have $q(2) = \pm 1 - 2m$, which is odd.

4.3 Theorem (Adams, Atiyah). *If there exists a map $S^{2m-1} \to S^m$ of the Hopf invariant 1, then $m = 2$, 4, or 8.*

Proof. We use the notations of (4.1) where $m = 2n$, and we calculate the following elements for k as any odd number.

$$\psi^2\psi^k(a_f) = \psi^2(k^n a_f + q(k)b_f)$$

$$= 2^n k^n a_f + k^n q(2)b_f + 2^{2n}q(k)b_f$$

$$\psi^k\psi^2(a_f) = \psi^k(2^n a_f + q(2)b_f)$$

$$= 2^n k^n a_f + 2^n q(k)b_f + k^{2n}q(2)b_f$$

From the relation $\psi^2\psi^k = \psi^{2k} = \psi^k\psi^2$ of 12(10.5), we have $k^n q(2) + 2^{2n}q(k) = 2^n q(k) + k^{2n}q(2)$ or $k^n(k^n - 1)q(2) = 2^n(2^n - 1)q(k)$. By (4.2), $q(2)$ is odd (this is the only place where we use $h_f = \pm 1$), and therefore we have 2^n divides $k^n - 1$ or $k^n \equiv 1 \bmod 2^n$ for all odd numbers k. By a well-known result from number theory (see Bourbaki [1, chap. VII, pp. 73–74]), we have $n \equiv 0 \bmod 2^{n-2}$. For $n \geq 5$ we have $0 < n < 2^{n-2}$, and the dimensions $n \geq 5$ are excluded. Finally, since $8 = 2^3$ does not divide $3^3 - 1 = 26$, the dimension $n = 3$ is excluded, leaving only $n = 1, 2$, and 4. This proves the theorem.

The above theorem could be proved using only the relation $\psi^2\psi^3 = \psi^3\psi^2$ and a different result in number theory. As a corollary we have the following result on H-space structures on spheres.

4.4 Corollary. *The only spheres with an H-space structure are S^1, S^3, and S^7.*

Proof. An H-space structure $\phi: S^{n-1} \times S^{n-1} \to S^{n-1}$ is a map of bidegree $(1, 1)$, and the Hopf construction yields a map $S^{2n-1} \to S^n$ of the Hopf invariant 1 by Theorem (3.5). By (4.3) we must have $n = 2, 4$, or 8, and by (3.7) there are H-space structures on S^1, S^3, and S^7.

Theorem 4.3 was first proved by Adams [5] using secondary operations in ordinary cohomology.

4.5 Corollary. *The only dimensions n for which we have multiplication $\mathbf{R}^n \times \mathbf{R}^n \to \mathbf{R}^n$, denoted $x \cdot y$, with $x \cdot y = 0$ implying either $x = 0$ or $y = 0$ are $n = 1, 2, 4$, and 8. These multiplications can be realized respectively by $\mathbf{R}, \mathbf{C}, \mathbf{H}$, and the Cayley numbers.*

CHAPTER 16

Vector Fields on the Sphere

In Chap. 12, Theorem (8.2), we saw that S^{n-1} has $\rho(n) - 1$ orthonormal tangent vector fields defined on it. The object of this chapter is to outline the steps required to prove that S^{n-1} does not have $\rho(n)$ orthonormal tangent vector fields defined on it; in fact, S^{n-1} does not have $\rho(n)$ linearly independent tangent vector fields, see also Adams [6].

To do this, we use several concepts: that of Thom space, S-duality, fibre homotopy equivalence, and the spectral sequence in K-theory. In this chapter we are able only to sketch the background material. Another way of stating the theorem on the nonexistence of vector fields is that there are no cross sections to the fibre map $V_{\rho(n)+1}(\mathbf{R}^n) \to V_1(\mathbf{R}^n) = S^{n-1}$, and Theorem 12(8.2) says that there is a cross section to $V_{\rho(n)}(\mathbf{R}^n) \to V_1(\mathbf{R}^n)$. An important reduction step in proving the nonexistence theorem is a statement that a cross section of $V_k(\mathbf{R}^n) \to V_1(\mathbf{R}^n)$ exists if and only if there is a map between shunted projective spaces with particular properties.

1. Thom Spaces of Vector Bundles

Let ξ be a real vector bundle with base space $B(\xi)$, associaed projective bundle $P(\xi)$, associated sphere bundle $S(\xi)$, and associated unit disk bundle $D(\xi)$ for some riemannian metric on ξ. If α is the associated principal $O(n)$-bundle to ξ, then $P(\xi)$ is the fibre bundle $\alpha[RP^{n-1}]$, $S(\xi)$ is $\alpha[S^{n-1}]$, and $D(\xi)$ is $\alpha[D^n]$. We consider the following diagram of cofibre maps f_1, f_2, and f_3, where f_2 and f_3 are inclusion maps and f_1 is the zero cross section $f_1(b) = (0, 1)$.

$$
\begin{array}{ccc}
B(\xi) & \xrightarrow{\ f_1\ } & S(\xi \oplus \theta^1) \\
\Big\uparrow{\scriptstyle p_1} & & \Big\downarrow{\scriptstyle q_1} \\
S(\xi) & \xrightarrow{\ f_2\ } & D(\xi) \\
\Big\downarrow{\scriptstyle p_2} & & \Big\downarrow{\scriptstyle q_2} \\
P(\xi) & \xrightarrow{\ f_3\ } & P(\xi \oplus \theta^1)
\end{array}
$$

The map p_1 is the projection in the bundle $S(\xi) \to B(\xi)$, $p_2(x)$ is the line through 0 and x for $x \in S(\xi)$, $q_1(x)$ is the vector $((1 - \|x\|)x, 2\|x\| - 1)$ divided by its length for $x \in D(\xi)$, and $q_2(x)$ is the line determined by 0 and $(x, 1 - \|x\|^2)$.

1.1 Proposition. *With the above notations the restrictions* $q_1 \colon D(\xi) - \operatorname{im} f_2 \to$ $S(\xi \oplus \theta^1) - \operatorname{im} f_1$ *and* $q_2 \colon D(\xi) - \operatorname{im} f_2 \to P(\xi \oplus \theta^1) - \operatorname{im} f_3$ *are homeomorphisms. These maps induce homeomorphisms* $D(\xi)/S(\xi) \to S(\xi \oplus \theta^1)/B(\xi)$ *and* $D(\xi)/S(\xi) \to P(\xi \oplus \theta^1)$.

Proof. A direct construction yields inverses for the restrictions of q_1 and q_2. The induced maps are continuous bijections which are bicontinuous except possibly at the base point. Observe that $q_1(V)$ is open in $S(\xi \oplus \theta^1)$ and $q_2(V)$ in $P(\xi \oplus \theta^1)$, where V is an open neighborhood of $S(\xi)$ in $D(\xi)$.

1.2 Definition. The Thom space of a real vector bundle, denoted $T(\xi)$, is the quotient space $D(\xi)/S(\xi)$.

Observe that (1.1) gives several models for the Thom space and it is independent of the riemannian metric used to define it, since $P(\xi \oplus \theta^1)/P(\xi)$ does not involve a metric.

1.3 Proposition. *If ξ is a real vector bundle with a compact base space, $T(\xi)$ is homeomorphic to the one-point compactification of $E(\xi)$, the total space of ξ.*

Proof. Observe that $D(\xi) - S(\xi)$ and $E(\xi)$ are homeomorphic. Then $D(\xi)/S(\xi)$, the one-point compactification of $D(\xi) - S(\xi)$, and the one-point compactification of $E(\xi)$ are homeomorphic.

1.4 Example. Let ξ be the (trivial) n-dimensional vector bundle over a point. Then we have $T(\xi) = S^n$. If θ^n is the trivial n-dimensional bundle over X, then $T(\theta^n)$ is homeomorphic to $S^n(X \cup \{\infty\})$. A Thom space is a generalized suspension.

1.5 Proposition. *Let ξ and η be two real vector bundles over a compact space. Then the Thom space $T(\xi \times \eta)$ and the space $T(\xi) \wedge T(\eta)$ are homeomorphic.*

Proof. The space $T(\xi \times \eta)$ is the one-point compactification of $E(\xi \times \eta)$ and $T(\xi) \wedge T(\eta)$ of $E(\xi) \times E(\eta)$. The result follows from $E(\xi \times \eta) = E(\xi) \times E(\eta)$.

The proposition holds more generally. One can construct a homeomorphism

$$f: (D(\xi \times \eta), S(\xi \times \eta)) \rightarrow (D(\xi) \times D(\eta), S(\xi) \times D(\eta) \cup D(\xi) \times S(\eta))$$

defined by the relation

$$f(x, y) = (\max(\|x\|, \|y\|))^{-1}(\|x\|^2 + \|y\|^2)^{-1/2}(x, y)$$

1.6 Corollary. *The Thom space* $T(\xi \oplus \theta^n)$ *is homeomorphic to the n-fold suspension* $S^n(T(\xi))$.

Proof. We have $\xi \oplus \theta^n$ isomorphic to $\xi \times \mathbf{R}^n$, where $\mathbf{R}^n$ is the n-dimensional vector bundle over a point.

The real k-dimensional projective space RP^k is the quotient space of S^k, modulo the relation: x is equivalent to $-x$ for $x \in S^k$. If ξ_k denotes the canonical line bundle on RP^k and if $m\xi_k$ denotes $\xi_k \oplus \overset{(m)}{\cdots} \oplus \xi_k$, the total space $E(m\xi_k)$ of $m\xi_k$ is the quotient space of $S^k \times \mathbf{R}^m$ modulo the relation: (x, y) is equivalent to $(-x, -y)$ for $x \in S^k$, $y \in \mathbf{R}^n$. Moreover, $D(m\xi_k)$ is the quotient space of $S^k \times D^m$ modulo the relation: (x, y) is equivalent to $(-x, -y)$ for $x \in S^k$, $y \in D^m$. Let $\langle x, y \rangle$ denote the class of (x, y) in $D(m\xi_k)$. Then we have $\langle x, y \rangle \in S(m\xi_k)$ if and only if $\|y\| = 1$.

1.7 Example. We define a map $f: S^k \times D^m \rightarrow S^{k+m}$ by the relation $f(x, y) = (y, (1 - \|y\|^2)x)$, where $f(S^k \times S^{m-1}) = S^{m-1}$ and $S^{m-1} \subset S^{k+m}$. Since $f(-x, -y) = -f(x, y)$, the map f defines a map $g: D(m\xi_k) \rightarrow RP^{k+m}$ such that $g(S(m\xi_k)) = RP^{m-1}$. Observe that $f: S^k \times \operatorname{int} D^m \rightarrow S^{k+m} - S^{m-1}$ and $g: D(m\xi_k) - S(m\xi_k) \rightarrow RP^{k+m} - RP^{m-1}$ are homeomorphisms. Therefore, the map g defines a quotient map

$$h: T(m\xi_k) \rightarrow RP^{m+k}/RP^{m-1}$$

which is a homeomorphism. More generally, we have the following theorem, using (1.6) and (1.7).

1.8 Theorem. *The Thom space* $T(m\xi_k \oplus \theta^n)$ *and the n-fold suspension of the stunted projective space* $S^n(RP^{m+k}/RP^{m-1})$ *are homeomorphic.*

2. *S*-Category

In the following discussion all spaces have base points, and all maps and homotopies preserve base points. The suspension factor S defines a function $S: [X, Y] \rightarrow [S(X), S(Y)]$. Moreover, $[S(X), S(Y)]$ is a group, and the function $S: [S^k(X), S^k(Y)] \rightarrow [S^{k+1}(X), S^{k+1}(Y)]$ is a group morphism.

2.1 Definition. We denote the direct limit of the sequence of abelian groups

$$[S^2(X), S^2(Y)] \to [S^3(X), S^3(Y)] \to \cdots \to [S^n(X), S^n(Y)] \to \cdots$$

by $\{X, Y\}$. An element of $\{X, Y\}$ is called an S-map from X to Y.

The suspension functor $S: \{X, Y\} \to \{S(X), S(Y)\}$ is easily seen to be an isomorphism. We state the following result of Spanier and Whitehead which is proved in a book by Spanier [3].

2.2 Theorem. *The natural function* $[X, Y] \to \{X, Y\}$ *is a bijection for X, a CW-complex of dimension n, and Y, an r-connected space, where $n - 1 < 2r - 1$.*

The idea of the proof is to show that $S: [X, Y] \to [S(X), S(Y)] = [X, \Omega S(Y)]$ is a bijection under the above hypotheses. If the above hypotheses hold for X and Y, they hold for $S(X)$ and $S(Y)$. The bijective character of $[X, Y] \to [X, \Omega S(Y)]$ is established using a Wang type of sequence over $S(Y)$ for the fibring $ES(Y) \to S(Y)$.

A composition $\{X, Y\} \times \{Y, Z\} \to \{X, Z\}$ is defined from the composition functions

$$[S^n(X), S^n(Y)] \times [S^n(Y), S^n(Z)] \to [S^n(X), S^n(Z)]$$

2.3 Proposition. *The composition function* $\{X, Y\} \times \{Y, Z\} \to \{X, Z\}$ *is biadditive.*

Proof. Using the coH-space structure of $S^k X$, we see that the function $[S^k X, S^k Y] \times [S^k Y, S^k Z] \to [S^k X, S^k Z]$ is additive in the first variable, and using the H-space structure on $\Omega S^k Z$, we see that the function $[S^{k-1} X, S^{k-1} Y] \times [S^{k-1} Y, \Omega S^k Z] \to [S^{k-1} X, \Omega S^k Z]$ is additive in the second variable. Now pass to the limit (with care).

The following proposition is useful in seeing how the Puppe sequence fits into the S-category.

2.4 Proposition. *Let $g: Z \to Y$ be a map such that $\alpha_f g$ is null homotopic, where $X \xrightarrow{f} Y \xrightarrow{\alpha_f} C_f$ is given by f. Then there exists a map $h: SZ \to SX$ such that Sg and $(Sf)h$ are homotopic.*

Proof. We consider the following mapping between Puppe sequences:

$$
\begin{array}{ccccccccc}
Z & \xrightarrow{g} & Y & \xrightarrow{\alpha_g} & C_g & \xrightarrow{\beta_g} & SZ & \xrightarrow{S_g} & SY \\
& & \downarrow{\scriptstyle 1} & & \downarrow & & \downarrow{\scriptstyle h} & & \downarrow{\scriptstyle S1} \\
X & \xrightarrow{f} & Y & \xrightarrow{\alpha_f} & C_f & \xrightarrow{\beta_f} & SX & \xrightarrow{S_f} & SY
\end{array}
$$

Then the map h above has the desired properties.

2.5 Theorem. *Let $f: X \to Y$ be a map, and form the Puppe sequence $X \xrightarrow{f}$ $Y \xrightarrow{\alpha_f} C_f \xrightarrow{\beta_f} SX \xrightarrow{S_f} SY$. Then for each space Z the following exact sequences hold:*

$$\{X, Z\} \leftarrow \{Y, Z\} \leftarrow \{C_f, Z\} \leftarrow \{SX, Z\} \leftarrow \{SY, Z\}$$

and

$$\{Z, X\} \to \{Z, Y\} \to \{Z, C_f\} \to \{Z, SX\} \to \{Z, SY\}$$

Proof. The first sequence is a limit of the usual Puppe sequence. Proposition (2.4) yields the exactness of $\{Z, X\} \to \{Z, Y\} \to \{Z, C_f\}$. The second sequence now follows by iteration.

3. S-Duality and the Atiyah Duality Theorem

We consider maps $u: X \wedge X' \to S^n$ which are referred to as n-pairings. Such a map defines two group morphisms $u_Z: \{Z, X'\} \to \{X \wedge Z, S^n\}$ and $u^Z: \{Z, X\} \to \{Z \wedge X', S^n\}$, by the relations $u_Z(\{f\}) = \{u(1 \wedge f)\}$ and $u^Z(\{g\}) = \{u(g \wedge 1)\}$. For $Z = S^k$ we write u_k and u^k, respectively. We propose the following definition of S-duality which has been considered independently by P. Freyd.

3.1 Definition. An n-pairing $u: X \wedge X' \to S^n$ is called an n-duality provided $u_k: \{S^k, X'\} \to \{X \wedge S^k, S^n\}$ and $u^k: \{S^k, X\} \to \{S^k \wedge X', S^n\}$ are group isomorphisms. If an n-duality map $u: X \wedge X' \to S^n$ exists, then X' is called an n-dual of X. We call X' an S-dual of X provided some suspension of X' is n-dual to some suspension of X for some n.

Observe that this definition of duality is closely related to the definition of duality in linear algebra where a morphism $V \oplus V' \to F$ is given. The non-degeneracy can be described by saying that a certain correlation morphism is an isomorphism. If $X \wedge X'$ is thought of as a tensor product, then u_k and u^k are correlation morphisms.

3.2 Example (Spanier and Whitehead [1]). Let X be a subcomplex of S^n that is finite, and let $g: X' \to S^n - X$ be an inclusion map that is a homotopy equivalence. Then there is a duality map $X \wedge X' \to S^n$.

3.3 Remark. Originally the situation in (3.2) was used as the definition of an n-duality. Then Spanier [2] modified the definition to consider n-pairings with a certain slant product property with its homology and cohomology. Duality in the Spanier sense is a duality in the sense of (3.2). In Sec. 8 we consider the converse.

3.4 Proposition. *Let* $u: X \wedge X' \to S^n$ *be an n-duality between finite CW-complexes. Then for each finite CW-complex Z the following group morphisms are isomorphisms:*

$$u_Z: \{Z, X'\} \to \{X \wedge Z, S^n\} \quad \text{and} \quad u^Z: \{Z, X\} \to \{Z \wedge X', S^n\}$$

For the proof, induction is used on the number of cells of Z and the Puppe sequence for the last cell attached to Z.

A duality between spaces defines a duality between maps.

3.5 Definition. Let $u: X \wedge X' \to S^n$ and $v: Y \wedge Y' \to S^n$ be two n-duality maps between finite CW-complexes. The induced n-duality between S-maps is the group isomorphism

$$D_n(u, v) = u_{Y'}^{-1} v^X: \{X, Y\} \to \{Y', X'\}$$

We have the following commutative diagram:

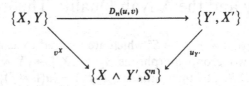

If $w: Z \wedge Z' \to S^n$ is a third n-duality map, we have

$$D_n(u, w)\{g\}\{f\} = (D_n(u, v)\{f\})(D_n(v, w)\{g\})$$

and $D_n(u, u)\{1\} = \{1\}$.

The following theorem is basic in the reduction of the vector field problem to one concerning shunted projective spaces. It is due to Atiyah [2] and is a generalization of a result of Milnor and Spanier [1]. Recall that v is a normal bundle to a manifold X with tangent bundle $\tau(X)$ provided $\tau(X) \oplus v$ is trivial.

3.6 Theorem. *Let* v *be a normal bundle of a compact manifold* X. *Then the Thom space* $T(v)$ *is an S-dual of* $X/\partial X$, *where* ∂X *is the boundary of* X.

This results by embedding X into a cube in $\mathbf{R}^n$ such that ∂X is the part of X carried into the boundary of the cube.

Since Thom spaces can be viewed as generalized suspensions, it is clear that Thom spaces should be related to S-duality. The next theorem of Atiyah is the most important result in this direction.

3.7 Theorem. *Let* ξ *and* η *be two vector bundles over a closed differentiable manifold* X *such that* $\xi \oplus \eta \oplus \tau(X)$ *is stably trivial on* X. *Then* $T(\xi)$ *and* $T(\eta)$ *are S-duals of each other.*

It is easy to see that (3.6) implies (3.7). It can be assumed that ξ is a differentiable vector bundle with a smooth riemannian metric. Then $D(\xi)$ is a compact differentiable manifold with $\partial D(\xi) = S(\xi)$. The tangent bundle to $D(\xi)$ is just $\pi^*(\xi \oplus \tau(X))$, and $\pi^*(\eta)$ is a normal bundle. Since the projection $\pi: D(\xi) \to X$ is a homotopy equivalence, $T(\xi) = D(\xi)/S(\xi)$ is an S-dual of $T(\eta)$.

4. Fibre Homotopy Type

Fibre homotopy is a general concept which could have been considered as part of the foundational material of Chaps. 1 and 2.

4.1 Definition. Let $p: E \to X$ and $p': E' \to X$ be two bundles over X. A homotopy $f_t: E \to E'$ is a fibre homotopy provided $p'f_t = p$ for all $t \in I$. Two bundle morphisms $f, g: E \to E'$ are fibre homotopic provided there is a fibre homotopy $f_t: E \to E'$ with $f_0 = f$ and $f_1 = g$. A bundle morphism $f: E \to E'$ is a fibre homotopy equivalence provided there exists a bundle morphism $g: E' \to E$ with gf and fg fibre homotopic to the identity. Two bundles E and E' have the same fibre homotopy type provided there exists a fibre homotopy equivalence $f: E \to E'$.

The reader can easily see that there is a category consisting of bundles and fibre homotopy equivalence classes of maps as morphisms. The isomorphisms in this category are the fibre homotopy equivalences.

For a general treatment of fibre homotopy equivalence see Dold [4]. We have need of a result from Dold [1] which we outline here. First, we consider the following lemma.

4.2 Lemma. *Let Y and Y' be two locally compact spaces. Let $f: D^n \times Y \to D^n \times Y'$ be a D^n-morphism of the bundles $D^n \times Y \to D^n$ and $D^n \times Y' \to D^n$. We assume the following properties for f.*

(1) *The map $f: \{x\} \times Y \to \{x\} \times Y'$ is a homotopy equivalence for each $x \in D^n$.*
(2) *There are a map $g': S^{n-1} \times Y' \to S^{n-1} \times Y$ of S^{n-1}-bundles and a fibre homotopy $h'_t: S^{n-1} \times Y \to S^{n-1} \times Y$ with $g'f = h'_0$ and $1 = h'_1$. Then there exist a prolongation $g: D^n \times Y' \to D^n \times Y$ of g' and a prolongation $h_t: D^n \times Y \to D^n \times Y$ of h'_t with $gf = h_0$ and $1 = h_1$.*

For a proof of this lemma, see Dold [1, pp. 118–120]. It consists of viewing g' as a map $S^{n-1} \to I(Y', Y) \subset Map(Y', Y)$, where $I(Y', Y)$ is the subspace of homotopy equivalences, and proving that it is null homotopic. The null homotopy defines g, and then with some care h_t is defined.

The following theorem of Dold is one of the first results on fibre homotopy equivalence.

4.3 Theorem. *Let $p: E \to B$ and $p': E' \to B$ be two locally trivial bundles over a finite CW-complex with locally compact fibres. Let $f: E \to E'$ be a map such that the restriction $f: E_b \to E'_b$ is a homotopy equivalence for all $b \in B$. Then f is a fibre homotopy equivalence.*

This theorem is proved inductively on the cells of B, using Lemma (4.2) and the fact that over a contractible space a locally trivial bundle is fibre homotopically equivalent to a product. We leave the details to the reader as an exercise.

4.4 Corollary. *Let $p: E \to B$ be a locally trivial bundle over a finite CW-complex and let $u: E \to Y$ be a map such that the restriction $u: E_b \to Y$ is a homotopy equivalence for all $b \in B$. Then $p: E \to B$ is fibre homotopically equivalent to the trivial bundle $B \times Y \to B$.*

Observe that the map $(p, u): E \to B \times Y$ satisfies the hypothesis of Theorem (4.3).

5. Stable Fibre Homotopy Equivalence

We are primarily interested in the fibre homotopy type of sphere bundles, that is, bundles whose fibre is a sphere. For example, the associated sphere bundle $S(\xi) \to B(\xi)$ of a vector bundle ξ. This leads to the next definition.

5.1 Definition. Let ξ and η be two vector bundles over X. The associated sphere bundles $S(\xi)$ and $S(\eta)$ are stable fibre homotopically equivalent provided $S(\xi \oplus \theta^n)$ and $S(\eta \oplus \theta^m)$ have the same fibre homotopy type for some n and m.

Clearly, stable fibre homotopy equivalence is an equivalence relation. We denote by $J(\xi)$ the stable fibre homotopy class determined by ξ and by $J(X)$ the set of all stable fibre homotopy classes of vector bundles on X. If every bundle is of finite type on X, there is a natural quotient surjection

$$\widetilde{KO}(X) \to J(X)$$

where $\widetilde{KO}(X)$ is viewed as the group of s-equivalence classes of vector bundles over X, by 8(3.8).

Let X be a space over which each real vector bundle is of the finite type.

5.2 Proposition. *The direct sum of vector bundles induces on $J(X)$ the structure of an abelian group, and the quotient function $\widetilde{KO}(X) \to J(X)$ is a group epimorphism.*

Proof. Everything will follow from the corresponding properties of $\widetilde{KO}(X)$ if we can prove that the sum operation on $J(X)$ induced from the sum operation on $\widetilde{KO}(X)$ is well defined; that is, for $J(\xi) = J(\xi')$ we shall prove that $J(\xi \oplus \eta) = J(\xi' \oplus \eta)$. A similar relation holds in the second summand. By replacing ξ by $\xi \oplus \theta^n$ and ξ' by $\xi' \oplus \theta^m$, we can assume that there are fibre homotopy equivalences $f: S(\xi) \to S(\xi')$ and $f': S(\xi') \to S(\xi)$ and fibre homotopies $h_t: S(\xi) \to S(\xi)$ and $h_t' = S(\xi') \to S(\xi')$ with $h_0 = f'f$, $h_1 = 1$, $h_0' = ff'$, and $h_1' = 1$. We define maps $g: S(\xi \oplus \eta) \to S(\xi \oplus \eta)$ and $g': S(\xi' \oplus \eta) \to S(\xi \oplus \eta)$ by the relations

$$g(x \cos \theta, y \sin \theta) = (f(x) \cos \theta, y \sin \theta)$$

$$g'(x' \cos \theta, y \sin \theta) = (f'(x') \cos \theta, y \sin \theta)$$

for $x \in S(\xi)$, $x' \in S(\xi')$, $y \in S(\eta)$, and $0 \leq \theta \leq \pi/2$. The homotopies

$$k_t(x \cos \theta, y \sin \theta) = (h_t(x) \cos \theta, y \sin \theta)$$

$$k'_t(x' \cos \theta, y \sin \theta) = (h'_t(x') \cos \theta, y \sin \theta)$$

are fibre homotopies with $k_0 = g'g$, $k_1 = 1$, $k'_0 = gg'$, and $k'_1 = 1$. This proves the proposition.

5.3 Remarks. Let X be a finite CW-complex. The group $J(X)$ is a subgroup of another group $\tilde{K}_{\text{Top}}(X)$. We outline its construction. Consider the set $F_n(X)$ of all fibre homotopy classes of fibrations with fibre S^{n-1}. The transition functions have values in the semigroup $H(n)$ of homotopy equivalences $S^{n-1} \to S^{n-1}$. As in Chaps. 3 and 4, the cofunctor $F_n(X)$ can be represented by $[X, B_{H(n)}]$, where $B_{H(n)}$ is a classifying space for the semigroup $H(n)$ (see Dold and Lashof [1]). The inclusion $H(n) \to H(n + r)$ defines a map $B_{H(n)} \to B_{H(n+r)}$. We denote by B_H the inductive limit of the sequence

$$B_{H(1)} \to B_{H(2)} \to \cdots \to B_{H(n)} \to \cdots$$

Then B_H admits a natural H-space structure, and there is an H-space map $B_0 \to B_H$ which is induced by the inclusion $O(n) \to H(n)$. By analogy with $\tilde{KO}(X)$, we view $[X, B_H]$ as $\tilde{K}_{\text{Top}}(X)$, the group of stable fibre homotopy classes of sphere bundles. The image of the natural group morphism induced by $B_0 \to B_H$

$$\tilde{KO}(X) = [X, B_0] \to [X, B_H] = \tilde{K}_{\text{Top}}(X)$$

is just $J(X)$. For a complete treatment of the construction of B_H, see Stasheff [1].

6. The Groups $J(S^k)$ and $\tilde{K}_{\text{Top}}(S^k)$

In this section we interpret the elements of $J(S^k)$ and $\tilde{K}_{\text{Top}}(S^k)$. In particular, we prove that $J(S^k)$ and $\tilde{K}_{\text{Top}}(S^k)$ are finite groups, and this in turn implies that $J(X)$ is a finite group.

6.1 Proposition. *For $2 \leq k \leq n - 2$ the groups $\pi_k(B_{H(n)})$, $\pi_{k-1}(H(n))$, and $\pi_{n+k-2}(S^{n-1})$ are isomorphic. Moreover, $\pi_1(B_{H(n)})$ and $\pi_0(H(n))$ are isomorphic to Z_2 for $n \geq 1$.*

Proof. The isomorphism between $\pi_k(B_{H(n)})$ and $\pi_{k-1}(H(n))$ is induced by the boundary operator in the exact homotopy sequence of the universal $H(n)$-bundle over $B_{H(n)}$. There are two components $H^+(n)$ and $H^-(n)$ of $H(n)$ corresponding to maps of degree $+1$ and -1, respectively. Consequently, $\pi_0(H(n)) = Z_2$.

Let M_n^d denote the component of $\mathrm{Map}_0(S^{n-1}, S^{n-1})$ consisting of maps of degree d. We have natural inclusions $M_n^1 \to H^+(n)$ and $M_n^{-1} \to H^-(n)$. The substitution map π from $f \in H^+(n)$ to $f(*)$ is a fibre map $\pi: H^+(n) \to S^{n-1}$ with fibre $\pi^{-1}(*)$ equal to M_n^1. From the exact sequence of homotopy groups with $k \leq n - 3$, we have

$$0 = \pi_{k+1}(S^{n-1}) \to \pi_k(M_n^1) \to \pi_k(H^+(n)) \to \pi_k(S^{n-1}) = 0$$

and the groups $\pi_k(M_n^1)$, $\pi_k(H^+(n))$, and $\pi_k(H(n))$ are isomorphic for $k \leq n - 3$. Since $\mathrm{Map}_0(S^{n-1}, S^{n-1})$ is homeomorphic to the H-space $\Omega^{n-1}(S^{n-1})$, there are isomorphisms between $\pi_k(H(n))$, $\pi_k(M_n^1)$, $\pi_k(M_n^0) = \pi_k(\Omega^{n-1}(S^{n-1})) = \pi_{n+k-1}(S^{n-1})$ for $1 \leq k \leq n - 3$. This proves the proposition.

6.2 Remark. Explicitly, the isomorphism θ between $\pi_k(M_n^1)$ and $\pi_{n+k-1}(S^{n-1})$ can be described as follows: For $[f] \in \pi_k(M_n^1)$ the map f is the translation by the identity of $h': S^k \times S^{n-1} \to S^{n-1}$, where $h'(x, y) = *$ if either $x = *$ or $y = *$. Then $\theta[f] = [g]$, where $g: S^k \wedge S^{n-1} \to S^{n-1}$ is given by $g(x \wedge y) = h'(x, y)$.

This isomorphism is closely related to the classical J-homomorphism of G. W. Whitehead.

6.3 Definition. We define the J-homomorphism $J: \pi_r(O(n)) \to \pi_{r+n}(S^n)$ by the requirement that $J[f] = [g]$, where $g: S^r * S^{n-1} \to SS^{n-1}$ is the Hopf construction [see 15(3.3)] applied to the map $(x, y) \mapsto f(x)y$ for $x \in S^r$ and $y \in S^{n-1}$.

We can view $J: \pi_r(SO(n)) \to \pi_{r+n}(S^n)$ by restriction.

6.4 Proposition. *The J-homomorphism $J: \pi_r(SO(n)) \to \pi_{r+n}(S^n)$ factors into the following three morphisms:*

$$\pi_r(SO(n)) \xrightarrow{\varepsilon_*} \pi_r(M_n^1) \xrightarrow{\theta} \pi_{r+n-1}(S^{n-1}) \xrightarrow{E} \pi_{r+n}(S^n)$$

where $\varepsilon: SO(n) \to M_n^1$ is the inclusion map and E is the suspension morphism.

Proof. The Hopf construction applied to the map $h(x, y) = f(x)y$ is a map $g\langle x, t, y \rangle = \langle f(x)y, t \rangle$. The map $x \mapsto h(x, y)$ viewed $S^r \to M_n^1$ can be translated by the identity on S^{n-1} to a map $S^r \to M_n^0$ which is of the form $h': (S^r \times S^{n-1}, S^r \vee S^{n-1}) \to (S^{n-1}, *)$ up to homotopy. Then we have $[h'] = \theta \varepsilon_*([h])$ and $E([h']) = [g] = J([f])$.

In the stable range $r + 2 \leq n$ we have the commutative diagram

where Π_r is the stable steam equal to $\pi_{r+n}(S^n)$ for $r + 2 \leq n$. Moreover, θ is an isomorphism defined as the composition $\pi_r(H) \to \pi_r(M_n^1) \xrightarrow{\theta} \pi_{r+n}(S^n)$ for $r + 2 \leq n$.

6.5 Remark. To the above picture we add $\tilde{K}O(X)$ and $\tilde{K}_{\text{Top}}(X)$ for $X = S^{r+1}$, and we have the following commutative diagram:

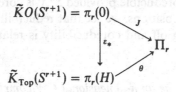

$$\tilde{K}O(S^{r+1}) = \pi_r(0)$$

$$\tilde{K}_{\text{Top}}(S^{r+1}) = \pi_r(H)$$

In particular, $J(S^{r+1})$ is isomorphic under θ to im J.

6.6 Theorem. *For a finite CW-complex X, the groups $\tilde{K}_{\text{Top}}(X)$ and $J(X)$ are finite groups.*

Proof. Since $\tilde{K}_{\text{Top}}(X)$ is a half-exact cofunctor, and since the $\tilde{K}_{\text{Top}}(S^m)$ are finite groups, we prove by induction on the number of cells, using the Puppe sequence, that $\tilde{K}_{\text{Top}}(X)$ is finite. Then $J(X)$ is finite because it is the subgroup of a finite group.

7. Thom Spaces and Fibre Homotopy Type

The relation between stable fibre homotopy equivalence and isomorphic spaces in the S-category is contained the next proposition.

7.1 Proposition. *Let ξ and η be two vector bundles over X. If $S(\xi)$ and $S(\eta)$ have the same fibre homotopy type, $T(\xi)$ and $T(\eta)$ have the same homotopy type. If $J(\xi) = J(\eta)$, then $T(\xi)$ and $T(\eta)$ are isomorphic in the S-category.*

Proof. Let $f: S(\xi) \to S(\eta)$ and $g: S(\eta) \to S(\xi)$ be fibre homotopy inverses of each other. Then f and g prolong radially to $f': D(\xi) \to D(\eta)$ and $g': D(\eta) \to D(\xi)$. The homotopies between fg and the identity prolong to a homotopy $(D(\xi), S(\xi)) \to (D(\xi), S(\xi))$ between $f'g'$ and the identity. Similarly, $g'f'$ is homotopic to the identity. By passing to quotients, we have maps $\bar{f}: T(\xi) \to T(\eta)$ and $\bar{g}: T(\eta) \to T(\xi)$ which are homotopy inverses of each other.

For the second statement, if $J(\xi) = J(\eta)$, then $S(\xi \oplus \theta^n)$ and $S(\eta \oplus \theta^m)$ have the same fibre homotopy type. From the first statement, $S^n T(\xi) = T(\xi \oplus \theta^n)$ and $S^m T(\eta) = T(\eta \oplus \theta^n)$ have the same homotopy type.

To consider the converse of the previous proposition, we need the following notions.

7.2 Definition. A space X (with a base point) is reducible provided there exists a map $f: S^n \to X$ such that $f_*: \tilde{H}_i(S^n) \to \tilde{H}_i(X)$ is an isomorphism for $i \geq n$. A space X is S-reducible provided $S^k X$ is reducible for some k.

The dual concept is introduced in the next definition.

7.3 Definition. A space X (with a base point) is coreducible provided there exists a map $g: X \to S^n$ such that $g^*: \tilde{H}^i(S^n) \to \tilde{H}^i(X)$ is an isomorphism for $i \leq n$. A space X is S-coreducible provided $S^k X$ is coreducible for some k.

In the next two propositions, we see that reducibility of a space is related to the top cell splitting off and coreducibility is related to the bottom cell splitting off.

7.4 Proposition. *Let X be an n-dimensional CW-complex with one n-cell, and let $v: X \to X/X^{n-1} = S^n$ be the natural projection map. The space X is reducible if and only if there is a map $f: S^n \to X$ such that vf is homotopic to the identity.*

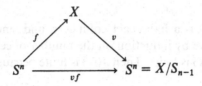

Proof. If X is reducible, there must be a reduction map $f: S^n \to X$ (in dimension n) by the homology condition. To prove that vf is homotopic to the identity, consider the following diagram:

$$\begin{array}{ccccccc} & & \tilde{H}_n(S^n) & & & & \\ & & \downarrow{\scriptstyle f_*} & \searrow{\scriptstyle (vf)_*} & & & \\ 0 = \tilde{H}_n(X^{n-1}) & \longrightarrow & \tilde{H}_n(X) & \overset{v_*}{\longrightarrow} & \tilde{H}_n(S^n) & \longrightarrow & \tilde{H}_{n-1}(X_{n-1}) \end{array}$$

Since $\tilde{H}_{n-1}(X^{n-1})$ is a free abelian group, v_* is an isomorphism, and, by hypothesis, f_* is an isomorphism. Consequently, $(vf)_*$ is an isomorphism and vf is homotopic to the identity. Conversely, the existence of such an f with vf of degree 1 yields a reduction of X since $\tilde{H}_n(X)$ has at most one generator.

7.5 Proposition. *Let X be a CW-complex with $X^n = S^n$, and let $u: S^n \to X$ be the natural inclusion map. The space X is coreducible if and only if there is a map $g: X \to S^n$ such that gu is homotopic to the identity.*

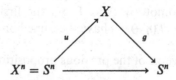

Proof. If X is coreducible, there exists a map $g: X \to S^n$ with $g^*: \tilde{H}^i(S^n) \to \tilde{H}^i(X)$ an isomorphism for $i \leq n$. Since $H^n(X)$ is a free abelian group, gu is homotopic to the identity, as in (6.4). The converse follows as in (6.4) since $\tilde{H}^n(X)$ has at most one generator.

7.6 Remark. Let ξ be a real vector bundle over X, and let $u_x\colon (D(\xi_x), S(\xi_x)) \to$ $(D(\xi), S(\xi))$ denote the natural inclusion map for $x \in X$. As in (7.5), the space $T(\xi) = D(\xi)/S(\xi)$ is coreducible if and only if there exists a map $g\colon$ $(D(\xi), S(\xi)) \to (S^n, *)$ such that $u_x^* g^*\colon \tilde{H}^n(S^n) \to \tilde{H}^n(D(\xi_x), S(\xi_x))$ is an isomorphism. The map g^* defines an orientation for ξ, and the Thom class of this oriented vector bundle will be $g^*(s)$, where s is a generator of $\tilde{H}^n(S^n)$. In effect, the Thom class determines the orientation preserving map $\mathbf{R}^n \to \xi_x \subset E(\xi)$ for each $x \in X$.

The next theorem is a partial converse of (6.1).

7.7 Theorem. *Let ξ be a vector bundle over a connected finite CW-complex X. Then the following statements are equivalent.*

(1) $J(\xi) = 0$ *in* $J(X)$
(2) *The space $T(\xi)$ is S-coreducible.*
(3) *The spaces $T(\xi)$ and $T(0) = X^+$ have the same S-type.*

Recall that X^+ is the space X plus an isolated discrete base point.

Proof. By Proposition (7.1), statement (1) implies (3). Assuming (3), we prove (2) by showing that X^+ is coreducible with the map $g\colon X^+ \to S^0$, where $g(X) = -1$ and $g(\infty) = +1$. Then $g^*\colon \tilde{H}^i(S^0) \to \tilde{H}^i(X^+)$ is an isomorphism for $i \leq 0$. Then $T(\xi)$ is S-coreducible.

To prove that statement (2) implies (1), we consider an integer m so large that $\eta = \xi \oplus \theta^m$ has the property that $T(\eta) = T(\xi \oplus \theta^m) = S^m T(\xi)$ is coreducible and $m > \dim X$. Let $g\colon (D(\eta), S(\eta)) \to (S^n, *)$ be the map defining a coreduction of $T(\eta)$. By (2.2) and by the first sequence in (2.5) we have the following commutative diagram with the horizontal arrows being isomorphisms:

$$
\begin{array}{ccc}
[S(\xi), S^{n-1}] & \longrightarrow & [D(\xi)/S(\xi), S^n] \\
\downarrow & & \downarrow \\
[S(\xi_x), S^{n-1}] & \longrightarrow & [D(\xi_x)/S(\xi_x), S^n]
\end{array}
$$

Consequently, the coreduction map defines a map $f\colon S(\eta) \to S^{n-1}$ such that $f|S(\eta_x)\colon S(\eta_x) \to S^{n-1}$ is a homotopy equivalence for each $x \in X$.

Now the theorem follows from (4.4).

8. S-Duality and S-Reducibility

Let $u\colon X \wedge X' \to S^n$ be a map. For two spaces W and Z we define a function $[W, Z \wedge X] \to [W \wedge X', S^n Z]$, where the image of $[f]\colon W \to Z \wedge X$ is $[(1 \wedge u)(f \wedge 1)]$. In the limit we have a morphism:

$$\delta^{W}(u)z: \{W, Z \wedge X\} \rightarrow \{W \wedge X', S^{r}Z\}$$

With an easy application of the Puppe sequence we can prove the following result by induction of the number of cells of W and Z.

8.1 Proposition. *Let $u: X \wedge X' \rightarrow S^{r}$ be an r-duality map. Then $\delta^{W}(u)_{Z}$: $\{W, Z \wedge X\} \rightarrow \{W \wedge X', S^{n}Z\}$ is an isomorphism for finite CW-complexes W and Z.*

8.2 Proposition. *Let $u: X \wedge X' \rightarrow S^{r}$ and $v: Y \wedge Y' \rightarrow S^{r}$ be two duality maps, and let $f: Y \rightarrow X$ and $g: X' \rightarrow Y'$ be two maps such that $D_{r}(v, u)\{f\} = \{g\}$. Then the following diagram is commutative where W and Z are finite CW-complexes:*

$$
\begin{array}{ccc}
\{W, Z \wedge Y\} & \xrightarrow{\ \delta\ } & \{W \wedge Y', S^{r}Z\} \\
{\scriptstyle f_{*}} \downarrow & & \downarrow {\scriptstyle g^{*}} \\
\{W, Z \wedge X\} & \xrightarrow{\ \delta\ } & \{W \wedge Y', S^{r}Z\}
\end{array}
$$

8.3 Homology and cohomology. Recall that there is an isomorphism between $\tilde{H}^{i}(X, G)$ and $[X, K(G, i)]_{0}$, and for X, a finite CW-complex, $K(G, i)$ can be replaced by one of its skeletons $K(G, i)^{*}$ which will be a finite CW-complex for G finitely generated.

Moreover, $H_{i}(X, G)$ is isomorphic to $\pi_{k+i}(K(G, k) \wedge X)$ for large k. Again, $K(G, k)$ can be replaced by a suitable skeleton $K(G, k)^{*}$. For the details, see G. W. Whitehead [5].

Now we bring together the discussion in (8.2) and (8.3). We have for $v: S^{n} \wedge S^{m} \rightarrow S^{r}$, with $r = n + m$ and v a homeomorphism, the following commutative diagram:

$$
\begin{array}{ccccc}
\tilde{H}_{k}(S^{n}) = \{S^{k+q}, K(\mathbf{Z}, q)^{*} \wedge S^{n}\} & \xrightarrow{\delta} & \{S^{k+q} \wedge S^{m}, S^{r}K(\mathbf{Z}, q)^{*}\} & = \tilde{H}^{r-k}(S^{m}) \\
{\scriptstyle f_{*}} \downarrow & & \downarrow {\scriptstyle g^{*}} \\
\tilde{H}_{k}(X) = \{S^{k+q}, K(\mathbf{Z}, q)^{*} \wedge X\} & \xrightarrow{\delta} & \{S^{k+q} \wedge X', S^{r}K(\mathbf{Z}, q)^{*}\} & = \tilde{H}^{r-k}(X')
\end{array}
$$

Therefore, $f_{*}: \tilde{H}_{i}(S^{n}) \rightarrow \tilde{H}_{i}(X)$ is an isomorphism for $i \leqq n$ if and only if g^{*}: $\tilde{H}^{j}(S^{m}) \rightarrow \tilde{H}^{j}(X')$ is an isomorphism for $m < j$. From this we get the following theorem.

8.4 Theorem. *Let X and X' be finite CW-complexes that are S-dual to each other. Then X is S-reducible if and only if X' is S-coreducible.*

9. Nonexistence of Vector Fields and Reducibility

We define a map $\theta: RP^{n-1} \rightarrow O(n)$ by the requirement that $\theta(L)$ be a reflection through the hyperplane perpendicular to L. In other words, $\theta(\{x, -x\})y = y - 2(x|y)x$ for $x \in S^{n-1}$ and $y \in \mathbf{R}^{n}$. The map θ is compatible with the following inclusions:

$$\begin{array}{ccc}
RP^{n-k-1} & \xrightarrow{\;\theta\;} & O(n-k) \\
\downarrow & & \downarrow \\
RP^{n-1} & \xrightarrow{\;\theta\;} & O(n) \\
\downarrow & & \downarrow \\
RP^{n-1}/RP^{n-k-1} & \longrightarrow & O(n)/O(n-k) = V_k(\mathbf{R}^n)
\end{array}$$

We have a map $\theta: RP^{n-1}/RP^{n-k-1} \to V_k(\mathbf{R}^n)$ making the above diagram commutative, where $\theta(L) = (v_1, \ldots, v_k)$ and v_i is the image of $\theta(\{x, -x\})e_{n-k+i}$ for $1 \leqq i \leqq k$ and $L = \{x, -x\}$.

9.1 Proposition. *The map θ defined $S^{n-1} = RP^{n-1}/RP^{n-2} \to V_1(\mathbf{R}^n) = S^{n-1}$ is a homeomorphism.*

Proof. We have $\theta(\{x, -x\})e_n = e_n - 2(x|e_n)x = e_n - 2x_n x = (-2x_1 x_n, \ldots, -2x_{n-1}x_n, 1 - x_n^2)$. If $y = (y_1, \ldots, y_n) \in S^{n-1}$ and $y_n \neq 1$, there is a unique x with $x_n > 0$ such that $\theta(\{x, -x\})e_n = y$. For $x_n = 0$ we have $\theta(\{x, -x\})e_n = (0, \ldots, 0, 1)$.

The following diagram is commutative:

$$\begin{array}{ccc}
RP^{n-2}/RP^{n-k-1} & \xrightarrow{\;\theta\;} & V_{k-1}(\mathbf{R}^{n-1}) \\
\downarrow{\scriptstyle u} & & \downarrow \\
RP^{n-1}/RP^{n-k-1} & \xrightarrow{\;\theta\;} & V_k(\mathbf{R}^n) \\
\downarrow & & \downarrow{\scriptstyle v} \\
RP^{n-1}/RP^{n-2} & \xrightarrow{\;\theta\;} & V_1(\mathbf{R}^n)
\end{array}$$

By 7(5.1), $V_{k-1}(\mathbf{R}^{n-1})$ is $(n-k-1)$-connected, and $V_1(\mathbf{R}^n)$ is $(n-2)$-connected. Since u is a cofibre map and v is a fibre map, we have an inductive argument on k leading to the following statement.

9.2 Proposition. *For $i < 2n - 2k - 1$ there is an isomorphism $\theta_*: \pi_i(RP^{n-1}/RP^{n-k-1}) \to \pi_i(V_k(\mathbf{R}^n))$.*
For this we use the Serre exact sequence and the Whitehead theorem (see Serre [1]).

9.3 Remark. Recall that $\rho(n) = 2^c + 8d$, where $n = (2a + 1)2^{c+4d}$ and $0 \leqq c \leqq 3$. For $n \neq 1, 2, 3, 4, 6, 8, 16$, we have $n - 1 < 2n - 2(\rho(n) + 1) - 1$ or $2\rho(n) + 2 < n$. By (9.2) this means that $\pi_{n-1}(RP^{n-1}/RP^{n-\rho(n)-2}) \to \pi_{n-1}(V_{\rho(n)+1}(\mathbf{R}^n))$ is an isomorphism. Therefore, we have the following result.

9.4 Theorem. *The projection $q\colon V_{\rho(n)+1}(\mathbf{R}^n) \to S^{n-1}$ has a cross section if and only if there is a map*

$$S^{n-1} \to RP^{n-1}/RP^{n-\rho(n)-2}$$

whose composition with $RP^{n-1}/RP^{n-\rho(n)-2} \to RP^{n-1}/RP^{n-2} = S^{n-1}$ is of degree 1.

Proof. If s is a cross section, then by (9.3) there is a map $S^{n-1} \to RP^{n-1}/RP^{n-\rho(n)-2}$ which when composed with θ is homotopic to s. After composition with the projection onto S^{n-1}, the resulting map is homotopic to the identity.

Conversely, a map $S^{n-1} \to RP^{n-1}/RP^{n-\rho(n)-2}$ when composed with θ yields a map s' such that gs' is homotopic to the identity on S^{n-1}. Since q is a fibre map, this homotopy lifts into $V_{\rho(n)+1}(R^n)$ which defines a cross section.

9.5 Remark. The nonexistence problem for vector fields reduces to proving a nonrecuction theorem for $RP^{n-1}/RP^{n-\rho(n)-2}$ for $n > 8$. In other words, it must be proved that the top cell of $RP^{n-1}/RP^{n-\rho(n)-2}$ does not split off. Unfortunately, in this form the problem is hard to solve with cohomology operations because a cohomology operation can be thought of as moving up a low dimensional cell and "hitting" a higher-dimensional cell nontrivially. This line of reasoning leads to the conclusion that the nonexistence of a coreduction may lend itself to the methods of cohomology operations.

10. Nonexistence of Vector Fields and Coreducibility

A reference for this section is the review article by Morin [1]. First, we make a calculation of an S-dual of RP^n/RP^{n-k}.

10.1 Proposition. *Let r denote the order of $J(\xi_{k-1})$ in $J(RP^{k-1})$. Then P^{n+rp}/P^{n-k+rp} and P^n/P^{n-k} have the same S-type, and for $rp > n + 1$ the space RP^{rp-k-2}/RP^{rp-n-2} is an S-dual of P^n/P^{n-k}.*

Proof. For the first part, by (1.8), RP^n/RP^{n-k} is homeomorphic to $T((n - k + 1)\xi_{k-1})$. By (1.6) and (7.1), the spaces $T((n - k + 1)\xi_{k-1} \oplus rp\xi_{k-1})$, $T((n - k + 1)\xi_{k-1} \oplus \theta^{rp})$, $S^{rp}T((n - k + 1)\xi_{k-1})$, and $S^{rp}(RP^n/RP^{n-k})$ are of the same homotopy type.

For the second part, we recall that $\tau(RP^{k-1}) \oplus \theta^1$ is isomorphic to $k\xi_{k-1}$ by 2(4.8). To apply the Atiyah duality theorem (3.7), we must find a vector bundle that equals $-(n - k + 1)\xi_{k-1} - k\xi_{k-1} = (-n - 1)\xi_{k-1}$ in $\widetilde{KO}(RP^{k-1})$. Up to S-type we have $T((-n + 1)\xi_{k-1}) = T((rp - n - 1)\xi_{k-1})$ for $rp > n + 1$. Since $T((rp - n - 1)\xi_{k-1})$ is the space RP^{rp-k-2}/RP^{rp-n-2}, we have the result by (7.1). This proves the proposition.

In the next theorem of Atiyah and James the problem of the nonexistence of vector fields is reduced to a problem in the nonexistence of a coreduction of a certain shunted projective space.

10.2 Theorem. *If there are $\rho(n)$ orthonormal vector fields on S^{n-1}, there exists an integer $m \geq 1$ with $\rho(m) = \rho(n)$ and such that $RP^{m+\rho(m)}/RP^{m-1}$ is coreducible.*

Proof. By the construction in 12(1.1), there are $\rho(n)$ orthonormal vector fields on S^{qn-1}, and by (9.4), if $qn \geq 2(\rho(n) + 1)$, the shunted projective space $RP^{qn-1}/RP^{qn-\rho(n)-2}$ is reducible.

Let r denote the order of $J(\xi_{\rho(n)})$ in $J(RP^{\rho(n)})$. For all integers p with $rp - qn = m \geq 1$, the shunted projective space $RP^{m+\rho(m)}/RP^{m-1}$ is an S-dual of $RP^{qn-1}/R^{qn-\rho(n)-2}$ and $RP^{m+\rho(m)}/RP^{m-1}$ is S-coreducible for $m \geq \rho(n) + 3$. By (2.2) for large m, the space $RP^{m+\rho(m)}/RP^{m-1}$ is S-coreducible if and only if it is coreducible.

Finally, if p is divisible by $2n$ and if q is odd, $m = tn$, where t is an odd number and $\rho(m) = \rho(n)$. This proves the theorem.

11. Nonexistence of Vector Fields and $J(RP^k)$

11.1 Remarks. We consider orthogonal pairings $R^{k+1} \times R^n \to R^n$ for k, or, equivalently, C_k-modules. Let N_k denote the Grothendieck group of C_k-modulus. By 12(2.4) and 12(6.5), if $R^{k+1} \times R^n \to R^n$ is an orthogonal pairing, then n is divisible by c_k, where $c_k = a_{k+1}$. We have the following table of values for c_k.

k	0	1	2	3	4	5	6	7	8
c_k	1	2	4	4	8	8	8	8	16

The numbers c_k satisfy the relation $c_{k+8} = 16c_k$. Finally, there exists a pairing $R^{k+1} \times R^{c_k} \to R^{c_k}$, and for $k = 3 \pmod 4$ there are two such pairings up to isomorphism, otherwise only one. The next proposition gives a characterization of c_k and relates it to $\rho(n)$ as in 12(8.2). We leave the proposition to the reader.

11.2 Proposition. *The number $c_k = 2^e$, where e is the number of m with $0 < m \leq k$ and $m = 0, 1, 2, 4 \pmod 8$. Moreover, $\rho(n)$ equals the maximum of all $k + 1$ such that $c_k | n$.*

For each orthogonal pairing $R^{k+1} \times R^m \to R^m$ or C_k-module M, we have defined a trivialization

$$\tau_M: m\xi_k \to \mathbf{R}^m$$

Here $m = \dim_R M$. The trivialization is defined by restricting the orthogonal pairing and factoring it, $E(m\xi_k) \to E(\theta^m)$.

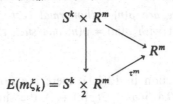

Clearly, $\tau_{M \oplus N} = \tau_M \oplus \tau_N$, and the bundle $c_k \xi_k$ is trivial on RP^k. In $J(RP^k)$ the order of the s-class $\xi_k - 1$ is a divisor of c_k.

11.3 Proposition. *For $k \leq l$ there is a group morphism $\theta'_{k,l}: N_k \to \widetilde{KO}(RP^l/RP^k)$ such that $\theta'_{k,l}(M) = (\dim M)\xi_l/\tau_M$ [see 10(1.2)] for each C_k-module M. Moreover, $\theta'_{k,l}$ is unique. We have a commutative diagram.*

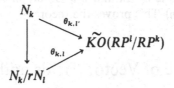

The vertical morphism is a quotient morphism, and the morphism $r: N_l \to N_k$ is induced by the inclusion $C_k \to C_l$.

Proof. The morphism $\theta'_{k,l}$ exists since $\tau_{M \oplus N} = \tau_M \oplus \tau_N$ and since $(\dim M \oplus N)\xi_l/\tau_{M \oplus N}$ and $((\dim M)\xi_l/\tau_M) \oplus ((\dim N)\xi_l/\tau_N)$ are isomorphic. Clearly, $\theta'_{k,l}$ is unique. Finally, for a module $M \in rN_l$ we have $c_l | \dim M$, and $(\dim M)\xi_l$ is trivial. Consequently, $\theta'_{k,l}(r(N_l)) = 0$, and $\theta_{k,l}$ is uniquely defined by the above diagram.

11.4 Remark. For $k = l - 1$ there is the group morphism

$$\theta_l = \theta_{l-1,l}: N_{l-1}/rN_k \to \widetilde{KO}(RP^l/RP^{l-1}) = \widetilde{KO}(S^l)$$

One form of the periodicity theorem says that θ_l is an isomorphism [see 14(13.3)]. We use this result in the remainder of this section.

This leads to the next proposition.

11.5 Proposition. *The morphisms defined above, $\theta_{k,l}: N/rN_l \to \widetilde{KO}(RP^l/RP^k)$, are epimorphisms.*

Proof. We prove this by induction on $l - k$. The case $l - k = 1$ is covered in (11.4). To prove that $\theta_{k,l}$ being an epimorphism implies that $\theta_{k,l+1}$ is an epimorphism, we use the following commutative diagram:

$$0 \longleftarrow N_k/rN_l \longleftarrow N_k/rN_{l+1} \longleftarrow N_l/rN_{l+1}$$

$$\downarrow {\scriptstyle \theta_{k,l}} \qquad\qquad \downarrow {\scriptstyle \theta_{k,l+1}} \qquad\qquad \downarrow {\scriptstyle \theta_{l,l+1}}$$

$$\widetilde{KO}(RP^l/RP^k) \longleftarrow \widetilde{KO}(RP^{l+1}/RP^k) \longleftarrow \widetilde{KO}(S^{l+1})$$

Since $\theta_{k,l}$ is an epimorphism and $\theta_{l,l+1}$ is an isomorphism, $\theta_{k,l+1}$ is an epimorphism by an easy 5-lemma type of argument.

11.6 Corollary. *The group $\widetilde{KO}(RP^l/RP^k)$ is a quotient of N_k/rN_l equal to $\mathbf{Z} \oplus Z_{2^e}$ for $k = 3$ (mod 4) and to Z_{2^e} for $k \neq 3$ (mod 4), where e is the number of $q = 0, 1, 2, 4$ (mod 8) with $k < q \leq l$.*

11.7 Remark. In the case of $k = 0$, $\widetilde{KO}(RP^l)$ is a quotient of Z_{c_l}. Consequently, it is a cyclic group generated by the class of the canonical line bundle ξ_l. This class has an order that is a divisor of c_l. In summary, there is a sequence of two epimorphisms

$$Z_{c_k} = \frac{N_0}{rN_k} \xrightarrow{\theta_{0,k}} \widetilde{KO}(RP^k) \longrightarrow J(RP^k)$$

If we can prove that $J(\xi_k)$ has order c_k, the above morphisms will be isomorphisms, and we shall have determined the structure of $\widetilde{KO}(RP^k) = J(RP^k)$.

The order of $J(RP^k)$ is related to the vector field problem.

11.8 Theorem. *If the order of $J(\xi_k)$ in $J(RP^k)$ is c_k for each k, then there do not exist $\rho(n)$ vector fields on S^{n-1} for each n.*

Proof. Suppose that there are $\rho(n)$ vector fields on S^{n-1} for some n. Then $RP^{m+\rho(m)}/RP^{m-1} = T(m\xi_{\rho(m)})$ is coreducible for some m by (10.2), and $J(m\xi_{\rho(m)}) = 0$ in $J(RP^{\rho(m)})$ by (7.7). By hypothesis, $c_{\rho(m)}$ divides m. This we want to show is impossible for all m.

Observe that $c_{\rho(1)} = c_1 = 2$, $c_{\rho(2)} = c_2 = 4$, $c_{\rho(4)} = c_4 = 8$, and $c_{\rho(8)} = c_8 = 16$. In each case, $c_{\rho(m)} = 2m$, and $c_{\rho(m)}$ does not divide m. In general, we write $m = (2a + 1)2^{c+4d} = (2a + 1)2^c 16^d$, where $0 \leq c \leq 3$. Then we have $\rho(m) = 2^c + 8d$ and $c_{\rho(m)} = 16^d c_{2^c} = 16^d 2(2^c)$ by the above calculation. Therefore, $c_{\rho(m)}$ has one power of 2 more than m, and $c_{\rho(m)}$ does not divide m. This proves the theorem.

12. Real K-Groups of Real Projective Spaces

Adams [6] made a calculation of the rings $K(CP^n/CP^m)$, $K(RP^n/RP^m)$, and $KO(RP^n/RP^m)$ and the action of the ψ^k operations on these rings. Finally, using these calculations, he was above to deduce that the coreduction described in Theorem (10.2) cannot exist. This calculation will not be reproduced here because it was developed in considerable detail in his paper. Instead, we shall consider only the calculation of the group $KO(RP^n)$ and in Sec. 13 prove that $KO(RP^n) = J(RP^n)$.

Up to this point, we have bypassed the generalized cohomology theories $K*$ and $KO*$ (see Atiyah and Hirzebruch [2]) because we have not needed them, but the cohomology theories generated by the K-functors are very important for certain considerations. We give a brief description of the generalized cohomology theories $K*$ and $KO*$ together with the spectral sequence relating them to the usual singular cohomology theory.

12.1 Definition. We define $\tilde{K}^{-p}(X) = \tilde{K}(S^p X)$ and $KO^{-p}(X) = KO(S^p X)$ for $p \geq 0$.

Using the Puppe sequence of Chap. 10, one proves that there is a sequence of cofunctors $\tilde{K}^{-p}$ and $\tilde{KO}^{-p}$ satisfying all the axioms for a cohomology theory on the category of finite CW-complexes except the dimension axiom (see Eilenberg and Steenrod [1]). These cohomology groups are defined for negative integers.

Using the periodicity theorems $\tilde{K}^{-p}(X) \cong \tilde{K}^{-p-2}(X)$ and $\tilde{KO}^{-p}(X)$ and $\tilde{KO}^{-p}(X) \cong \tilde{KO}^{-p-8}(X)$, we can systematically define $\tilde{K}^p$ and $\tilde{KO}^p$ for all integers p.

12.2 Theorem. *Let X be a finite CW-complex. There is a spectral sequence $E_r^{p,q}$ with the following properties:*

(1) *The term $E_2^{p,q}$ equals $\tilde{H}^p(X, KO^q(*))$.*

(2) *We have $E_\infty^{p,q} \Rightarrow \tilde{KO}^{p+q}(X)$.*

The filtration on $\tilde{KO}^{p+q}(X)$ which has $E_\infty^{p,q}$ as its associated graded groups is given by $\ker(\tilde{KO}^n(X) \to \tilde{KO}^n(X_{p-1}))$, where X_{p-1} is the $(p-1)$-skeleton of X.

A similar theorem holds for K. This spectral sequence is natural with respect to maps.

12.3 Coefficient groups. By the periodicity theorem, $K^p(*) = \mathbf{Z}$ for p even and 0 for p odd. In KO theory, there is the following table:

p	0	1	2	3	4	5	6	7	8
$\tilde{KO}^{-p}(*)$	$\mathbf{Z}$	$\mathbf{Z}_2$	$\mathbf{Z}_2$	0	$\mathbf{Z}$	0	0	0	$\mathbf{Z}$
$\pi_{p+8m}(0)$	$\mathbf{Z}_2$	$\mathbf{Z}_2$	0	$\mathbf{Z}$	0	0	0	$\mathbf{Z}$	$\mathbf{Z}_2$

We recall that $H^i(CP^n, \mathbf{Z}) = \mathbf{Z}$ for i even and $0 \leq i \leq 2n$ and 0 otherwise. For the space $X = CP^n$, the spectral sequence in (12.2) collapses. From this we easily get the next proposition.

12.4 Proposition. *The ring $K(CP^n)$ is $\mathbf{Z}[v]$, where $v = \xi_n(\mathbf{C}) - 1$ with one relation $v^{n-1} = 0$.*

For the multiplicative structure of $K(CP^n)$, one uses the ring morphism ch: $K(CP^n) \to H^{ev}(CP^n, \mathbf{Q})$ which we consider at length in Chap. 20.

There is a natural map $q: RP^{2n+1} \to CP^n$. One proves that $q^*(\xi_n(\mathbf{C}))$ is the complexification of $\xi_{2n+1}(\mathbf{R})$, where $\xi_k(F)$ denotes the canonical line bundle over FP^k.

Again the spectral sequence for $\tilde{K}(RP^{2n+1})$ collapses. Using this fact and the exact sequence

$$\mathbf{Z} \leftarrow \tilde{K}(RP^{2n}) \leftarrow \tilde{K}(RP^{2n+1}) \leftarrow 0$$

we have the next proposition, after looking at the spectral sequence.

12.5 Proposition. *The group $\tilde{K}(RP^n)$ is the finite cyclic of order $a(n) = 2^e$, where e is the integral part of $n/2$, and with generator w, where $w = q'(v)$ for $n = 2t$ or $2t + 1$, and v, the ring generator of $K(CP^n)$.*

Next we consider $\varepsilon_U: \tilde{KO}(RP^n) \to \tilde{K}(CP^n)$. Since the generator w of (12.5) has the property that $\varepsilon_U(u) = w$, by 14(11.4), where $u = \xi_n(\mathbf{R}) - 1$, we have the next proposition.

12.6 Proposition. *The morphism $\varepsilon_U: \tilde{KO}(RP^n) \to \tilde{K}(CP^n)$ is an epimorphism, and for $n = 6, 7, 8 \pmod 8$ ε_U is an isomorphism.*

Proof. For the isomorphism statement, consider the E_2 term in the spectral sequence for $\tilde{KO}(RP^n)$. It has at most $a(n)$ elements in it.

For $n = 6, 7, 8 \pmod 8$, we have $a(n) = c_n$, where c_n was defined in (11.1). Finally, with the above information and a look at the spectral sequence for $\tilde{KO}(RP^n)$, we have the following result.

12.7 Theorem. *The group $\tilde{KO}(RP^k)$ is cyclic of order c_k with generator $u = \xi_k(\mathbf{R}) - 1$.*

13. Relation Between $KO(RP^n)$ and $J(RP^n)$

In this section, we outline a solution to the vector field problem on spheres. This solution follows a general schema of Bott [5]. Our special calculations follow ideas developed by F. Hirzebruch in a lecture at the Summer Institute in Seattle in 1963. The reader is referred to Atiyah, Bott, and Shapiro [1] for further details on the following construction.

13.1 Notations. Let ξ be a Spin$(8m + 1)$ real vector bundle, that is, Spin$(8m + 1)$ is a structure group. The associated sphere bundle $S(\xi)$ can be thought of as Spin($8m + 1$)/Spin($8m$) bundle. As in Chap. 14, Sec. 13, we are able to define a morphism

$$\tilde{\alpha}_\xi^*: RO\,\text{Spin}(8m)/RO\,\text{Spin}(8m + 1) \to KO(S(\xi))$$

using the principal Spin($8m + 1$) bundle α_ξ^* resulting from lifting up α_ξ, the associated principal bundle of ξ.

The following theorem of Bott (see Bott [5]) is a generalization of 14(13.3). It is proved by a Mayer-Vietoris argument using the fact that it reduces to the case of 14(13.3) on each fibre.

13.2 Theorem. *Let ξ be a* Spin($8m + 1$) *real vector bundle. Let α_ξ^* denote the induced bundle of α_ξ on $S(\xi)$. Then α_ξ^* is a* Spin($8m$) *principal bundle. Then $KO(S(\xi))$ is a free $KO(B(\xi))$-module with two generators 1 and $\tilde{\alpha}_\xi^*(\Delta_{8m}^+)$. Moreover, the $KO(B(\xi))$-module $KO(S(\xi))$ has involution with $\tilde{\alpha}_\xi^*(\Delta_{8m}^\pm) \mapsto \tilde{\alpha}_\xi^*(\Delta_{8m}^\mp)$.*

13.3 Notations. We denote $\tilde{\alpha}_\xi^*(\Delta_{8m}^\pm)$ simply by $\Delta_{8m}^\pm$ in $KO(S(\xi))$. From (13.2), we have

$$\psi^k \Delta_{8m}^\pm = \theta_k(\xi)\Delta_{8m}^\pm + b$$

We are interested in calculating the $\theta_k(\xi)$ characteristic classes. We make use of the following multiplicative property of θ_k.

13.4 Proposition. *We have $\theta_k(\xi \oplus \eta) = \theta_k(\xi)\theta_k(\eta)$.*

We calculate θ_k using the morphism $\tilde{\alpha}_\xi^*$ and the character formula for $\Delta_{8m}^\pm$.

13.5 Character Calculation. We have $\psi^k \Delta_{8m}^\pm = \theta_k(\xi)\Delta_{8m}^\pm + b$, and subtracting these two relations, we have

$$\psi^k(\Delta_{8m}^+ - \Delta_{8m}^-) = \theta_k(\xi)(\Delta_{8m}^+ - \Delta_{8m}^-)$$

In $RO\,\text{Spin}(8m)$, we have

$$\Delta_{8m}^+ - \Delta_{8m}^- = \prod_{1 \leq j \leq 4m} (\alpha_j^{1/2} - \alpha_j^{-1/2})$$

and

$$\psi^k(\Delta_{8m}^+ - \Delta_{8m}^-) = \prod_{i \leq j \leq 4m} (\alpha_j^{k/2} - \alpha_j^{-k/2})$$

Therefore, we have the following formula for $r = (k - 1)/2$:

$$\begin{aligned}
\theta_k(\xi) &= \prod_{1 \leq j \leq 4m} \frac{\alpha_j^{k/2} - \alpha_j^{-k/2}}{\alpha_j^{1/2} - \alpha_j^{-1/2}} \\
&= \prod_{1 \leq j \leq 4m} (\alpha_j^r + \alpha_j^{r-1} + \cdots + 1 + \cdots + \alpha_j^{-r}) \\
&= \prod_{1 \leq j \leq 4m} (1 + \psi^1(\alpha_j) + \cdots + \psi^r(\alpha_j))
\end{aligned}$$

Observe that these formulas involve only ψ^k operations of α_j and α_j^{-1}. Therefore, for any real oriented $2n$-dimensional vector bundle η, these formulas can be used to define $\theta_k(\eta)$.

13.6 θ_k of a 2-Plane Bundle. We calculate $\theta_k(2\zeta)$, where ζ is a line bundle and $\zeta^2 = 1$.

$$\theta_k(2\zeta) = 1 + \psi^1(2\zeta) + \cdots + \psi^r(2\zeta)$$

$$= 1 + 2(\zeta + 1 + \zeta + \cdots + \zeta + 1)$$

$$= (2r + 1) + r(\zeta - 1) \qquad \text{for } r = (k - 1)/2 \text{ and } r \text{ even}$$

13.7 Calculation of $\theta_k(2n\zeta)$. Let ζ be a line bundle, and recall that $(\zeta - 1)^2 = -2(\zeta - 1)$. Then we compute the following class.

$$\theta_k(2n\zeta) = [(2r + 1) + r(\zeta - 1)]^n$$

$$= (2r + 1)^n + a(\zeta - 1)$$

We substitute $\zeta - 1 = -2$ in this expression and get the relation

$$1^n = (2r + 1)^n - 2a$$

where $r = (k - 1)/2$ and $2r + 1 = k$. Then we have the following relation $\theta_k(2n\zeta) = k^n + [(k^n - 1)/2](\zeta - 1)$.

13.8 Remark. The class $\theta_k(\xi)$ is a fibre homotopy type invariant of $S(\xi)$. Consequently, if $J(2n\zeta) = 0$, we have $[(k^n - 1)/2](\zeta - 1) = 0$.

We are now in a position to prove the main theorem of this section.

13.9 Theorem. *Let X be a finite CW-complex such that $\widetilde{KO}(X)$ is generated by a line bundle. Then the canonical epimorphism $J: \widetilde{KO}(X) \to J(X)$ is an isomorphism.*

Proof. By the classification theorem and (11.8), the order of $\zeta - 1$ is a power of 2 for the generating line bundle ζ. Let 2^r equal the order of $\zeta - 1$ in $\widetilde{KO}(X)$ and $J(2n\zeta) = 0$ in $J(X)$. Then, by (13.8), $(k^n - 1)/2 = 0 \pmod{2^r}$. For $k = 5$, we have $5^n - 1 = 0 \pmod{2^{r-1}}$. Since the units in $Z_{2^{r+1}}$ have the structure of $Z_2 \oplus Z_{2^{r-1}}$, where 5 generates the second factor, we have $2^{r-1} | n$ or $2^r | 2n$. Therefore, we have $2n(\zeta - 1) = 0$ in $\widetilde{KO}(X)$.

Combining the results of (11.8) and (13.9), we have the following theorem of Adams.

13.10 Theorem. *On the sphere S^{n-1}, there are at most $\rho(n) - 1$ orthonormal vector fields.*

14. Remarks on the Adams Conjecture

In the previous sections we studied the natural morphism $J: \widetilde{KO}(X) \to J(X)$ using K-theory characteristic classes and proved that it was an isomorphism for $X = RP^n$. This in turn led to a solution of the vector field problem.

Now we consider $J(X)$ and the morphism J for X a sphere. In (6.5) we saw that $J(S^n)$ can be viewed as a subgroup of the stable stem $\sigma_{n-1} = \pi_{i+n-1}(S^i)$ (for $i \geq n + 1$). By (12.3) the possible nonzero morphisms J are in the following degrees.

$$J: KO(S^{8m+i}) = \mathbf{Z}/2 \to J(S^{8m+i}) \subset \sigma_{8m+i-1} \qquad \text{for } i = 1, 2,$$

$$J: KO(S^{4k}) = \mathbf{Z} \to J(S^{4k}) \subset \sigma_{4k-1}.$$

For example $J(S^4) = \sigma_3 = \mathbf{Z}/24$ and J is an epimorphism of $\mathbf{Z} = KO(S^4)$ onto $\mathbf{Z}/24$. In a series of four papers [7] Adams considers the groups $J(X)$ by using K-theory characteristic classes in order to detect elements of $\widetilde{KO}(X)$ whose image in $J(X)$ is nonzero. A key question centered around the nature of the image of $\psi^k(x) - x$ in $J(X)$ for $x \in \widetilde{KO}(X)$ and in [7, I, p. 183] Adams posed the following conjecture.

14.1 Adams Conjecture. *For an integer k and $x \in \widetilde{KO}(X)$ there exists an integer $e = e(k, x)$ such that $k^e(\psi^k(x) - x)$ has image zero in $J(X)$.*

Note that since $\psi^{mn} = \psi^m \psi^n$, it suffices to prove that $\psi^p(x) - x$ is a p-torsion element in $J(X)$.

Adams proved in [7, I] that the conjecture holds for elements x which are linear combinations of $0(1)$ and $0(2)$ bundles over a finite complex X. From this he deduced in the same paper that it holds for elements $x \in im(r: K(S^{2n}) \to KO(S^{2n}))$. This together with the e-invariant $e: \sigma_{4k-1} \to \mathbf{Q}/\mathbf{Z}$ that Adams defined in [7, II] led Adams to the following theorem about the stable homotopy groups of spheres.

First recall that the Bernoulli numbers B_k are the rational numbers defined by

$$\frac{x}{e^x - 1} = 1 - \frac{x}{2} + \Sigma_{1 \leq k}(-1)^{k-1} B_k \frac{x^k}{(2k)!}$$

and that m_r denotes the denominator of $B_r/4r$ (expressed in lowest terms).

14.2 Theorem. (Adams) *The J-homomorphism on spheres has the following properties*:

(1) $J: \widetilde{KO}(S^{8k+i}) = \mathbf{Z}/2 \to \sigma_{8k+i-1}$ *is a monomorphism for $i = 1$ or 2,*

(2) $J: \widetilde{KO}(S^{8k+4}) = \mathbf{Z} \to \sigma_{8k+3}$ *has as image a direct summand which is cyclic of order m_{2k+1}, and*

(3) $J: \widetilde{KO}(S^{8k}) = \mathbf{Z} \to \sigma_{8k-1}$ *has an image a cyclic subgroup of order either m_{2k} or $2m_{2k}$. In the first case the image $J(S^{8k})$ is a direct summand.*

14.3 Remark. Adams observed that if the Adams conjecture were true for $X = S^{8k}$, then we can combine (2) and (3) of 14.2 into the single statement: $J: KO(S^{4m}) \to \sigma_{4m-1}$ has image a cyclic direct summand of order m_k. Thus one of the byproducts of the proof of the Adams conjecture is a precise determination of $J(S^m)$. Apart from finiteness of σ_n and the determination of the summand $J(S^{4k})$ in σ_{4k-1}, we have little global information about the graded abelian group σ_*.

Up to now there have been four separate proofs of the Adams conjectures which we now describe briefly. The mathematics that they have generated has far outshadowed the tying up of the "loose end" in part 3 of 4.2. To prove the Adams conjecture, it suffices to prove it for the universal bundle over $BO(n)$ and ψ^p where p is a prime. In the first two proofs one uses the fact that $BO(n)$ can be approximated by Grassmannians which are algebraic varieties defined over any field.

Proof 1. (Quillen, Friedlander) Quillen in [1] made the observation that for algebraic vector bundles E with class $[E]$ in $K(X)$ that the following formula holds

$$\psi^p[E] = [\text{Frob}^* E]$$

where Frob: $X \to X$ is the Frobenius endomorphism. Using the étale homotopy theory of Artin, Mazur [1], he was able to show that the Adams conjecture for *complex* vector bundles reduced to a conjecture on the etale properties of sphere bundles which Friedlander proved in [1]. Since this proof was only for complex K-theory, it did not resolve the question of m_{2k} or $2m_{2k}$ in 14.2(3).

Proof 2. (Sullivan) Sullivan uses etale homotopy type considerations for BO as a limit of algebraic varieties $BO(m, n) = 0(m + n)/0(m) \times 0(n)$ defined over $\mathbf{Q}$, the rational numbers. The Galois group $G = \text{Gal}(\overline{\mathbf{Q}}/\mathbf{Q})$ acts on BO_l the l-adic completion of BO, and a certain element of G induces the l-adic completion of the action of $\psi^p: BO \to BO$ for $p \neq l$. This in turn means that $\psi^p: BO(n)_l \to BO(n)_l$ is well defined and functorial in n. The l-adic completion $F(n)_l$ of the homotopy fibre $F(n)$ of $BO(n - 1) \to BO(n)$ has an action of ψ^p on it making the following diagram homotopy commutative.

$$
\begin{array}{ccccc}
F(n)_l & \longrightarrow & BO(n - 1)_l & \longrightarrow & BO(n)_l \\
\psi^p\downarrow & & \psi^p\downarrow & & \psi^p\downarrow \\
F(n)_l & \longrightarrow & BO(n - 1)_l & \longrightarrow & BO(n)_l
\end{array}
$$

Since $F(n)$ is just the associated sphere bundle of the universal vector bundle V over $BO(n)$, it follows that V and $\psi^p V$ are spherically homotopy equivalent over $BO(n)_l$ for each prime $l \neq p$, and so the difference in the finite groups $J(BO(n, m))$ is a p torsion element. Sullivan also shows that there is a decomposition $\Omega^\infty S^\infty = \text{Im}(J) \times \text{Coker}(J)$ as spaces from his methods.

For references to this proof and related questions see Sullivan [1], [2], Atiyah and Tall [1] and Atiyah and Segal [1].

Proof 3. (Quillen) In [2] Quillen gives a proof of the Adams conjecture by first showing the conjecture is true for vector bundles with finite structure group. Then using modular character theory for finite groups, one produces enough examples of virtual representations of finite groups to define maps

$$BGL(k) \to BU \quad \text{and} \quad BO(k) \to BO$$

where k is an algebraic closure of the field of p elements. These maps are homology isomorphisms over $\mathbf{Z}/d$ where d is prime to p. Then the general Adams conjecture is deduced from the special case of finite structure group using standard methods of algebraic topology. The J-morphism for vector bundles with finite structure group was also considered by Atiyah and Tall [1].

Proof 4. (Becker, Gottlieb) Let $p\colon E \to B$ be a fibre bundle with fibre F a compact smooth manifold, structure group G a compact Lie group acting smoothly on F, and B a finite complex. Becker and Gottlieb construct an S-map $p^{tr}\colon B \cup \{\infty\} \to E \cup \{\infty\} = E^{+}$ such that for any cohomology theory $h^{*}(p^{tr})h^{*}(p)$ is multiplication by $e(F)$, the Euler number of F. For a real $2n$-dimensional vector bundle ξ over B, there is a map $f\colon X \to B$ such that $h^{*}(f)$ is a split monomorphism for any general cohomology theory and $f^{*}(\xi) = \eta$ has structure group $N(T)$, the normalizer of the maximal torus $SO(2)^{n} \subset O(2n)$. In fact $f\colon X \to B$ is an $O(2n)/N(T)$ fibre bundle and $e(O(2n)/N(T)) = 1$. Bundles η of this form are treated by methods similar to those of Quillen in the previous proof.

CHARACTERISTIC CLASSES

PART III

CHARACTERISING CLASSES

CHAPTER 17

Chern Classes and Stiefel-Whitney Classes

We consider Chern classes, Stiefel-Whitney classes, and the Euler class from an axiomatic point of view. The uniqueness of the classes follows from the splitting principle, and the existence is derived using the bundle of projective spaces associated with a vector bundle and the Leray-Hirsch theorem. These results could be obtained by using obstruction theory or the cohomology of the classifying space. Finally, we consider the relation between characteristic classes and the Thom isomorphism.

1. The Leray-Hirsch Theorem

In this section all cohomology groups have coefficients in a principal ring K (usually equal to $\mathbf{Z}$, $\mathbf{Z}_p$, where p is a prime, or $\mathbf{Q}$).

1.1 Theorem (Leray-Hirsch). *Let $p: E \to B$ be a bundle which is of finite type, that is, trivial over a finite covering, let E_0 be an open subspace of E, and let (F, F_0) be an open pair of spaces such that for each $b \in B$ there is a homeomorphism $j_b: (F, F_0) \to (p^{-1}(b), p^{-1}(b) \cap E_0) \subset (E, E_0)$. Let $a_1, \ldots, a_r \in H^*(E, E_0)$ be homogeneous elements such that $j_b^*(a_1), \ldots, j_b^*(a_r)$ is a K-base of $H^*(F, F_0)$ for each $b \in B$. Then $H^*(E, E_0)$ is a free $H^*(B)$-module with base $a_1, \ldots, a_r$ under the action defined by $p^*: H^*(B) \to H^*(E, E_0)$.*

Proof. For an open subset U of B, let E_U denote $p^{-1}(U)$, let $j_U: E_U \to E$ be the natural inclusion, and let $p_U: E_U \to U$ be the restriction of p. If there exists a homeomorphism $(U \times F, U \times F_0) \to (E_U, E_U \cap E_0)$ preserving the projections onto U, then $p^*: H^*(U) \to H^*(E_U, E_U \cap E_0)$ is a monomorphism and $j_U^*(a_1), \ldots, j_U^*(a_r)$ is a base of the $H^*(U)$-module $H^*(E_U, E_U \cap E_0)$. This follows directly from the Künneth formula. Consequently, the theorem is true over such an open set U.

Finally, it suffices to prove that if the theorem is true over open sets U, V, and $U \cap V$ it is true over $U \cup V$. To do this, we define two functors $K^n(U)$ and $L^n(U)$ on the open subsets U of B. Let $n(i)$ denote the degree of a_i, and let x_i denote an indeterminant of degree $n(i)$. Let $K^n(U)$ be the direct sum $\sum_{1 \leq i \leq r} H^{n-n(i)}(U)x_i$, let $L^n(U)$ denote $H^n(E_U, E_U \cap E_0)$, and the $\theta_U : K^n(U) \to L^n(U)$ be the morphism defined by the relation $\theta_U \left(\sum_i c_i x_i \right) = \sum_i p^*(c_i)a_i$, where $c_i \in H^{n-n(i)}(U)$. Observe that the theorem is true over U if and only if θ_U is an isomorphism.

Since $L^n(U)$ is constructed from a functor for which the Mayer-Vietoris sequence exists and is exact, and since $K^n(U)$ is a direct sum of functors for which the Mayer-Vietoris sequence exists and is exact, we have the following commutative diagram with exact rows.

$$
\begin{array}{ccccccc}
\cdots \longleftarrow & K^n(U \cap V) & \longleftarrow & K^n(U) \oplus K^n(V) & \longleftarrow \\
& \Big\downarrow \theta_1 & & \Big\downarrow \theta_2 = \theta \oplus \theta & \\
\longleftarrow & L^n(U \cap V) & \longleftarrow & L^n(U) \oplus L^n(V) & \longleftarrow \\
\end{array}
$$

$$
\begin{array}{ccccccc}
K^n(U \cup V) & \longleftarrow & K^{n-1}(U \cap V) & \longleftarrow & K^{n-1}(U) \oplus K^{n-1}(V) & \longleftarrow \\
\Big\downarrow \theta_3 & & \Big\downarrow \theta_4 & & \Big\downarrow \theta_5 & \\
L^n(U \cup V) & \longleftarrow & L^{n-1}(U \cap V) & \longleftarrow & L^{n-1}(U) \oplus L^{n-1}(V) & \longleftarrow \\
\end{array}
$$

If the theorem is true for $U \cap V$, U, and V, then θ_1, θ_2, θ_4, and θ_5 are isomorphisms. By the "5-lemma," θ_3 is an isomorphism, and the theorem is true over $U \cap V$. The proof of the theorem follows now by an easy induction on n where $B = U_1 \cup \cdots \cup U_n$ and $p : (E, E_0) \to B$ is trivial over U_i for $1 \leq i \leq n$ as a bundle and subbundle pair. This proves the theorem.

This theorem plays a fundamental role in the construction of Stiefel-Whitney classes and Chern classes.

1.2 Remark. The above theorem holds for arbitrary fibrations and not only for bundles of finite type by a well-known spectral sequence argument. Again one must be given elements $a_i \in H^{n(i)}(E, E_0)$ such that $j_b^*(a_i)$ is a K-base of $H^*(F, F_0)$.

We shall make use of this generalized version later in calculating $BU(n) = G_n(\mathbf{C}^\infty)$, for example.

2. Definition of the Stiefel-Whitney Classes and Chern Classes

We work with real and complex vector bundles. Let c denote 1, K_1 denote Z_2, and F denote **R** for real vector bundles, and let c denote 2, K_2 denote **Z**, and F denote **C** for complex vector bundles.

For an n-dimensional vector bundle $\xi = (E, p, B)$, let E_0 denote the non-zero vectors in E, and let p_0 denote the restriction $p|E_0$. Let E' be the quotient space of E_0 where two vectors in the same line in a given fibre of ξ are identified, and let $q: E' \to B$ be the factorization of $p_0: E_0 \to B$ through E' by the natural projection $E_0 \to E'$.

2.1 Definition. The bundle (E', q, B) is called the projective bundle associated with ξ and is denoted $P\xi$.

The fibre of $P\xi^n$ is FP^{n-1}, and the bundle is locally trivial. For each $b \in B$, the inclusion $F^n \to p^{-1}(b) \subset E$ defines a natural inclusion $j_b: FP^{n-1} \to q^{-1}(b) \subset E'$. A point in E' is a line L in the fibre of ξ over $q(L)$.

The induced bundle $q^*(\xi)$ has a canonical line bundle λ_ξ as a subbundle where a point in the total space $E(\lambda_\xi)$ of λ_ξ over L is a pair (L, x), where $q(L) = b = p(x)$, or, equivalently, $x \in L$. There is an exact sequence over E' where σ_ξ is the quotient of $q^*(\xi)$ by λ_ξ:

$$0 \to \lambda_\xi \to q^*(\xi) \to \sigma_\xi \to 0$$

2.2 Proposition. Let $j_b: FP^{n-1} \to E(\Gamma(\zeta))$ be the inclusion map onto the fibre. Then $j_b^*(\lambda_\xi)$ is isomorphic to the canonical line bundle on FP^{n-1}.

Proof. A point in the total space of $j_b^*(\lambda_\xi)$ is a pair (L, y), where L is a line in F^n and $y \in j_b(L)$. Then y has a unique representation of the form (L, x), where $x \in L$ and the function that assigns the pair (L, x) to (L, y) is the desired isomorphism.

2.3 The Cohomology Rings of Projective Spaces. We recall a basic result of algebraic topology, namely, that the cohomology ring $H^*(FP^\infty, K_c)$ is a polynomial ring $K_c[z]$, where $\deg z = c$. The inclusion $FP^n \to FP^\infty$ induces an epimorphism $H^*(FP^\infty, K_c) \to H^*(FP^n, K_c)$ where the image of z is z_n, a generator of the ring $H^*(FP^n, K_c)$ with the one relation $(z_n)^{n+1} = 0$. Then z^i has as image z_n^i for $i \leq n$ and z^i has as image 0 for $i \geq n + 1$.

2.4 Henceforth, we assume that $E(P\xi)$ is a paracompact space. This is true if B is paracompact. Or we could require that ξ be a numerable vector bundle from which we deduce that λ_ξ is numerable. Then there exists, by the classification theorem 3(7.2) (or in Chap. 4), a map $f: E(P\xi) \to FP^\infty$ such that $f^*(\gamma_1) \cong \lambda_\xi^*$, where λ_ξ^* is the conjugate bundle to λ_ξ and γ_1 is the universal line bundle. Let a_ξ denote the class $f^*(z)$. Since f is unique up to homotopy, the class a_ξ is well defined. The basic properties of the class a_ξ are contained in the next theorem.

2.5 Theorem. *For an n-dimensional vector bundle ξ, the classes $1, a_\xi, \ldots, a_\xi^{n-1}$ form a base of the $H^*(B(\xi), K_c)$-module $H^*(E(P(\xi)), K_c)$. Moreover, q^*: $H^*(B(\xi)) \to H^*(E(P(\xi)))$ is a monomorphism.*

Proof. Clearly we need only check the hypothesis of Theorem (1.1). For this, we note that $j_b^*(f^*(\gamma_1))$ is the conjugate of the canonical line bundle over FP^{n-1}. Therefore, up to an automorphism of FP^{n-1}, the map fj_b is homotopic to the inclusion $FP^{n-1} \to FP^\infty$. This means that $j_b^*(1), \ldots, j_b^*(a_\xi^{n-1})$ is a K_c-base of $H^*(FP^{n-1}, K_c)$. The theorem follows from (1.1).

In view of (2.5), there exists unique $x_i(\xi) \in H^{ci}(B, K_c)$ such that

$$a_\xi^n = -\sum_{1 \le i \le n} x_i(\xi) a_\xi^{n-i}$$

We write $x(\xi) = 1 + x_1(\xi) + \cdots + x_n(\xi)$.

2.6 Definition. For a real vector bundle ξ, the ith Stiefel-Whitney class of ξ, denoted $w_i(\xi)$, is $x_i(\xi) \in H^i(B(\xi), Z_2)$. For a complex vector bundle ξ, the ith Chern class of ξ, denoted $c_i(\xi)$, is $x_i(\xi) \in H^{2i}(B(\xi), Z)$. In addition, $w(\xi) = 1 + w_1(\xi) + \cdots + w_n(\xi)$ and $c(\xi) = 1 + c_1(\xi) + \cdots + c_n(\xi)$ are called the total Stiefel-Whitney class and total Chern class, respectively.

3. Axiomatic Properties of the Characteristic Classes

We write several properties of the Stiefel-Whitney classes and of the Chern classes. Then we consider relations between these properties and verify all of them except the Whitney sum property. In Sec. 5 we shall prove that these properties uniquely characterize these classes, and in Sec. 6 we shall verify the Whitney sum property.

3.1 Properties of Stiefel-Whitney Classes. For each real vector bundle ξ over a space B there is a class $w(\xi) \in H^*(B, Z_2)$ with the following properties:

(SW$_0$) We have $w(\xi) = 1 + w_1(\xi) + \cdots + w_n(\xi)$, where $w_i(\xi) \in H^i(B, Z_2)$ and $w_i(\xi) = 0$ for $i > \dim \xi = n$.

(SW$_1$) If ξ and η are B-isomorphic, it follows that $w(\xi) = w(\eta)$, and if $f: B_1 \to B$ is a map, then we have $f^*(w(\xi)) = w(f^*(\xi))$.

(SW$_2$) (Whitney sum formula) For two vector bundles ξ and η over B, the relation $w(\xi \oplus \eta) = w(\xi)w(\eta)$ (cup multiplication) holds.

(SW$_3$) For the canonical line bundle λ over $S^1 = RP^1$, the element $w_1(\lambda)$ is nonzero in $H^1(S^1, Z_2) = Z_2$.

(SW$'_3$) For the canonical line bundle γ_1 over RP^∞, the element $w_1(\gamma_1)$ is the generator of the polynomial ring $H^*(RP^\infty, Z_2)$.

Using property (SW$_1$) and the inclusion $RP^1 \to RP^\infty$, we find that (SW$_3$) and (SW$'_3$) are equivalent to each other.

We choose a generator of $H^2(S^2, Z)$ which in turn defines a generator of $H^2(CP^n, Z)$ for each n with $1 \leq n \leq +\infty$. This element z will generate the polynomial ring $H^*(CP^\infty, Z)$.

3.2 Properties of Chern Classes. For each complex vector bundle ξ over a space B there is a class $c(\xi) \in H^*(B, Z)$ with the following properties:

(C$_0$) We have $c(\xi) = 1 + c_1(\xi) + \cdots + c_n(\xi)$, where $c_i(\xi) \in H^{2i}(B, Z)$ and $c_i(\xi) = 0$ for $i > \dim \xi$.

(C$_1$) If ξ and η are B-isomorphic, it follows that $c(\xi) = c(\eta)$, and if $f: B_1 \to B$ is a map, then we have $f^*(c(\xi)) = c(f^*(\xi))$.

(C$_2$) For two vector bundles ξ and η over B, the relation $c(\xi \oplus \eta) = c(\xi)c(\eta)$ (cup multiplication) holds.

(C$_3$) For the canonical line bundle λ over $S^2 = CP^1$, the element $c_1(\lambda)$ is the given generator of $H^2(S^2, Z)$.

(C$'_3$) For the canonical line bundle γ_1 over CP^∞, the element $c_1(\gamma_1)$ is the generator z of the polynomial ring $H^*(CP^\infty, Z)$.

Using (C$_1$) and the inclusion $CP^1 \to CP^\infty$, we find that (C$_3$) and (C$'_3$) are equivalent to each other.

From the parallel character of properties (SW$_0$) to (SW$_3$) and properties (C$_0$) to (C$_3$) it is clear that the two sets of characteristic classes have many formal properties in common.

3.3 Proposition. *Properties* (SW$_0$), (SW$_1$), *and* (SW$'_3$) *hold for Stiefel-Whitney classes, and properties* (C$_0$), (C$_1$), *and* (C$'_3$) *hold for Chern classes.*

Proof. Property 0 in both cases follows from (2.6) immediately. For a map $f: B_1 \to B$ there is an f-morphism of bundles $u: P(f^*(\xi)) \to P(\xi)$ which is an isomorphism on the fibres. From the definition of λ_ξ it is clear that $u^*(\lambda_\xi)$ and $\lambda_{f^*(\xi)}$ are isomorphic. Therefore, in the cohomology ring $H^*(E(P(f^*(\xi))))$ we have $u^*(a_\xi) = a_{f^*(\xi)}$. From the two relations

$$a_{f^*(\xi)}^n = - \sum_{1 \leq i \leq n} x_i(f^*(\xi))(a_{f^*(\xi)})^{n-i}$$

and

$$u^*(a_\xi^n) = - \sum_{1 \leq i \leq n} f^*(x_i(\xi))(u^*(a_{f^*(\xi)}))^{n-i}$$

we have the result $x_i(f^*(\xi)) = f^*(x_i(\xi))$. Clearly, for isomorphic bundles ξ and η we have $x_i(\xi) = x_i(\eta)$.

As for axiom (C$'_3$), we observe that if ξ is a line bundle the total space of $P\xi$ is $B(\xi)$ and λ_ξ is ξ. The defining relation $a_\xi = -x_1(\xi)$ becomes $x_1(\xi^*) =$

a_ξ. But to define a_ξ, we used a map $f: FP^\infty \to FP^\infty$ with $f^*(\gamma_1) = \xi^*$ and $f^*(z) = a_\xi$. In other words, $x_1(\gamma_1)$ is the canonical generator of $H^*(FP^\infty, K_c)$. The above discussion applies to both Stiefel-Whitney and Chern classes. This proves the proposition.

Let $L_F(B)$ denote the group of isomorphism classes of line bundles over B with tensor product $\xi \otimes \eta$ as group operation and ξ^* as the inverse of ξ. Then $L_F(-)$ is a group-valued cofunctor defined on the category of paracompact spaces and homotopy classes of maps. The classification theorem 3(7.2) says that the function that assigns to a homotopy class $[f]: B \to FP^\infty$ the isomorphism class of $f^*(\gamma)$ defines an isomorphism of cofunctors $[-, FP^\infty] \to L_F(-)$. Then FP^∞ has an H-space structure corresponding to the group operation on $L_F(-)$.

The spaces FP^∞ are $K(\pi, n)$-spaces, and $K(\pi, n)$-spaces have a unique H-space structure. (This can easily be seen from the prolongation theorems of Chap. 1.) Since the cofunctors $H^c(-, K_c)$ and $[-, FP^\infty]$ are isomorphic in such a way that $[i] \in [FP^\infty, FP^\infty]$ corresponds to the generator of $H^c(FP^\infty, K_c)$, and since $x_1(\gamma_1)$ is the generator of $H^c(FP^\infty, K_c)$, the function $x_1: L_F(B) \to H^c(B, K_c)$ is an isomorphism for $c = 1, 2$.

3.4 Theorem. *The functions* $w_1: L_R(B) \to H^1(B, Z_2)$ *and* $c_1: L_C(B) \to H^2(B, \mathbf{Z})$ *define isomorphisms of cofunctors.*

We have $w_1(\xi \otimes \eta) = w_1(\xi) + w_1(\eta)$ and $c_1(\xi \otimes \eta) = c_1(\xi) + c_1(\eta)$ from this theorem. The cohomology ring $H^*(RP^\infty, Z_2)$ is generated by $w_1(\lambda_1)$ and $H^*(CP^\infty, \mathbf{Z})$ by $c_1(\lambda_1)$. Consequently, the characteristic classes of line bundles are uniquely defined by their axiomatic properties.

4. Stability Properties and Examples of Characteristic Classes

In this section we assume the existence and uniqueness of the characteristic classes.

4.1 Proposition. *If* ξ *is a trivial bundle over* B, *then* $w_i(\xi) = 0$ *for* $i > 0$ *in the real case and* $c_i(\xi) = 0$ *for* $i > 0$ *in the complex case.*

Proof. The statement is true for ξ over a point because the cohomology in nonzero dimensions is zero, and every trivial bundle is isomorphic to the induced bundle by a map to a point. By property 1 we have the result.

In addition, using property 2, we have the following result.

4.2 Theorem. *Let* ξ *and* η *be two s-equivalent vector bundles. Then the relation* $w(\xi) = w(\eta)$ *holds in the real case and* $c(\xi) = c(\eta)$ *in the complex case.*

Proof. For some n and m, there is an isomorphism between $\xi \oplus \theta^n$ and $\eta \oplus \theta^m$. From this we have the following equalities in the real case: $w(\xi) = w(\xi)w(\theta^n) = w(\eta \oplus \theta^m) = w(\eta)w(\theta^m) = w(\eta)^1 = w(\eta)$, or we have $w(\xi) = w(\eta)$. Similarly, in the complex case we have $c(\xi) = c(\eta)$.

4.3 Remark. We denote by $G^*(B, K_c)$ the subset of those elements in $H^*(B, K)$ of the form $1 + a_i + \cdots + a_n$, where $a_i \in H^{ci}(B, K_c)$ with the cup product as group operation for a finite complex B. For the two cases $K_1 = Z_2$ and $K_2 = Z$ this is a commutative group. Moreover, $G^*(-, K_c)$ is a cofunctor from the category of spaces and homotopy classes of maps with values in the category of abelian groups. Properties 1, 2, and 3 allow us to view

$$w: KO(-) \to G^*(-, Z_2) \quad \text{and} \quad c: K(-) \to G^*(-, Z)$$

as morphisms of cofunctors. Observe that characteristic classes cannot distinguish between s-equivalent bundles.

4.4 Proposition. *For the tangent bundle $\tau(S^n)$ to the sphere, $w(\tau(S^n)) = 1$.*

Proof. Since $\tau(S^n) \oplus \theta^1$ and θ^{n+1} are isomorphic by 2(4.7), the tangent bundle $\tau(S^n)$ is s-trivial. Therefore, by (4.1) and (4.2), $w(\tau(S^n)) = 1$.

By the calculation of 2(4.8), $\tau(RP^n)$ is s-equivalent to $\lambda \oplus \overset{(n+1)}{\cdots} \oplus \lambda = (n+1)\lambda$, where λ is the canonical line bundle over RP^n. This holds in both the real and complex cases. Therefore, we have the next proposition.

4.5 Proposition. *For the tangent bundle $\tau(RP^n)$ there are the relations $w(\tau(RP^n)) = (1 + z)^{n+1}$, where z is the generator of $H^1(RP^n, Z_2)$, and $c(\tau(CP^n)) = (1 + z)^{n+1}$, where z is the generator of $H^2(CP^n, Z)$.*

Every vector field on RP^n defines a vector field on S^n. A nonexistence statement for vector fields on S^n is stronger than a nonexistence statement for vector fields on RP^n, but as an application we include the next proposition which is really an easy consequence of 12(1.4).

4.6 Proposition. *Every tangent vector field on RP^{2k} has at least one zero.*

Proof. Observe that $w_{2k}(\tau(RP^{2k})) = (2k + 1)z^{2k} = z^{2k} \neq 0$. If $\tau(RP^{2k})$ had a cross section that was everywhere nonzero, we would have $\tau(RP^{2k}) = \xi \oplus \theta^1$. Then $w_{2k}(\tau(RP^{2k})) = w_{2k-1}(\xi)w_1(\theta^1) = 0$, which is a contradiction.

5. Splitting Maps and Uniqueness of Characteristic Classes

The following notation allows the reduction of general considerations of vector bundles to those of line bundles, i.e., uniqueness of characteristic classes.

5.1 Definition. Let ξ be a vector bundle over B. A splitting map of ξ is a map $f\colon B_1 \to B$ such that $f^*(\xi)$ is a sum of line bundles and $f^*\colon H^*(B, K_c) \to H^*(B_1, K_c)$ is a monomorphism.

The next proposition says that splitting maps exist.

5.2 Proposition. *If ξ is a vector bundle over B, there exists a splitting map for ξ.*

Proof. We prove this by induction on the dimension of ξ. For a line bundle, the identity on the base space is a splitting map. In general, let $q\colon E(P\xi) \to B$ be the associated projective bundle. Then $q^*\colon H^*(B, K_c) \to H^*(E(P\xi), K_c)$ is a monomorphism, and $q^*(\xi) = \lambda_\xi \oplus \sigma_\xi$. By inductive hypothesis there exists a splitting map $g\colon B_1 \to E(P\xi)$ for σ_ξ. Then $f = qg$ from B_1 to B is a splitting map for ξ.

5.3 Corollary. *Let $\xi_1, \ldots, \xi_r$ be r vector bundles over B which are either all real or all complex. Then there exists a map $f\colon B_1 \to B$ such that f is a splitting map for each ξ_i, where $1 \leqq i \leqq r$.*

5.4 Theorem. *The properties* (SW_0) *to* (SW_3) *completely determine the Stiefel-Whitney classes, and the properties* (C_0) *to* (C_3) *completely determine the Chern classes.*

Proof. Let w_i and $\bar{w}_i$ be two sets of Stiefel-Whitney classes, and let ξ^n be a vector bundle with splitting map $f\colon B_1 \to B$. Since w_1 is uniquely determined for line bundles λ_i, we have $f^*(w(\xi)) = w(f^*(\xi)) = (1 + w_1(\lambda_1)) \cdots (1 + w_1(\lambda_n)) = (1 + \bar{w}_1(\lambda_1)) \cdots (1 + \bar{w}(\lambda_n)) = \bar{w}(f^*(\xi)) = f^*(\bar{w}(\xi))$, where $f^*(\xi) = \lambda_1 \oplus \cdots \oplus \lambda_n$. Since f^* is a monomorphism, we have $w(\xi) = \bar{w}(\xi)$. The same proof applies to Chern classes.

This theorem is a good illustration of the use of splitting maps.

6. Existence of the Characteristic Classes

In (2.5) and (2.6) we defined the characteristic classes of vector bundles. We have verified properties 0, 1, and 3 for these classes. In this section we verify the Whitney sum formula. The next proposition is the key result.

6.1 Proposition. *Let $\xi = \lambda_1 \oplus \cdots \oplus \lambda_n$ be a Whitney sum of line bundles. Then the relation $x(\xi) = (1 + x_1(\lambda_1)) \cdots (1 + x_1(\lambda_n))$ holds. Again x denotes Stiefel-Whitney classes or the Chern classes.*

Proof. Let $q\colon E(P\xi) \to B$ be the projective bundle associated with ξ. Then there are two exact sequences

$$0 \to \lambda_\xi \to q^*(\xi) = q^*(\lambda_1) \oplus \cdots \oplus q^*(\lambda_n) \to \sigma \to 0$$

and

$$0 \to \theta^1 = \lambda_\xi^* \otimes \lambda_\xi \to [\lambda_\xi^* \otimes q^*(\lambda_i)] \oplus \cdots \oplus [\lambda_\xi^* \otimes q^*(\lambda_n)] \to \lambda_\xi^* \otimes \sigma \to 0$$

Therefore, $\lambda_\xi^* \otimes q^*(\xi)$ has a cross section s which is everywhere nonzero and projects to a cross section s_i in $\lambda_\xi^* \otimes q^*(\lambda_i)$. Let V_i be the open subset over which $s_i \neq 0$. The image of $x_1(\lambda_\xi^* \otimes q^*(\lambda_i))$ is zero in $H^c(V_i, K_c)$, and therefore it can be pulled back to $H^c(E(P\xi), V_i; K_c)$. Moreover, the cup product $x_1(\lambda_\xi^* \otimes q^*(\lambda_1)) \cdots x_1(\lambda_\xi^* \otimes q^*(\lambda_n))$ is an element of $H^*(E(P\xi), \bigcup_{1 \leq i \leq n} V_i; K_c) = H^*(E(P\xi), E(P\xi); K_c) = 0$. Therefore, we have the relation

$$\prod_{1 \leq i \leq n} [q^*(x_1(\lambda_i)) + x_1(\lambda_\xi^*)] = 0$$

For $a_\xi = x_1(\lambda_\xi^*)$, this is just the equation $a_\xi^n + q^* x_1(\xi) a_\xi^{n-1} + \cdots + q^* x_n(\xi) = 0$. Therefore, we have $x(\xi) = x(\lambda_1) \cdots x(\lambda_n)$. This proves the proposition.

6.2 Theorem. *For ξ and η, two vector bundles over B, there is the relation $x(\xi \oplus \eta) = x(\xi)x(\eta)$.*

Proof. Let $f: B_1 \to B$ be a splitting map for ξ and η, where $f^*(\xi) = \lambda_1 \oplus \cdots \oplus \lambda_n$ and $f^*(\eta) = \lambda_{n+1} \oplus \cdots \oplus \lambda_{n+m}$. Then we have

$$f^*(x(\xi \oplus \eta)) = x(f^*(\xi \oplus \eta)) = x(\lambda_1 \oplus \cdots \oplus \lambda_{n+m}) = x(\lambda_1) \cdots x(\lambda_{n+m})$$

$$= x(f^*(\xi)) x(f^*(\eta)) = f^*(x(\xi)x(\eta))$$

Since f^* is a moonomorphism, $x(\xi \oplus \eta) = x(\xi)x(\eta)$. This completes the proof.

7. Fundamental Class of Sphere Bundles. Gysin Sequence

For a vector bundle ξ with total space $E(\xi)$, let $E_0(\xi)$ denote the open subspace of nonzero vectors. For $b \in B$, let $j_b: (\mathbf{R}^n, \mathbf{R}^n - \{0\}) \to (E(\xi), E_0(\xi))$ denote the inclusion onto the fibre of ξ over $b \in B$. Each complex n-dimensional vector bundle restricts to a real $(2n)$-dimensional real vector bundle. In Secs. 1 to 6, we developed characteristic classes, using $P(\xi)$; now we use $E_0(\xi)$ to define other classes.

7.1 Definition. A vector bundle is orientable provided its structure group restricts from $O(n)$ to $SO(n)$. An orientation of a vector bundle is a particular restriction of the structure group to $SO(n)$. An oriented vector bundle is a pair consisting of a vector bundle and an orientation on the bundle.

In other words, a vector bundle has an atlas of charts where the linear transformations changing from one chart to another have strictly positive determinants.

7.2 Example. Every restriction of a complex vector bundle to a real vector bundle is orientable and has a natural orientation because $U(n) \subset SO(2n) \subset O(2n)$.

The next theorem contains the fundamental construction of this section.

7.3 Theorem. *Let ξ be a real vector bundle. The cohomology groups have coefficients in $\mathbf{Z}$ if the bundle is oriented and in $\mathbf{Z}_2$ in general. The following statements hold:*

(1) *There exists a unique $U_\xi \in H^n(E, E_0)$ such that $j_b^*(U_\xi)$ is a fixed generator of $H^n(\mathbf{R}^n, \mathbf{R}^n - \{0\})$.*
(2) *For $i < n$, there is the relation $H^i(E, E_0) = 0$.*
(3) *The function $a \mapsto p^*(a)U_\xi$ (cup product) of $H^i(B) \to H^{i+n}(E, E_0)$ is an isomorphism.*

Proof. As with Theorem (1.1), a spectral sequence argument is needed if ξ is not of finite type. This we leave to the reader and given an elementary argument when ξ is of finite type. Statements (1) and (2) hold over open sets V where $\xi|V$ is trivial. Let U_V denote the resulting class.

If statements (1) and (2) hold over V_1, V_2, and $V_1 \cap V_2$, they hold over $V_1 \cup V_2$ by a Mayer-Vietoris sequence argument. Let $E_1 = p^{-1}(V_1)$, $E_2 = p^{-1}(V_2)$, $E_3 = p^{-1}(V_1 \cap V_2)$, $E_4 = p^{-1}(V_1 \cup V_2)$, and $E_i^0 = E_i \cap E_0$. Then we have the following exact sequence:

$$H^i(E_3, E_3^0) \leftarrow H^i(E_1, E_1^0) \oplus H^i(E_2, E_2^0) \leftarrow H^i(E_4, E_4^0) \leftarrow H^{i-1}(E_3, E_3^0)$$

Statement (2) follows immediately since $0 \leftarrow H^i(E_4, E_4^0) \leftarrow 0$ is exact for $i < n$. The classes $U_1 \in H^n(E_1, E_1^0)$ and $U_2 \in H^n(E_2, E_2^0)$ restrict to the same class in $H^n(E_3, E_3^0)$, and there is a unique class U_4 in $H^n(E_4, E_4^0)$ that restricts to U_1 and U_2. By induction on the number of charts in a finite atlas, we have the proof of statements (1) and (2).

Finally, statement (3) is a direct application of Theorem (1.1). This proves the theorem.

In summary, we consider the following isomorphisms for a vector bundle ξ of dimension n.

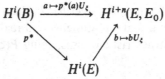

For the inclusion $j: E \to (E, E_0)$ there is the morphism $j^*: H^i(E, E_0) \to H^i(E)$. The morphism $p^*: H^i(B) \to H^i(E)$ is an isomorphism.

7.4 Definition. The Euler class of a real vector bundle ξ^n over B, denoted $e(\xi)$, is $p^{*-1}j^*(U_\xi)$, where $p: E \to B$ is the projection of ξ. The term "Euler class" is usually used only in the oriented case and with integral coefficients. Moreover, we have $e(\xi) \in H^n(B)$. The class U_ξ is called the fundamental class.

Let ψ denote the composition of the coboundary $H^i(E_0) \to H^{i+1}(E, E_0)$ and $\phi^{-1}: H^{i+1}(E, E_0) \to H^{i+1-n}(B)$, where $\phi(a) = p^*(a)U_\xi$. There results the following exact sequence which relates the cohomology of B and E_0.

7.5 Theorem (Gysin). *For an n-dimensional real vector bundle ξ^n there is the following exact sequence of cohomology groups where the coefficients are in Z_2 in general or Z for oriented bundles.*

$$\cdots \to H^i(B) \xrightarrow{\text{mult. by } e(\xi)} H^{i+n}(B) \xrightarrow{p^*} H^{i+n}(E_0) \xrightarrow{\psi} H^{i+1}(B) \to \cdots$$

Proof. We have the following commutative diagram with top row exact.

$$
\begin{array}{ccccccc}
\longrightarrow & H^{i+n}(E, E_0) & \xrightarrow{j^*} & H^{i+n}(E) & \longrightarrow & H^{i+n}(E_0) & \longrightarrow & H^{i+n+1}(E, E_0) & \longrightarrow \\
& \uparrow \phi & & \uparrow p^* & & \uparrow 1 & & \uparrow \phi & \\
\longrightarrow & H^i(B) & \longrightarrow & H^{i+n}(B) & \xrightarrow{p^*} & H^{i+n}(E_0) & \xrightarrow{\psi} & H^{i+1}(B) & \longrightarrow
\end{array}
$$

We have only to calculate

$$p^{*-1}j^*\phi(a) = p^{*-1}j^*(p^*(a)U_\xi) = p^{*-1}(p^*(a)j^*(U_\xi)) = a[p^{*-1}j^*(U_\xi)] = ae(\xi)$$

This proves the theorem.

The sequence in (7.5) is called the Gysin sequence of the bundle (E_0, p, B).

7.6 Proposition. *The expression $e(\xi) = \phi^{-1}(U_\xi^2)$ holds for the Euler class.*

Proof. We calculate $\phi(e(\xi)) = p^*(p^{*-1}j^*(U_\xi))U_\xi = j^*(U_\xi)U_\xi = U_\xi^2$ since $j: E \to (E, E_0)$ is the inclusion, and the cup product is a natural operation.

7.7 Corollary. *The Euler class of an odd-dimensional oriented bundle ξ has the property that $2e(\xi) = 0$.*

7.8 Proposition. *Let ξ be a vector bundle over B, and let $f: B' \to B$ be a map. Then $f^*(e(\xi)) = e(f^*(\xi))$.*

Proof. There is a map $g: (E', E_0') \to (E, E_0)$ inducing f, where E is the total space of ξ and E' is the total space of $f^*(\xi)$. By the uniqueness property of Theorem (7.3), $f^*(U_\xi)$ is equal to $U_{f^*(\xi)}$, since they are equal on each fibre of $f^*(\xi)$. The proposition follows now from the commutative diagram

$$
\begin{array}{ccccc}
H^*(E, E_0) & \longrightarrow & H^*(E) & \longrightarrow & H^*(B) \\
\downarrow g^* & & \downarrow g^* & & \downarrow f^* \\
H^*(E', E_0') & \longrightarrow & H^*(E') & \longrightarrow & H^*(B')
\end{array}
$$

As with (4.1), there is the following corollary.

7.9 Corollary. *If ξ is a trivial bundle of dimension $n \geq 1$, then $e(\xi) = 0$.*

8. Multiplicative Property of the Euler Class

Let ξ^n be a real vector bundle with total space E', space of nonzero vectors E'_0, and fibre F'_b over b; let η^m have total space E'', etc., E''_0, and F''_b; and let $\xi \oplus \eta$ have total space E, etc., E_0, and F_b. Let $F_{0,b}$ denote $E_0 \cap F_b$, etc.

Next, let E_1 be the union of $F'_{0,b} \times F''_b$ for all $b \in B$, and E_2 the union of $F'_b \times F''_{0,b}$ for all $b \in B$, where $F''_{0,b} = E''_0 \cap F''_b$, etc. We have the following commutative diagrams where the rows are inclusions:

$$
\begin{array}{ccccc}
E_1 \longrightarrow & E & \longrightarrow (E, E_1) & E_2 \longrightarrow & E & \longrightarrow (E, E_2) \\
\downarrow r_1 & \downarrow p_1 & \downarrow q_1 & \downarrow r_2 & \downarrow p_2 & \downarrow q_2 \\
E'_0 \longrightarrow & E' & \longrightarrow (E', E'_0) & E''_0 \longrightarrow & E'' & \longrightarrow (E'', E''_0)
\end{array}
$$

The vertical maps are projections, and r_1, r_2, p_1, and p_2 are homotopy equivalences. Finally, we have $E_0 = E_1 \cup E_2$, and the following functions are isomorphisms.

$$H^*(E', E'_0) \xrightarrow{q_1^*} H^*(E, E_1) \qquad \text{and} \qquad H^*(E'', E''_0) \xrightarrow{q_2^*} H^*(E, E_2)$$

Let U' denote $U_\xi \in H^*(E', E'_0)$, let U'' denote $U_\eta \in H^*(E'', E''_0)$, and let U denote $U_{\xi \oplus \eta} \in H^*(E, E_0)$. The next proposition describes the multiplicative character of U_ξ, using these notations.

8.1 Proposition. *The relation $U = q_1^*(U')q_2^*(U'')$ holds.*

Proof. From the uniqueness character of U, it suffices to prove $j_b^*(q_1^*(U')q_2^*(U''))$ is the standard generator $j_b^*(U)$ of $H^{n+m}(F_b, F_{b,0})$. For this, we consider the following diagram:

$$
\begin{array}{ccc}
U' \otimes U'' \in H^n(E', E'_0) \otimes H^m(E'', E''_0) & \xrightarrow{q_1^* \otimes q_2^*} & \\
\downarrow j_1^* \otimes j_2^* & & \\
H^n(F', F'_0) \otimes H^m(F'', F''_0) & \longrightarrow & \\
H^n(E, E_1) \otimes H^m(E, E_2) & \xrightarrow{\text{cup}} & H^{n+m}(E, E_0) \\
\downarrow & & \downarrow j^* \\
H^n(F, F_1) \otimes H^m(F, F_2) & \xrightarrow{\text{cup}} & H^{n+m}(F, F_0)
\end{array}
$$

The unnamed arrows are induced by obvious inclusion maps. All fibres are over $b \in B$. Then $j_b^*(q_1^*(U')q_2^*(U''))$ is the composition of the upper row and right vertical map. The standard generator in $H^n(F', F'_0) \otimes H^m(F'', F''_0)$ is carried into the standard generator in $H^{n+m}(F, F_0)$. This proves the proposition.

8.2 Theorem. *For the Euler class, the relation* $e(\xi \oplus \eta) = e(\xi)e(\eta)$ *holds.*

Proof. By Definition (7.4), we have $e(\xi \oplus \eta) = p^{*-1}j^*(U)$, using the above notation. Then we calculate $p^{*-1}j^*(U) = p^{*-1}j^*(q_1^*(U')q_2^*(U'')) = p^{*-1}[p_1^* j_1^*(U')p_2^* j_2^*(U'')] = e(\xi)e(\eta)$. This proves the theorem.
   The following result is useful.

8.3 Theorem. *If a vector bundle ξ has an everywhere-nonzero cross section, then* $e(\xi) = 0$.

Proof. A vector bundle with an everywhere-nonzero cross section splits off a line bundle, that is, $\xi = \theta^1 \oplus \eta$. Then $e(\xi) = e(\theta^1)e(\eta) = 0e(\eta) = 0$. The Euler class of a trivial bundle is zero.

9. Definition of Stiefel-Whitney Classes Using the Squaring Operations of Steenrod

Let $\phi: H^i(B) \to H^{i+n}(E, E_0)$ denote the isomorphism $\phi(a) = p^*(a)U_\xi$. We work with Z_2 coefficients only in this section [see (7.3) and (7.4)].

9.1 Theorem (Thom). *The Stiefel-Whitney class $w_i(\xi)$ is given by $\phi^{-1}(Sq^iU_\xi)$, where Sq^i denotes the ith Steenrod square.*

Proof. From the uniqueness character of w_i it suffices to prove that $w^*(\xi) = \phi^{-1}SqU_\xi$ satisfies statements (SW$_0$) to (SW$_3$).
   First, (SW$_0$) follows from the property that $Sq^0 = 1$ and $Sq^iU_\xi = 0$ for $i > n$. For (SW$_1$), let $f: B' \to B$ be a map. Then we have a commutative diagram:

$$
\begin{array}{ccc}
H^i(B) & \xrightarrow{\phi} & H^{i+n}(E, E_0) \\
\downarrow{\scriptstyle f^*} & & \downarrow{\scriptstyle g^*} \\
H^i(B') & \longrightarrow & H^{i+n}(E', E_0')
\end{array}
$$

The vector bundle map (g, f) is defined $f^*(\xi) \to \xi$. Then $f^*(w^*(\xi)) = f^*\phi^{-1}(SqU_\xi) = \phi^{-1}(Sqg^*(U_\xi)) = \phi^{-1}(Sq(U_{f^*(\xi)})) = w^*(f^*(\xi))$. This proves (SW$_1$).
   For (SW$_2$), we use the results and notations of the previous section. We have $U = q_1^*(U')q_2^*(U'')$ by (8.1), and $SqU = Sq(q_1^*(U'))Sq(q_2^*(U'')) = q_1^*(Sq(U'))q_2^*(Sq(U''))$. By the formula for w^*, we have $SqU = [p^*w^*(\xi \oplus \eta)]U$, $SqU' = [p^*w^*(\xi)]U'$, and $SqU'' = [p^*w^*(\eta)]U''$.
   Now we calculate

$$SqU = q_1^*(Sq(U'))q_2^*(Sq(U''))$$

$$= (q_1^*([p^*w^*(\xi)]U'))(q_2^*([p^*w^*(\eta)]U''))$$

$$= [p_1^*p^*w^*(\xi)]q_1^*(U')[p_2^*p^*w^*(\eta)]q_2^*(U'')$$

$$= [p_1^*p^*w^*(\xi)][p_2^*p^*w^*(\eta)]q_1^*(U')q_2^*(U'')$$

$$= p^*(w^*(\xi)w^*(\eta))U$$

But $w^*(\xi \oplus \eta)$ is uniquely defined by this relation. Therefore, we have the Whitney sum formula $w^*(\xi \oplus \eta) = w^*(\xi)w^*(\eta)$.

Finally, for (SW_3), Sq^1 carries the nonzero class of $H^1(E, E^0)$ to the nonzero class of $H^2(E, E_0)$, where $(E(\lambda), E_0(\lambda))$ over $S^1 = RP^1$ is, up to homotopy type, the Moebius band modulo its boundary. The Moebius band equals RP^2 minus an open two-cell. Then Sq^1U_λ is nonzero, and $w_1^*(\lambda)$ is nonzero. This proves the theorem.

10. The Thom Isomorphism

Let ξ be a vector bundle. If ξ is oriented, then **Z**-coefficients are used in the cohomology groups; otherwise Z_2 is used. By Theorem (7.3), there is an isomorphism $\phi(a) = p^*(a)U_\xi$ of $H^i(B) \to H^{i+n}(E, E_0)$ or $H^*(B) \to H^*(E, E_0)$.

10.1 Definition. We assume that ξ has a riemannian metric. Then we define the associated disk bundle, denoted $D(\xi)$, as the subbundle of $x \in E$ with $\|x\| \leq 1$, and the associated sphere bundle, denoted $S(\xi)$, as the subbundle of $x \in E$ with $\|x\| = 1$.

For $D(\xi)_0 = D(\xi) \cap E_0$, there is the following sequence of isomorphisms.

$$H^i(B) \overset{\phi}{\to} H^{i+n}(E, E_0) \to H^{i+n}(D(\xi), D(\xi)_0) \to H^{i+n}(D(\xi), S(\xi))$$

The first is given by (7.3), the second is given by excision, and the third by the homotopy equivalence $S(\xi) \to D(\xi)_0$. The class U_ξ is carried into $H^{i+n}(D(\xi), S(\xi))$, and composition of the above three maps is an isomorphism $\phi': H^i(B) \to H^{i+n}(D(\xi), S(\xi))$, where $\phi'(a) = p^*(a)U_\xi$. This leads to the following definition.

10.2 Definition. The Thom space of a vector bundle ξ, denoted ξ^B, is the quotient $D(\xi)/S(\xi)$.

The projection $\sigma: H^{i+n}(D(\xi), S(\xi)) \to \tilde{H}^{i+n}(\xi^B)$ is an isomorphism, and we define the Thom morphism $\psi: H^i(B) \to \tilde{H}^{i+n}(\xi^B)$ to be $\psi = \sigma\phi'$. The following theorem is immediate.

10.3 Theorem. *The Thom morphism* $\psi: H^i(B) \to \tilde{H}^{i+n}(\xi^B)$ *is an isomorphism.*

The reader should refer to the exercises for an alternative approach to the Thom isomorphism.

11. Relations Between Real and Complex Vector Bundles

We have considered the operation of conjugation ξ^* of a complex vector bundle ξ, see 5(7.6). We can also restrict the scalars of ξ to $\mathbf{R}$. This yields a group homomorphism

$$\varepsilon_0: K_C(X) \to K_R(X)$$

The process of tensoring a real vector bundle η with $\mathbf{C}$ yields a complex vector bundle $\eta \otimes \mathbf{C}$, called the complexification of η. This yields a ring morphism

$$\varepsilon_U: K_R(X) \to K_C(X)$$

Clearly, there are the relations, see 13(11.3),

$$\varepsilon_0(\varepsilon_U(\eta)) = 2\eta \quad \text{and} \quad \varepsilon_U(\varepsilon_0(\xi)) = \xi + \xi^*$$

Observe that for a real vector bundle η the complex vector bundles $\eta \otimes \mathbf{C}$ and $(\eta \otimes \mathbf{C})^*$ are isomorphic.

11.1 Proposition. *For a complex vector bundle ξ, the relation $c_i(\xi^*) = (-1)^i c_i(\xi)$ holds.*

Proof. The proposition is true for line bundles. Let $f: B_1 \to B$ be a splitting map, where $f^*(\xi) = \lambda_1 \oplus \cdots \oplus \lambda_n$. Then

$$c(f^*(\xi^*)) = c(\lambda_1^*) \cdots c(\lambda_n^*) = (1 - c_1(\lambda_1)) \cdots (1 - c_1(\lambda_n)) = \sum_{0 \leq i} (-1)^i c_i(f^*(\xi))$$

This proves the result.

11.2 Corollary. *If a complex vector bundle ξ is isomorphic to ξ^*, then $2c_{2i+1}(\xi) = 0$ for $0 \leq i$.*

Proof. We have $c_{2i+1}(\xi) = -c_{2i+1}(\bar{\xi}) = -c_{2i+1}(\xi)$ or $2c_{2i+1}(\xi) = 0$.
    The above corollary applies to the complexification $(\eta \otimes \mathbf{C})^*$ of a real vector bundle.

11.3 Definition. The ith Pontrjagin class of a real vector bundle ξ, denoted $p_i(\xi)$, is $(-1)^i c_{2i}(\xi \otimes \mathbf{C})$ which is a member of $H^{4i}(B, \mathbf{Z})$.
    We define $p(\xi) = 1 + p_1(\xi) + \cdots \in H^*(B, \mathbf{Z})$ to be the total Pontrjagin class of the vector bundle ξ. The Whitney sum theorem holds only in the following modified form:

$$2(p(\xi \oplus \eta) - p(\xi)p(\eta)) = 0$$

Let $q: RP^{2n-1} \to CP^{n-1}$ be the map that assigns to each real line determined by $\{z, -z\}$ for $z \in S^{2n-1}$ the complex line determined by z.

11.4 Theorem. *Let ξ be the canonical (real) line bundle on RP^{2n-1} and η the canonical (complex) line bundle on CP^{n-1}. Then the following statements apply.*

(1) *The bundle $q^*(\eta)$ is isomorphic to $\varepsilon_U(\xi)$.*
(2) *The class $w_2(\varepsilon_0(\eta))$ is the mod 2 reduction of $c_1(\eta)$.*
(3) *$c_1(\eta) = e(\varepsilon_0(\eta))$.*

Proof. For (1), we recall that $H^2(RP^{2n-1}, \mathbf{Z}) = Z_2$ for $n \geqq 2$. Since complex line bundles are classified up to isomorphism by their first Chern class c_1, we need show only that $c_1(q^*(\eta)) \neq 0$ and $c_1(\varepsilon_U(\xi)) \neq 0$. Since $c_1(\eta)$ generates $H^2(CP^{n-1})$ and since $q^*: H^2(CP^{n-1}) \to H^2(RP^{2n-1})$ is an epimorphism, we have $c_1(q^*(\eta)) = q^*(c_1(\eta)) \neq 0$. Since $\varepsilon_0\varepsilon_U(\xi) = \xi \oplus \xi$, it suffices to show that $\xi \oplus \xi$ is nontrivial. But we have $w(\varepsilon_0\varepsilon_U(\xi)) = w(\xi \oplus \xi) = w(\xi)w(\xi) = (1 + x)^2 = 1 + x^2$ and $x^2 \neq 0$. Therefore, $\varepsilon_U(\xi)$ is nontrivial, and we have $c_1(\varepsilon_U(\xi)) \neq 0$.

For (2), we observe that $q^*(\varepsilon_0(\eta))$ equals $\varepsilon_0(\varepsilon_U(\xi))$, and therefore $q^*(w(\varepsilon_0(\eta))) = w(\varepsilon_0(\varepsilon_U(\xi))) = 1 + x^2$. Since $q^*: H^2(CP^{n-1}) \to H^2(RP^{2n-1})$ is an epimorphism, $w(\varepsilon_0(\eta))$ is equal to the nonzero element of $H^2(CP^{n-1}, Z_2)$.

For (3), if $n = 2$, then $c_1(\eta)$ is by definition the canonical generator which comes by double suspension from $\alpha_0 \in \tilde{H}^0(S^0)$. From the construction of $e(\varepsilon_0(\eta))$, it is also the double suspension of α_0. In general, the restriction $H^2(CP^n, \mathbf{Z}) \to H^2(CP^1, \mathbf{Z})$ is a monomorphism with both $c_1(\eta)$ and $e(\varepsilon_0(\eta))$ having as image the canonical generator of $H^2(CP^1, \mathbf{Z})$. This proves the theorem.

11.5 Corollary. *Over a paracompact space, $w_2(\varepsilon_0(\eta))$ is equal to the mod 2 restriction of $c_1(\eta)$ for any complex line bundle η.*

Proof. The corollary is true for the universal bundle by (2) in (11.4) and therefore for all complex line bundles.

11.6 Corollary. *Over a paracompact space, $e(\varepsilon_0(\eta))$ is equal to $c_1(\eta)$ for any complex line bundle η.*

Both corollaries are true for the universal bundles by (11.4), and therefore they are verified for all line bundles, using the classifying maps.

12. Orientability and Stiefel-Whitney Classes

12.1 Theorem. *Let ξ be a real vector bundle over a space B. Then ξ is orientable if and only if $w_1(\xi) = 0$.*

Proof. First, we consider the case where $\xi = \lambda_1 \oplus \cdots \lambda_n$ is a Whitney sum of line bundles. Then the line bundle $\Lambda^n \xi$ has as coordinate transformations the determinant of the coordinate transformations of ξ. Therefore, ξ is orientable if and only if $\Lambda^n \xi$ is trivial, that is, $w_1(\Lambda^n \xi) = 0$. But $w_1(\Lambda^n \xi) = w_1(\lambda_1 \otimes \cdots \otimes \lambda_n) = w_1(\lambda_1) + \cdots + w_1(\lambda_n) = w_1(\xi)$ by 5(6.10). Therefore, ξ is orientable if and only if $w_1(\xi) = 0$.

For the general case, let $f: B_1 \to B$ be a splitting map for ξ. Then ξ is orientable if and only if $f^*(\xi)$ is orientable, because $f^*(w_1(\Lambda^n \xi)) = w_1(\Lambda^n f^*(\xi))$ holds and f^* is a monomorphism. Finally, we have $w_1(\xi) = 0$ if and only if $w_1(f^*(\xi)) = 0$. This proves the theorem.

Exercises

1. Prove that if a vector bundle ξ^n has an everywhere-nonzero cross section s, then $w_n(\xi) = 0$ in the real case and $c_n(\xi) = 0$ in the complex case.

2. Construct a real vector bundle ξ^n with no everywhere-nonzero cross sections but with $w_n(\xi^n) = 0$.

3. Let ξ be an n-dimensional complex vector bundle. Prove that $w_{2n}(\varepsilon_0(\xi))$ is the mod 2 reduction of $c_n(\xi)$ and that $c_n(\xi) = e(\varepsilon_0(\xi))$.

4. Prove that the Thom space of a bundle ξ over a compact space B is homeomorphic to the one-point compactification of $E(\xi)$.

5. For a vector bundle ξ, define a map $f: D(\xi) \to P(\xi \oplus \theta^1)$ by the relation that $f(x)$ is the line generated by $x + (1 - \|x\|^2) 1$ in $E(\xi \oplus \theta^1)$. Prove that the restriction $f: D(\xi) - S(\xi) \to P(\xi \oplus \theta^1) - P(\xi)$ is a homeomorphism and that it induces a homeomorphism $\xi^B \to P(\xi \oplus \theta^1)/P(\xi)$.

6. Prove that the sequence of spaces $P(\xi) \to P(\xi \oplus \theta^1) \to B^\xi$ associated with a vector bundle ξ defines an exact sequence.

$$0 \leftarrow H^*(P(\xi)) \xleftarrow{\alpha} H^*(P(\xi \oplus \theta)) \xleftarrow{\beta} H^*(B^\xi) \leftarrow 0$$

Also, prove that $\alpha(a_{\xi \oplus \theta^1}) = a_\xi$ and that the ideal im β^* is generated by $a \in H^*(P(\xi \oplus \theta^1))$, where $a = \sum\limits_{i \leq i \leq n} x_i(\xi) a_{\xi \oplus \theta^1}^{n-i}$ with $n = \dim \xi$ and $\alpha(a_{\xi \oplus \theta^1}) = 0$.

7. Define a natural inclusion $B(\xi) \to S(E \oplus \theta^1)$ such that $S(\xi \oplus \theta^1)/B(\xi)$ is homeomorphic to ξ^B.

8. If λ is a real line bundle and if $H^2(X, \mathbf{Z})$ has no 2-torsion, prove that $\lambda \otimes \mathbf{C}$ is a trivial complex line bundle.

9. Prove Wu's formula for the Stiefel-Whitney classes w_i of a vector bundle

$$Sq^k w_i = \sum_{0 \leq r \leq k} \binom{i + r - k - 1}{r} w_{k-r} w_{i+r}$$

Hint: See Wu-Chung Hsiang, on Wu's formula of Steenrod squares on Stiefel-Whitney classes, *Boletin de la Sociedad Matematica Mexicana*, 8:20–25 (1963).

CHAPTER 18

Differentiable Manifolds

Fibre bundles first arose as bundles associated with the geometry of a manifold. In this chapter we consider topics in the topology of manifolds connected with vector bundles and the orientability of manifolds. Using Stiefel-Whitney classes we derive elementary nonexistence theorems for immersions of manifolds into euclidean space.

1. Generalities on Manifolds

We begin by recalling some ideas from differential calculus. Let U be an open subset of $\mathbf{R}^n$ and V of $\mathbf{R}^m$. The ith partial derivative of a function $f: U \to V$ is denoted by $D_i f: U \to \mathbf{R}^m$. The derivative of f, denoted Df, is a function $Df: U \to \mathbf{L}(\mathbf{R}^n, \mathbf{R}^m)$ defined by the relation $Df(x)a = \sum_{1 \leqq i \leqq n} D_i f(x) a_i$ for $a \in \mathbf{R}^n$. Recall that $\mathbf{L}(\mathbf{R}^n, \mathbf{R}^m)$ denotes the space of all linear transformations $\mathbf{R}^n \to \mathbf{R}^m$.

Let $A \subset \mathbf{R}^n$, and let $B \subset \mathbf{R}^m$. A function $u: A \to B$ is of class C^r provided there exist open sets U and V with $A \subset U$ and $B \subset V$ and a function $f: U \to V$ such that all partial derivatives $D_{i(1)} \cdots D_{i(s)} f$ exist and are continuous on U for $s \leqq r$ and $f|A = u$. If $u: A \to B$ and $v: B \to C$ are of class C^r, then $vu: A \to C$ is of class C^r and u is of class C^s for $s \leqq r$. It is said that $u: A \to B$ is a C^r isomorphism provided $u^{-1}: B \to A$ exists and is of class C^r.

If $f: U \to V$ and $g: V \to W$ are two maps of class C^1, we have $D(gf)(x) = D(g)(f(x)D(f)(x)$.

1.1 Definition. A chart on a space X is a pair (U, ϕ), where U is an open subset of X and $\phi: U \to \phi(U)$ is a homeomorphism onto an open subset $\phi(U)$ of H^n. Here H^n is the upper half space of all $x \in \mathbf{R}^n$ with $x_n \geqq 0$.

1.2 Definition. A C^r-atlas **A** on a space X is a family of charts on X such that the following two properties are satisfied.

(1) The set X is the union of all U for $(U, \phi) \in \mathbf{A}$.
(2) For (U, ϕ), $(V, \psi) \in \mathbf{A}$, the restriction $\psi\phi^{-1}: \phi(U \cap V) \to \psi(U \cap V)$ is a C^r-map (thus a C^r-isomorphism with inverse $\phi\psi^{-1}$).

Two C^r-atlases **A** and **A**′ are equivalent provided $\mathbf{A} \cup \mathbf{A}'$ is a C^r-atlas. A maximal C^r-atlas **A** is an atlas such that for every C^r-atlas $\mathbf{A}' \supset \mathbf{A}$ there is the relation $\mathbf{A} = \mathbf{A}'$. The reader can easily verify that equivalence of C^r-atlases is an equivalence relation and that every C^r-atlas is included in a unique maximal C^r-atlas. Two C^r-atlases are equivalent if and only if they determine the same maximal C^r-atlas.

1.3 Definition. A manifold M is a pair consisting of a Hausdorff space M and a maximal C^r-atlas (or an equivalence class of C^r-atlases).

1.4 Example. If N is an open subset of a manifold M of class C^r, then N has a natural C^r-manifold structure where the charts on N are of the form $(U \cap N, \phi|(U \cap N))$ and (U, ϕ) is a chart on M. Then N is referred to as an open submanifold of M.

The spaces H^n and $\mathbf{R}^n$ are manifolds whose maximal atlas is determined by a single chart, the identity $H^n \to H^n$ for H^n and $(x_1, \ldots, x_n) \mapsto (x_1, \ldots, x_{n-1}, \exp x_n)$ of $\mathbf{R}^n \to H^n$ for $\mathbf{R}^n$.

1.5 Definition. A C^r-morphism between two C^r-manifolds is a map $f: M \to N$ such that for each chart (U, ϕ) on M and (V, ψ) on N the composition of restrictions $\psi f \phi^{-1}: \phi^{-1}(U \cap f^{-1}(V)) \to \psi(V)$ is of class C^r.

If $f: M \to N$ and $g: N \to V$ are two C^r-morphisms, then $gf: M \to V$ is a C^r-morphism. For this statement, we need the result that a function $f: A \to B$ for $A \subset \mathbf{R}^n$, $B \subset \mathbf{R}^m$ is of class C^r if and only if f is of class C^r near each point of A. Therefore, the category of manifolds of class C^r is defined.

Let ∂H^n denote the $x \in H^n$ with $x_n = 0$. Then $\partial H^n = \mathbf{R}^{n-1}$ in a natural way. For a manifold M and for $x \in U \cap V$, where (U, ϕ) and (V, ψ) are two charts on M, we have $\phi(x) \in \partial H^n$ if and only if $\psi(x) \in \partial H^n$. This follows because $(B(a, r) - \{a\}) \cap H^n$ is contractible for $a \in \partial H^n$ and has the homotopy type of S^{n-1} otherwise.

1.6 Definition. The boundary of a manifold M, denoted ∂M, is the subspace of $x \in M$ with $\phi(x) \in \partial H^n$ for some chart (U, ϕ). Moreover, ∂M is a manifold with charts $(U \cap \partial M, \phi|(U \cap \partial M))$, where (U, ϕ) is any chart of M.

If M has dimension n, that is, the range of each chart is in H^n, then ∂M is either empty or of dimension $n - 1$. When ∂M is empty, M is referred to as a manifold without boundary.

1.7 Convention. In the subsequent discussion we incorporate paracompactness into the definition of a manifold. This property is equivalent to each component being separable or being a countable union of compact sets.

2. The Tangent Bundle to a Manifold

Historically, fibre bundle theory arose in connection with the study of the geometry of manifolds (see also Seifert [1].) Associated with every n-dimensional manifold M is an n-dimensional vector bundle $\tau(M)$ called the tangent bundle. Most fibre bundles of interest in manifold theory arise from the tangent bundle. As we shall see, the tangent bundle reflects many of the geometric properties of the base manifold.

We define a tangent bundle using the existence theorem of Chap. 5; namely, we describe a system of transition functions relative to an open covering of the manifold.

2.1 Definition. Let M^n be an n-dimensional manifold with maximal atlas **A**. The tangent bundle $\tau(M^n)$ is the n-dimensional vector bundle defined by transition functions $D(\psi\phi^{-1})(\phi(x)) \in GL(n, \mathbf{R})$ for $x \in U \cap V$ and (U, ϕ), $(V, \psi) \in \mathbf{A}$.

For $x \in U \cap V \cap W$ and $(U, \phi), (V, \psi), (W, \eta) \in \mathbf{A}$ the relation $D(\eta\phi^{-1})(\phi(x)) = D(\eta\psi^{-1})(\psi(x))D(\psi\phi^{-1})(\phi(x))$ holds by the "chain rule." From this calculation, the set of functions $D(\psi\phi^{-1})(x))$ is a family of transition functions for a vector bundle.

A vector field on a manifold M is just a cross section of $\tau(M)$.

The total space E of the tangent bundle $\tau(M)$ of a C^r-manifold is a C^{r-1}-manifold. The coordinate charts of E are of the form $(U \times \mathbf{R}^n, \phi \times 1)$, where (U, ϕ) is a chart of M. The coordinate change from $(U \times \mathbf{R}^n, \phi \times 1)$ to $(V \times \mathbf{R}^n, \psi \times 1)$ is given by $(x, a) \mapsto (\psi\phi^{-1}(x), D(\psi\phi^{-1})(\phi(x)a)$ and is of class C^{r-1}. The projection is a map of class C^{r-1}.

2.2 Example. The tangent bundle to M, an open subset of $\mathbf{R}^n$, is the trivial product bundle $M \times \mathbf{R}^n \to M$. See also Exercise 4.

2.3 Definition. Let $f: M \to N$ be a C^r-map between two manifolds. The induced map C^{r-1}-map of tangent vector bundles $(\tau(f), f): \tau(M) \to \tau(N)$ follows from the requirement that $D(\psi f\phi^{-1}) = h_V\tau(f)h_U^{-1}$, where $h_U: \tau(M)|U \to U \times \mathbf{R}^m$ and $h_V: \tau(N)|V \to V \times \mathbf{R}^n$ are charts for the respective tangent bundles, (U, ϕ) for M, and (V, ψ) for N.

Note the relation $D(\psi f\phi^{-1}) = h_V\tau(f)h_U^{-1}$ holds over $U \cap f^{-1}(V)$. Since $D(\psi f\phi^{-1})$ is linear, $\tau(f)$ is linear on the fibres. The character of the transition functions assures that $\tau(f)$ exists and $(\tau(f), f)$ is a morphism of vector bundles.

For two C^r-maps $f: M \to N$ and $g: N \to V$ the relation $\tau(gf) = \tau(g)\tau(f)$ holds. Consequently, τ is a functor from C^r-manifolds and C^r-maps to vector bundles and vector bundle morphisms.

The inverse function theorem says that for a C^r-map $f: M \to N$ such that $\tau(f): \tau(M)_z \to \tau(N)_{f(x)}$ is an isomorphism there are open submanifolds U of M and V of N such that the restriction $f|U: U \to V$ is a C^r-isomorphism and $x \in U$.

2.4 Definition. A subspace X of a manifold M is called a submanifold provided each $x \in X$ has a chart (U, ϕ) of M such that $X \cap U = \phi^{-1}((0 \times \mathbf{R}^q) \cap H^n)$.

Then the restrictions $(X \cap U, \phi|(X \cap U))$ determine a C^r-structure on X.

2.5 Definition. A C^r-map $f: M \to N$ is an immersion provided $\tau(f)$ is a monomorphism on each fibre. A C^r-map $f: M \to N$ is an embedding provided f is an immersion and $f: M \to f(M)$ is a homeomorphism.

An application of the inverse function theorem says that the restriction of an immersion to some neighborhood of each point yields an embedding. Moreover, when $f: M \to N$ is an immersion, there exists a monomorphism of vector bundles over M, namely $\tau(M) \to f^*(\tau(N))$.

2.6 Definition. The quotient bundle, denoted v_f, of $f^*(\tau(N))$ by the image of $\tau(M)$ is called the normal bundle of the immersion.

By 3(8.2), the normal bundle always exists, and there is the following exact sequence:

$$0 \to \tau(M) \to f^*(\tau(N)) \to v_f \to 0$$

If M is paracompact, then by 3(9.6) the above exact sequence splits and $f^*(\tau(N)) = \tau(M) \oplus v_f$.

2.7 Remark. If M^m is an m-dimensional manifold and if $N = \mathbf{R}^n$ is euclidean space, then for an immersion $f: M^m \to \mathbf{R}^n$ we have the exact sequence

$$0 \to \tau(M) \to f^*\tau(\mathbf{R}^n) \to v_f \to 0$$

and the splitting $\tau(M) \oplus v_f = \theta^n$, the trivial n-dimensional vector bundle since $\tau(\mathbf{R}^n)$ is trivial. Therefore, the normal bundles v_f of immersions f arise as stable inverses of $\tau(M)$ in $\tilde{K}_{\mathbf{R}}(M)$. The converse also holds, that every s-inverse of $\tau(M)$ is the normal bundle of some immersion; this is Hirsch's theorem (see Hirsch [1]).

The next result is useful in finding relations between a manifold and its boundary. For it we use the fact that, if M is a manifold with boundary ∂M, there exists an embedding $h: \partial M \times [0, 1] \to M$ such that $h(x, 0) = x$.

2.8 Proposition. *The relation* $\tau(M)|\partial M = \tau(\partial M) \oplus \theta^1$ *holds.*

Proof. This is true locally, and the above diffeomorphism $h: \partial M \times [0,1) \rightarrow M$ arising from the local situation yields the isomorphism over ∂M.

3. Orientation in Euclidean Spaces

Using both homology theory and the tangent bundle of a manifold, we wish to investigate the notion of orientation. The relation between positive and negative, clockwise and counterclockwise, and right- and left-handed systems are all manifestations of this general concept.

3.1 Definition. Let $f: U \rightarrow V$ be a diffeomorphism of class C^r for $r \geqq 1$ of open connected subsets of $\mathbf{R}^n$. It is said that f preserves orientation, denoted $O(f) = +1$, provided $\det(Df)(x) > 0$ for $x \in U$ and that f reverses orientation, denoted $O(f) = -1$, provided $\det(Df)(x) < 0$ for $x \in U$.

3.2 Examples. The identity, a translation $x \mapsto x + a$, and a stretching $x \mapsto bx$ with $b > 0$ all preserve orientation. A reflection through a hyperplane reverses orientation. If $f: \mathbf{R}^n \rightarrow \mathbf{R}^n$ is linear, then $O(f)$ is the sign of $\det f$.

3.3 Proposition. *Let* $f: U \rightarrow V$ *and* $g: V \rightarrow W$ *be two diffeomorphisms between connected open sets in* $\mathbf{R}^n$. *Then* $O(gf) = O(g)O(f)$. *Let* U' *be an open connected subset of* U *and* V' *of* V *such that* $f(U') = V'$. *Then we have* $O(f) = O(f|U')$.

Proof. For the first statement, we observe that $\det[D(gf)(x)] = \det[D(g)(f(x))] \det[D(f)(x)]$ and $O(gf) = O(g)O(f)$. The second statement is immediate.

Next, we develop a homological determination of $O(f)$. We choose generators $\alpha_n \in \tilde{H}_n(S^n)$ and $\beta_n \in \tilde{H}^n(S^n)$ (integral coefficients) such that $\sigma(\alpha_n) = \alpha_{n+1}$ and $\sigma(\beta_n) = \beta_{n+1}$, where σ is the suspension isomorphism, and such that $\langle \alpha_n, \beta_n \rangle = 1$ for the pairing $H_n(X) \times H^n(X) \rightarrow \mathbf{Z}$.

We have the following natural isomorphisms for $x \in U$ and $\overline{B(x,r)} \subset U$, where U an open subset of $\mathbf{R}^n$:

$$\tilde{H}_{n-1}(S^{n-1}) \overset{(1)}{\leftarrow} H_n(B^n, S^{n-1}) \overset{(2)}{\rightarrow} H_n(B^n, B^n - 0)$$

$$\overset{(3)}{\rightarrow} H_n(B(x,r), B(x,r) - x) \overset{(4)}{\rightarrow} H_n(U, U - x)$$

Similarly for cohomology there is a natural isomorphism $H^{n-1}(S^{n-1}) \rightarrow H^n(U, U - x)$. Isomorphism (1) comes from the exact sequence, (2) comes from the fact that $S^{n-1} \rightarrow B^n - 0$ is a homotopy equivalence, (3) comes from translation and stretching homeomorphisms, and (4) is excision where $\overline{B(x,r)} \subset U$. The image of $\alpha_n \in \tilde{H}_{n-1}(S^{n-1})$ in $H_n(U, U - x)$, denoted α_x, is a canonical generator of $H_n(U, U - x)$. The image of $\beta_n \in \tilde{H}^{n-1}(S^{n-1})$ in $H^n(U, U - x)$, denoted β_x, is a canonical generator of $H^n(U, U - x)$.

3.4 Theorem. *Let $f: U \to V$ be a diffeomorphism between two open connected subsets of $\mathbf{R}^n$. Then for $x \in U$ there are the relations $f_*(\alpha_x) = O(f)\alpha_{f(x)}$ and $f^*(\beta_{f(x)}) = O(f)\beta_x$.*

Proof. By translations, we can assume that $f(0) = 0 = x$. Let $f(x) = Lx + g(x)$, where Lx is linear and $D(g)(0) = 0$, $g(0) = 0$. As maps defined $(U, U - 0) \to (\mathbf{R}^n, \mathbf{R}^n - 0)$, there is a homotopy $h_t(x) = Lx + (1 - t)g(x)$, where $h_0(x) = f(x)$ and $h_1(x) = Lx$. We have reduced the problem to the case where $f: U \to V$ is linear.

By a second homotopy we can change f into an orthogonal transformation. Then $f = r_1 \cdots r_q$ is a reflection through an $(n - 1)$-subspace of $\mathbf{R}^n$. Then $(r_i)_*(\alpha_0) = -\alpha_0$, $(r_i)^*(\beta_0) = -\beta_0$, and $O(r_i) = -1$. By the rule for composition, we have $f_*(\alpha_0) = O(f)\alpha_0$ and $f^*(\beta_0) = O(f)\beta_0$. This proves the theorem.

This theorem allows extension of the notion of orientation to topological maps.

3.5 Definition. Let $f: U \to V$ be a homeomorphism of subsets of $\mathbf{R}^n$. Then the orientation number of f at x, denoted $O_x(f)$, is defined by the relation $f_*(\alpha_x) = O_x(f)\alpha_{f(x)}$ or, equivalently, $f^*(\beta_{f(x)}) = O_x(f)\beta_x$.

The number $O_x(f) \in Z_2$ is locally a constant. By (3.4) it agrees with (3.1). Proposition (3.3) applies to this definition.

4. Orientation of Manifolds

The coefficient ring will be either $\mathbf{Z}$ or Z_2 in the following discussion. If (ϕ, U) is a local coordinate chart on a manifold M, consider the following isomorphisms for $x \in U \subset M$. (The first is an excision isomorphism.)

$$H_n(M, M - x) \leftarrow H_n(U, U - x) \to H_n(\phi(U), \phi(U) - \phi(x))$$

Therefore, $H_n(M, M - x)$, and similarly $H^n(M, M - x)$, equals the coefficient ring. Orientation arises as a problem because $H_n(M, M - x)$ has two generators for coefficients in $\mathbf{Z}$.

4.1 Definition. An orientation ω of a manifold M^n with ∂M empty is a function that assigns, to each $x \in M$, a generator $\omega_x \in H_n(M, M - x)$ such that each $x \in M$ has a neighborhood W and a class $\omega_W \in H_n(M, M - W)$ whose image in $H_n(M, M - y)$ is ω_y for each $y \in W$.

Implicit in the above definition is the notion of orientation in two senses Z_2 and $\mathbf{Z}$.

4.2 Definition. A manifold M is orientable provided it has an orientation with integer coefficients.

Before we investigate homological properties of an orientation ω, we derive some characterization of the notion.

4.3 Theorem. *Every manifold M has a unique Z_2-orientation. For a manifold M the following statements are equivalent.*

(1) *M is orientable.*
(2) *There exists an atlas $\mathbf{A}$ of M such that for (U, ϕ), $(V, \psi) \in \mathbf{A}$ the relation $\det D(\psi \phi^{-1}) > 0$ on $\phi(U \cap V)$ holds.*
(3) *The tangent bundle $\tau(M)$ has $SO(n)$ as its structure group.*

Finally, if M is connected, M has precisely two orientations.

Proof. Let ω_x be the nonzero element of $H_n(M, M - x; Z_2)$. Then for a ball W around x in the domain of a chart we have $H_n(M, M - W; Z_2) = Z_2$, and the nonzero element is mapped onto ω_y for each $y \in W$. This proves the first statement.

For the equivalence of statements (1) to (3), we begin by assuming (2). Then for a Z-orientation of M we define $\omega_x = (\phi^{-1})_*(\alpha_n)$, where $\alpha_n \in H_n(\phi(U), \phi(U) - \phi(x))$ is the canonical class used in (3.4), and $(U, \phi) \in \mathbf{A}$. By (3.4) and (2), change of coordinate is orientation preserving, and ω is a well-defined orientation. Conversely, let $\mathbf{A}$ be the atlas of those charts such that $\omega_x = (\phi^{-1})_*(\alpha_n)$ for $x \in U$ and $\alpha_n \in H_n(\phi(U), \phi(U) - \phi(x))$. This verifies the equivalence of (1) and (2).

To see that (2) implies (3), observe that the transition functions $D(\psi \phi^{-1})(\phi(x))$, where $x \in U \cap V$, have a positive determinant and are homotopic to transition functions in $SO(n)$. Conversely, if $\tau(M)$ has a family of transition functions, $D(\psi \phi^{-1})(\phi(x))$ can be chosen with a positive determinant for some atlas of M.

For the final statement, it should be noted that the maximal atlas divides into two atlases each with property (2). This holds at a point and then along any path to another point. This proves the theorem.

In the next theorem we derive homological properties of an orientation ω which apply both to coefficients in Z and in Z_2. We shall not have to distinguish in the proof between the two kinds of coefficients.

For two subsets $K \subset L$ of M, let $r: H_i(M, M - L) \to H_i(M, M - K)$ be the homomorphism induced by inclusion.

4.4 Theorem. *Let M^n be an n-dimensional manifold, and let K be a compact subset of M^n. We have $H_i(M, M - K) = 0$ for $i > n$, and for $a \in H_n(M, M - K)$ the relation $a = 0$ holds if and only if $r(a) = 0$ in $H_n(M, M - x)$ for each $x \in K$. Finally, if ω is an orientation for M, there exists $\omega_K \in H_n(M, M - K)$ such that $r(\omega_K) = \omega_x$ in $H_n(M, M - x)$ for each $x \in K$. Moreover, ω_K is unique, and it is called the orienting homology class of M.*

Proof. We consider for which compact sets $K \subset M$ the theorem is true. First, we prove that if the theorem is true for K_1, K_2, and $K_1 \cap K_2$ it is true for $K_1 \cup K_2$. Let K denote $K_1 \cup K_2$ and L denote $K_1 \cap K_2$. The Mayer-Vietoris sequence has the following form (see Eilenberg and Steenrod [1]).

$$\cdots \to H_{i+1}(M, M - L) \xrightarrow{\partial} H_i(M, M - K)$$
$$\xrightarrow{u} H_i(M, M - K_1) \oplus H_i(M, M - K_2) \xrightarrow{v} \cdots$$

Recall that $u(a) = (r_1(a), r_2(a))$, and $v(a, b) = r_1'(a) - r_2'(b)$. Then, clearly, we have $H_i(M, M - K) = 0$ for $i > n$. For the next statement, if each $r(a) = 0$ in $H_n(M, M - x)$, we have $u(a) = 0$. Since $H_{n+1}(M, M - L) = 0$, we have $a = 0$. Finally, the existence of ω_K follows from the fact that $v(\omega_{K_1}, \omega_{K_2}) = 0$, and the uniqueness of ω_K follows from the previous statement.

Next we suppose that $K \subset U$, where (U, ϕ) is a coordinate chart with $\phi(K)$ a convex set in $\phi(U) \subset \mathbf{R}^n$. Since $\phi(U) - \phi(K) \to \phi(U) - \phi(x)$ is a homotopy equivalence, the induced homomorphisms $H_i(M, M - K) \to H_i(M, M - x)$ are isomorphisms. For these K the theorem follows. Using the result of the first paragraph, the theorem holds for compact sets $K \subset U$, where (U, ϕ) is a coordinate chart and $\phi(K)$ is a union of a finite number of convex sets.

Now we consider the case where K is an arbitrary compact set with $K \subset U$ and (U, ϕ) is a coordinate chart. We consider $\phi_*(a) \in H_i(\phi(U), \phi(U) - \phi(K))$ which is represented by a chain c in $\mathbf{R}^n$ where all simplexes in ∂c are disjoint from $\phi(K)$. Then this is true for a compact neighborhood L of $\phi(K)$. We cover $\phi(K)$ with finitely many balls $B_1, \ldots, B_m$ so that $B_1 \cup \cdots \cup B_m \subset L$ with $B_i \cap \phi(K) \neq \phi$. There exists $b \in H_i(\phi(U), \phi(U) - L)$ with $r(b) = \phi_*(a)$. For $i > n$ we have $r(b) = 0$ in $H_i(\phi(U), \phi(U) - (B_1 \cup \cdots \cup B_m))$ from the above discussion. For $i = n$ we have $r(b) = 0$ in $H_i(\phi(U), \phi(U) - x)$ for $x \in B_1 \cup \cdots \cup B_m$ by the above discussion. Therefore, we have $a = 0$ since $H_i(U, U - K) = H_i(M, M - K)$ by excision. Finally, we define $\omega_K = \phi_*^{-1}(r(\omega_{B_1 \cup \cdots \cup B_m}))$, and therefore the theorem is true for small compact sets.

Finally, the theorem follows from the fact that every compact K is of the form $K = K_1 \cup \cdots \cup K_m$, where K_i is a subset of some U and (U, ϕ) is a coordinate chart.

4.5 Remark. For a manifold M with boundary ∂M, Theorems (4.3) and (4.4) hold with appropriate modifications.

We leave this as an exercise for the reader.

5. Duality in Manifolds

Let K be a compact subset and V an open subset of an n-dimensional manifold M with $K \subset V$. Then there is the following diagram involving the cup product:

$$H^q(M) \otimes H^{n-q}(M, M - K) \xrightarrow{\text{cup}} H^n(M, M - K)$$

$$j^* \otimes 1 \downarrow \qquad\qquad\qquad\qquad \downarrow j^*$$

$$H^q(V) \otimes H^{n-q}(M, M - K) \xrightarrow{\text{cup}} H^n(V, V - K)$$

Observe that $j: (V, V - K) \to (M, M - K)$ is an excision map inducing an isomorphism in both singular homology and cohomology theory. Next there is the following diagram involving the cap product:

$$H^q(M) \otimes H_n(M, M - K) \xrightarrow{\text{cap}} H_{n-q}(M, M - K)$$

$$j^* \otimes (j_*^{-1}) \downarrow \qquad\qquad\qquad\qquad \downarrow 1$$

$$H^q(V) \otimes H_n(V, V - K) \xrightarrow{\text{cap}} H_{n-q}(M, M - K)$$

If $\langle , \rangle: H^*(M, A) \otimes H_*(M, A) \to \mathbf{Z}$ or $\mathbf{Z}_2$ is the canonical pairing induced by the substitution pairing $C^*(M, A) \otimes C_*(M, A) \to \mathbf{Z}$ or $\mathbf{Z}_2$, there is the following relation between cup products $c_1 \cup c_2$ and cap products $c \cap u$.

5.1 Cup-Cap Relation. Let $u \in H_n(V, V - K)$, let $c_1 \in H^k(V)$, and let $c_2 \in H^{n-k}(V)$. Then

$$\langle c_1 \cup c_2, u \rangle = \langle c_1, c_2 \cap u \rangle$$

5.2 Notations. Let ω be an orientation of M, and let $\omega_K \in H_n(M, M - K)$ be the corresponding class given by Theorem (4.4) for each compact set K in M. Let ω_K^V equal $j_*^{-1}(\omega_K)$, where $j_*: H_n(V, V - K) \to H_n(M, M - K)$ is the excision isomorphism. We define $D_K^V: H^q(V) \to H_{n-q}(M, M - K)$ by $D_K^V(c) = j_*(c \cap \omega_K^V)$.

Next, let $\bar{H}^q(K)$ be the direct limit of $H^q(V)$, where V is an open set with $V \supset K$. Then the direct limit of $D_K^V: H^q(V) \to H_{n-q}(M, M - K)$ is denoted by $D_K: \bar{H}^q(K) \to H_{n-q}(M, M - K)$.

The main duality theorem is the following statement.

5.3 Theorem. *With the above notations, $D_K: \bar{H}^q(K) \to H_{n-q}(M, M - K)$ is an isomorphism for all compact subsets K of M.*

Proof. First, we prove that if the theorem holds for compact subsets K, L, and $K \cap L$ of M it holds for $K \cup L$. Let V and W be open subsets of M with $K \subset V$ and $L \subset W$. Then $J = K \cap L \subset V \cap W$ and $I = K \cup L \subset V \cup W$. The following pair of Mayer-Vietoris sequences is connected by cap products with ω_K, ω_L, $\omega_{K \cap L}$, or $\omega_{K \cup L}$ for $p = n - q$.

$$H^{q-1}(V) \oplus H^{q-1}(W) \longrightarrow H^{q-1}(V \cap W) \longrightarrow$$

$$DK^V \oplus DL^W \downarrow \qquad\qquad\qquad\qquad \downarrow D_J^{V \cap W}$$

$$H_{p+1}(M, M - K) \oplus H_{p+1}(M, M - L) \longrightarrow H_{p+1}(M, M - J) \longrightarrow$$

$$H^q(V \cup W) \longrightarrow H^q(V) \oplus H^q(W)$$

$$\downarrow D_I^{V \cup W} \qquad\qquad\qquad \downarrow D_K^V \oplus D_L^W$$

$$H_p(M, M - I) \longrightarrow H_p(M, M - K) \oplus H_p(M, M - L)$$

In the limit there is the following morphism between two exact sequences.

$$\bar{H}^{q-1}(K) \oplus \bar{H}^{q-1}(L) \longrightarrow \bar{H}^{q-1}(K \cap L) \longrightarrow$$

$$\downarrow D_K \oplus D_L \qquad\qquad\qquad\qquad \downarrow D_{K \cap L}$$

$$H_{p+1}(M, M - K) \oplus H_{p+1}(M, M - L) \longrightarrow H_{p+1}(M, M - J) \longrightarrow$$

$$\bar{H}^q(K \cup L) \longrightarrow \bar{H}^q(K) \oplus \bar{H}^q(L)$$

$$\downarrow D_{K \cup L} \qquad\qquad\qquad\qquad \downarrow D_K \oplus D_L$$

$$H_p(M, M - I) \longrightarrow H_p(M, M - K) \oplus H_p(M, M - L)$$

Since $D_{K \cap L}$ and $D_K \oplus D_L$ are all isomorphisms by hypothesis, the 5-lemma implies that $D_{K \cup L}$ is an isomorphism.

Secondly, observe that the theorem holds for compact, convex K or a ball K, where $K \subset U$ for a coordinate chart (U, ϕ). Again, that K is convex means that $\phi(K)$ is convex in euclidean space. If V is a ball and open neighborhood of K in U, then D_K^V is an isomorphism. This follows easily from the relation in (5.1) in the nontrivial dimension.

Next, we see that the theorem holds for finite simplicial complexes K embedded in M since each simplex, after subdivision, can be assumed to be in an open set U for a chart (U, ϕ).

If K is a compact subset of an open set U, where (U, ϕ) is a chart of M, then K is the intersection of K_i which are simplicial complexes. Then $\bar{H}^*(K)$ is the limit of the $\bar{H}^*(K_i)$, $H_*(M, M - K)$ is the limit of the $H_*(M, M - K_i)$, and D_K is the limit of the D_{K_i}. Consequently, D_K is an isomorphism.

Finally, for a general compact subset K we write K as the union of K_i with $1 \leq i \leq m$, where $K_i \subset U_i$ and (U_i, ϕ_i) is a coordinate chart. Using the result of the first paragraph, we have the proof in the general case.

The first corollary is the classical Poincaré duality theorem.

5.4 Corollary. *Let M be a compact manifold with an orientation. Then D_M: $H^q(M) \to H_{n-q}(M)$ given by the cap product is an isomorphism.*

The next corollary is the Alexander duality theorem.

5.5 Corollary. *Let K be a compact subset of $\mathbf{R}^n$. Then the composition of D_K and the boundary morphism is an isomorphism for $n > q$:*

$$\bar{H}^q(K) \to \tilde{H}_{n-q-1}(\mathbf{R}^n - K)$$

5.6 Remark. The reader may compare $\bar{H}^q(K)$ with $H^q(K)$, the singular cohomology of K, and $\check{H}^q(K)$, the Čech cohomology group of K. In general, $\bar{H}^q(K)$ and $\check{H}^q(K)$ are isomorphic, and $\bar{H}^q(K)$ and $H^q(K)$ are isomorphic if K has a neighborhood base of open sets each of which K is a deformation retract.

6. Thom Class of the Tangent Bundle

Let M^n denote a closed (compact with no boundary) connected manifold and let τ denote the tangent bundle to M.

6.1 Notations. With a riemannian metric on τ, we define for each $x \in M$ a smooth map $\exp_x: E(\tau)_x \to M$ by the condition that $\exp_x(v)$ is the value at $t = 1$ of the unique geodesic through x with tangent vector v for $t = 0$. By adjusting the metric on τ with a positive constant, we can assume that the homotopy $h_t: (D(\tau), D_0(\tau), M) \to (M \times M, M \times M - \Delta, \Delta)$ defined by the relation $h_t(x, v) = (\exp_x(-tv), \exp_x(v))$ is a diffeomorphism for each $t \in [0, 1]$ of $D(\tau)$ onto a closed manifold neighborhood of Δ in M.

6.2 Notations. Let R denote a ring which we assume contains Z_2 if M is not oriented; require that all cohomology groups have their coefficients in R. Let U_τ or U denote the Thom class of τ where $U \in H^n(D(\tau), S(\tau))$. By excision $(h_t)^*: H^*(M \times M, M \times M - \Delta) \to H^*(D(\tau), S(\tau))$ is an isomorphism, which is independent of t, and we denote the class $(h_t^*)^{-1}(U)$ by U'. The inclusion

$$j: M \times M \to (M \times M, M \times M - \Delta)$$

defines an induced morphism

$$j^*: H^*(M \times M, M \times M - \Delta) \to H^*(M \times M)$$

Let U_M denote the class $j^*(U')$. We call U_M the fundamental class of M.

In the next propositions, we derive several properties of the class U_M.

6.3 Propositions. *For $a, b \in H^*(M)$ the relation $U_M(a \times b) = (-1)^{d(a)d(b)} U_M(b \times a)$ holds where $d(a)$ is the degree of a and $d(b)$ of b.*

Proof. Consider the maps $\alpha: M \times M \to M \times M$ given by $\alpha(x, y) = (y, x)$ and $\beta: D(\tau) \to D(\tau)$ given by $\beta(x, v) = (x, -v)$. They are related by $h_1\beta = \alpha h_1$. By the uniqueness property of the Thom class $\beta^*(U) = (-1)^n U$. Since $\phi: H^i(M) \to H^{n+i}(D(\tau), S(\tau))$ given by $\phi(a) = aU$ is an isomorphism, the morphism $\beta^*: H^*(D(\tau), S(\tau)) \to H^*(D(\tau), S(\tau))$ is multiplication by $(-1)^n$. From the relation $\beta^* h_1^* = h_1^* \alpha^*$ it follows that $\alpha^*: H^*(M \times M, M \times M - \Delta) \to H^*(M \times M, M \times M - \Delta)$ is multiplication by $(-1)^n$. For two classes a, $b \in H^*(M)$ we have $\alpha^*(a \times b) = (-1)^{d(a)d(b)}(b \times a)$ in $H^*(M \times M)$. Now we calculate

$$U_M(a \times b) = j^*[(-1)^n \alpha^*(U'(a \times b))]$$
$$= (-1)^{d(a)d(b)} j^*[U'(b \times a)] = (-1)^{d(a)d(b)} U_M(b \times a)$$

This proves the proposition.

We denote the natural substitution pairing $H^k(M) \otimes H_k(M) \to R$ by $a \otimes b \mapsto \langle a, b \rangle$. Then we have $\langle \overline{\omega}_M, \omega_M \rangle = 1$ for the orientation classes $\omega_M \in H_n(M)$ and $\overline{\omega}_M \in H^n(M)$.

6.4 Proposition. *The following relation holds between the fundamental class U_M and orientation class ω_M of M:* $\langle U_M, 1 \times \omega_M \rangle = 1$.

Proof. We have the following commutative diagram where $f_x : M \to M \times M$ is defined by the relation $f_x(y) = (x, y)$.

$$
\begin{array}{ccccc}
H^*(D(\tau), D_0(\tau)) & \xleftarrow[\approx]{h_t^*} & H^*(M \times M, M \times M - \Delta) & \xrightarrow{j^*} & H^*(M \times M) \\
\Big\downarrow{j_x^*} & & & & \Big\downarrow{f_x^*} \\
H^*(D(\tau)_x, D_0(\tau)_x) & \xleftarrow[\exp_x^*]{\approx} & H^*(M, M - \{x\}) & \xrightarrow{j^*} & H^*(M)
\end{array}
$$

In $H^*(D(\tau)_x, S(\tau)_x)$ we have $\exp_x^*(\overline{\omega}_x) = j_x^*(U)$. Consequently, we have $\langle (\exp_x^*)^{-1} j_x^*(U), \omega_x \rangle = 1$ and $\langle j^*(\exp_x^*)^{-1} j_x^*(U), \omega_M \rangle = 1$ for the pairing $H^n(M) \otimes H_n(M) \to R$. Therefore, we have $\langle U_M, 1 \times \omega_M \rangle = 1$ since $(f_x)_*(\omega_M) = 1 \times \omega_M$, which is a homology cross product. This proves the proposition.

Recall the duality morphism $D: H^k(M) \to H_{n-k}(M)$ is defined by $D(a) = a \cap \omega_M$.

6.5 Proposition. *Let U_M be any class in $H^n(M \times M)$ satisfying the properties of the fundamental class in (6.3) and (6.4). For $a \in H^k(M)$ and $b \in H_k(M)$ the relation $\langle a, b \rangle = (-1)^{n+d(a)d(b)} \langle U_M, b \times Da \rangle$ holds.*

Proof. We have $U_M(a \times 1) = U_M(1 \times a)$ by (6.3) since $d(1)d(a) = 0$. First, we calculate using (5.1) and $a \cap b = \langle a, b \rangle 1$.

$$\langle U_M(a \times 1), b \times \omega_M \rangle = \langle U_M, (a \times 1) \cap (b \times \omega_M) \rangle$$
$$= \langle U_M, (a \cap b) \times \omega_M \rangle$$
$$= \langle a, b \rangle \langle U_M, 1 \times \omega_M \rangle = \langle a, b \rangle$$

Therefore we can make the next calculation

$$\langle a, b \rangle = (-1)^n \langle U_M(1 \times a), b \times \omega_M \rangle = (-1)^n \langle U_M, (1 \times a) \cap (b \times \omega_M) \rangle$$
$$= (-1)^{n+d(a)d(b)} \langle U_M, b \times (a \cap \omega_M) \rangle$$
$$= (-1)^{n+d(a)d(b)} \langle U_M, b \times Da \rangle$$

This proves the proposition.

6.6 Remark. The formula in (6.5) can be used to prove that the duality morphism $D: H^k(M) \to H_{n-k}(M)$ is an isomorphism for field coefficients. First, observe that D is a monomorphism; for if $a \in H^k(M)$ with $a \neq 0$, we have $b \in H_k(M)$ with $\langle a, b \rangle \neq 0$. By (6.5) the relation $Da \neq 0$ follows. Since $\dim H^k(M) = \dim H_k(M)$ and $\dim H^{n-k}(M) = \dim H_{n-k}(M)$, we deduce from the two monomorphisms

$$H^k(M) \overset{D}{\to} H_{n-k}(M) \qquad H^{n-k}(M) \overset{D}{\to} H_k(M)$$

that $\dim H^k(M) = \dim H_{n-k}(M)$. Consequently, D is an isomorphism.

7. Euler Characteristic and Class of a Manifold

In this section M is a closed, connected manifold and the coefficient ring R is the rational numbers.

7.1 Definition. The Euler characteristic $\chi(X)$ of a space X is the sum $\sum_{0 \leq i} (-1)^i \dim H^i(X, \mathbf{Q})$ where we assume $H^i(X, \mathbf{Q}) = 0$ for i large and $\dim H^i(X, \mathbf{Q})$ is finite.

The main result of this section is contained in the next theorem. Observe that $\chi(M)$ is defined for a closed manifold M.

7.2 Theorem. *Let M^n be a closed, connected manifold with $\mathbf{Z}$-orientation ω_M. Then the Euler class of the tangent bundle is related to the Euler characteristic by the relation $e(\tau(M)) = \chi(M)\overline{\omega}_M$.*

Proof. For n odd, we use the relation $e(\tau(M)) = \phi^{-1}(U^2)$ given in 17(7.6) and observe that $e(\tau(M))$ is an element of order 2 or is zero in $H^n(M, \mathbf{Z}) = \mathbf{Z}$. Consequently, it follows that $e(\tau(M)) = 0$. Since $\dim H^i(M, \mathbf{Q}) = \dim H^{n-i}(M, \mathbf{Q})$ by (6.6), we have $\chi(M) = 0$ for an odd dimension manifold M. Therefore, the above formula holds.

Henceforth, we assume n to be even. Let $s: M \to D(\tau)$ be the zero cross section. We have the following commutative diagram.

$$
\begin{array}{ccccc}
H^*(D(\tau), S(\tau)) & \overset{(ht)^*}{\longleftarrow} & H^*(M \times M, M \times M - \Delta) & \overset{j^*}{\longrightarrow} & H^*(M \times M) \\
\downarrow{\scriptstyle j^*} & & & & \downarrow{\scriptstyle \Delta^*} \\
H^*(D(\tau)) & & \overset{s^*}{\longrightarrow} & & H^*(M)
\end{array}
$$

From this diagram we deduce that $e(\tau(M)) = \Delta^*(U_M)$. Consequently, to prove the theorem we must show that

$$\langle e(\tau(M)), \omega_M \rangle = \langle \Delta^*(U_M), \omega_M \rangle = \chi(M)$$

For this, let $e_i \in H^{r(i)}(M)$ such that the set of e_i with $r(i) = k$ is a basis of $H^k(M) = H^k(M, \mathbf{Q})$. Let $e_i^* \in H_{r(i)}(M)$ such that $\langle e_i, e_j^* \rangle = \delta_{i,j}$. Then we have $U_M = \Sigma a_{i,j}(e_i \times e_j)$ and $e(\tau) = \Delta^* U_M = \Sigma a_{i,j} e_i e_j$. Now we calculate with (6.5).

$$(-1)^{d(e_k)} = (-1)^{d(e_k)} \langle e_k, e_k^* \rangle$$

$$= \Sigma a_{i,j} \langle e_i \times e_j, e_k^* \times De_k \rangle$$

$$= \Sigma a_{i,j} \langle e_i, e_k^* \rangle \langle e_j, e_k \cap \omega_M \rangle$$

$$= \sum_j a_{k,j} \langle e_j e_k, \omega_M \rangle$$

Now adding over the index k we have

$$\chi(M) = \Sigma(-1)^{d(ek)} = \Sigma a_{k,j} \langle e_j e_k, \omega_M \rangle$$

$$= \langle \Delta^* \alpha^*(U_M), \omega_M \rangle$$

$$= \langle \Delta^*(U_M), \omega_M \rangle$$

The last equality follows from $\alpha \Delta = \Delta$ where $\alpha(x, y) = (y, x)$. This proves the theorem.

7.3 Corollary. *Let M be an orientable manifold with an everywhere-nonzero vector field. Then we have $\chi(M) = 0$.*

Proof. By 16(8.3), we have $e(\tau(M)) = 0$, and by (7.2), we have $\chi(M) = 0$.

7.4 Remark. The relations $\chi(S^{2n}) = 2$ and $\chi(S^{2n+1}) = 0$. This yields another proof of part of 12(1.4): that is, an even sphere cannot have an everywhere-nonzero vector field.

8. Wu's Formula for the Stiefel-Whitney Classes of a Manifold

All coefficients in this section are in Z_2 and the results will apply to non-orientable closed connected manifolds.

8.1 Definition. The Stiefel-Whitney class of a manifold M, denoted $\omega(M)$, is the Stiefel-Whitney class $w(\tau(M))$ of its tangent bundle.

Let $Sq = \sum_{0 \leq i} Sq^i$ denote the Steenrod operation in Z_2 cohomology, and let $Sq^T : H_*(X) \to H_*(X)$ be the transpose in homology; that is, Sq^T is defined by the relation $\langle Sqa, b \rangle = \langle a, Sq^T(b) \rangle$. As before, $D : H^k(M) \to H_{n-k}(M)$ is the duality morphism.

8.2 Theorem (Wu). *Let M^n be a closed manifold. Then the class $v = D^{-1} Sq^T(\omega_M)$, called the Wu class of M, has the property that $\langle av, \omega_M \rangle = \langle Sqa, \omega_M \rangle$ for each $a \in H^*(M)$, and moreover we have $w(M) = Sqv$.*

Proof. For the first relation observe that $Dv = Sq^T \omega_M$ implies the following

$$\langle a, Sq^T \omega_M \rangle = \langle a, Dv \rangle$$

or $\qquad\qquad\qquad \langle Sqa, \omega_M \rangle = \langle a, v \cap \omega_M \rangle$

$$= \langle av, \omega_M \rangle$$

This proves the first relation.

For the second, recall the $SqU = \phi(w(M)) = (\pi^* w(M))U$ by 16(9.1). Then we have $SqU' = (h_t^*)^{-1}[(\pi^* w(M))U] = [(h_t^*)^{-1}\pi^* w(M)][(h_t^*)^{-1}U]$ and $SqU_M = U_M(w(M) \times 1)$ using the notation of (6.2). By (6.5), we derive the following relations:

$$\langle w(M), b \rangle = \langle U_M(w(M) \times 1), b \times \omega_M \rangle = \langle SqU_M, b \times \omega_M \rangle$$

$$= \langle U_M, Sq^T(b \times \omega_M)) \rangle = \langle U_M, Sq^T b \times Sq^T \omega_M \rangle$$

$$= \langle U_M, Sq^T b \times Dv \rangle$$

$$= \langle v, Sq^T b \rangle$$

$$= \langle Sqv, b \rangle$$

In the above relation, we used the transpose of the Cartan formula $Sq^T(b_1 \times b_2) = Sq^T(b_1) \times Sq^T(b_2)$. Since $\langle w(M), b \rangle = \langle Sqv, b \rangle$ holds for all $b \in H_*(M)$, we deduce the formula $w(M) = Sqv$. This proves the theorem.

8.3 Corollary. *The Stiefel-Whitney classes of closed manifolds are homotopy invariants of the manifold.*

8.4 Corollary. *The Wu class $v = \sum v_i$ of M where $v_i \in H^i(M)$ has the property that $v_i = 0$ for $2i > \dim M$.*

Proof. Recall that $Sq^i a_k = 0$ for $a_k \in H^k(M)$ and $i > k$. Now we use the relation $\langle av, \omega_M \rangle = \langle Sqa, \omega_M \rangle$.

9. Stiefel-Whitney Numbers and Cobordism

Let $w(M)$ denote the total Stiefel-Whitney class and ω the orientation of a manifold M. A monomial $w_1^{r(1)} \cdots w_n^{r(n)}$ is of degree $r(1) + 2r(2) + \cdots + nr(n)$.

9.1 Definition. The Stiefel-Whitney number of a manifold M corresponding to a monomial $\mu = w_1^{r(1)} \cdots w_n^{r(n)}$ of degree n is the number $\langle \mu, \omega \rangle \in Z_2$.

An n-dimensional manifold M^n has one Stiefel-Whitney number for each sequence $r(1), \ldots, r(n)$ with $n = r(1) + 2r(2) + \cdots + nr(n)$. For calculations of Stiefel-Whitney numbers we use the next theorem of Pontrjagin.

9.2 Theorem. *Let M^n be a compact manifold which is the boundary of a compact manifold W^{n+1}. Then all the Stiefel-Whitney numbers of M^n are zero.*

Proof. Let ω be the Z_2-orientation class of W, where $\omega \in H_{n+1}(W, M)$. Then $\partial\omega$ is the Z_2-orientation class in $H_n(M)$. For a monomial $\mu = w_1^{r(1)} \cdots w_n^{r(n)}$ there is the relation $\langle \mu, \partial w \rangle = \langle \delta\mu, \omega \rangle$. Note this is true on the chain and cochain level and extends to boundary and coboundary morphisms.
   Since $\tau(W)|M = \tau(M) \oplus \theta^1$, by (2.8), we have $i^*(w_i(\tau(W))) = w_i(\tau(M)) = w_i(M)$. Since $\delta i^* = 0$ in the exact sequence $H^n(W) \overset{i^*}{\to} H^n(M) \overset{\delta}{\to} H^{n+1}(W, M)$, we have $\delta(w_1^{r(1)} \cdots w_n^{r(n)}) = 0$. Therefore, all the Stiefel-Whitney numbers of M are zero.

   The next proposition provides a useful criterion for a manifold to be the boundary of another manifold.

9.3 Proposition. *Let $T: M \to M$ be a fixed point free differentiable involution of M, that is, $T^2 = 1$ and $T(x) \neq x$ for all $x \in M$. Then there exists a manifold W with $\partial W = M$.*

Proof. Form $M \times [-1, +1]$, and identify (x, t) with $(T(x), -t)$. The resulting space W is a manifold, and $\partial W = M$. The manifold structure of W arises from that of $M \times [-1, +1]$ and the fact that T is a local diffeomorphism.

9.4 Corollary. *The sphere S^n equals ∂B^{n+1}, and all Stiefel-Whitney numbers of S^n are zero. Also, $x \mapsto -x$ is an involution of the type in (9.3).*

9.5 Corollary. *Odd-dimensional real projective RP^{2n+1} equals ∂W, and all Stiefel-Whitney numbers are zero.*

Proof. For $z = (z_0, \ldots, z_n) \in S^{2n+1}$ we define $R(z) = iz$, where $i^2 = -1$. Then $R^2(z) = -z$, and R induces $T: RP^{2n+1} \to RP^{2n+1}$, where $T^2 = 1$ and $T(x) \neq x$ for all $x \in RP^{2n+1}$ because $R(-z) = -R(z)$. This yields the corollary.

9.6 Definition. Two manifolds M^n and N^n are in the same cobordism class, provided $M^n \cup N^n = \partial W^{n+1}$.
   The relation of manifolds in the same cobordism class is an equivalence relation. Theorem (9.2) now takes the following form.

9.7 Theorem. *Two manifolds in the same cobordism class have the same Stiefel-Whitney numbers.*

Proof. The Stiefel-Whitney numbers of $M^n \cup N^n$ are the sum of the Stiefel-Whitney numbers of M and N because $\omega_{M \cup N} = \omega_M + \omega_N$. This proves the theorem.

9.8 Remark. Thom proved the converse of (9.2) and (9.7), which means that the cobordism class of a manifold is determined by its Stiefel-Whitney numbers.

10. Immersions and Embeddings of Manifolds

10.1 Definition. For a real vector bundle ξ, the dual Stiefel-Whitney class $\bar{w}(\xi)$ of ξ is the unique element of $H^*(B(\xi), Z_2)$ with the property that $w(\xi)\bar{w}(\xi) = 1$. The dual Stiefel-Whitney class of a manifold M, denoted $\bar{w}(M)$, is $\bar{w}(\tau(M))$.

The notion of the dual Stiefel-Whitney class is useful, because if v is the normal bundle for an immersion $M \to \mathbf{R}^p$ then $w(v) = w(M)$.

10.2 Theorem. *Let M^n be a manifold. If M^n can be immersed in $\mathbf{R}^{n+k}$, then $\bar{w}_i(M) = 0$ for $i > k$. If M^n can be embedded in $\mathbf{R}^{n+k}$, then $\bar{w}_i(M) = 0$ for $i \geq k$, and if, in addition, M^n is embedded in $\mathbf{R}^{n+k}$ with an orientable normal bundle v, then $e(v) = 0$ [besides $w_i(v) = 0$ for $i \geq k$].*

Proof. For the first part, the immersion yields a normal bundle v of dimension k, where $\tau(M) \oplus v$ is trivial. Therefore, we have $0 = w_i(v) = \bar{w}_i(M)$ for $i > k$.

For the statement about embeddings, recall that $e(\xi) = s^*i^*(U_\xi)$ or $w^k(\xi) = s^*i^*(U_\xi)$ mod 2 for dim $\xi = k$, where $B \xrightarrow{s} E(\xi) \xrightarrow{i} (E(\xi), E_0(\xi))$ by 17(7.4) and 17(9.1). Choosing a riemannian metric on ξ and a prolongation $f: E(\varepsilon) \to \mathbf{R}^{n+k}$ of the embedding $f: M \to \mathbf{R}^{n+k}$, we have the following diagram of maps.

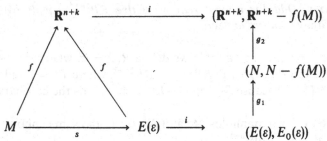

Here g_1 is a diffeomorphism, g_2 is excision, $g = g_2 g_1$, and s is the zero cross section. Then $e(v)$ equals $s^*i^*(U_v) = f^*i^*g^{*-1}(U_v)$. Since $H^k(\mathbf{R}^{n+k}) = 0$, we have $f^* = 0$ and $e(v) = 0$ or $w^k(v) = 0$. This proves the theorem.

Let x be the generator of $H^*(RP^n, Z_2)$. Then $w(RP^n) = (1 + x)^{n+1}$ and $w_i(RP) = \binom{n+1}{i}_2 x^i$ for $0 \leq i \leq n$ and $w_i(RP^n) = 0$ for $i > n$. Note that $w(RP^n) = 1$ if and only if $n + 1 = 2^r$ for some r. Then we have $\bar{w}(RP^n) = 1$. If $\bar{w}(RP^n) = 1 + x$, then $(1 + x)^{n+2} = 1$ mod 2 or $w(RP^n) = 1 + x + \cdots + x^n$. Note that $\bar{w}(RP^n) = 1 + x$ if and only if $n + 2 = 2^r$ for some r.

When $n = 2^r$, then $w(RP^n) = (1 + x)^{2^r}(1 + x) = 1 + x + x^n$ and $\tilde{w}(RP^n) = (1 + x + x^n)^{-1} = (1 + x^n)(1 + x + \cdots + x^n) = 1 + x + \cdots + x^{n-1}$. Therefore, $\tilde{w}_i(RP^n) \neq 0$ for $0 \leq i \leq n - 1$ and $\tilde{w}_i(RP^n) = 0$ for $n \leq i$.

In summary, we have the following theorem.

10.3 Theorem. *If there is an immersion of RP^n into $\mathbf{R}^{n+1}$, then $n = 2^r - 1$ or $n = 2^r - 2$ for some r. If there is an embedding of RP^n into $\mathbf{R}^{n+1}$, then $n = 2^r - 1$ for some r. For $n = 2^r$, there is no immersion of RP^n into $\mathbf{R}^{2n-2}$ and no embedding into $\mathbf{R}^{2n-1}$.*

10.4 Corollary. *Let $n = 2^r + q$, where $0 \leq q < 2^r$. Then there is no immersion of RP^n into $\mathbf{R}^m$ for $m = 2^{r+1} - 2$ and no embedding into $\mathbf{R}^k$ for $k = 2^{r+1} - 1$.*

Proof. Apply (10.3) to $RP^{2^r} \subset RP^n$.

10.5 Remark. A theorem of Whitney [1] says for every M^n there exists an immersion of M into $\mathbf{R}^{2n-1}$ and there exists an embedding of M into $\mathbf{R}^{2n}$.

Exercises

1. Prove that if $\tau(RP^n)$ is trivial then $n - 1 = 2^r$ for some r.

2. Calculate directly the Stiefel-Whitney numbers of RP^n for $n \leq 10$, $n = 2^r$, and $n = 2r + 1$.

3. Define Pontrjagin numbers for an orientable real manifold. Define orientable cobordism, and prove that all the Pontrjagin numbers of a manifold M^n are zero if $M^n = \partial W^{n+1}$. Observe that M^n has nonzero Pontrjagin numbers only if $n = 0$ (mod 4).

4. Prove that the tangent bundles to S^n and RP^n as described in 2(2.1) and 2(2.6) are tangent bundles in the sense of definition (2.1).

5. Prove that $w_1(M) = \cdots = w_r(M) = 0$ implies $v_1 = \cdots = v_r = 0$ and $w_{r+1}(M) = v_{r+1}$ for the Wu class $v = \Sigma v_i$ of a closed manifold.

6. Prove that $w_1(M) = \cdots = w_r(M) = 0$ where dim $M = 2r$ or $2r + 1$ implies that $w(M) = 0$ for a closed manifold M.

7. For RP^n prove that $v_i = \binom{n-i}{i} x^i$ where $H^*(RP^n) = Z_2[x]$. From this calculate $w(RP^n)$.

CHAPTER 19

Characteristic Classes and Connections

Apart from the previous chapter, the theory of fibre bundles in this book is a theory over an arbitrary space. Even the relation to manifolds in Chapter 18 is treated from a topological point of view, but in the context of smooth manifolds and vector bundles we can approach Chern classes using constructions from analysis. This idea, which goes back to a letter from A. Weil (see A. Weil *Collected papers*, Volume III, pages 422–36 and 571–574), involves choosing a connection or covariant derivative on the complex vector bundle, defining the curvature 2-form of the connection, and representing the characteristic class as closed $2q$-form which is a polynomial in the curvature form. This proceedure is outlined in this chapter.

1. Differential Forms and de Rham Cohomology

The characteristic class of smooth bundles on a smooth manifold can be described as analytic representatives of cohomology classes. This is the theory of de Rham, and we give a brief explanation of those parts of the theory needed for an analytic approach to characteristic classes.

1.1 Notation. Let M denote a smooth manifold. To denote the fact that it has dimension n, we will write M^n. For a smooth vector bundle E over M we denote the space of smooth cross sections by $\Gamma(E)$. The tangent bundle to M is denoted $T(M)$ and the cotangent bundle by $T^*(M)$. Then $\Lambda^q T^*(M)$ denotes the qth exterior power of the cotangent bundle, and the space of q-forms with values in a smooth vector bundle is denoted $A^q(M, E)$. It equals the space of cross sections $\Gamma(\Lambda^q T^*(M) \otimes E)$. When E is the trivial line bundle, then

$A^q(M)$ denotes $A^q(M, E) = \Gamma(\Lambda^q T^*(M))$, the space of q-forms on M. On the vector space of q-forms with values in the trivial bundle we have a differential operator which is called exterior differentiation.

1.2 Definition. For a smooth manifold M the exterior derivative $d: A^q(M) \to A^{q+1}(M)$ is given locally by the formula

$$d(a(x_1, \ldots, x_n) \, dx_{n(1)} \wedge \cdots \wedge dx_{n(q)})$$

$$= \sum_{1 \le i \le n} \frac{\partial}{\partial x_i} a(x_1, \ldots, x_n) \, dx_i \wedge dx_{n(1)} \wedge \cdots \wedge dx_{n(q)}.$$

As usual, the relation $dx_i \wedge dx_j = -dx_j \wedge dx_i$ is incorporated in the formula for the exterior derivative. There is a global axiomatic characterization of the exterior derivative d.

1.3 Proposition. *For a smooth manifold M there is a unique sequence of linear $d: A^q(M) \to A^{q+1}(M)$ such that*

(1) *$d(\omega\omega') = (d\omega)\omega' + (-1)^q \omega(d\omega')$ where $\omega \in A^q(M)$, $\omega \in A^{q'}(M)$,*
(2) *the composite $dd = 0$, and*
(3) *$d: A^0(M) \to A^1(M)$ is given by $df(\xi) = \xi(f)$ for any $\xi \in \mathrm{Vect}(M)$, the Lie algebra of vector fields on M.*

The proof comes from the defining formula and the local version of condition (3), namely, that $df = \sum_{1 \le i \le n} \frac{\partial f}{\partial x_i}$.

1.4 Proposition. *Let $g: M \to N$ be a smooth mapping of manifolds. For each natural number q there is a unique linear map $A^q(g): A^q(N) \to A^q(M)$ such that*

(1) *$A^{q+q}(g)(\omega\omega') = A^q(g)(\omega)A^{q'}(g)(\omega')$ where $\omega \in A^q(N)$, $\omega \in A^{q'}(N)$,*
(2) *$dA^q(g)(\omega) = A^{q+1}(g)(d\omega)$ for $\omega \in A^q(N)$,*
(3) *$A^0(g)(f) = fg$ for a smooth function f on N.*

The proof comes from the defining formula for differential forms and defining $A^q(g)(\omega)$ locally using (3) and the relation $A^1(g)(df) = d(fg)$ for a smooth coordinate function f on N coming from (2) and (3). If $q > 1$, then we use (1). The idea of the proof of the proposition is similar to that of (1.3).

1.5 Definition. The de Rham cohomology vector space is

$$H^q_{DR}(M) = \ker(d: A^q(M) \to A^{q+1}(M))/\mathrm{im}(d: A^{q-1}(M) \to A^q(M)).$$

We can use either real valued forms giving $H^*_{DR}(M, \mathbf{R})$ or complex valued forms giving $H^*_{DR}(M, \mathbf{C})$. Let $g: M \to N$ be a smooth mapping. The induced mapping $H^*_{DR}(g): H^*_{DR}(N) \to H^*_{DR}(M)$ is defined by the class of $H^*_{DR}(g)(\omega)$ equals the class of $A^*(\omega)$ where $d\omega = 0$.

There is a direct argument that shows $H^*_{DR}(g): H^*_{DR}(N) \to H^*_{DR}(M)$ is a well-defined linear mapping. In general, de Rham cohomology is isomorphic to singular cohomology, sheaf and Čech cohomology, and for a compact, orientable manifold M it satisfies Poincaré duality. All of this for coefficients in $\mathbf{R}$ or in $\mathbf{C}$.

1.6 Comparison with Singular Cohomology. This is done by Stokes' formula

$$\int_{\partial c} \omega = \int_c d\omega$$

where c is a smooth $(q + 1)$-chain with boundary ∂c and ω is a q-form. Integration defines a pairing $(c, \omega) \mapsto \int_c \omega$, and Stokes' formula is just the assertion that ∂ on chains and d on forms are adjoint to each other under the integration pairing of chains and forms. It is a theorem: This pairing given by integration induces a perfect pairing $H_q(M) \otimes H^q_{DR}(M) \to \mathbf{R}$ which implies that $H^q_{DR}(M) = H_q(M)^\vee$, the dual to singular homology which is singular cohomology.

1.7 Poincaré Duality for de Rham Cohomology. Let M be a closed oriented manifold which means, firstly, that integration over the manifold is well defined. If $\dim(M) = n$, then given a p-form ω and a q-form ω' with $p + q = n$ so that $\omega \wedge \omega'$ is an n-form, and thus we can define a pairing $A^p(M) \otimes A^q(M) \to R$ given by

$$(\omega, \omega') \mapsto \int_M \omega \wedge \omega' = [\omega, \omega'].$$

Since the boundary of M is empty, we have $[d\theta, \theta'] = (-1)^{p+1}[\theta, d\theta']$ for a p-form θ and a q-form θ' with $p + q = n - 1$. Since d is either selfadjoint or skewadjoint with respect to this pairing, it follows that a pairing is induced on de Rham cohomology

$$H^p_{DR}(M) \otimes H^{n-p}_{DR}(M) \to \mathbf{R}.$$

We conclude this section with an integration formula on the product manifold $M \times [0, 1]$ for a closed manifold M. This is used to prove a homotopy formula as well as in the following sections to study homotopy properties of connections.

1.8 Remark. Each differential form $\omega \in A^q(M \times [0, 1])$ can be decomposed as $\omega = \alpha + \beta \wedge dt$ where $\alpha \in A^q(M \times [0, 1])$, $\beta \in A^{q-1}(M \times [0, 1])$ each without any dt factor. Moreover, d decomposes as

$$d = d_x + d_t : A^q(M \times [0, 1]) \to A^{q+1}(M \times [0, 1])$$

where $d\omega = (d_x + d_t)(\alpha + \beta \wedge dt) = d_x\alpha + (d_x\beta) \wedge dt + d_t\alpha$ since $dt \wedge dt = 0$. We define a linear operator $Q: A^q(M \times [0, 1]) \to A^{q-1}(M)$ to be integration in t from $t = 0$ to $t = 1$ with a sign by the formula

$$Q(\omega) = Q(\alpha + \beta \wedge dt) = (-1)^{q-1} \int_0^1 \beta\, dt.$$

1.9 Proposition. (Homotopy formula) *With the above notations we have*

$$dQ + Qd = j_1 - j_0$$

as maps $A^q(M \times [0,1]) \to A^q(M)$ *where* $j_s(\alpha + \beta\, dt) = \alpha|_{t=s}$ *for* $s \in [0,1]$.

Proof. Since d_x and $\int dt$ commute, it follows that the sign in the definition of Q gives $Qd_x + d_x Q = 0$. Now we consider $d_t Q + Qd_t$ on each summand of $\omega = \alpha + \beta\, dt$. Firstly, $(d_t Q + Qd_t)(\beta\, dt) = 0 = j_1(\beta\, dt) - j_0(\beta\, dt)$. Now we decompose $\alpha = \sum_I a_I(x,t)\, dx_I$ with $I = \{i(1) < \cdots < i(q)\}$ and $dx_I = dx_{i(1)} \wedge \cdots \wedge dx_{i(q)}$. Using the fundamental theorem of calculus, we calculate the following expression

$$(d_t Q + Qd_t)(\alpha) = Qd_t(\alpha)$$

$$= Q\left(\sum_I \frac{\partial}{\partial t} a_I(x,t)\, dt\, dx_I\right)$$

$$= \sum_I \int_0^1 \left\{\frac{\partial}{\partial t}\{a_I(x,t)\}\, dx_I\right\} dt$$

$$= \sum_I a_I(x,1)\, dx_I - \sum_I a_I(x,0)\, dx_I$$

$$= j_1(\alpha) - j_0(\alpha).$$

Adding up the various results, we obtain the homotopy formula

$$dQ(\omega) + Qd(\omega) = j_1(\omega) - j_0(\omega).$$

This proves the proposition.

1.10 Corollary. *If* $f_0, f_1: M \to N$ *are two smooth maps such that there is a smooth homotopy* $h: M \times [0,1] \to N$ *with* $h(x,0) = f_0(x)$ *and* $h(x,1) = f_1(x)$, *then* f_0 *and* f_1 *induce the same map on de Rham cohomology, that is,* $H^*_{DR}(f_0) = H^*_{DR}(f_1): H^*_{DR}(N) \to H^*_{DR}(M)$.

2. Connections on a Vector Bundle

Now we study the possibility of defining an operator

$$A^q(M, E) \to A^{q+1}(M, E)$$

which is like the exterior derivative when E is the trivial bundle. As with the exterior derivative, we start with $q = 0$.

2.1 Definition. A connection ∇ on a smooth vector bundle E over a smooth manifold M is a complex linear map $\nabla: A^0(M, E) = \Gamma(E) \to \Gamma(T^*(M) \otimes E) = A^1(M, E)$ such that

$$\nabla(fs) = f\nabla(s) + df \otimes s \qquad \text{for all } s \in A^0(M, E), f \in A^0(M).$$

2.2 Theorem. *Let ∇ be a connection on a smooth vector bundle E over M. Then there exists a unique extension of ∇ to a linear map $\nabla: A^q(M, E) \to A^{q+1}(M, E)$ for each $q \geq 0$ and a two form, called the curvature form, $K \in A^2(M, \operatorname{End}(E))$ such that*

(1) $\nabla(\alpha \otimes s) = d\alpha \otimes s + (-1)^p \alpha \wedge \nabla s$ *for all $\alpha \in A^p(M), s \in A^q(M, E)$, and*
(2) $\nabla\nabla(s) = Ks$ *for any $s \in A^q(M, E)$.*

Proof. The existence of the extensions will follow the lines of ideas in (1.2) and (1.3). We calculate locally over an open subset U of M where E has a basis of cross sectons $s_1, \ldots, s_q$ and the connection takes the values

$$\nabla s_i = \sum_{1 \leq j \leq q} \theta_{i,j} \otimes s_j \qquad \text{where } \theta_{i,j} \in A^1(M).$$

This leads to the value of $\nabla\nabla$ on these cross sections locally on U

$$\nabla\nabla(s_i) = \sum_{1 \leq k \leq q} \nabla(\theta_{i,k} \otimes s_k)$$

$$= \sum_{1 \leq k \leq q} \left\{ d\theta_{i,k} - \sum_{1 \leq j \leq q} \theta_{i,j} \wedge \theta_{j,k} \right\} \otimes s_k$$

$$= \sum_{1 \leq k \leq q} K_{i,k} \otimes s_k.$$

This defines K on open sets with a chart for the bundle as a matrix of differential forms

$$K_{i,j} = d\theta_{i,j} - \sum_{1 \leq k \leq q} \theta_{i,k} \wedge \theta_{k,j}.$$

In order to see that K is globally defined we must consider a second basis of cross sections $s'_1, \ldots, s'_q$ and express them in terms of the first basis by a relation

$$s'_i = \sum_{1 \leq j \leq q} a_{i,j} s_j \qquad \text{where } (a_{i,j}) = A \in M_q(A^0(M)).$$

In the language of matrices we have the next four relations

$$\nabla s = \theta_s, \quad \nabla s' = \theta' s', \quad s' = As, \quad \text{and} \quad \theta' = A\theta A^{-1} + (dA)A^{-1}. \quad (1)$$

Only the last relation needs some explanation, as it follows from the first two by the computation

$$\theta' As = \theta' s' = \nabla s' = \nabla(As) = A\nabla s + (dA)s = (A\theta + dA)s.$$

Thus, the last relation follows by multiplying by A^{-1} on the right.

Locally, the curvature form is either

$$K = d\theta - \theta \wedge \theta \qquad \text{or} \qquad K' = d\theta' - \theta' \wedge \theta'$$

and the relation between two matrices of two forms is given by

$$K' = AKA^{-1}. \tag{2}$$

To establish this relation, which shows that K is a well-defined element in $A^2(M, \text{End}(E))$, we use $d(A^{-1}) = -A^{-1}(dA)A^{-1}$ and calculate the curvature expression K' in terms of K using formula (1)

$$
\begin{aligned}
K' &= d\theta' - \theta' \wedge \theta' \\
&= d(A\theta A^{-1} + (dA)A^{-1}) - (A\theta A^{-1} + (dA)A^{-1}) \wedge (A\theta A^{-1} + (dA)A^{-1}) \\
&= (dA)\theta A^{-1} + A(d\theta)A^{-1} - A\theta(dA^{-1}) - (dA) \wedge (dA^{-1}) - [A\theta \wedge \theta A^{-1} \\
&\quad + (dA) \wedge \theta A^{-1} + A\theta A^{-1}(dA)A^{-1} + (dA^{-1})A^{-1}(dA)A^{-1}] \\
&= A(d\theta - \theta \wedge \theta)A^{-1} \\
&= AKA^{-1}.
\end{aligned}
$$

This establishes the conjugate relation between two local curvature matrices. This proves the theorem.

3. Invariant Polynomials in the Curvature of a Connection

The relation $K' = AKA^{-1}$ is the motivation for considering polynomials $P(x_{1,1}, \ldots, x_{n,n}) = P(X)$ in n by n matrices X such that

$$P(AXA^{-1}) = P(X),$$

for such polynomial expressions in the locally-defined curvature forms will define global forms on the manifold.

3.1 Elementary symetric functions. The permutation group Sym_n on n objects permutes the varibles of $k[x_1, \ldots, x_n]$ preserving the k-submodule $k[x_1, \ldots, x_n]_q$ of homogeneous polynomials of degree q. The subalgebra of polynomials invariant under the symmetric group is a polynomial algebra on certain elementary sysmmetric functions denoted σ_q of degree q. Thus, we have

$$k[x_1, \ldots, x_n] \supset k[x_1, \ldots, x_n]^{\text{Sym}_n} = k[\sigma_1, \ldots, \sigma_n]$$

where $\sigma_q(x) \in k[x_1, \ldots, x_n]$ is defined by either of the following relations:

$$Q_x(t) = \prod_{1 \le j \le n} (1 + x_j t) = \sum_{0 \le q \le n} \sigma_q(x_1, \ldots, x_n) t^q, \qquad \text{or}$$

$$\sigma_q(x_1, \ldots, x_n) = \sum_{i(1) < \cdots < i(q)} x_{i(1)} \cdots x_{i(q)}.$$

3.2 Notation. Again, using the above polynomial $Q_x(t)$, we introduce some new polynomials $e_q(\sigma_1, \ldots, \sigma_n)$ by the following formal relation in characteristic zero

$$-t\frac{d}{dt}\log Q_x(t) = \sum_{q \geq 1} e_q(\sigma_1, \ldots, \sigma_n)(-t)^q.$$

See also 14(1.8), where the notation is slightly different.

By conjugating with matrices, we define an action of the general linear group $GL_n(k)$ on the algebra of polynomials in the n^2 matrix elements $k[x_{1,1}, \ldots, x_{n,n}]$.

3.3 Remark. The subalgebra of polynomials invariant under conjugation by $GL_n(k)$ is a polynomial algebra over a field of characteristic 0 in elements $c_1(x), \ldots, c_n(x)$, that is,

$$k[x_{1,1}, \ldots, x_{n,n}] \supset k[x_{1,1}, \ldots, x_{n,n}]^{GL_n(k)} = k[c_1(x), \ldots, c_n(x)]$$

where the polynomials $c_q(x)$ are given by the following relations:

$$R_x(t) = \det(I + Xt) = \sum_{0 \leq q \leq n} c_q(x)t^q.$$

In the case where X is diagonal with $x_{i,j} = \delta_{i,j}\lambda_i$, the expression for $c_q(x)$ is an elementary symmetric function, that is,

$$c_q(x_{1,1}, \ldots, x_{n,n}) = \sum_{i(1) < \cdots < i(q)} \lambda_{i(1)} \cdots \lambda_{i(q)}.$$

3.4 Remark. Assume that k is a field of characteristic zero. We have the relations

$$-t\frac{d}{dt}\log(\det(I + Xt)) = \sum_{1 \leq q} \mathrm{Tr}(X^q)(-t)^q$$

or equivalently

$$\det(I + Xt) = \exp\left(-\sum_{1 \leq q} \mathrm{Tr}(X^q)\frac{(-t)^q}{q}\right).$$

The following algebras of polynomials are equal

$$k[c_1(x), \ldots, c_n(x)] = k[\mathrm{Tr}(X), \mathrm{Tr}(X^2), \ldots, \mathrm{Tr}(X^n)].$$

We denote this algebra by $\mathrm{Inv}(n)$, and the subspace of elements homogeneous of degree q by $\mathrm{Inv}_q(n)$. The algebra $\mathrm{Inv}(n)$ is a direct sum of the subspaces $\mathrm{Inv}_q(n)$, and we have $c_q(x)$, $\mathrm{Tr}(X^q) \in \mathrm{Inv}_q(n)$.

With the invariant polynomials $c_q \in k[x_{1,1}, \ldots, x_{n,n}]$ we introduce the characteristic differential forms of a smooth vector bundle E with connection ∇.

3.5 Definition. The characteristic Chern forms $c_q(E, \nabla)$ of a smooth vector bundle E with connection ∇ are defined to be

$$c_q(E, \nabla) = \frac{1}{(2\pi i)^q} c_q(K).$$

For any invariant polynomial $\phi(x) \in \text{Inv}(n) = k[x_{1,1}, \ldots, x_{n,n}]^{GL_n(k)}$ we denote the characteristic form with K substituted into ϕ by $\phi(K)$.

By (2.2)(2) for two curvature matrices K and K' in two local coordinate systems, we have $K' = AKA^{-1}$ and thus $c_q(K) = c_q(K')$. Also more generally $\phi(K) = \phi(K')$ is a well-defined global form for any invariant polynomial. To prove that the characteristic Chern forms and more generally $\phi(K)$ are closed, we use the following formula.

3.6 Proposition. (Bianchi identity) *If θ is a local connection form with curvature $K = d\theta - \theta \wedge \theta$, then we have*

$$dK = \theta \wedge K - K \wedge \theta = [\theta, K].$$

Proof. We calculate

$$dK = d\, d\theta - d\theta \wedge \theta + \theta \wedge d\theta = -(K + \theta \wedge \theta) \wedge \theta + \theta \wedge (K + \theta \wedge \theta)$$

$$= \theta \wedge K - K \wedge \theta = [\theta, K].$$

This proves the proposition.

3.7 Proposition. *The characteristic Chern forms $c_q(E, \nabla)$ of a smooth n-dimensional vector bundle E with connection ∇ are closed forms. Also $\phi(K)$ is a closed form for each invariant polynomial $\phi(x) \in \text{Inv}(n)$.*

Proof. Firstly, we calculate using the derivation property of $\alpha \mapsto [\theta, \alpha]$ on even forms α

$$d\,\text{Tr}(K^q) = \sum_{i+j=q-1} \text{Tr}(K^i(dK)K^j)$$

$$= \sum_{i+j=q-1} \text{Tr}(K^i[\theta, K]K^j)$$

$$= \text{Tr}([\theta, K^q])$$

$$= 0.$$

Since $\text{Inv}(n) = k[c_1(x), \ldots, c_n(x)] = k[\text{Tr}(X), \text{Tr}(X^2), \ldots, \text{Tr}(X^n)]$ is the algebra of all invariant polynomials, we see that the characteristic Chern forms and all $\phi(K)$ are in a polynomial algebra which is generated by closed forms $\text{Tr}(K^q)$, and therefore, they are also closed forms. This proves the proposition.

3.8 Example. In the algebra of conjugation invariant polynomial functions on n by n matrices

$$k[c_1(X), \ldots, c_n(X)] = k[\text{Tr}(X), \text{Tr}(X^2), \ldots, \text{Tr}(X^n)],$$

we have the relation $c_1(X) = \text{Tr}(X)$ corresponding to the sum of the eigenvalues. Another relation is

$$\text{Tr}(X^2) = c_1(X)^2 - 2c_2(X).$$

This can be seen by squaring the sum of the eigenvalues and subtracting off the sum of products of distinct eigenvalues. In the case of $n = 2$, the Hamilton-Cayley theorem says that

$$X^2 - \text{Tr}(X)X + \det(X)I = 0,$$

and taking the trace of this relation, we obtain

$$\text{Tr}(X^2) = \text{Tr}(X)^2 - 2\det(X)$$

which is another version of this relation. This leads to another interesting characteristic class of bundles

$$a_2(E) = c_1(E)^2 - 2c_2(E).$$

4. Homotopy Properties of Connections and Curvature

4.1 Definition. Let E be a smooth vector bundle over a smooth manifold, and let ∇_0 and ∇_1 be two connections on E. A homotopy from ∇_0 to ∇_1 is a connection ∇ on $E \times [0, 1]$ over $M \times [0, 1]$ such that ∇ restricts to ∇_i on $E \times \{i\}$ over $M \times \{i\}$ for $i = 0, 1$.

4.2 Remark. We use the notations of the previous definition and consider an invariant polynomial $\phi \in \text{Inv}$. The form $\phi(K)$ coming from the curvature K of ∇ on $E \times [0, 1]$ over $M \times [0, 1]$ restricts to $\phi(K_i)$ on $E \times \{i\}$ over $M \times \{i\}$. In particular $\phi(K_0)$ and $\phi(K_1)$ determine the same de Rham cohomology class.

There is a standard way of constructing a homotopy between any two connections.

4.3 Definition. Let E be a smooth vector bundle over a smooth manifold M. Let ∇_0 and ∇_1 be two connections on E with local connection forms. The affine homotopy from ∇_0 to ∇_1 is the connection

$$\nabla = (1 - t)\nabla_0 + t\nabla_1$$

defined on the smooth bundle $E \times [0, 1]$ over the product $M \times [0, 1]$.

If the local connection forms for ∇_0 and ∇_1 are θ_0 and θ_1 respectively, then $\theta = (1 - t)\theta_0 + \theta_1$ is the local connection form for the affine homotopy ∇. The restriction of ∇ to $E \times \{t\}$ over $M \times \{t\}$ is denoted ∇_t and of θ and K to $B \times \{t\}$ are denoted θ_t and K_t.

4.4 Proposition. *The curvature form K_t for the affine homotopy $\nabla_t = (1 - t)\nabla_0 + t\nabla_1$ is given by*

$$K_t = (1 - t)K_0 + tK_1 + t(1 - t)(\theta_0 - \theta_1)^2 + (\theta_0 - \theta_1)\,dt$$

where θ_0 or θ_1 is the local connection form from ∇_0 to ∇_1.

Proof. Now calculate using $(1 - t) - (1 - t)^2 = t - t^2 = t(1 - t)$ from the definition

$$K_t = d\theta_t - \theta_t \wedge \theta_t$$

$$= (1 - t)\,d\theta_0 + t\,d\theta_1 + dt(-\theta_0 + \theta_1) - (1 - t)^2\theta_0 \wedge \theta_0 - t^2\theta_1 \wedge \theta_1$$

$$- t(1 - t)\{\theta_0 \wedge \theta_1 + \theta_1 \wedge \theta_0\}$$

$$= (1 - t)K_0 + tK_1 + t(1 - t)\{\theta_0 \wedge \theta_0 - \theta_0 \wedge \theta_1 - \theta_1 \wedge \theta_0 + \theta_1 \wedge \theta_1\}$$

$$+ (\theta_0 - \theta_1)\,dt$$

$$= (1 - t)K_0 + tK_1 + t(1 - t)(\theta_0 - \theta_1)^2 + (\theta_0 - \theta_1)\,dt.$$

This proves the proposition.

Now we wish to apply the homotopy formula $dQ + Qd = j_1 - j_0$ of (1.7) to a form coming from an invariant polynomial $\phi \in \text{Inv}(n)$ with K on $M \times [0, 1]$ substituted into ϕ. By (3.7) the form $\phi(K)$ is a closed form on $M \times [0, 1]$.

4.5 Definition. Let $\phi(x) \in \text{Inv}_q$ be an invariant homogeneous polynomial of degree q. Then for two connections ∇_0 and ∇_1 with curvatures K_0 and K_1 repectively, we form ∇ with curvature K and define

$$\phi(K_0, K_1) = Q(\phi(K)) \in A^{2q-1}(M) \qquad \text{as in (1.6).}$$

This leads to a more explicit version of (4.4).

4.6 Proposition. *Let ∇_0 and ∇_1 be two connections on a smooth n-dimensional vector bundle with curvature forms K_0 and K_1 respectively, and let $\phi \in \text{Inv}_q(n)$ be a homogeneous polynomial of degree q. Then the two 2q-forms $\phi(K_0)$ and $\phi(K_1)$ define the same de Rham cohomology class. In particular, the Chern forms $c_q(E, \nabla)$ define de Rham cohomology classes independent of the connection ∇ used to define these classes.*

Proof. Using the formula $dQ + Qd = j_1 - j_0$ of (1.7) applied to $\phi(K_0, K_1)$, we obtain $dQ(\phi(K_0, K_1)) = j_1(\phi(K_0, K_1)) - j_0(\phi(K_0, K_1)) = \phi(K_1) - \phi(K_0)$. Thus the difference between $\phi(K_0)$ and $\phi(K_1)$ is an exact form so that the de Rham cohomology classes $[\phi(K_0)] = [\phi(K_1)]$. In the special case of the Chern forms we have $[c_q(E, \nabla_0)] = [c_q(E, \nabla_1)]$. This proves the proposition.

4.7 Remark. We can denote $c_q(E)_{DR} = [c_q(E, \nabla)] \in H^{2q}_{DR}(M, \underline{C})$ where ∇ is any connection on the smooth vector bundle E. There is another Chern class $c_q(E) \in H^{2q}(M, \underline{Z})$ defined axiomatically in 16(3.2). There is a coefficient morphism $H^{2q}(M, \underline{Z}) \to H^{2q}_{DR}(M, \underline{C})$ in cohomology, and under this coefficient morphism $c_q(E)$ is mapped to $c_q(E)_{DR}$.

4.8 Reference. There are further considerations in the theory of connections. For example, there is the notion of universal connections. For this and related topics, see M. S. Narasimhan and S. Ramanan, *Existence of universal connections I and II*, American Journal of Mathematics, 83, 1961, pp. 563–572 and 85, 1963, pp. 223–231.

5. Homotopy to the Trivial Connection and the Chern-Simons Form

5.1 Remark. The curvature form K_t of (4.2) for $\nabla_t = (1 - t)\nabla_0 + t\nabla_1$ in the case of $\nabla_0 = 0$ takes the following form in terms of the connection form θ_1:

$$K_t = tK_1 + t(1 - t)\theta_1^2 + \theta_1 \, dt$$
$$= t \, d\theta_1 - t\theta_1^2 + t\theta_1^2 - t^2\theta_1^2 - \theta_1 \, dt$$
$$= t \, d\theta_1 - t^2\theta_1 \wedge \theta_1 - \theta_1 \, dt.$$

5.2 Proposition. *For* $K_t = t \, d\theta - t^2\theta \wedge \theta - \theta \, dt$ *on the product manifold* $M \times [0, 1]$ *the trace of the square is*

$$\mathrm{Tr}(K_t^2) = \mathrm{Tr}(2t\theta \wedge d\theta + 2t^2\theta \wedge \theta \wedge \theta) \, dt - \mathrm{Tr}(2t^3\theta \wedge \theta \wedge d\theta),$$

and when integrated from $t = 0$ to $t = 1$ we obtain

$$\int_0^1 \mathrm{Tr}(K_t^2) = \mathrm{Tr}(\theta \wedge d\theta) + \frac{2}{3}\mathrm{Tr}(\theta \wedge \theta \wedge \theta).$$

Proof. The first expression comes from the cross terms in the trace of a square. The integral dt gives contribution from the first two terms of the expression for $\mathrm{Tr}(K_t^2)$. This proves the proposition.

5.3 Definition. The three form $\mathrm{Tr}(\theta \, d\theta + \frac{2}{3}\theta^3)$ associated with a connection ∇ with connection form θ is called the Chern-Simon form of the connection.

There are other versions associated with $\mathrm{Tr}(K_t^n)$, but it is this 3-form that has many applications in the theory of three dimensional manifolds since its integral over the manifold is an invariant of the manifold mod **Z**.

The above considerations are special to the three-dimensional case. In the general case, the Chern-Simon's form for a flat vector bundle lives on the principal bundle.

5.4 Definition. Let (E, ∇) be a flat vector bundle over a manifold M, and let $q: \Pr(E) \to M$ denote the associated principal bundle of E over M. We consider two connections on the induced vector bundle $q^*(E)$ over the total space $\Pr(E)$ of the principal bundle: the induced flat connection ∇_0 from ∇ on E, and the canonical trivial connection ∇_1 on $q^*(E)$ coming from the canonical trivialization of $q^*(E)$ over the total space of the principal bundle. Again, the affine homotopy (4.3) defines a standard homotopy, and for each $\phi \in \mathrm{Inv}_q$ we have a canonical Chern-Simons form $CS(E, \nabla) \in A^{2q-1}(\Pr(E))$ on the principal bundle $\Pr(E)$ of E with

$$\phi(K_0) = d(CS(E, \nabla)).$$

Observe that the curvature form $K_0 = q^*(K)$ where K is the curvature form on M for the flat bundle (E, ∇) on M.

5.5 Remark. When the flat bundle (E, ∇) in the previous definition is also trivial, or equivalently, when the principal bundle $\Pr(E)$ has a cross section s, the Chern-Simons form can be pulled back to $s^*(CS(E, \nabla))$ of the base space M. This form depends on the fibre homotopy type of s.

5.6 Remark. In the case of an oriented 3 dimensional manifold M and its tangent bundle $E = T(M)$, we know that $T(M)$ is trivial and hence for each flatness structure ∇ on $T(M)$ we have a curvature class on $\Pr(E) = M \times SO(3)$. Now, consider the class

$$a_2(E) - c_1(E)^2 - 2c_2(E)$$

on the principal bundle. It is of the form $d(CS(M))$ where $CS(M)$ is a three form on the principal bundle. Now we induce $s^*(CS(M))$ to the base 3 dimensional manifold M giving a three form on M. By integrating this over M we obtain a real number which depends on s. Since $[M, SO(3)] = \mathbf{Z}$, we see that the exponential

$$\exp\left\{ 2\pi i \int_M s^*(CS(M)) \right\},$$

called the Chern-Simon invariant of the flatness structure on the 3 dimensional manifold, is a well-defined numerical invariant.

6. The Levi-Civita or Riemannian Connection

Every Riemannian manifold has a unique connection on its tangent bundle called the Levi-Civita or Riemannian connection. In this section we show that it exits and is unique.

6.1 Definition. A pseudo-riemannian metric g on a smooth real vector bundle E over a manifold M is an $A^0(M)$-bilinear map $g: \Gamma(E) \times \Gamma(E) \to A^0(M)$ such that

(1) $g(s, s') = g(s', s)$ for all $s, s' \in \Gamma(E)$, and
(2) there exists an isomorphism $\gamma: \Gamma(E) \to \Gamma(E^\vee)$ such that the section $\gamma(s)$ the dual $E^\vee$ to E applied to the section s' of E gives $\gamma(s)(s') = g(s, s')$.

Note that in 3(9.1) an inner product on topological vector bundles was considered fibre by fibre on the bundle itself. If such an inner product is smooth, then it defines a pseudo-riemannian metric on $\Gamma(E)$.

6.2 Definition. The covariant derivative asociated with a connection $\nabla: \Gamma(E) \to \Gamma(T^*(M) \otimes E)$ is the map $\xi \mapsto \nabla_\xi$ defined $\Gamma(T(M)) \to \mathrm{Hom}(\Gamma(E), \Gamma(E))$ such that $\nabla_\xi(s) = (\nabla(s))(\xi \otimes id)$.

6.3 Remark. The covariant derivative satisfies the following relations as a' function of $(\xi, s) \mapsto \nabla_\xi s$:

(1) For $\xi \in \Gamma(T(M))$ and $s \in \Gamma(E)$ the function $\nabla_\xi s$ is C-linear, and
(2) for a smooth function f we have $\nabla_{f\xi} s = f \nabla_\xi s$ and

$$\nabla_\xi(fs) = f\nabla_\xi s + \xi(f)s.$$

Conversely, given a covariant derivative $\nabla_\xi s$ we can define a connection $\nabla: \Gamma(E) \to \Gamma(T^*(M) \otimes E)$ by the relation $\nabla_\xi(s) = (\nabla(s))(\xi \otimes id)$. One way to see this is in local coordinates $x_1, \ldots, x_n$ giving local sections $\partial_1, \ldots, \partial_n$ of $T(M)$ and local sections $s_1, \ldots, s_q$ of E where $\nabla(s_i) = \sum_{1 \le j \le q} \theta_{i,j} s_j$ with $\theta_{i,j} = \sum_{1 \le k \le n} \Gamma_{i,j,k} dx_k$. Then the covariant derivative is given by the relation

$$\nabla_{\partial_k}(s_i) = \sum_{1 \le j \le q} \Gamma_{i,j,k} s_j.$$

Also, the covariant derivative determines the connection form.

6.4 Definition. A pseudo-riemannian metric g on a smooth bundle is compatible with a connection ∇ provided for the covariant derivative ∇_ξ associated with ∇ and three elements $\xi, \eta, \zeta \in \Gamma(T(M))$

$$\zeta g(\xi, \eta) = g(\nabla_\xi \xi, \eta) + g(\xi, \nabla_\zeta \eta).$$

6.5 Theorem. *Let M be a smooth manifold with a pseudo-riemannian metric g on the tangent bundle $T(M)$. There exists a unique connection ∇ compatible with g such that*

$$\nabla_\xi \eta - \nabla_\eta \xi = [\xi, \eta]. \tag{T}$$

Proof. The compatibility relation can be written three ways

$$\zeta g(\xi, \eta) = g(\nabla_\zeta \xi, \eta) + g(\xi, \nabla_\zeta \eta),$$

$$\eta g(\zeta, \xi) = g(\nabla_\eta \zeta, \xi) + g(\zeta, \nabla_\eta \xi)$$

$$\xi g(\eta, \zeta) = g(\nabla_\xi \eta, \zeta) + g(\eta, \nabla_\xi \zeta)$$

$$= g(\nabla_\xi \eta, \zeta) + g(\eta, \nabla_\zeta \xi) + g(\eta, [\xi, \zeta]).$$

Now we form an alternating sum of these three expressions to obtain

$$\zeta g(\xi, \eta) - \xi g(\eta, \zeta) + \eta g(\zeta, \xi)$$

$$= 2g(\xi, \nabla_\zeta \eta) - g(\eta, [\xi, \zeta]) + g([\eta, \zeta], \xi) + g(\zeta, [\eta, \xi]).$$

This is a formula for $g(\xi, \nabla_\zeta \eta)$ as one half of a six-term formula involving g and brackets of vector fields. This proves the uniqueness, and further the six term expression is a formula for $\nabla_\zeta \eta$ since it is determined by knowing $g(\xi, \nabla_\zeta \eta)$ for each $\xi \in \Gamma(T(M))$. It is easy to see that $\nabla_\zeta \eta$ is C-bilinear in ζ and η. The $A^0(M)$-homogeneity properties are a direct calculation left to the reader. Using the formula to calculate $2(g(\xi, \nabla_\zeta \eta) - g(\xi, \nabla_\eta \zeta))$, we get the term $2g(\xi, [\zeta, \eta])$ showing that condition (T) holds. Similiarly, ∇ is compatible with g. This proves the theorem.

6.6 Remark. The condition (T) of the previous theorem says that the torsion tensor of the connection $T(\xi, \eta) = 0$ where

$$T(\xi, \eta) = \nabla_\xi \eta - \nabla_\eta \xi - [\xi, \eta].$$

6.7 Remark. The previous existence theorem is the beginning of the subject of Riemannian geometry. With respect to a connection there is the notion of parallel transport along a curve in the manifold of a vector in the fibre at the initial point to a vector in the fibre at the final point of the curve. With respect to a positive definite Riemannian metric there is the notion of the curve of minimal length between two points called a geodesic. The relation between the two is that for a geodesic the tangent vector field along the curve is a parallel transport for the Levi-Civita connection. This condition of parallel transport is a first order differential equation for a general vector, while the fact that tangent vector to a curve is a parallel transport is a second order differential equation in the coordinates of the curve. The local existence of geodesics follows from theory of differential equations.

For a clear introduction of these ideas, together with application to geometry and Bott periodicity, we recommend the book by J. Milnor, *Morse theory*, Annals of Mathematics Studies.

CHAPTER 20

General Theory of Characteristic Classes

Using vector bundles over a space X, we are able to associate with X various sets which reflect some of the topological properties of X, for example, $\mathrm{Vect}_F(X)$, the semigroup of isomorphism classes of F-vector bundles; $\mathrm{Vect}_F^n(X)$, the set of isomorphism classes of n-dimensional vector bundles over X; and $K_F(X)$, the group completion of $\mathrm{Vect}_F(X)$. We view a characteristic class as a morphism defined on one of the cofunctors Vect_F, Vect_F^n, or K_F with values in a cohomology cofunctor. In several important cases, we are able to give a complete description of all characteristic classes. We conclude with a discussion of properties of the Chern character.

1. The Yoneda Representation Theorem

We begin with a general result in category theory that is basic for calculating characteristic classes and cohomology operations. In view of our area of application, the result is stated for cofunctors. By duality there is a corresponding statement for functors.

1.1 Proposition. *Let* A *be a category, and let* K *be an object in* A*. For each morphism* $f: Y \to X$*, let* $[f, K]$ *denote the function that assigns to each morphism* $u \in [X, K]$ *the morphism* $uf \in [Y, K]$*. Let* $[-, K]$ *denote the function that assigns to each* X *in* A *the set* $[X, K]$ *and to each morphism* $f: Y \to X$ *the function* $[f, K]: [X, K] \to [Y, K]$*. Then* $[-, K]:$ A $\to$ ens *is a cofunctor.*

Proof. We have only to check the axioms. If $u \in [X, K]$ and if $g: Z \to Y$ and $f: Y \to X$ are morphisms, we have $[fg, K]u = ufg = [g, K]uf = [g, K][f, K]u$ and $[1, K]u = u$.

The next theorem is the Yoneda representation theorem, and it concerns the calculation of a certain set of morphisms between cofunctors.

1.2 Theorem. *Let $T: A \to$ ens be a cofunctor, and let $\beta: [[-, K], T] \to T(K)$ be the function defined by the relation $\beta(\phi) = \phi(K)1_K$ for each morphism $\phi: [-, K] \to T$. Then β is a bijection.*

Proof. We define an inverse function α of β by the relation $\alpha(x)(X)u = T(u)x$ for $x \in T(K)$ and $u \in [X, K]$. To verify that $\alpha(x): [-, K] \to T$ is a morphism of cofunctors, we make the following calculation for $f: Y \to X$:

$$T(f)\alpha(x)(X)u = T(f)T(u)x = T(uf)x = \alpha(x)(Y)uf = \alpha(x)(T)[f, K]u$$

To show that α and β are inverses to each other, we calculate $\beta(\alpha(x)) = \alpha(x)(K)1_K = T(1_K)x = x$ and $\alpha\beta(\phi)(X)u = \alpha(\phi(K)1_K)(X)u = T(u)\phi(K)1_K = \phi(X)u1_K = \phi(X)u$. Consequently, we have $\beta\alpha = 1$ and $\alpha\beta = 1$. This proves the theorem.

1.3 Corollary. *For two objects K and L in a category A, let $\beta: [[-, K], [-, L]] \to [K, L]$ be the function defined by the relation $\beta(\phi) = \phi(K)1_K$ for a morphism $\phi: [-, K] \to [-, L]$. Then β is a bijection.*

1.4 Remark. Theorem (1.2) is useful for calculating the set of all morphisms $[F, T]$ for two set-valued cofunctors F and T, where F is isomorphic to $[-, K]$ for some $K \in A$.

2. Generalities on Characteristic Classes

2.1 Definition. Let H be a cofunctor defined on a category of spaces and maps with values in the category of sets. A characteristic class for n-dimensional bundles with values in H is a morphism $\mathrm{Vect}_{F^n} \to H$.

In most cases, we shall discuss the situation where Vect_{F^n} is isomorphic to some $[-, K_n]$ and the set of all characteristic classes in dimension n is in natural bijection with the set $H(K_n)$.

2.2 Definition. Let H be a cofunctor defined on a category of spaces and maps with values in the category of commutative semigroups (or semirings). A characteristic class for F-vector bundles with values in H is a morphism $\mathrm{Vect}_F \to H$. A characteristic class ϕ is stable provided there is a factorization $\phi': K_F \to H$ of ϕ by the natural morphism $\mathrm{Vect}_F \to K_F$.

The following diagram is commutative, and ϕ uniquely determines ϕ' with respect to this property.

There is the following simple criterion for the stability of a characteristic class.

2.3 Proposition. *A characteristic class ϕ with values in a cofunctor H is stable if and only if $\phi(\xi)$ is invertible for each bundle class ξ. If ϕ is defined over the category of finite CW-complexes and if $\phi(\theta^q)$ is the neutral element in $H(X)$ for each trivial bundle class θ^q, then ϕ is stable.*

Proof. The first statement says that ϕ factors through the group completion K_F if and only if its image is included in a subgroup of $H(X)$ for each X. The second statement follows from the first, because each bundle ξ has an s-inverse η, where $\xi \oplus \eta$ is trivial. Then $\phi(\xi \oplus \eta) = \phi(\xi)\phi(\eta)$ is the neutral element of $H(X)$.

2.4 Example. Let $H(X)$ be $H^{ev}(X, \mathbf{Z}) = \sum_{k \geq 0} H^{2k}(X, \mathbf{Z})$ with the cup product as the commutative semigroup operation. Then the total Chern class $c: \mathrm{Vect}_{\mathbf{C}} \to H^{ev}(-, \mathbf{Z})$ is a characteristic class. Since $c(\xi) = 1$ for a trivial bundle ξ, c is stable by (2.3), and we can view c as defined by $K \to H^{ev}(-, \mathbf{Z})$.

2.5 Example. Let $H(X)$ be $H^*(X, \mathbf{Z}_2) = \sum_{k \geq 0} H^k(X, \mathbf{Z}_2)$ with the cup product as the commutative semigroup operation. Then the total Stiefel-Whitney class $w: \mathrm{Vect}_{\mathbf{R}} \to H^*(-, \mathbf{Z}_2)$ is a characteristic class. Since $w(\xi) = 1$ for a trivial bundle ξ, w is stable by (2.3), and we can view w as defined by $KO \to H^*(-, \mathbf{Z}_2)$.

2.6 Example. Let $H(X)$ be $H^{ev}(X, \mathbf{Z})$ with the cup product as the commutative semigroup operation. Then the Euler class $e: \mathrm{Vect}_{\mathbf{R}}^+ \to H^{ev}(-\mathbf{Z})$ is an example of a nonstable characteristic class. The cofunctor $\mathrm{Vect}_{\mathbf{R}}^+$ is defined by requiring $\mathrm{Vect}_{\mathbf{R}}^+(X)$ to be the semiring of isomorphism classes of even-dimenion real orientable vector bundles.

3. Complex Characteristic Classes in Dimension n

We consider complex characteristic classes with values in $H^*(X, \mathbf{Z})$ and $H^{ev}(X, \mathbf{Z})$. (We denote $\mathrm{Vect}_{\mathbf{C}}$ by Vect and $K_{\mathbf{C}}$ by K.) The following result is useful in determining the n-dimensional characteristic classes.

3.1 Proposition. *Let* $h_n: CP^\infty \times \overset{(n)}{\cdots} \times CP^\infty \to G_n(\mathbb{C}^\infty)$ *be the classifying map for the bundle* $\gamma \times \overset{(n)}{\cdots} \times \gamma$, *where* γ *is the canonical line bundle over* CP^∞. *Then* h_n *is a splitting map for the canonical bundle* γ_n *over* $G_n(\mathbb{C}^\infty)$.

Proof. Since $h_n^*(\gamma_n)$ is isomorphic to $\gamma \times \overset{(n)}{\cdots} \times \gamma$, it suffices to show that $h_n^*: H^*(G_n(\mathbb{C}^\infty)) \to H^*(CP^\infty \times \overset{(n)}{\cdots} = CP^\infty)$ is a monomorphism. For this, let $f: X \to G_n(\mathbb{C}^\infty)$ be any splitting map of γ_n, where $f^*(\gamma_n) = \lambda_1 \oplus \cdots \oplus \lambda_n$. Let $g_i: X \to CP^\infty$ be a classifying map for λ_i, where λ_i is isomorphic to $g_i^*(\gamma)$. Then for $g = (g_1, \ldots, g_n)$ the bundle $\lambda_1 \oplus \cdots \oplus \lambda_n$ is isomorphic to $g^*(\gamma \times \overset{(n)}{\cdots} \times \gamma)$. Therefore, by 3(6.2) the maps f and $h_n g$ are homotopic. Then as morphisms of cohomology we have $f^* = (h_n g)^* = g^* h_n^*$. Since f^* is a monomorphism, h_n^* is a monomorphism.

For some expositions of characteristic classes the next theorem is used as the starting point.

3.2 Theorem. *Let* c_i *denote* $c_i(\gamma_n)$, *where* γ_n *is the universal n-dimensional vector bundle. Then the cohomology ring* $H^*(G_n(\mathbb{C}^\infty), \mathbb{Z}) = \mathbb{Z}[c_1, \ldots, c_n]$ *and the classes* $c_1, \ldots, c_n$ *are algebraically independent.*

Proof. By (3.1), $h_n: CP^\infty \times \overset{(n)}{\cdots} \times CP^\infty \to G_n(\mathbb{C}^\infty)$ is a splitting map. Since $\gamma \times \overset{(n)}{\cdots} \times \gamma$ is invariant under the action of the symmetric group in n letters, the image of h_n^* in $H^*(CP^\infty \times \overset{(n)}{\cdots} \times CP^\infty, \mathbb{Z}) = \mathbb{Z}[y_1, \ldots, y_n]$ is a subring of the ring of symmetric polynomials in the variables $y_1, \ldots, y_n$. If $pr_i: CP^\infty \times \overset{(n)}{\cdots} \times CP^\infty \to CP^\infty$ is the projection on the ith factor, then y_i equals $c_1(pr_i^*(\gamma))$. Since $h_n^*(\gamma_n) = pr_1^*(\gamma) \oplus \cdots \oplus pr_n^*(\gamma)$, there is the following relation for the total Chern class $c(\gamma_n)$, that is, $h_n^*(c(\gamma_n)) = c(h_n^*(\gamma_n)) = c(pr_1^*(\gamma) \oplus \cdots \oplus pr_n^*(\gamma)) = (1 + y_1) \cdots (1 + y_n)$. Consequently, $h_n^*(c_i(\gamma_n))$ is the ith elementary symmetric function σ_i in the variables $y_1, \ldots, y_n$. Since $\mathbb{Z}[\sigma_1, \ldots, \sigma_n]$ is a polynomial ring and since $h_n^*: H^*(G_n(\mathbb{C}^\infty), \mathbb{Z}) \to \mathbb{Z}[\sigma_1, \ldots, \sigma_n]$ is an isomorphism with $h_n^*(c_i) = \sigma_i$, we have proved the theorem.

3.3 Remarks. From the relation of vector bundles and general fibre bundles [see 5(2.8)], we have $G_n(\mathbb{C}^\infty) = BU(n)$ and $CP^\infty \times \overset{(n)}{\cdots} \times CP^\infty = BT(n)$, where $T(n) = S_1 \times \overset{(n)}{\cdots} \times S^1$. The inclusion of $T(n) \subset U(n)$ as a maximal torus induces the splitting map $BT(n) \to BU(n)$ or $CP^\infty \times \cdots \times CP^\infty \to G_n(\mathbb{C}^\infty)$.

From Theorems (3.2) and (1.2) we have the following theorem immediately.

3.4 Theorem. *Each characteristic class has the form* $q(c_1(\xi), \ldots, c_n(\xi))$, *where* $q(y_1, \ldots, y_n)$ *is a polynomial uniquely determined by* ϕ *and* $c_i(\xi)$ *is the ith Chern class of* ξ.

3.5 Remark. Since the Chern classes are all even-dimensional classes, each characteristic class $\phi: \text{Vect}_{\mathbb{C}^n} \to H^*(-, \mathbb{Z})$ can be viewed as a characteristic class with values in $H^{ev}(-, \mathbb{Z})$.

4. Complex Characteristic Classes

Let R denote a commutative ring with a unit. Associated with the ring $R[[t]]$ of formal series in t with coefficients in R are several commutative semigroups. Let $R[[t]]_+$ denote the additive group of $R[[t]]$, let $R[[t]]_\times$ denote the multiplicative semigroup of nonzero elements of $R[[t]]$, and let $R[[t]]_\times^*$ denote the group of units in $R[[t]]$.

4.1 Notation. Let H be a ring-valued cofunctor. Let $[\mathrm{Vect}_F, H]_\mu$ denote the set of all characteristic classes $\phi: \mathrm{Vect}_F \to H$ which preserve the following algebraic structures:

(1) For $\mu = (+, +)$, $\phi(\xi \oplus \eta) = \phi(\xi) + \phi(\eta)$.
(2) For $\mu = (+, \times)$, $\phi(\xi \oplus \eta) = \phi(\xi)\phi(\eta)$.
(3) For $\mu = r$, each $\phi(X)$ is required to be a semiring morphism.

4.2 Remarks. The natural morphism $\mathrm{Vect}_F \to K_F$ defines an injection $[K_F, H]_\mu \to [\mathrm{Vect}_F, H]_\mu$. In the two sets $[\mathrm{Vect}_F, H]_{+,+}$ and $[\mathrm{Vect}_F, H]_{+,\times}$ there is a natural semigroup structure, where $(\phi + \psi)(\xi) = \phi(\xi) + \psi(\xi)$ in the first case and $(\phi\psi)(\xi) = \phi(\xi)\psi(\xi)$ in the second case.

In the next theorem, we consider a category of paracompact spaces which contains CP^n for $0 \leqq n \leqq +\infty$.

For a ring R, we denote by $c_1(\xi)$ the image under $H^2(X, \mathbf{Z}) \to H^2(X, R)$ of the first Chern class of ξ. Finally, we denote Vect_C by Vect and K_C by K.

4.3 Theorem. *There exist a function $a: [\mathrm{Vect}, H^{ev}(-, R)]_{+,+} \to R[[t]]_+$ such that $a(\phi)(c_1(\lambda)) = \phi(X)\lambda$ for each line bundle λ over X and a function $m: [\mathrm{Vect}, H^{ev}(-, R)]_{+,\times} \to R[[t]]_\times$ such that $m(\psi)(c_1(\lambda)) = \psi(X)(\lambda)$ for each line bundle λ over X. The functions a and m are unique with respect to this property, and a and m are semigroup isomorphisms.*

Additional properties of a and m are contained in the following statements.

(1) *The natural injection $[K, H^{ev}(-, R)]_{+,+} \to [\mathrm{Vect}, H^{ev}(-, R)]_{+,+}$ is a bijection.*
(2) *After restriction of m, there is a commutative diagram where the horizontal functions are bijections.*

$$
\begin{array}{ccc}
[K, H^{ev}(-, R)]_{+,+} & \xrightarrow{\ m\ } & R[[t]]_\times^* \\
\downarrow & & \downarrow \\
[\mathrm{Vect}, H^{ev}(-, R)]_{+,\times} & \xrightarrow{\ m\ } & R[[t]]_\times
\end{array}
$$

(3) *The restriction of a to $[\mathrm{Vect}, H^{ev}(-, R)]_r$ is a bijection onto the set of $\{0, e^{bt}\}$, where $b \in R$ in the case that R is a ring with $\mathbf{Q} \subset R$.*
(4) *The conjugate $\bar{\phi}$ of a characteristic class ϕ is defined by the relation $\bar{\phi}(\xi) = \phi(\bar{\xi})$, where $\bar{\xi}$ is the conjugate bundle. Then $a(\bar{\phi})(t) = a(\phi)(-t)$ and $m(\bar{\psi})(t) = m(\psi)(-t)$.*

Proof. We construct a as follows: Let $a(\phi)(t) \in R[[t]]$ be such that $a(\phi)(c_1(\gamma)) = \phi(CP^n)\gamma$ in $H^{ev}(CP^n, R) = R[x] \bmod x^{n+1}$. This yields an inductive definition of the first n coefficients of $a(\phi)(t)$ and displays the unicity of a. If λ is a line bundle over X, there is a map $f: X \to CP^n$ such that $f^*(\gamma)$ and λ are X-isomorphic. Then we have $\phi(X)(\lambda) = f^*(\phi(CP^n)\gamma) = f^*[\alpha(\phi)(c_1(\gamma))] = a(\phi)[f^*(c_1(\gamma))] = \alpha(\phi)(c_1(\lambda))$. Similarly, m is constructed in this unique way.

Next, for two characteristic classes ϕ and ϕ' we have $a(\phi + \phi')(c_1(\gamma)) = (\phi + \phi')(CP^n)\gamma = \phi(CP^n)\gamma + \phi'(CP^n)\gamma$ in $H^{ev}(CP^n, R) = R[x] \bmod x^{n+1}$. Consequently, we have $a(\phi + \phi')(t) \equiv a(\phi)(t) + a(\phi')(t) \bmod t^{n+1}$ for each n and $a(\phi + \phi') = a(\phi) + a(\phi')$. Similarly, the relation $m(\psi\psi') = m(\psi)m(\psi')$ holds.

To prove that a is injective, we suppose that $a(\phi_1) = a(\phi_2)$. Then for a line bundle λ over X we have $\phi_1(X)\lambda = a(\phi_1)(c_1(\lambda)) = a(\phi_2)(c_1(\lambda)) = \phi_2(X)\lambda$. For a bundle ξ over Y, let $u: X \to Y$ be a splitting map with $u^*(\xi) = \lambda_1 \oplus \cdots \oplus \lambda_n$. Then the following relation holds:

$$u^*(\phi_1(Y)\xi) = \phi_1(X)\lambda_1 + \cdots + \phi_1(X)\lambda_n = \phi_2(X)\lambda_1 + \cdots + \phi_2(X)\lambda_n$$

$$= u^*(\phi_2(T)\xi)$$

Since u^* is a morphism, we have $\phi_1 = \phi_2$. Similarly, if $m(\psi_1) = m(\psi_2)$, we have $\psi_1 = \psi_2$, and a and m are injections.

To prove that a is surjective, we take $f(t) \in R[[t]]$ and define $\phi(X)\lambda$ to be $f(c_1(\lambda))$ for a line bundle λ and X. For an arbitrary bundle ξ over X, let $u: X' \to X$ be a splitting map with $u^*(\xi) = \lambda_1 \oplus \cdots \oplus \lambda_n$. Then we require $\phi(X')u^*(\xi)$ to be $\phi(X')\lambda_1 + \cdots + \phi(X')\lambda_n$. Observe that $\phi(X')\lambda_1 + \cdots + \phi(X')\lambda_n$ is a polynomial in $c_i(u^*(\xi)) = u^*(c_i(\xi))$, the elementary symmetric functions of the classes $c_1(\lambda_1), \ldots, c_1(\lambda_n)$. Since $u^*: H^{ev}(X, R) \to H^{ev}(X', R)$ is a monomorphism, there is a unique element $\phi(X)\xi$ such that $u^*(\phi(X')\xi) = \phi(X)u^*(\xi)$. The above construction contains implicitly the fact that ϕ is a morphism Vect $\to H^{ev}(-, R)$. To prove that m is surjective, we use the above argument and require $\psi(X)u^*(\xi)$ to be the $\psi(X')\lambda_1 \cdots \psi(X')\lambda_n$. The above construction is independent of u.

Statement (1) follows from the fact that $H^{ev}(X, R)$ is an additive group, and statement (2) follows from the criterion in Proposition (2.3). For (3), from the splitting principle, a characteristic class $\phi: \text{Vect} \to H^{ev}(-, R)$ is a semiring morphism if and only if we have

$$a(\phi)(c_1(\lambda) + c_1(\lambda')) = a(\phi)(c_1(\lambda \otimes \lambda')) = \phi(X)(\lambda \otimes \lambda') = (\phi(X)\lambda)(\phi(X)\lambda')$$

$$= a(\phi)(c_1(\lambda))a(\phi)(c_1(\lambda'))$$

that is, we must have $a(\phi)(t \oplus t') = a(\phi)(t)a(\phi)(t')$. For (4), we have $\bar{\phi}(\lambda) = -\phi(\lambda)$ for line bundles, and therefore $a(\bar{\phi})(t) = a(\phi)(-t)$ and $m(\bar{\psi})(t) = m(\psi)(-t)$. This proves the theorem.

Under m the total Chern class corresponds to $1 + t$.

5. Real Characteristic Classes Mod 2

The results of Secs. 3 and 4 on complex characteristic classes carry over to real characteristic classes by replacing Chern classes with Stiefel-Whitney classes and restricting to coefficient rings R containing Z_2. We state the results and leave it to the reader to supply the parallel verifications.

5.1 Proposition. *Let $h_n: RP^\infty \times \overset{(n)}{\cdots} \times RP^\infty \to G_n(\mathbf{R}^\infty)$ be the classifying map for the bundle $\gamma \times \overset{(n)}{\cdots} \times \gamma$, where γ is the canonical line bundle over RP^∞. Then h_n is a splitting map for the canonical bundle γ_n over $G_n(\mathbf{R}^\infty)$ with respect to Z_2 cohomology.*

5.2 Theorem. *Let w_i denote $w_i(\gamma_n)$. Then the cohomology ring $H^*(G_n(\mathbf{R}^\infty), Z_2)$ $= Z_2[w_1, \ldots, w_n]$ and the classes $w_1, \ldots, w_n$ are algebraically independent.*

Statements (5.1) and (5.2) are parallel to (3.1) and (3.2).

5.3 Remarks. We have $G_n(\mathbf{R}^\infty) = BO(n)$ and $RP^\infty \times \overset{(n)}{\cdots} \times RP^\infty = B(Z_2)^n$. The inclusion of diagonal matrices $(Z_2)^n$ into $O(n)$ induces the splitting map $B(Z_2)^n \to BO(n)$ or $RP^\infty \times \overset{(n)}{\cdots} \times RP^\infty \to G_n(\mathbf{R}^\infty)$.

5.4 Theorem. *Each characteristic class $\phi: \mathrm{Vect}_{\mathbf{R}^n} \to H^*(-, Z_2)$ is of the form $\phi(X)\xi = q(w_1(\xi), \ldots, w_n(\xi))$, where $q(y_1, \ldots, y_n)$ is a polynomial uniquely determined by ϕ and $w_i(\xi)$ is the ith Stiefel-Whitney class of ξ.*

In keeping with the notations of (4.1), we have the following theorem which is similar to (4.3).

5.5 Theorem. *Let Z_2 be the subring of R generated by 1. There exist a function $a: [\mathrm{Vect}_{\mathbf{R}}, H^*(-, R)]_{+,+} \to R[[t]]_+$ such that $a(\phi)(w_1(\lambda)) = \phi(X)\lambda$ for each line bundle λ over X and a function $m: [\mathrm{Vect}_{\mathbf{R}}, H^*(-, R)]_{+,\times} \to R[[t]]_\times$ such that $m(\psi)(w_1(\lambda)) = \psi(X)\lambda$ for each line bundle λ over X. The functions a and m are unique with respect to this property, and a and m are semigroup isomorphisms.*

Additional properties of a and m are contained in the following statements.

(1) *The natural injection $[KO, H^*(-, R)]_{+,+} \to [\mathrm{Vect}, H^*(-, R)]_{+,+}$ is a bijection.*

(2) *After restriction of m, there is a commutative diagram where the horizontal functions are bijections.*

$$
\begin{array}{ccc}
[KO, H^*(-, R)]_{+,\times} & \xrightarrow{\ m\ } & R[[t]]_\times^* \\
\downarrow & & \downarrow \\
[\mathrm{Vect}, H^*(-, R)]_{+,\times} & \xrightarrow{\ m\ } & R[[t]]_\times
\end{array}
$$

In the above theorem, $w_i(\xi) \in H^i(B(\xi), R)$ denotes the image of the ith Stiefel-Whitney class of ξ under $H^i(B(\xi), Z_2) \to H^i(B(\xi), R)$ which is induced by $Z_2 \to R$.

6. 2-Divisible Real Characteristic Classes in Dimension n

The theory of real characteristic classes divides naturally in two parts: the theory for rings R with $2R = 0$ (see Sec. 5) and the theory for rings R with $1/2 \in R$, that is, 2-divisible rings which we consider in this and the following section.

The next result on principal Z_2-bundles is very useful in the following discussion.

6.1 Proposition. *Let $q: E \to B$ be a locally trivial principal Z_2-bundle, let R be a 2-divisible ring, and let τ denote the involution of $H^*(E, R)$ induced by the action of $-1 \in Z_2$ on E. Then $q^*: H^*(B, R) \to H^*(E, R)$ is a monomorphism, and the image of q^* is the fixed elements of τ.*

Proof. Let U_i for $i \in I$ be an open covering of B such that the restriction $q^{-1}(U_i) \to U_i$ is a trivial Z_2-bundle. By a standard result in singular cohomology theory (see Eilenberg and Steen rod [1, p. 197]), it suffices to consider cochains which are defined only on singular simplexes with images in some U_i or $q^{-1}(U_i)$. Then, as a cochain complex, $C^*(E, R) = C^*_+(E, R) \oplus C^*_-(E, R)$, where $C^*_\pm(E, R)$ are the ± 1 eigenspaces of the involution induced by the action of -1 on E. Then $q^*: C^*(B, R) \to C^*_+(E, R)$ is a cochain isomorphism, and this proves the proposition.

6.2 Remarks. Since $SO(2) = U(1) = S^1$, oriented real plane bundles η have a natural complex line bundle structure compatible with their orientation. These bundles over X are classified by elements of $H^2(X, Z)$, using the characteristic class $e(\eta)$ which is the first Chern class $c_1(\eta)$ of the complex line bundle or the Euler class of the real plane bundle. The complexification $\eta \otimes C$ of the real plane bundle η equals $\eta \oplus \bar{\eta}$ as complex vector bundles. The Chern classes are given by $c_1(\eta \otimes C) = c_1(\eta) + c_1(\bar{\eta}) = c_1(\eta) - c_1(\eta) = 0$ and $-c_2(\eta \otimes C) = -c_1(\eta)c_1(\bar{\eta}) = c_1(\eta)^2 = p_1(\eta)$, the first Pontrjagin class of η. If η is any real plane bundle, we have $c_1(\eta \otimes C) = c_1(\overline{\eta \otimes C}) = -c_1(\eta \otimes C)$ and $2c_1(\eta \otimes C) = 0$. If the coefficient ring is 2-divisible, then the relations $c_1(\eta \otimes C) = 0$ and $p_1(\eta) = -c_2(\eta \otimes C)$ hold.

The constructions in the previous remarks yield the following commutative diagram:

$$
\begin{array}{ccccc}
CP^\infty & \xrightarrow{\cong} & G_2^+(R^\infty) & \longrightarrow & CP^\infty \times CP^\infty \\
 & \searrow & \downarrow & & \downarrow \\
 & & G_2(R^\infty) & \longrightarrow & G_2(C^\infty)
\end{array}
$$

The two horizontal maps on the right refer to complexification, $G_2^+(R^\infty) \to G_2(R^\infty)$ is a locally trivial principal Z_2-bundle defined by forgetting orientation, and $CP^\infty \times CP^\infty \to G_2(C^\infty)$ is the splitting map considered in (3.1).

6.3 Proposition. *The cohomology of the above spaces with values in a 2-divisible ring R is the following:* $H^*(CP^\infty) = R[c_1(\gamma)]$, $H^*(G_2^+(\mathbf{R}^\infty)) = R[e(\gamma_2^+)]$, $H^*(G_2(\mathbf{R}^\infty)) = R[p_1(\gamma_2)]$, $H^*(CP^\infty \times CP^\infty) = R[c_1(\gamma'), c_1(\gamma'')]$, *and* $H^*(G_2(\mathbf{C}^\infty)) = R[c_1(\gamma_2), c_2(\gamma_2)]$. *Moreover, in the following commutative diagram,* $w_2(e(\gamma_2^+)) = c_1(\gamma)$, $v_2(p_1(\gamma_2)) = e(\gamma_2^+)^2$, $v_1(c_1(\gamma_2)) = 0$, $v_1(c_2(\gamma_2)) = p_1(\gamma_2)$, $u_2(c_1(\gamma')) = e(\gamma_2^+)$, *and* $u_2(c_1(\gamma'')) = -e(\gamma_2^+)$.

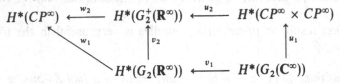

Proof. The statements about CP^∞, $CP^\infty \times CP^\infty$, and $G_2(\mathbf{C}^\infty)$ follow from (3.1) and (3.2). Since u_2 is induced by the map corresponding to $\eta \mapsto \eta \oplus \bar{\eta}$, we have determined u_2. The character of w_2 and $H^*(G_2^+(\mathbf{R}^\infty))$ follows from the fact that $CP^\infty \to G_2^+(\mathbf{R}^\infty)$ is an isomorphism. The statement about v_2 and $H^*(G_2(\mathbf{R}^\infty))$ follows from (6.1) and about v_1 from (6.2). This proves the proposition.

6.4 Diagram. The calculations $H^*(BSO(2), R) = R[e]$ and $H^*(BO(2), R) = R[p_1]$ can be extended to the case of any $n \geq 2$. We have the following commutative diagram where $n = 2r$ or $n = 2r + 1$:

$$
\begin{array}{ccc}
[G_2^+(\mathbf{R}^\infty)]^r & \xrightarrow{\ v\ } & [CP^\infty]^n \\
\downarrow{\scriptstyle f} & & \downarrow{\scriptstyle g} \\
G_n^-(\mathbf{R}^\infty) & \xrightarrow{\ u\ } & G_n(\mathbf{C}^\infty)
\end{array}
$$

The horizontal arrows are the classifying maps for the complexification of real universal bundles, and the vertical maps are the classifying maps for the sum of real plane bundles or complex line bundles.

6.5 Theorem. *Let R be a 2-divisible ring. If e denotes the Euler class of the universal bundle, and if p_i denotes the ith Pontrjagin class of the universal bundle, then there are the following polynomial rings:*

$$H^*(BO(2r), R) = R[p_1, \ldots, p_r] \qquad H^*(BSO(2r), R) = R[p_1, \ldots, p_{r-1}, e]$$

$$H^*(BO(2r + 1), R) = R[p_1, \ldots, p_r] \qquad H^*(BSO(2r + 1), R) = R[p_1, \ldots, p_r]$$

Proof. We begin by observing that (6.1) allows us to deduce the result for $BO(n)$ from $BSO(n)$ and the relation $e^2 = p_r$.

We have the following maps:

$$
\begin{array}{ccc}
E_0(\gamma_n^+) & \xrightarrow{\ \pi\ } & G_{n-1}^+(\mathbf{R}^\infty) \\
\downarrow{\scriptstyle q} & & \\
G_n^+(\mathbf{R}^\infty) & &
\end{array}
$$

which are defined by the relations $q(W, x) = W$ and $\pi(W, x) = W'$, where $W = W' \perp \mathbf{R}x$ and W' has the induced orientation. In addition, there is an isomorphism between $q^*(\gamma_n^+)$ and $\pi^*(\gamma_{n-1}^+) \oplus \theta^1$. The Gysin sequence [see 17(7.5)] has the following form:

$$
\cdots \to H^i(G_n^+(\mathbf{R}^\infty)) \overset{\underset{\text{mult.}}{\text{by } e}}{\to} H^{i+n}(G_n^+(\mathbf{R}^\infty)) \overset{q^*}{\to} H^{i+n}(E_0) \overset{\psi}{\to} H^{i+1}(G_n^+(\mathbf{R}^\infty))
$$

$$
\pi^{*-1}q^* \searrow \qquad \pi^* \uparrow
$$

$$
H^{i+1}(G_n^+(\mathbf{R}^\infty))
$$

From the naturality of $p(\gamma_n^+)$ and $p(\gamma_{n-1}^+)$ we see that $\pi^{*-1}q^*(p_i(\gamma_n^+)) = p_i(\gamma_{n-1}^+)$. By induction on n we see that $H^*(BSO(n), R)$ is generated by $p_1, \ldots, p_r$ and e. We recall, if $n = 2r + 1$, that $e = 0$; otherwise, we have $e^2 = p_r$.

Finally, we must show that the set $p_i, \ldots, p_{r-1}, e$ for $n = 2r$ and $p_1, \ldots, p_r$ for $n = 2r + 1$ is algebraically independent. For this, there is the following diagram where $e_{(j)} = e(pr_j^*(\gamma_2^+))$ and $c_{(i)} = c_1(pr_i^*(\gamma))$:

$$
H^*(G_2^+(\mathbf{R}^\infty)^r) = \quad R[e_{(1)}, \ldots, e_{(r)}] \quad \overset{v^*}{\longleftarrow} H^*((CP^\infty)^n) = R[c_{(1)}, \ldots, c_{(n)}]
$$

$$
\uparrow f^* \qquad\qquad\qquad \uparrow g^*
$$

$$
;H^*(G_n^+(\mathbf{R}^\infty)) = \begin{cases} R[p_1, \ldots, p_r], n = 2r+1 \\ R[p_1, \ldots, p_{r-1}, e], n = 2r \end{cases} \overset{u^*}{\longleftarrow} H^*(G_n(\mathbf{C}^\infty)) = R[c_1, \ldots, c_n]
$$

We consider the formal sum $\sum_{1 \leq i \leq n} c_i x^i$. We have

$$
g^*\left(\sum_{1 \leq i \leq n} c_i x^i \right) = \prod_{1 \leq i \leq n} (1 + c_{(i)}x)
$$

and $v^* g^*\left(\sum_i c_i x^i \right) = \prod_{i \leq j \leq r} (1 - e_{(j)}^2 x)$. But we have, in addition,

$$
f^*\left(\sum_{1 \leq j \leq r} (-1)^j p_j x^{2j} \right) = \prod_{1 \leq j \leq r} (1 - e_{(j)}^2 x)
$$

Therefore, the elements $f^*(p_1), \ldots, f^*(p_r)$ are the elementary symmetric functions in $e_{(1)}^2, \ldots, e_{(r)}^2$, and therefore the set $p_1, \ldots, p_r$ is algebraically independent. In the case $n = 2r$, let $f(p_1, \ldots, p_{r-1}, e) = 0$ be a polynomial relation. Then $0 = f(p_1, \ldots, p_{r-1}, e) = f_1(p_1, \ldots, p_r) + f_2(p_1, \ldots, p_r)e$, where f_1 and f_2 are polynomials. This means $f_1(p_1, \ldots, p_r) = f_2(p_1, \ldots, p_r) = 0$ and $f_1 = f_2 = 0$. This proves the theorem.

6.6 Remark. In the process of proving Theorem (6.5) we demonstrated that $f: [G_2^+(\mathbf{R}^\infty)]^r \to G_n^+(\mathbf{R}^\infty)$ is a splitting map, in the sense that $f^*: H^*(G_n^+(\mathbf{R}^\infty), R) \to H^*([G_2^+(\mathbf{R}^\infty)]^r, R)$ is a monomorphism and that $f^*(\gamma_n^+)$ splits into a sum of oriented plane bundles and oriented line bundles. A similar statement holds for the classifying map $[G_2(\mathbf{R}^\infty)]^r \to G_n(\mathbf{R}^\infty)$.

From Theorems (6.5) and (1.2) we have the following theorem immediately.

6.7 Theorem. *Let R be a 2-divisible ring. Each characteristic class ϕ: $\mathrm{Vect}_R^n \to$ $H^*(-,R)$ is of the form $\phi(X)\xi = q(p_1(\xi),\dots,p_r(\xi))$, where $q(y_1,\dots,y_r)$ is a polynomial uniquely determined by ϕ, $p_i(\xi)$ is the ith Pontrjagin class, and n equals $2r$ or $2r + 1$. If Vect_+^n denotes the cofunctor of oriented n-dimensional real vector bundles, each characteristic class ϕ: $\mathrm{Vect}_+^n \to H^*(-,R)$ is of the form $\phi(X)\xi = q(p_1(\xi),\dots,p_{r-1}(\xi),e(\xi))$ for $n = 2r$ or $q(p_1(\xi),\dots,p_r(\xi))$ for $n = 2r + 1$, where q is a polynomial uniquely determined by ϕ.*

7. Oriented Even-Dimensional Real Characteristic Classes

Let $R^+[[t]]$ denote the subgroup of $R[[t]]_+$ consisting of all $f(t)$ with $f(-t) = f(t)$, that is, $f(t) = g(t^2)$ for some $g \in R[[t]]$, and let $R^\pm[[t]]$ denote the subsemigroup of $f(t) \in R[[t]]_\times$ with $f(-t) = f(t)$ or $f(-t) = -f(t)$, that is, $f(t) = g(t^2)$ or $tg(t^2)$ for some $g \in R[[t]]$.

Let $\mathrm{Vect}^+(X)$ denote the semiring of isomorphism class of even-dimensional oriented real vector bundles.

7.1 Theorem. *Let R be a 2-divisible ring. There exist a function a: $[\mathrm{Vect}^+, H^{\mathrm{ev}}(-,R)]_{+,+} \to R^+[[t]]$ such that $a(\phi)(e(\eta)) = \phi(X)\eta$ for each oriented plane bundle η over X and a function m: $[\mathrm{Vect}^+, H^{\mathrm{ev}}(-,R)]_{+,\times} \to R^\pm[[t]]$ such that $m(\psi)(e(\eta)) = \psi(X)\eta$ for each oriented plane bundle η over X. The functions a and m are unique with respect to this property, and a and m are semigroup isomorphisms.*

Additional properties of a and m are contained in the following statements:

(1) *The following squares are commutative.*

$$
\begin{array}{ccc}
[\mathrm{Vect}^+, H^{\mathrm{ev}}(-,R)]_{+,+} & \xrightarrow{\;a\;} & R^+[[t]] \\
{\scriptstyle u_1}\Big\downarrow\Big\uparrow{\scriptstyle u_2} & & {\scriptstyle v_1}\Big\downarrow\Big\uparrow{\scriptstyle v_2} \\
[\mathrm{Vect}, H^{\mathrm{ev}}(-,R)]_{+,+} & \xrightarrow{\;a\;} & R[[t]]
\end{array}
$$

$$
\begin{array}{ccc}
[\mathrm{Vect}^+, H^{\mathrm{ev}}(-,R)]_{+,\times} & \xrightarrow{} & R^\pm[[t]] \\
{\scriptstyle u_1}\Big\downarrow\Big\uparrow{\scriptstyle u_2} & & {\scriptstyle v_1}\Big\downarrow\Big\uparrow{\scriptstyle v_2} \\
[\mathrm{Vect}, H^{\mathrm{ev}}(-,R)]_{+,\times} & \xrightarrow{} & R[[t]]
\end{array}
$$

The morphism u_1 is induced by ε_0: $\mathrm{Vect} \to \mathrm{Vect}^+$ and u_2 by ε_U: $\mathrm{Vect}^+ \to \mathrm{Vect}$. The morphism v_1 is inclusion. The morphism v_2 assigns to $f(t)$ the series $f(t) + f(-t)$, and v_3 assigns to $f(t)$ the series $f(t)f(-t)$.

(2) *The restriction of a to* $[\text{Vect}^+, H^{ev}(-, R)]_r$ *is a bijection onto the set* $\{0, e^{at} + e^{-at}\}$ *for* $a \in R$, *where* $\mathbf{Q} \subset R$.

Proof. The construction of a and m proceeds as in paragraph 1 of the proof of (4.3) and the monomorphic character of a and m as semigroup morphisms as in paragraphs 2 and 3.

Let η be the oriented 4-plane bundle over $CP^n \times CP^n$ which is the Whitney sum $\eta_1 \oplus \eta_2$, where $\eta_i = pr_i^*(\varepsilon_0(\gamma))$ for $i = 1, 2$. Observe that $\varepsilon_0(\gamma) = \gamma_2^+$ over $CP^n = G_2^+(R^{2n+2})$. Then $H^*(CP^n \times CP^n, R)$ is the polynomial ring $R[x, y]$ modulo the relations $x^{n+1} = 0$ and $y^{n+1} = 0$. Then we have $a(\phi)(x) + a(\phi)(y) = \phi(\eta_1 \oplus \eta_2) = \phi(\eta_1^* \oplus \eta_2^*) = a(\phi)(-x) + a(\phi)(-y)$ and $m(\phi)(x)m(\phi)(y) = \phi(\eta_1 \oplus \eta_2) = \phi(\eta_1^* \oplus \eta_2^*) = m(\phi)(-x)m(\phi)(-y)$ since reversing the orientation of both η_1 and η_2 leaves the orientation of $\eta_1 \oplus \eta_2$ unchanged. By writing a power series $f(t) = g_1(t^2) + tg_2(t^2)$, we see that $a(\phi)(t) \in R^+[[t]]$ and $m(\phi)(t) \in R^{\pm}[[t]]$.

To show that a and m are surjective, let $f(t) = g(t^2) \in R^+[[t]]$. Define ϕ by $\phi(X)(\eta)) = f(c_1(\eta)) = g(p_1(\eta))$ for an oriented 2-plane bundle η over X. For an arbitrary oriented real bundle ξ over X, let $u: X_1 \to X$ be a splitting map $u^*(\xi) = \eta_1 \oplus \cdots \oplus \eta_r$ in the sense of (6.6), where η_i is an oriented 2-plane bundle. Then $\phi(X_1)(u^*(\xi))$ is required to be $\phi(X_1)(\eta_1) + \cdots + \phi(X_1)(\eta_r)$. Note that $\phi(X_1)(\eta_1) + \cdots + \phi(X_1)(\eta_r)$ is a polynomial in the elementary symmetric functions evaluated on the classes $p_1(\eta_1), \ldots, p_1(\eta_r)$. The ith elementary symmetric function in $p_1(\eta_j)$ is $p_i(u^*(\xi)) = u^*(p_i(\xi))$ because $1/2$ exists in R. Since $u^*: H^{ev}(X, R) \to H^{ev}(X_1, R)$ is a monomorphism, there is a unique element $\phi(X)\xi$ such that $u^*(\phi(X)\xi) = \phi(X_1)(u^*(\xi))$. The above construction demonstrates that ϕ is a morphism $\text{Vect}_R^+ \to H^{ev}(-, R)$. To prove m is surjective, the above argument is used, but $\psi(X_1)(u^*(\xi))$ is required to be $\psi(X_1)(\eta_1) \cdots \psi(X_1)(\eta_r)$. If $f(t) = tg(t^2)$, this is $c_1(\eta_1)g(p_1(\eta_1)) \cdots c_1(\eta_r)g(p_1(\eta_r))$ which is the Euler class $e(u^*(\xi))$ times a polynomial in $p_i(u^*(\xi)) = u^*(p_i(\xi))$. If $f(t) = g(t^2)$, only Pontrjagin classes appear.

The commutativity of the above diagrams in statement (1) is immediate. For $\phi \in [\text{Vect}_R^+, H^{ev}(-, \mathbf{Q})]_{+,+}$ with $a(\phi) = f$, the morphism ϕ preserves the ring structure if and only if $\phi(\eta_1 \otimes_R \eta_2) = \phi(\eta_1)\phi(\eta_2)$, where η_1 and η_2 are two plane bundles by the splitting principle of (6.6). We can take η_i to be $pr_i^*(\varepsilon_0(\gamma))$ for $i = 1, 2$ on $CP^n \times CP^n$. Then we have $f(x)f(y) = \phi(\eta_1)\phi(\eta_2) = \phi(\eta_1 \otimes_R \eta_2) = \phi(\eta_1 \otimes_C \eta_2) + \phi(\eta_1 \otimes_C \eta_2) = f(x + y) + f(x - y)$. By differentiating twice with respect to x and setting $y = 0$, we see that $f(t) = 0$ or $f(t) = e^{at} + e^{-at}$ are the only solutions of this functional relation. This proves the theorem.

7.2 Remark. For fibre maps of finite type we were able to give an "elementary" proof of the Leray-Hirsch theorem 17(1.1) from which the splitting principle follows. The reader is invited to determine under what hypothesis the results of Secs. 2 to 7 can be developed in an elementary framework.

8. Examples and Applications

8.1 Example. Under the bijection $m\colon [\mathrm{Vect}(-), H^{ev}(-,\mathbf{Z})]_{+,\times} \to \mathbf{Z}[[t]]$, the total Chern class corresponds to $1+t$, and under bijection $m\colon [\mathrm{Vect}_{\mathbf{R}}(-), H^*(-,\mathbf{Z}_2)]_{+,\times} \to \mathbf{Z}_2[[t]]$, the total Stietel-Whitney class corresponds to $1+t$. The top Chern class and top Stiefel-Whitney class correspond to t in each of the respective cases. The above statements are clearly true for line bundles, and therefore they are true in general.

8.2 Definition. Under the bijection $a\colon [\mathrm{Vect}(-), H^{ev}(-,\mathbf{Q})]_{\mathrm{ring}} \to \{0, e^{at}\} \subset \mathbf{Q}[[t]]_+$, the Chern character ch: $\mathrm{Vect}(-) \to H^{ev}(-,\mathbf{Q})$ is defined to be the morphism corresponding to e^t.

The Chern character is stable, and it can be viewed as a morphism ch: $K(-) \to H^{ev}(-,\mathbf{Q})$.

8.3 Example. Under the bijection $m\colon [\mathrm{Vect}_{\mathbf{R}}^+(-), H^{ev}(-,\mathbf{Z})]_{+,\times} \to \mathbf{Z}^{\pm}[[t]]$, the Euler class corresponds to t. By the uniqueness statement, this need be checked only on 2-plane bundles where it is clear. Similarly, the total Pontrjagin class corresponds to $1+t^2$ and the top Pontrjagin class to t^2.

If the first vertical arrow represents the function induced by ε_0, restriction of scalars from $\mathbf{C}$ to $\mathbf{R}$, there is the following commutative diagram from Theorem (7.1).

$$
\begin{array}{ccc}
[\mathrm{Vect}(-), H^{ev}(-,\mathbf{Z})]_{+,\times} & \xrightarrow{\ m\ } & \mathbf{Z}[[t]]_{\times} \\
\Big\downarrow & & \Big\downarrow \\
[\mathrm{Vect}_{\mathbf{R}}^+(-), H^{ev}(-,\mathbf{Z})]_{+,\times} & \xrightarrow{\ m\ } & \mathbf{Z}^+[[t]]_{\times}
\end{array}
$$

The second vertical arrow refers to the function that assigns to $f(t)$ the series $f(t)f(-t)$. Since $1-t^2 = (1+t)(1-t)$, there follows the relation $\tilde{p}(\varepsilon_0(\eta)) = c(\eta)\tilde{c}(\eta)$, where $\tilde{p}_i = (-1)^i p_i$ and $\tilde{c}_i = (-1)^i c_i$ are augmentations of the usual Pontrjagin and Chern classes. Here $\tilde{p}$ corresponds to $1-t^2$ under m and $\tilde{c}$ to $1-t$. Translating the above relation into a relation between Pontrjagin class and Chern classes, we have the following result.

8.4 Proposition. *If $\eta_{\mathbf{R}}$ denotes the real bundle associated with the complex vector bundle η, there exists the relation* $(-1)^i p_i(\eta_{\mathbf{R}}) = \sum_{j+k=2i} (-1)^k c_j(\eta) c_k(\eta)$.

If $p(M^n)$ denotes $p(\tau(M)_{\mathbf{R}})$ for a complex manifold, there is the following result for CP^n, using the fact that $c(\tau(CP^n)) = (1+u)^{n+1}$, where u is the canonical generator of $H^2(CP^n, \mathbf{Z})$.

8.5 Proposition. *The relation $p(CP^n) = (1+u^2)^{n+1}$ holds.*

Proof. Since $\tilde{p}(CP^n) = c(\tau(CP^n))\tilde{c}(\tau(CP^n)) = (1-u^2)^{n+1}$, then we have $p(CP^n) = (1+u^2)^{n+1}$.

8.6 Examples. For complex vector bundles, the L-genus is defined to be the characteristic class corresponding to $t/\tanh t$, the A-genus is defined to be the characteristic class corresponding to $t/\sinh t$, and the Todd class is the characteristic class corresponding to $t/1 - e^{-t}$. For real oriented vector bundles, the $\hat{A}$-genus is the characteristic class corresponding to $(t/2)/(\sinh t/2)$. For manifolds these characteristic classes are evaluated on the top class of the manifold to give a number, also called the L-genus, A-genus and $\hat{A}$-genus respectively.

9. Bott Periodicity and Integrality Theorems

In this section we relate the Chern character to the complex periodicity and prove the Bott integrality theorem. In doing this, we given an alternative proof of the complex periodicity theorem.

Let γ denote the canonical line bundle (or hyperplane bundle) over $S^2 = CP^1$. Then, from the results in Chap. 11, 1 and $\gamma - 1$ form an additive base for $K(S^2)$. All K-groups are complex K-groups.

9.1 Proposition. *The ring homomorphism* ch: $K(S^2) \to H^*(S^2, \mathbf{Z})$ *is an isomorphism.*

Proof. Clearly ch(1) = 1, where $c_1(1) = 0$ for the trivial line bundle 1 and ch(γ 1) $= c_1(\gamma)$ which generates $H^2(S^2, \mathbf{Z})$.

9.2 Corollary. *The relation* $(\gamma - 1)^2 = 0$ *holds on* S^2.

Proof. Since ch is a ring homomorphism, ch$(\gamma - 1)^2 = $ ch$(\gamma - 1)$ ch$(\gamma - 1) = c_1(\gamma)^2 = 0$. Since ch is an isomorphism, $(\gamma - 1)^2 = 0$.

9.3 Corollary. *The relation* $\gamma^m = 1 + m(\gamma - 1)$ *holds on* S^2.

Proof. From (9.2), we have $\gamma^m = (1 + (\gamma - 1))^m = 1 + m(\gamma - 1)$.

9.4 Definition. The Bott homomorphism $\beta: \tilde{K}(X) \to \tilde{K}(X \wedge S^2)$ is defined by $\beta(\xi) = \xi \otimes (\gamma - 1)$.

The justification for this definition results from the following split exact sequence of 9(3.4).

$$0 \leftarrow \tilde{K}(X \vee S^2) \leftarrow \tilde{K}(X \times S^2) \leftarrow \tilde{K}(X \wedge S^2) \leftarrow 0$$

Since the image of $\xi \otimes (\gamma - 1)$ in $\tilde{K}(X \vee S^2)$ is zero, we can view $\xi \otimes (\gamma - 1) \in \tilde{K}(X \wedge S^2)$.

9.5 Remark. By elementary methods one can prove that the Bott homomorphism $\beta: \tilde{K}(X) \to \tilde{K}(X \wedge S^2)$ is surjective if X is compact.

We give Hirzebruch's version of the Bott periodicity and integrality theorem.

9.6 Theorem. *For each sphere* X *the Bott homomorphism* $\beta: \tilde{K}(X) \to \tilde{K}(X \wedge S^2)$ *is an isomorphism. For each* $k \geq 1$, ch *carries* $\tilde{K}(S^{2k})$ *isomorphically onto* $H^{2k}(S^{2k}, \mathbf{Z}) \subset H^{2k}(S^{2k}, \mathbf{Q})$.

Proof. We begin by proving the Bott map β is an isomorphism for $X = S^{2k}$ and the integrality statement, both by induction on k.

For this, we consider the following diagram:

$$
\begin{array}{ccc}
\tilde{K}(S^{2k}) & \xrightarrow{\hspace{2cm}} & \tilde{K}(S^{2k+2}) \\
\end{array}
$$

$$
\tilde{H}^*(S^{2k}, \mathbf{Z}) \xrightarrow{\hspace{1cm}} \quad \downarrow \text{ch} \quad \xrightarrow{\text{cup with } u} \quad \downarrow \quad \xrightarrow{\hspace{1cm}} \tilde{H}^*(S^{2k+2}, \mathbf{Z})
$$

$$
\tilde{H}^*(S^{2k}, \mathbf{Q}) \xrightarrow{\text{cup with } u} \tilde{H}^*(S^{2k+2}, \mathbf{Q})
$$

Here $u \in H^2(S^2, Z)$ is the canonical generator. The above diagram commutes since $\text{ch}(\gamma - 1) = u$. The bottom two horizontal arrows are isomorphisms. If ch factors through $\tilde{H}^*(S^{2k}, \mathbf{Z})$ by an isomorphism, then β is injective, and, by (9.5), β is bijective. If β is an isomorphism, then ch factors through $\tilde{H}^*(S^{2k+2}, \mathbf{Z})$ by an isomorphism. Since the induction starts at $k = 1$, the integrality theorem holds.

Since $\tilde{K}(S^1) = [S^0, U]_0$ and U is connected, and since β is an epimorphism, we have $\tilde{K}(S^{2k+1}) = 0$ for all k. This proves the theorem.

9.7 Remark. The Bott morphism $\beta: \tilde{K}(X) \to \tilde{K}(X \wedge S^2)$ is an isomorphism for each sphere $X = S^i$. By a standard comparison theorem for half-exact cofunctors, β is a cofunctor isomorphism.

9.8 Corollary. *Let* $a \in H^{2n}(S^{2n}, \mathbf{Z})$ *be a generator. Then for each complex vector bundle* ξ *over* S^{2n}, *the nth Chern class* $c_n(\xi)$ *is a multiple of* $(n-1)!a$, *and for each* m *with* $m \equiv 0 \bmod (n-1)!$ *there exists a unique* $\xi \in \tilde{K}(S^{2n})$ *with* $c_n(\xi) = ma$.

Proof. Since $\text{ch}\,\xi = rk(\xi) \pm nc_n(\xi)/n!$ in $H^0(S^{2k}, \mathbf{Z}) \oplus H^{2k}(S^{2k}, \mathbf{Z})$ we have the first statement. The second follows immediately from the above formula and Theorem (9.6). Note that we used the relation $s_n(\sigma_1, \ldots, \sigma_n) = \pm n\sigma_n + \cdots$ and $s_n(0, \ldots, 0, \sigma_n) = \pm n\sigma_n$.

10. Comparison of K-Theory and Cohomology Definitions of Hopf Invariant

10.1 Let $f: S^{2n-1} \to S^n$ be a map, and form C_f, the cell complex, with one 0 cell, one n cell, and one $2n$ cell. From the sequence of spaces $S^{2n-1} \xrightarrow{f} S^n \to C_f \to SS^{2n-1} = S^{2n}$ there are the following isomorphisms in integral cohomology: $H^n(S^n) \leftarrow H^n(C_f)$ and $H^{2n}(C_f) \leftarrow H^{2n}(S^{2n})$. There are well classes $c_n \in H^n(S^n)$ such that $\mathrm{ch}_n \beta_n = c_n$ for n even where ch_n is the nth component of ch. In this way there are two classes $a^* \in H^n(C_f)$ and $b^* \in H^{2n}(C_f)$ with $\mathrm{ch}_n a_f = a^*$ and $\mathrm{ch}_{2n} b_f = b^*$ for n even. The cohomology Hopf invariant h_f^* is defined by $(a^*)^2 = h_f^* b^*$. By (graded) commutativity of the cup product, $h_f^* = 0$ for n odd. The main result of this section is contained in the next proposition.

10.2 Proposition. *The K-theory Hopf invariant h_f equals the cohomology Hopf invariant h_f^*.*

Proof. The Chern character $\mathrm{ch}: K(X) \to H^{ev}(X, \mathbf{Q})$ is a ring homomorphism. Therefore, we have $\mathrm{ch}\, a_f^2 = \mathrm{ch}\, a_f \, \mathrm{ch}\, a_f = (a^* + rb^*) = (a^*)^2$, where $r \in \mathbf{Q}$. Then from $a_f^2 = h_f b_f$ we derive the relation $h_f b_f^* = \mathrm{ch}\, h_f b_f = \mathrm{ch}\, a_f^2 = (a^*)^2 = h_f^* b^*$ or $h_f = h_f^*$. This proves the proposition.

11. The Borel-Hirzebruch Description of Characteristic Classes

11.1 Notations. Let α be a locally trivial principal G-bundle over X, let T^r be a maximal torus of G of dimension r, and let M be an m-dimensional complex G-module. By restricting the action of G on M to T^r, we get m T^r-modules $M_1, \ldots, M_m$ defined byy m elements of $\mathrm{Hom}(T^r, T^1) = H^1(T^r, \mathbf{Z})$. We wish to calculate the Chern class $c(\alpha[M])$ and the Chern character $\mathrm{ch}(\alpha[M])$ in terms of $M_1, \ldots, M_m$. Here $T^1 = S^1$ is the circle.

11.2 The transgression. Let γ be a principal T^r-bundle and define the transgression $\tau_\gamma: H^1(T^r, \mathbf{Z}) \to H^2(B(\gamma), \mathbf{Z})$ by the following commutative diagram:

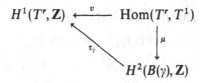

The isomorphism v is defined by the relation $v(h) = h^*(s)$, where s is a generator of $H^1(T, \mathbf{Z})$, and the morphism μ is defined by the relation $\mu(h)$ equals the first Chern class $c_1(\gamma[\mathbf{C}_h])$, where $\mathbf{C}_h$ is the simple T^r-module $\mathbf{C}$ and scalar multiplication tz is equal to $h(t)z$ for $t \in T^r$ and $z \in \mathbf{C}$.

11.3 Proposition. *If γ is the universal T^r-bundle over $(CP^x)^r$, then τ_y: $H^1(T^r, \mathbf{Z}) \to H^2(B(\gamma), \mathbf{Z})$ is an isomorphism.*

Proof. For $r = 1$ the free generator of $H^1(S^1, \mathbf{Z})$ corresponds to the canonical line bundle over CP^∞, and its first Chern class is the free generator of $H^2(CP^\infty, \mathbf{Z})$. For general r the transgression τ_y is the product of r transgressions, where $r = 1$, and consequently τ_y is an isomorphism.

Using the notation (11.1), we associate with a principal G-bundle α a principal T-bundle α_T, where $E(\alpha) = E(\alpha_T)$ and $B(\alpha_T) = E(\alpha) \bmod T$. We have the following commutative diagram:

$$E(\alpha_T) = E(\alpha)$$

$$B(\alpha_T) = E(\alpha) \bmod T \xrightarrow{\quad q_T \quad} B(\alpha) = X$$

The bundle can be proved to be locally trivial, using a local cross section of $G \to G \bmod T$, and q_T is the natural quotient map. The map q_T plays the role of a splitting map, where α_T and $q_T^*(\alpha)$ are $B(\alpha_T)$-isomorphic.

11.4 Theorem. *Let M be an m-dimensional complex G-module, where $M = M_1 \oplus \cdots \oplus M_m$ as T^r-modules. Then*

$$q_T^*(c(\alpha[M])) = \prod_{1 \leq j \leq m} (1 + \tau(M_j))$$

and

$$q_T^*(\mathrm{ch}(\alpha[M])) = \sum_{1 \leq j \leq m} e^{\tau(M_j)}$$

Proof. By the naturality of c and the Whitney sum formula for c, we have

$$q_T^*(c(\alpha[M])) = c(q_T^*(\alpha[M]))$$
$$= c(\gamma[M_1 \oplus \cdots \oplus M_m])$$
$$= c(\gamma[M_1]) \cdots c(\gamma[M_m])$$
$$= (1 + \tau(M_1)) \cdots (1 + \tau(M_m))$$

and

$$q_T^*(\mathrm{ch}(\alpha[M])) = \mathrm{ch}(q_T^*(\alpha[M]))$$
$$= \mathrm{ch}(\gamma[M_1 \oplus \cdots \oplus M_m])$$
$$= \mathrm{ch}(\gamma[M_1]) + \cdots + \mathrm{ch}(\gamma[M_m])$$
$$= e^{\tau(M_1)} + \cdots + e^{\tau(M_m)}$$

This proves the theorem.

Now we state the following theorem of Borel [3].

11.5 Theorem. *For the universal G-bundle* α, *the morphism* $q_T^*: H^*(B_G, \mathbf{Q}) \to H^*(B_T, \mathbf{Q})$ *is a monomorphism and the image of* q_T^* *consists of the subgroup of* $H^*(B_T, \mathbf{Q})$ *of elements left elementwise fixed by the action of the Weyl group on* $H^*(B_T, \mathbf{Q})$.

The statement of (11.5) should be compared with the relation between $R(G)$ and elements of $R(T)$ elementwise fixed by the action of the Weyl group.

11.6 Theorem. *Let* α *be the universal G-bundle, where G is a compact, connected Lie group. There is a commutative triangle*

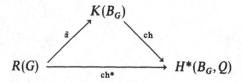

The morphism ch *is the Chern character, and* ch* *has the property that* $\mathrm{ch}^* M = e^{-\tau(M)}$ *for M, a complex T-module.*

This theorem follows from (11.5) by comparing $R(G)$ with $R(T)$ and $H^*(B_G, \mathbf{Q})$ with $H^*(B_T, \mathbf{Q})$.

APPENDIX 1

Dold's Theory of Local Properties of Bundles

In this appendix we state the main results in the important paper of A. Dold [4]. These results are concerned with the relation between local and global properties of bundles.

A.1 Definition. A covering $\{U_i\}$ of a space X is numerable provided there exists a partition of unity $\{\eta_j\}$ on X such that the covering $\{\eta_j^{-1}((0,1])\}$ refines $\{U_i\}$.

We consider four properties of bundles. The main results of Dold are that if these properties hold locally they hold globally. That a property holds locally means it holds over each set in a numerable covering of the space.

A.2 Definition. A halo of a subset A in a space X is an open set V with $A \subset V$ such that there is a map $f: X \to [0,1]$ with $A \subset f^{-1}(1)$ and $f^{-1}((0,1]) \subset V$.

A.3 Definition. The section extension property holds for a bundle $p: E \to X$ provided each cross section s over $A \subset X$ which extends to a halo of A also extends to all of X.

The next statement refers to the local character of the section extension property.

A.4 Theorem. Let $p: E \to X$ be a bundle. If $p: E \to X$ has the section extension property, its restriction to an open subset of X has the section extension property. If $p: E \to X$ has the section extension property over the subsets U_i of a numerable covering $\{U_i\}$, then $p: E \to X$ has the section extension property.

The following definition is stronger than that used in Chap. 1.

A.5 Definition. A bundle $p: E \to X$ is a fibration provided the homotopy lifting property holds for all spaces.

A.6 Theorem. *If a bundle $p \ E \to X$ is a fibration when restricted to the subsets U_i of a numerable covering $\{U_i\}$, then $p: E \to X$ is a fibration.*

The next concept is preserved under fibre homotopy equivalence.

A.7 Definition. A bundle $p: E \to X$ is a weak fibration provided for each homotopy $h_t: W \to X$ and map $f: W \to E$ with $pf = h_0$ there is a homotopy $f_t: W \to E$ such that $pf_t = h_t$. Moreover, it is assumed that f_0 and f are homotopic with a homotopy $k_t: W \to E$ such that $pk_t = f_0 = f$ for all $t \in [0, 1]$ and $k_0 = f_0, k_1 = f$.

A.8 Theorem. *If a bundle $p: E \to X$ is a weak fibration when restricted to the subsets U_i of a numerable covering $\{U_i\}$, then $p: E \to X$ is a weak fibration.*

We used a mild form of the next theorem in Chap. 16.

A.9 Theorem. *Let $f: E \to E'$ be an X-morphism of bundles $p: E \to X$ and $p': E' \to X$ over X. If the restriction of f over subsets U_i of a numerable covering $\{U_i\}$ of X is a fibre homotopy equivalence, then $f: E \to E'$ is a fibre homotopy equivalence.*

In the three statements (A.6), (A.8), and (A.9), the proofs are achieved by reducing the statements to an application of (A.4) by mapping space construction.

On the Double Suspension

We analyze the pair $(\Omega^2 S^{2n+1}, S^{2n-1})$ considered by Toda [2] using the fibre space techniques of John Moore introduced in exposé 22 of the Cartan Séminaire 1954/55 and the process of localizing a space at a prime p. As in the approach of Toda, we consider a filtration $S^{2n-1} \subset \Omega F_n \subset \Omega^2 S^{2n+1}$. In our case the analysis of the pair $(\Omega F_n, S^{2n-1})$ depends on the nonexistence of elements of mod p Hopf invariant one for $n > 1$. The boundary map $\pi_i(\Omega^2 S^{2n+1}, \Omega F_n) \to \pi_{i-1}(\Omega F_n, S^{2n-1})$ is exhibited as an induced map of the homotopy groups of spaces and related to the double suspension.

We develop the necessary properties of ΩSX and $S\Omega SX$ in the first few sections and then give an exposition of the results related to the mod p Hopf invariant. The single suspension sequence of I James (see Annals of Math., vol. 63, 1956, pp. 407–429) is included.

1. $H_*(\Omega S(X))$ as an Algebraic Functor of $H_*(X)$

Let R be a commutative ring. In order to calculate $H_*(\Omega S(X), R)$ in terms of $H_*(X, R)$, we use the following preliminaries on the path space fibration $\pi: E(Y) \to Y$ where $\pi(\delta) = \delta(1)$ and $\pi^{-1}(*) = \Omega(Y)$. We define μ to be the path product making the following diagram commutative

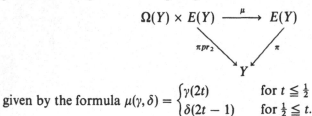

given by the formula $\mu(\gamma, \delta) = \begin{cases} \gamma(2t) & \text{for } t \leq \frac{1}{2} \\ \delta(2t - 1) & \text{for } \frac{1}{2} \leq t. \end{cases}$

For $E(Y) \times E(Y)$ the subspace of $(\delta_1, \delta_2) \in E(Y) \times E(Y)$ with $\pi(\delta_1) = \pi(\delta_2)$, we define the path product $v: E(Y) \underset{\pi}{\times} E(Y) \to \Omega(Y)$ by the formula

$$v(\delta_1, \delta_2) = \begin{cases} \delta_1(2t) & t \leq \frac{1}{2} \\ \delta_2(2 - 2t) & \frac{1}{2} \leq t \end{cases}$$

Let $S_+(X)$ (resp. $S_-(X)$) denote the subspace of all $[x, t] \in S(X)$ where $\frac{1}{2} \leq t$ (resp. $t \leq \frac{1}{2}$), and identify X with $S_+(X) \cap S_-(X)$. Restrict the pathspace fibration $\pi: E = E(\Omega(X)) \to S(X)$ to $\pi_\pm: E_\pm \to S_\pm(X)$ where $E_\pm = \pi^{-1}(S_\pm(X))$ over the cones $S_\pm(X)$.

Now define maps $f_\pm$ and $g_\pm$ mapping the following diagrams commutative.

$$\Omega S(X) \times S_\pm(X) \underset{g_\pm}{\overset{f_\pm}{\rightleftarrows}} E_\pm$$

with pr_2 and π mapping to $S_\pm(X)$.

Let $u_\pm: S_\pm(X) \to E_\pm$ be the canonical cross sections given by $u_-[x, s](t) = [x, st]$ and $u_+[x, s](t) = [x, 1 - (1 - s)t]$. Then we define $f_\pm(\gamma, [x, s]) = \mu(\gamma, u_\pm([x, s]))$ and $g_\pm(\delta) = (v(\delta, u_\pm(\pi(\delta))), \pi(\delta))$. An easy construction will show that $f_\pm$ and $g_\pm$ are fibre homotopy inverses of each other.

Let $\beta(X) = \beta: X \to \Omega S(X)$ be the natural map where $\beta(X)(t) = [x, t]$. We denote $H_*(X, x_0) = \bar{H}_*(X)$ for a pointed space X.

1.1 Proposition. *Let X be a pointed space. If $H_*(X) = H_*(X, R)$ is R-flat, then the following composite is an isomorphism where ϕ is the multiplication on the algebra $H_*(\Omega S(X))$.*

$$H_*(\Omega S(X)) \otimes \bar{H}_*(X) \xrightarrow{1 \otimes H_*(\beta)} H_*(\Omega S(X)) \otimes \bar{H}_*(\Omega S(X))$$

$$\downarrow$$

$$\bar{H}_*(\Omega S(X))$$

Proof. We have the following commutative diagram where the top horizontal morphism and the morphisms in the last column are isomorphisms.

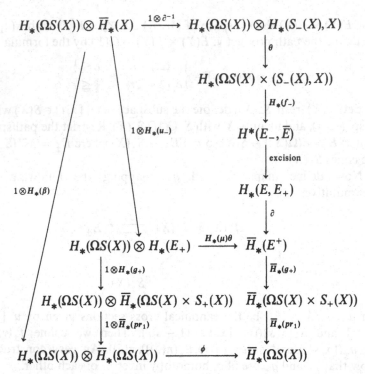

Since $\beta: X \to \Omega S(X)$ is the composite

$$X \xrightarrow{u_-} E^+ \xrightarrow{g_+} \Omega S(X) \times S_-(X) \xrightarrow{pr_1} \Omega S(X)$$

the triangle on the left is commutative. The large rectangle is commutative from the relation between u_- and f_- and the naturality of boundary morphisms. Since $\phi(1 \otimes H_*(\beta))$ is a composite of isomorphisms, it is an isomorphism and this proves the proposition.

1.2 Proposition. *Let C be a connected graded R-coalgebra, let A be a connected graded R-Hopf algebra, and let $i: C \to A$ be a morphism of coalgebras such that the following composite is an isomorphism*

$$A \otimes I(C) \xrightarrow{A \otimes I(i)} A \otimes I(A) \xrightarrow{\phi(A)} I(A)$$

Then the unique Hopf algebra morphism $f: T(C) \to A$, of the tensor Hopf algebra $T(C)$ into A induced by $i: C \mapsto A$, is an isomorphism.

Proof. Since $A = R \oplus I(A)$ as modules, we can iterate the above given isomorphism to an isomorphism

$$\coprod_{1 \leq k < m} I(C)^{k \otimes} \oplus (A \otimes I(C)^{m \otimes}) \to I(A)$$

whose restriction $f_m: \coprod_{1 \leq k < m} I(C)^{k \otimes} \to I(A)$ is just the restriction of f to this part of the tensor algebra. By the connectivity hypothesis f_m is an isomorphism in degrees $<m$ and hence $f: T(C) \to A$ is an isomorphism because $I(C)^{m\otimes}$ is zero in degrees $<m$. This proves the proposition.

Putting together (1.1) and (1.2), we have the following theorem which was pointed out to us by John Moore.

1.3 Theorem. *Let X be a path connected space such that $H_*(X) = H_*(X, R)$ is R-flat. Then the coalgebra morphism $H_*(\beta): H_*(X) \to H_*(\Omega S(X))$ is factored by a natural morphism of Hopf algebras $TH_*(X) \to H_*(\Omega S(X))$ which is an isomorphism.*

Concerning cohomology, the following result comes from duality.

1.4 Theorem. *Let X be a path connected space such that $H_*(X)$ is R-projective of finite type. Then the algebra morphism $H_*(\beta): H^*(\Omega S(X)) \to H^*(X)$ is factored by a natural morphism of Hopf algebras $H^*(\Omega S(X)) \to T'H^*(X)$ which is an isomorphism where T' is the tensor Hopf algebra on the algebra $H_*(X)$.*

Moreover, if $R = \mathbf{F}_p$ the field with p elements, then the action of the mod p Steenrod algebra on $H^*(X)$ induces an action on $T'H^*(X)$ and the above isomorphism preserves this action.

Proof. The first statement follows from (1.3) by duality. The second statement follows by observing that the dual of the composite isomorphism of (1.1) defined $\overline{H}^*(\Omega S(X)) \to H^*(\Omega S(X)) \otimes \overline{H}^*(X)$ is an isomorphism preserving the Steenrod algebra. This is carried through in the dual to (1.2).

The above theorems yield immediately the first suspension theorem.

1.5 Theorem. *Let X be an $(n-1)$-connected space for $n > 1$. The pair $(\Omega SX, X)$ is $(2n-1)$ connected and $\beta: X \to \Omega SX$ induces $\pi_i(X) \to \pi_i(\Omega SX) \simeq \pi_{i+1}(SX)$ which is an isomorphism for $i \leq 2n - 2$ and an epimorphism for $i \leq 2n - 1$.*

Proof. Since $H_*(\Omega SX, R) = T(\overline{H}_*(X, R))$ and $\overline{H}_i(X, R) = 0$ for $i < n$ and R a field, it follows that β induces an isomorphism $H_i(X, R) \to H_i(\Omega SX, R)$ for $i \leq 2n - 1$ and so $H_i(\Omega SX, X; R) = 0$ for $i \leq 2n - 1$. This holds over any field R and thus over any ring R by the universal coefficient theorem. Since X is simply connected, the pair $(\Omega SX, X)$ is also simply connected and by the Whitehead theorem $\pi_i(\Omega SX, X) = 0$ for $i \leq 2n - 1$ and $(\Omega SX, X)$ is $(2n-1)$-connected. The second statement follows from the exact homotopy sequence of a pair, and this proves the theorem.

1.6 Corollary. *For S^n the n sphere where $n \geq 2$, the pair $(\Omega S^{n+1}, S^n)$ is $(2n-1)$-connected and the induced morphism $\pi_i(S^n) \to \pi_i(\Omega S^{n+1}) = \pi_{i+1}(S^{n+1})$ is an isomorphism for $i \leq 2n - 2$ and an epimorphism for $i \leq 2n - 1$.*

Proof. Since S^n is $(n-1)$-connected the corollary follows from the theorem.

The induced morphism $\pi_i(\beta)$: $\pi_i(S^n) \to \pi_i(\Omega S^{n+1}) = \pi_{i+1}(S^{n+1})$ is just the morphism which assigns to the homotopy class of u: $S^i \to S^n$ the homotopy class of its suspension Su: $S^{i+1} \to S^{n+1}$. Hence the terminology suspension morphism while the notation E: $\pi_i(S^n) \to \pi_{i+1}(S^{n+1})$ for $\pi_i(\beta)$ comes from the fact that suspension in German is Einhängung.

2. Connectivity of the Pair $(\Omega^2 S^{2n+1}, S^{2n-1})$ Localized at p

The connectivity of the localized pair $(\Omega^2 S^{2n+1}, S^{2n-1})$ will be derived from the following calculation, due to J. C. Moore, of $H_*(\Omega^2 S^{2n+1}, k)$ where k will denote in this section a field of characteristic $p > 2$. In [2] J. C. Moore did this calculation as a structure theorem on Hopf algebras, while here we sketch another argument of Moore which is an easy consequence of the induced fibration spectral sequence of Eilenberg and Moore.

2.1 Theorem. *As a Hopf algebra over k of characteristic $p > 2$*

$$H_*(\Omega^2 S^{2n+1}, k) = \bigotimes_{0 \le i} E(u_i, 2np^i - 1) \otimes \bigotimes_{1 \le j} S(v_j, 2np^j - 2)$$

Proof. As a Hopf algebra $H^*(\Omega S^{2n+1}, k) = S'(w, 2n)$ by (1.4) and as an algebra for $w_j = \gamma_{p^j}(w)$ we have

$$H^*(\Omega S^{2n+1}, k) = \bigotimes_{1 \le j} S(w_j, 2np^j)/(w_j^p)$$

To see that the spectral sequence going from the cohomology of the base ΩS^{2n+1} to the fibre $\Omega^2 S^{2n+1}$ collapses, we recall that it is a spectral sequence of Hopf algebras with

$$E_2 = \mathrm{Tor}^{H^*(\Omega S^{2n+1}, k)}(k, k)$$

or

$$E^2 = \mathrm{Cotor}^{H_*(\Omega S^{2n+1}, k)}(k, k)$$

This Hopf algebra E^2 is a tensor product of polynomial algebras generated by elements on the -2 filtration line and exterior algebras generated by elements on the -1 filtration line which are homology suspensions. Since the image of the homology suspension is stationary in the spectral sequence and since the other factor of the spectral sequence is in even total degree, the spectral sequence collapses. Since $E^\infty = E^0 H_*(\Omega^2 S^{2n+1}, k)$ is free commutative, the theorem follows.

2.2 Corollary. *For* $i < 2pn - 2$ *we have* $\bar{H}_i(\Omega^2 S^{2n+1}, S^{2n-1}, k) = 0$ *and* $\pi_i(\Omega^2 S^{2n+1}, S^{2n-1})_{(p)} = 0$.

2.3 Corollary. *The double suspension*

$$\pi_i(S^{2n-1})_{(p)} \to \pi_i(\Omega^2 S^{2n+1})_{(p)} = \pi_{i+2}(S^{2n+1})_{(p)}$$

is an isomorphism for $i < 2np - 2$ and an epimorphism for $i \leq 2np - 2$.

Note that corollaries (2.2) and (2.3) can be derived from a simple application of the total space (Serre) spectral sequence, and they should be regarded as refinements of (1.6).

We will make several uses of the following calculation.

2.4 Proposition. *For an odd sphere S^{2n-1} the pair $(\Omega^2 S^{2n+1}, S^{2n-1})$ localized at a prime p has the following first nonzero homology group $H_{2np-2}(\Omega^2 S^{2n+1}, S^{2n-1}) = \mathbf{Z}/p\mathbf{Z}$.*

Proof. We make the calculation in cohomology over $\mathbf{Z}_{(p)}$. The total space spectral sequence for the path space fibration over ΩS^{2n+1} has $E_2^{*,*} = S'(x, 2n) \otimes H^*(\Omega^2 S^{2n+1})$ and $E_\infty^{*,*} = \mathbf{Z}_{(p)}$. Thus $H^{2n-1}(\Omega^2 S^{2n+1}) = \mathbf{Z}_{(p)} y$, $E_2^{*,*} = E_{2n}^{*,*}$, and $d_{2n}(y) = x$. Hence $d_{2n}(y\gamma_m(x)) = (m+1)\gamma_{m+1}(x)$ and $E_{2n+1}^{a,b} = 0$ for $(a,b) \neq (0,0)$ and $a < 2np$ or $b < 2np - 1$. Since $d_{2n}(y\gamma_{p-1}(x)) = p\gamma_p(x)$, $E_{2n+1}^{2np,0} = \cdots = E_{2np}^{2np,0} = \mathbf{Z}/p$, and because $d_{2np}: E_{2np}^{0,2np-1} \to E_{2np}^{2np,0}$ must be an isomorphism for $E_\infty^{*,*} = \mathbf{Z}_{(p)}$, it follows that $H^{2np-1}(\Omega^2 S^{2n+1}, \mathbf{Z}_{(p)}) = \mathbf{Z}/p$. Hence by the universal coefficient theorem $H_{2np-1}(\Omega^2 S^{2n+1}, \mathbf{Z}_{(p)}) = H_{2np-1}(\Omega^2 S^{2n+1}, S^{2n-1}, \mathbf{Z}_{(p)}) = \mathbf{Z}/p$ and this proves the proposition.

Proposition (2.4) holds for $p = 2$ while (2.1) does not as it stands. The reader is invited to reformulation and prove (2.1) for $p = 2$.

3. Decomposition of Suspensions of Products and $\Omega S(X)$

3.1 Proposition. *For two points spaces X and Y there is a homotopy equivalence $S(X \times Y) \to S(X \wedge Y) \vee S(X) \vee S(Y)$.*

Proof. We give two arguments for the proof. The projections $X \times Y \to X \wedge Y$, $X \times Y \to X$, and $X \times Y \to Y$ induce the maps $S(X \times Y) \to S(X \wedge Y)$, $S(X \times Y) \to S(X)$ and $S(X \times Y) \to S(Y)$. The coHspace structure of the suspension defines a map $S(X \times Y) \to \bigvee_3 S(X \times Y)$ which composed with the wedge of these three maps yields the desired map $S(X \times Y) \to S(X \wedge Y) \vee S(X) \vee S(Y)$. Now this map is seen to be a homology isomorphism for any ring of coefficients, and this proves the proposition at least in the case when the spaces X and Y are connected and have the homotopy type of a CW complex. Alternatively for any space W the following is a split exact sequence of groups

$$[S(X \wedge Y), W]_0 \qquad [S(X \times Y), W]_0 \qquad [S(X \vee Y), W]_0$$

$$0 \longrightarrow \qquad \| \qquad \longrightarrow \qquad \| \qquad \underset{s}{\overset{\longrightarrow}{\longleftarrow}} \qquad \| \qquad \longrightarrow 0$$

$$[X \wedge Y, \Omega W]_0 \qquad [X \times Y, \Omega W]_0 \qquad [X \vee Y, \Omega W]_0$$

$$= [X, \Omega W]_0 \times [Y, \Omega W]_0$$

Again the projections $X \times Y \to X$ and $X \times Y \to Y$ induce the splitting s. Hence we have an isomorphism of group valued functors of the pointed space W

$$[S(X \wedge Y), W]_0 \times [S(X), W]_0 \times [S(Y), W]_0$$
$$= [S(X \wedge Y) \vee S(X) \vee S(Y), W]_0 \to [S(X \times Y), W]_0$$

which must be defined by a homotopy equivalence

$$S(X \times Y) \to S(X \wedge Y) \vee S(X) \vee S(Y)$$

This proves the proposition.

Since $\beta = \beta(X): X \to \Omega S(X)$ is a map into an H-space, it defines by taking products $\beta_n: X^n \to \Omega S(X)$ and we suspend this map and apply inductively the splitting of the suspension of a product

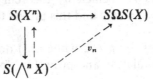

3.2 Theorem. The maps $v_n: S(\bigwedge_n X) \to S\Omega S(X)$ collect to define a map $v: \bigvee_{1 \leq n} S(\bigwedge_n X) \to S\Omega S(X)$ which is a weak homotopy equivalence for path connected X.

Proof. By (1.3), $H_*(\Omega S(X)) = TH_*(X)$ over any field. Since $\bar{H}_*(\bigwedge_n X) = \bar{H}_*(X)^{\otimes n}$ over a field, we see that

$$H_*(v_n): S\bar{H}_*(X)^{\otimes n} = \bar{H}_*(S(\bigwedge_n X)) \to \bar{H}_*(S\Omega S(X))$$

maps isomorphically onto the suspension of the nth summand in $\bar{H}_*(\Omega S(X)) = \coprod_{1 \leq n} \bar{H}_*(X)^{\otimes n}$ and the collection $v: \bigvee_{1 \leq n} S(\bigwedge_n X) \to S\Omega S(X)$ induces an isomorphism of homology with any field coefficients. Since the spaces are simply connected being suspensions of path connected spaces, v is a weak homotopy equivalence. This proves the theorem.

3.3 Corollary. The maps $v_n: S^{nr+1} \to S\Omega S^{r+1}$ collect to define a map $v: \bigvee_{1 \leq n} S^{nr+1} \to S\Omega S^{r+1}$ which is a homotopy equivalence.

Proof. The proof follows from (3.2) and the Whitehead theorem.

3.4 Remark. With the above notations for a map $f: S^r \to S^r$ of degree q, it follows that $\beta_n(f \times \overset{(n)}{\cdots} \times f) = \Omega S(f)$,

$$\Omega S(f)_*: H_{nr}(\Omega S^{r+1}) \to H_{nr}(\Omega S^{r+1})$$

is multiplication by q^r, and the following diagram is commutative up to homotopy

$$
\begin{array}{ccc}
\displaystyle\bigvee_{1 \le n} S^{nr+1} & \overset{v}{\longrightarrow} & S\Omega(S^{r+1}) \\
\Big\downarrow & & \Big\downarrow{\scriptstyle S\Omega S(f)} \\
\displaystyle\bigvee_{1 \le n} S^{nr+1} & \overset{v}{\longrightarrow} & S\Omega(S^{r+1})
\end{array}
$$

$S(f) \wedge f \wedge \overset{(n-1)}{\cdots} \wedge f$

Let $u: S\Omega S^{n+1} \to \bigvee_{1 \le i} S^{ni+1}$ be a homotopy inverse of $v: \bigvee_{1 \le i} S^{ni+1} \to S\Omega S^{n+1}$ considered in (3.3), and let $p_{(q)}: \bigvee_{1 \le i} S^{ni+1} \to S^{nq+1}$ be the restriction to the wedge of the projection onto the qth factor. Form the composite $p_{(q)}u: S\Omega S^{n+1} \to S^{nq+1}$, take the adjoint map $\Omega S^{n+1} \to \Omega S^{nq+1}$, and make it into a fibre map which we will denote $f_{(q)}: \Omega S^{n+1} \to \Omega S^{nq+1}$ with $f_{(q)}(S^n) = *$. Let $F_n(q)$ denote the fibre of the map $f_{(q)}$. In fact $f_{(q)}: \Omega SX \to \Omega S(\bigwedge^q X)$ can be defined for any space X where the map v in (3.2) is homotopy equivalence. This would be the case for any CW complex.

For odd spheres the cohomology algebra $H^*(\Omega S^{2n+1}, R)$ is the divided power algebra $S'(x, 2n)$ by (1.4) and $H^*(f_{(q)})$ is described in the next proposition.

3.5 Proposition. *With the above notations the map* $f_{(q)}: \Omega S^{2n+1} \to \Omega S^{2nq+1}$ *induces* $f_{(q)}^*: H^{2knq}(\Omega S^{2nq+1}, R) \to H^{2knq}(\Omega S^{2n+1}, R)$ *which is multiplication by* $[k!(q!)^k/(qk)!]$ *on modules of rank 1.*

Proof. For $H^*(\Omega S^{2n+1}, R) = S'(x, 2n)$ and $H^*(\Omega S^{2nq+1}, R) = S'(y, 2nq)$,

$$f_{(q)}^*: H^{2nq}(\Omega S^{2nq+1}, R) = Ry \to H^{2nq}(\Omega S^{2n+1}, R) = R\gamma_q(x)$$

is an isomorphism and we choose y such that $f_{(q)}^*(y) = \gamma_q(x)$. Since the calculation is independent of R, we can assume R is the integers and use $m! \gamma_m(x) = x^m$. We calculate

$$f_{(q)}^*(\gamma_k(y)) = k! f_{(q)}^*(y^k) = k! [\gamma_q(x)]^k = k!(q!)^k x^{qk}$$

$$= [k!(q!)^k/(qk)!] \gamma_{qk}(x)$$

This proves the proposition because $H^{2knq}(\Omega S^{2nq+1}, R) = R\gamma_k(y)$ and $H^{2knq}(\Omega S^{2n+1}, R) = R\gamma_{kq}(x)$.

In the next section we study the fibre map $f_{(2)}: \Omega S^{n+1} \to \Omega S^{2n+1}$ for both even and odd spheres. In order to study whether or not $f_{(q)}^*$ is an isomorphism we use the notation $i = \text{ord}_p x$ where x is a nonzero rational number of the form $x = p^i \frac{a}{b}$ and a, b are integers not divisible by the prime p. If $[x]$ denotes the largest integer $\leq x$, then one shows easily that

$$\mathrm{ord}_p(k!) = \left[\frac{k}{p}\right] + \left[\frac{k}{p^2}\right] + \cdots + \left[\frac{k}{p^i}\right] + \cdots$$

which is a finite sum terminating when $k < p^i$. When $q = p^j$ in (3.5) we calculate that

$$\mathrm{ord}_p[k!(q!)^k/(qk)!] = \mathrm{ord}_p(k!) + k\,\mathrm{ord}_p(p^j!) - \mathrm{ord}_p(p^j k!)$$

$$= \left[\frac{k}{p}\right] + \left[\frac{k}{p^2}\right] + \cdots + k(1 + p + \cdots + p^{j-1})$$

$$- \left(\left[\frac{p^j k}{p}\right] + \cdots + \left[\frac{p^j k}{p}\right] + \cdots\right) = 0$$

This means that $[k!(q!)^k/(qk)!]$ is a unit over any $\mathbf{Z}_{(p)}$ − algebra R and $(f_{(p^j)})^*$: $H^{2knq}(\Omega S^{2nq+1}, R) \to H^{2knq}(\Omega S^{2n+1}, R)$ is an isomorphism. Moreover

$$\mathrm{Tor}^{H^*(\Omega S^{2nq+1}, R)}(R, H^*(\Omega S^{2n+1}, R)) = R \otimes_{H^*(\Omega S^{2nq+1}, R)} H^*(\Omega S^{2n+1}, R)$$

and this is the cohomology of the fibre $F_n(q)$ of $f_{(q)}: \Omega S^{2n+1} \to \Omega S^{2nq+1}$ localized at p for the case $q = p^j$. From this discussion the following theorem results immediately using the relation

$$(mp^i, k) \equiv 1 \ (\mathrm{mod}\, p) \qquad \text{for } 0 < k < mp^i.$$

3.6 Theorem. *The fibre $F_n(q)$ of $\Omega S^{2n+1} \to S^{2nq+1}$ localized at p for $q = p^j$ has cohomology $H^*(F_n(q), R) = S'(x, 2n)$ truncated with $\gamma_i(x) = 0$ for $i \geq q$ and homology $H_*(F_n(q), R)$ the subcoalgebra of $T(x, 2n) = S(x, 2n)$ consisting of those elements in degrees $\leq 2nq - 1$.*

4. Single Suspension Sequences

In order to investigate further the terms $\pi_*(\Omega S^{n+1}, S^n)$, considered first in (1.6), we study the map $f_{(2)}: \Omega S^{n+1} \to \Omega S^{2n+1}$ introduced in section 3 and examined in (3.5). We have already used the relation $H^*(\Omega S^{2n+1}, R) = S'(2n, R)$ for an odd sphere and now we remark that for an even sphere $H^*(\Omega S^m, R) = E(y, m - 1) \otimes S'(z, 2m - 2)$ where $E(y, m - 1)$ is the exterior algebra on the generator y in degree $m - 1$.

4.1 Proposition. *For the fibre map $f_{(2)}: \Omega S^{n+1} \to \Omega S^{2n+1}$ with fibre F_n and $n \geq 2$, the inclusion $j: S^n \to F_n$ is a homotopy equivalence for n odd and localized at 2 the map $j: S^n \to F_n$ is a homotopy equivalence for n even.*

Proof. For n odd, $n + 1$ is even and $H^*(\Omega S^{n+1}, R) = E(y, n) \otimes S'(z, 2n)$ and $f_{(2)}^*: H^*(\Omega S^{2n+1}, R) = S'(x, 2n) \to H^*(\Omega S^{n+1}, R)$ carries $H^*(\Omega S^{2n+1}, R)$ isomorphically onto the factor $S'(z, 2n)$, and $H^*(F_n, R) = E(y, n)$ so that $j^*: H^*(F_n, R) \to H^*(S^n, R)$ and $j_*: H_*(S^n, R) \to H_*(F_n, R)$ are isomorphisms for all R. This proves the first statement.

For n even, $n + 1$ is odd and $j_*: H_*(S^n, R) \to H_*(F_n, R)$ is an isomorphism for each $\mathbf{Z}_{(2)}$-algebra by (3.6). This proves the theorem.

The fibre map $f_{(2)}: (\Omega S^{n+1}, F_n(2)) \to (\Omega S^{2n+1}, *)$ induces an isomorphism $\pi_*(\Omega S^{n+1}, F_n(2)) \to \pi_*(\Omega S^{2n+1}, *)$ and leads to the following diagram with an exact triangle

$$\pi_*(S^n) \longrightarrow \pi_*(F_n(2)) \longrightarrow \pi_*(\Omega S^{n+1})$$

$$\overset{\text{deg}=1}{\swarrow} \qquad \searrow$$

$$\pi_*(\Omega S^{n+1}, F) \underset{\sim}{\longrightarrow} \pi_*(\Omega S^{2n+1})$$

From this and (4.1) we deduce the single suspension sequences immediately. These are also called the *EHP* sequences.

4.2 Theorem. *For n odd the following triangle is exact.*

$$\pi_*(S^n) \overset{E}{\longrightarrow} \pi_*(\Omega S^{n+1})$$

$$\overset{P}{\nwarrow} \qquad \swarrow H = f_{(2)*}$$

$$\pi_*(\Omega S^{2n+1})$$

For n even the following triangle of 2 torsion groups is exact.

$$\pi_*(S^n)_{(2)} \longrightarrow \pi_*(\Omega S^{n+1})_{(2)}$$

$$\overset{P}{\nwarrow} \qquad \swarrow$$

$$\pi_*(\Omega S^{2n+1})_{(2)}$$

Here P is the boundary morphism of degree -1.

Since $\pi_i(\Omega S^n, S^{n-1}) = 0$ for $i < 2n - 2$, the Hurwicz morphism $h: \pi_{2n-2}(\Omega S^n, S^{n-1}) \to H_{2n-2}(\Omega S^n, S^{n-1}) = \mathbf{Z}$ is an isomorphism.

4.3 Definition. The (integral) Hopf invariant $H: \pi_{2n-1}(S^n) \to \mathbf{Z}$ is the composite morphism

$$\pi_{2n-1}(S^n) = \pi_{2n-2}(\Omega S^n) \overset{H}{\to} \pi_{2n-2}(\Omega S^n, S^{n-1}) \overset{h}{\to} H_{2n-2}(\Omega S^n, S^{n-1}) = \mathbf{Z}$$

Because of the ambiguity in the identification of $\mathbf{Z}$ with $H_{2n-2}(\Omega S^n, S^{n-1})$ the morphism $H: \pi_{2n-1}(S^n) \to \mathbf{Z}$ is well defined only up to sign. From the following commutative diagram with indicated isomorphisms

$$\pi_{2n-1}(S^n) = \pi_{2n-2}(\Omega S^n) \longrightarrow \pi_{2n-2}(\Omega S^n, S^{n-1})$$

$$\downarrow h \qquad\qquad\qquad \wr \downarrow h$$

$$\mathbf{Z} \cdot \gamma_2(x) = H_{2n-2}(\Omega S^n) \xrightarrow{\ \sim\ } H_{2n-2}(\Omega S^n, S^{n-1})$$

we see that $h(\alpha) = H(\alpha)\gamma_2(x)$ up to sign for any $\alpha \in \pi_{2n-1}(S^n) = \pi_{2n-2}(\Omega S^n)$ and this provides an alternative formulation for the Hopf invariant H: $\pi_{2n-1}(S^n) \to \mathbf{Z}$.

4.4 Theorem. (*Serre*) *Let* $f: S^{2n-1} \to S^n$ *be a map and let* C_f *denote its mapping cone. Then* $H^n(C_f) = \mathbf{Z}v$, $H^{2n}(C_f) = \mathbf{Z}w$, *and* $v^2 = \pm H([f])w$ *where* $[f]$ *is the homotopy class of* f.

Proof. Let $S^{2n-1} \to M_f \to S^n$ be the mapping cylinder factorization of f where $M_f/S^{2n-1} = C_f$ the mapping cone and $S^n \to M_f$ is a homotopy equivalence. Then $H^n(C_f) = H^n(M_f, S^{2n-1}) = \mathbf{Z}v$ and $H^{2n}(C_f) = H^{2n}(M_f, S^{2n-1}) = \mathbf{Z}w$ are the only nonzero cohomology groups in this two cell complex. If $v^2 = hw$, then we must show $h = \pm H([f])$.

Let $E \to M_f$ be the path space fibration and let $E' \to S^{2n-1}$ be its restriction to S^{2n-1}. The fibre can be taken to be $\Omega S^n \subset E' \subset E$ because $S^n \to M_f$ is a homotopy equivalence. Now the groups $H^{2n-1}(E') \gtrsim H^{2n}(E, E')$ isomorphic, and we will prove that $\pm h = H([f])$ by showing that $H^{2n-1}(E') = \mathbf{Z}/H([f])$ and $H^{2n}(E, E') = \mathbf{Z}/h$.

For $H^{2n}(E, E')$ we consider the spectral sequence E_r of the pair of fibre spaces (E, E') where $E_2^{*,*} = H^*(M_f, S^{2n-1}) \otimes H^*(\Omega S^n)$. So $E_2^{a,b} = 0$ for $a \neq 0$, n, $2n$ or $b \not\equiv 0 \bmod(n-1)$. Thus $E_n = E_2$ and $E_n^{n,n-1} = E_2^{n,n-1}$ is generated by $v \otimes x$ and $E_n^{2n,0} = E_2^{2n,0}$ is generated by $w \otimes 1$. Calculate

$$d_n(v \otimes x) = (v \otimes 1)d_n(1 \otimes x) = \pm(v \otimes 1)^2 = \pm h(w \otimes 1)$$

using the action $E \times (E, E') \to (E, E')$ to see that $d_n((v \otimes 1)(1 \otimes x)) = (v \otimes 1)d_n(1 \otimes x)$. Hence it follows that $H^{2n}(E, E') = E_{n+1}^{2n,0} = \mathbf{Z}/h$.

If $g: S^{2n-2} \to \Omega S^n$ denotes the adjoint of $f: S^{2n-1} \to S^n$, then g_*: $H_{2n-2}(S^{2n-2}) \to H_{2n-2}(\Omega S^n)$ is multiplication by $h' = H([f])$. In the following commutative diagram the image of α is generated by $[g]$

$$\pi_{2n-1}(S^{2n-1}) \longrightarrow \pi_{2n-2}(\Omega S^n) \longrightarrow \pi_{2n-2}(E') \longrightarrow 0$$

$$\wr\downarrow \qquad\qquad\qquad \downarrow \qquad\qquad\qquad \downarrow$$

$$H_{2n-1}(S^{2n-1}) \longrightarrow H_{2n-2}(\Omega S^n) \longrightarrow H_{2n-2}(E') \longrightarrow 0$$

Hence $H_{2n-2}(E') = \mathbf{Z}/h'$ and in cohomology $H^{2n-1}(E') = \mathbf{Z}/h'$ since it is the cokernel of $H^{2n-2}(\Omega S^n) \to H^{2n-1}(S^{2n-1})$. This proves the theorem.

4.5 Corollary. *For* n *odd and* $\alpha \in \pi_{2n-1}(S^n)$ *the Hopf invariant* $H(\alpha) = 0$.

Proof. The cup square of an odd dimensional class is zero.

4.6 Remark. If $f: S^{2n-1} \to S^n$ is a map, if $v: S^n \to S^n$ has degree a, and if $u: S^{2n-1} \to S^{2n-1}$ has degree b, then $H([vfu]) = a^2 b H([f])$.

The Hopf construction on a map $g: X \times Y \to Z$ is a map $g^0: X * Y \to SZ$ where $X * Y$ is the join of X and Y defined as the quotient of $X \times [0,1] \times Y$ where $(x, 0, y)$ and $(x', 0, y)$ are identified and $(x, 1, y)$ and $(x, 1, y')$ are identified, and g^0 is given by $g^0(x, t, y) = \langle g(x, y), t \rangle \in S(Z)$. Note that $S^p * S^q$ is homeomorphic to S^{p+q+1}.

The bidegree of a map $g: S^{n-1} \times S^{n-1} \to S^{n-1}$ is (d_1, d_2) provided $x \mapsto g(x, y)$ has degree d_1 and $y \mapsto g(x, y)$ has degree d_2.

4.7 Remark. If $g: S^{n-1} \times S^{n-1} \to S^{n-1}$ is a map of bidegree (d_1, d_2), then for the Hopf construction $g^0: S^{2n-1} \to S^n$ the Hopf invariant $H([g^0]) = d_1 d_2$, see 14(3.5).

4.8 Example. In 8(10.2) we prove the existence of a map $S^{n-1} \times S^{n-1} \to S^{n-1}$ of bidegree $(+1 + (-1)^n, -1)$. Hence for $n = 2m$, there is a map $S^{4m-1} \to S^{2m}$ of Hopf invariant 2 by (4.7).

4.9 Example. The restriction to the unit spheres $S^{n-1} \times S^{n-1} \to S^{n-1}$ of multiplication in the complex numbers, quaternions, and Cayley numbers yields H-space structures on S^{n-1} for $n = 2$, 4, and 8 respectively. These maps of bidegree $(1, 1)$ give maps of Hopf invariant 1 defined $S^3 \to S^2$ for $n = 2$, $S^7 \to S^4$ for $n = 4$, and $S^{15} \to S^8$ for $n = 8$.

In section 6 we will see that these are the only integers m for which there is a map $S^{4m-1} \to S^{2m}$ of Hopf invariant 1. The infinite real projective space $\mathbf{P}_\infty(\mathbf{R})$ is the classifying space for the H-space S^0, the infinite complex projective space $\mathbf{P}_\infty(\mathbf{C})$ for S^1, and the infinite quaternionic projective space $\mathbf{P}_\infty(\mathbf{H})$ for S^3. In section 6 we will prove that S^7 has no classifying space, in fact, it is not homotopy associative.

4.10 Remark. The existence of a map $S^{n-1} \times S^{n-1} \to S^{n-1}$ of bidegree $(2, -1)$ for any odd sphere shows that the localization at p of an odd sphere $S^{2m+1}_{(p)}$ is an H-space. In section 6 we will see which of these H-spaces have classifying spaces.

4.11 Remark. Let $f: S^{4n-1} \to S^{2n}$ be a map of Hopf invariant h. Then a straightforward argument shows that the EHP exact triangle in (4.2) tensored with $\mathbf{Z}\left[\dfrac{1}{h}\right]$ is decomposed into a split short sequence.

$$0 \to \pi_+(S^{2n-1}) \otimes \mathbf{Z}\left[\frac{1}{h}\right] \xrightarrow{E} \pi_*(\Omega S^{2n}) \otimes \mathbf{Z}\left[\frac{1}{h}\right] \underset{\rightleftarrows}{\overset{H}{}} \pi_*(\Omega S^{4n-1}) \otimes \mathbf{Z}\left[\frac{1}{h}\right] \to 0$$

For $n = 1$, 2, and 4 we have a map of Hopf invariant 1 and hence an isomorphism

$$\pi_*(S^{2n-1}) \oplus \pi_*(\Omega S^{4n-1}) \to \pi_*(\Omega S^{2n}).$$

For any n we have a map of Hopf invariant 2 and hence an isomorphism of the localizations at an odd prime

$$\pi_*(S^{2n-1})_{(p)} \oplus \pi_*(\Omega S^{4n-1})_{(p)} \overset{\approx}{\rightarrow} \pi_*(\Omega S^{2n})_{(p)}.$$

Thus the homotopy localized at an odd prime p of an even sphere is determined by the homotopy of the odd spheres.

5. Mod p Hopf Invariant

The integral Hopf invariant considered in the previous section used the Hurwicz isomorphism $h: \pi_{2n-2}(\Omega S^n, S^{n-1}) \rightarrow H_{2n-2}(\Omega S^n, S^{n-1}) = \mathbf{Z}$. The mod p Hopf invariant uses the Hurwicz isomorphism

$$h: \pi_{2pn-2}(\Omega^2 S^{2n+1}, S^{2n-1})_{(p)} \rightarrow H_{2pn-2}(\Omega^2 S^{2n+1}, S^{2n-1})_{(p)} = \mathbf{Z}/p$$

for the pair $(\Omega^2 S^{2n+1}, S^{2n-1})$ localized at p.

5.1 Definition. The mod p Hopf invariant $H_{(p)}: \pi_{2pn}(S^{2n+1})_{(p)} \rightarrow \mathbf{Z}/p$ is the composite

$$\pi_{2pn}(S^{2n+1})_{(p)} \overset{\approx}{\rightarrow} \pi_{2pn-2}(\Omega^2 S^{2n+1})_{(p)}$$

$$\rightarrow \pi_{2pn-2}(\Omega^2 S^{2n+1}, S^{2n-1})_{(p)}$$

$$\rightarrow H_{2pn-2}(\Omega^2 S^{2n+1}, S^{2n-1})_{(p)}$$

$$= \mathbf{Z}/p.$$

The integral Hopf invariant H involves the first degree i where $E: \pi_i(S^{m-1}) \rightarrow \pi_i(\Omega S^m)$ is not necessarily an isomorphism. This is made precise with the following exact sequence

$$\cdots \rightarrow \pi_{2m-2}(S^{m-1}) \overset{E}{\rightarrow} \pi_{2m-2}(\Omega S^m) \overset{H}{\rightarrow} \mathbf{Z} \rightarrow \pi_{2m-3}(S^{m-1}) \rightarrow \pi_{2m-3}(\Omega S^m) \rightarrow 0$$

and isomorphism $E: \pi_i(S^{m-1}) \rightarrow \pi_i(\Omega S^m)$ for $i < 2m - 3$. The mod p Hopf invariant $H_{(p)}$ involves the first degrees i where the double suspension $E^2: \pi_i(S^{2n-1})_{(p)} \rightarrow \pi_i(\Omega^2 S^{2n+1})_{(p)} = \pi_{i+2}(S^{2n+1})_{(p)}$ is not necessarily an isomorphism. This is made precise with the following exact sequence

$$\cdots \rightarrow \pi_{2pn-2}(S^{2n-1})_{(p)} \rightarrow \pi_{2pn-2}(\Omega^2 S^{2n+1})_{(p)} \overset{H_{(p)}}{\rightarrow} \mathbf{Z}/p$$

$$\rightarrow \pi_{2pn-3}(S^{2n-1})_{(p)} \overset{E^2}{\rightarrow} \pi_{2pn-3}(\Omega^2 S^{2n+1})_{(p)} \rightarrow 0$$

and isomorphism $\pi_i(S^{2n-1})_{(p)} \rightarrow \pi_i(\Omega^2 S^{2n+1})_{(p)}$ for $i < 2n - 3$.

The integral Hopf invariant $H: \pi_{4n-1}(S^{2n}) \rightarrow \mathbf{Z}$ is compared to the mod 2 Hopf invariant by the following commutative diagram.

$$\pi_{4n-2}(\Omega S^{2n})_{(2)} \xrightarrow{\;H'\;} \pi_{2n-2}(\Omega S^{2n}, S^{2n-1})_{(2)} = \mathbf{Z}$$

$$\Big\downarrow E$$

$$\pi_{4n-2}(\Omega^2 S^{2n+1})_{(2)} \xrightarrow{\;H'_{(2)}\;} \pi_{4n-2}(\Omega^2 S^{2n+1}, S^{2n-1})_{(2)} = \mathbf{Z}/2$$

$$\pi_{4n-2}(\Omega^2 S^{2n+1}, \Omega S^{2n})_{(2)}$$

If $E(\alpha') = \beta' \in \pi_{4n-2}(\Omega^2 S^{2n+1})_{(2)}$ where $\alpha' \in \pi_{4n-2}(\Omega S^{2n})_{(2)}$, then $H'_{(2)}(\beta) = 1$ if and only if $H'(\alpha') = 2k + 1$ some integer k. Let $H'(\alpha_0) = 2$ as in (4.8). Thus $H(\alpha) = 1$ for $\alpha = \alpha' - k\alpha_0$ and we have the following proposition.

5.2 Proposition. *There exists an element of Hopf invariant 1 in $\pi_{4n-1}(S^{2n})$ if and only if $H_{(2)} \colon \pi_{4n}(S^{2n+1}) \to \mathbf{Z}/2$ is nonzero.*

In view of (4.9) the mod 2 Hopf invariant $H_{(2)} \colon \pi_{4n}(S^{2n+1}) \to \mathbf{Z}/2$ is nonzero for $n = 1, 2$, and 4 and in the next section we see that these are the only cases of nonzero mod 2 Hopf invariant.

Recall from (3.6) that the fibre $F_n(p)$ of $f_{(p)} \colon \Omega S^{2n+1} \to \Omega S^{2np+1}$ has cohomology $H^*(F_n(p), \mathbf{Z}_{(p)}) = S(x, 2n)/(x^p)$. As for the homology and cohomology of $\Omega F_n(p)$, we have the following result using the same argument as in the proof of (2.1).

5.3 Proposition. *Let R be any algebra over $\mathbf{Z}_{(p)}$. As Hopf algebras over R, we have*

$$H^*(\Omega F_n(p), R) = E(v^*, 2n - 1) \otimes S'(w^*, 2pn - 2)$$

and

$$H_*(\Omega F_n(p), R) = E(v, 2n - 1) \otimes S(w, 2pn - 2).$$

In the next proposition we consider several criterions for the nontriviality of the mod p Hopf invariant which includes (5.2) for $p = 2$ since $F_n(2) = S^{2n}$.

5.4 Proposition. *The following are equivalent for $n \geq 1$.*

(1) *The mod p Hopf invariant $H_{(p)} \colon \pi_{2np}(S^{2n+1})_{(p)} \to \mathbf{Z}/p$ is nontrivial.*
(2) *The double suspension $E^2 \colon \pi_i(S^{2n-1})_{(p)} \to \pi_i(\Omega^2 S^{2n+1})_{(p)}$ is not an epimorphism for $i \leq 2np - 2$ (only an epimorphism for $i \leq 2np - 3$).*
(3) *There exists a complex $X = F_n(p) \cup_u e^{2np}$ and $u \in H^{2n}(X, \mathbf{F}_p)$ with $u^p \neq 0$ in $H^{2pn}(X, \mathbf{F}_p)$.*
(4) *There exists a complex $Y = S^i \cup_w e^{i+2n(p-1)}$ such that*

$$\mathscr{P}^n H^i(Y, \mathbf{F}_p) \neq 0 \qquad for\ p > 2$$

and

$$Sq^{2n}(H^i(Y, \mathbf{F}_2)) \neq 0 \qquad for\ p = 2$$

Proof. The equivalence of (1) and (2) follow immediately from the exact sequence involving $H_{(p)}$ mentioned after (5.1).

For (1) implies (3) observe that the pair $(\Omega S^{2n+1}, F_n(p))$ localized at p is $(2pn - 1)$-connected and by (1) we have elements $[u] \in \pi_{2pn-1}(F_n(p))_{(p)}$ and $[v] \in \pi_{2pn-2}(\Omega F_n(p))_{(p)}$ related by the usual isomorphism and mapping to a nonzero element of $\mathbf{Z}/p$ in the following commutative diagram.

$$[u] \in \pi_{2pn-1}(F_n(p))_{(p)} \to \pi_{2pn-1}(\Omega S^{2n+1})_{(p)} \to \pi_{2pn-1}(\Omega S^{2n+1}, F_n(p))_{(p)} = 0$$

$$[v] \in \pi_{2pn-2}(\Omega F_n(p))_{(p)} \to \pi_{2pn-2}(\Omega^2 S^{2n+1})_{(p)}$$

$$\pi_{2pn-2}(\Omega^2 S^{2n+1}, S^{2n-1})_{(p)} \xrightarrow{H_{(p)}} \mathbf{Z}/p$$

Form $X = F_n(p) \cup_u e^{2pn}$ and view $S^{2n} \subset F_n(p) \subset X$. Then we have maps $S^{2n-1} \to \Omega F_n(p) \cup_v e^{2pn-1} \to \Omega X$. By (5.3) and the homological properties of u and v it follows that the composite is an isomorphism.

$$H_i(S^{2n-1}) \to H_i(\Omega F_n(p) \cup_v e^{2pn-1}) \to H_i(\Omega X)$$

over $\mathbf{Z}_{(p)}$ for $i < 2(p + 1)n - 3$. By the argument used to calculate $H_*(\Omega X)$ in (5.3), if the pth power $H^{2n}(X, \mathbf{F}_p) \to H^{2np}(X, \mathbf{F}_p)$ is zero, then $H^{2np-2}(\Omega X, \mathbf{F}_p) \neq 0$ which contradicts the fact that $H_i(S^{2n-1}, \mathbf{F}_p) \xrightarrow{\sim} H_i(\Omega X, \mathbf{F}_p)$ for $i < 2(p + 1)n - 3$. This proves (3).

Assuming (3), we use the inclusion $SF_n(p) \subset S\Omega S^{2n+1}$ and the map $\overline{w}: S\Omega S^{2n+1} \to S^{2n+1}$ given in (3.2) to map

$$g: S(F_n(p) \cup_u e^{2np}) \to SS^{2n} \cup_{\overline{w}Su} e^{2np+1} = S^{2n+1} \cup_w e^{2np+1} = Y.$$

Note that $g^*: H^*(Y, \mathbf{F}_p) \to H^*(SX, \mathbf{F}_p)$ is a monomorphism. Since $\mathscr{P}^n$ or Sq^{2n} (for $p = 2$) is the pth power in the cohomology of $X = F_n(p) \cup_u e^{2np}$ and hence nonzero, it is nonzero defined $H^{2n+1}(SX, \mathbf{F}_p) \to H^{2np+1}(SX, \mathbf{F}_p)$ and $H^{2n+1}(Y, \mathbf{F}_p) \to H^{2np+1}(Y, \mathbf{F}_p)$ because this cohomology operation commutes with suspensions and induced morphisms of maps.

Finally for (4) implies (2), assume that (2) is false, that is, $E^2: \pi_i(S^{2n-1}) \to \pi_{i+2}(S^{2n+1})$ is an epimorphism for all $i \leq 2np - 2$. Consider a two cell complex $Y = S^t \cup_w e^{t+2n(p-1)}$ where $[w] \in \pi_{2n(p-1)-1+t}(S^t)$. Suspending Y is necessary, we can assume t is odd. Then $[w] = E^{2r}[w_0]$ where $[w_0] \in \pi_{2n(p-1)-1+(2n-1)}(S^{2n-1})$ for $2r + (2n - 1) = t$, and Y and $S^{2r}(Y_0)$ have the same homotopy type. Since $\mathscr{P}^n$ or Sq^n is zero $H^{2n-1}(Y_0, \mathbf{F}_p) \to H^{2np-1}(Y_0, \mathbf{F}_p)$ when the class has such a low dimension, it is also zero $H^t(Y, \mathbf{F}_p) \to H^{t+2n(p-1)}(Y, \mathbf{F}_p)$ because the cohomology operation commutes with suspension. Thus (4) is false when (2) is false and we have proved the proposition.

5.5 Remark. In the next section we will see that the mod p Hopf invariant

$$H_{(p)}: \pi_{2pn}(S^{2n+1})_{(p)} \to \mathbf{Z}/p$$

is zero for $n > 1$ when $p > 2$ (result originally of Luilevicius, Shimada, and Yamoshita) and for $n \neq 1, 2, 4$ when $p = 2$ (Adams). The case $p = 2$ is contained in (5.2) and (4.9). Now we show that $H_{(p)} \colon \pi_{2p}(S^3)_{(p)} \to \mathbf{Z}/p$ is nonzero (the case $n = 1$). This can be seen in different ways and we consider two. Let P_n denote the complex projective n space.

Proof (1). Map $g_m \colon S^2 \times \overset{(m)}{\cdots} \times S^2 \to P_m$ by the relation

$$g_m((a_1, b_1), \dots, (a_m, b_m)) = \prod_{j=1}^{m} (a_j z + b_j)$$

where the coordinates of P_m are coefficients of nonzero polynomials of degree $\leq m$ up to scalar factor. The map g_m has degree $m!$, and $S(g_m) \colon S(S^2 \times \overset{(m)}{\cdots} \times S^2) \to SP_m$ restricts to the top cell S^{2m+1} in the wedge decomposition given by (3.1) to a map $S^{2m+1} \to SP_m$ of degree $m!$. Hence for $m < p$, localized at p we have a homotopy equivalence $SP_{m-1} \vee e^{2m+1} \to SP_m$. Again localized at p, we can split and map

$$SP_p \simeq (S^3 \vee S^5 \vee \cdots \vee S^{2p-1}) \cup_\alpha e^{2p+1} \overset{f}{\to} S^3 \cup_\alpha e^{2p+1}$$

where f^* is a monomorphism on cohomology. The pth power $= \mathscr{P}^1$ for $H^2(P_p, \mathbf{F}_p) \to H^{2p}(P_p, \mathbf{F}_p)$ and is $\neq 0$. Using properties of $\mathscr{P}^1$, we deduce that $\mathscr{P}^1 \colon H^3(SP_p, \mathbf{F}_p) \to H^{2p+1}(SP_p, \mathbf{F}_p)$ and therefore

$$\mathscr{P}^1 \colon H^3(S^3 \cup_\alpha e^{2p+1}, \mathbf{F}_p) \to H^{2p+1}(S^3 \cup_\alpha e^{2p+1}, \mathbf{F}_p)$$

is nonzero. By (4)(5.4) the mod p Hopf invariant is nonzero.

Proof (2). Map $F_1(p) \to P_\infty$ by representing the two dimensional cohomology generator of $H^2(F_1(p), \mathbf{Z})$ for $P_\infty = K(\mathbf{Z}, 2)$ such that it factors $F_1(p) \to P_{p-1}$. Localized at p, this is a homotopy equivalence and P_p is of the same homotopy type as $F_1(p) \cup_u e^{2p}$ for some u localized at p. Since the pth power $H^2(P_p, \mathbf{F}_p) \to H^{2p}(P_p, \mathbf{F}_p)$ is nonzero, we deduce by (3)(5.4) that the mod p Hopf invariant is nonzero.

6. Spaces Where the pth Power Is Zero

In order to describe the elementary number theory needed we use the notation $i = \mathrm{ord}_p x$ where x is a nonzero rational number of the form $x = p^i \dfrac{a}{b}$ and a, b are integers not divisible by the prime p. Clearly

$$\mathrm{ord}_p(xy) = \mathrm{ord}_p(x) + \mathrm{ord}_p(y).$$

6.1 Remark. Let k be an integer whose class in $(\mathbf{Z}/2p^2)^*/(\pm 1)$ generates this cyclic group. For $p = 2$, we can let $k = 5$ and for $p > 2$, $k = (1 + p)k'$ where k' generates $(\mathbf{Z}/p)^*$ is such an example. Then $\mathrm{ord}_p(k^n - 1) = 0$ if $p - 1 \nmid n$ and

$$\mathrm{ord}_p(k^{m(p-1)}-1) = \begin{cases} \mathrm{ord}_p(m)+1 & \text{for } p > 2 \\ 1 & \text{for } p = 2,\ m \text{ odd} \\ \mathrm{ord}_2(m)+2 & \text{for } p = 2,\ m \text{ even} \end{cases}$$

Moreover $\mathrm{ord}_p(k^{m(p-1)}-1) \geqq m$ if and only if $m = 1$ for $p > 2$ and $m = 1, 2,$ or 4 for $p = 2$.

6.2 Theorem. *Let X be a finite complex with nonzero mod p homology only in dimensions ni for $0 \leq i \leq p$.*

(1) *If $p > 2$, and if $x^p \neq 0$ for some $x \in H^n(X, \mathbf{F}_p)$, then $n = 2m$ is even and m divides $p - 1$.*
(2) *If $n = 2m$, if $x \in H^{2m}(X, \mathbf{Z}_{(p)})$ and $y \in K(X) \otimes \mathbf{Z}_{(p)}$ such that the Chern character $ch(y) = x + cx'$ where $x' \in H^{2mp}(X, \mathbf{Z}_{(p)})$, if each $H^{2mi}(X, \mathbf{Z}_{(p)})$ has $\mathbf{Z}_{(p)}$-rank 1, and if $x^p \neq 0$ in $H^{2mp}(X, \mathbf{F}_p)$, then $m = 1$ for $p > 2$ and $m = 1, 2,$ or 4 for $p = 2$.*
(3) *For $p = 2$, if n is odd and $x^2 \neq 0$ for $x \in H^n(X, \mathbf{F}_2)$, then $n = 1$.*

Proof. (1) We make use of the projections $\pi_i \colon \tilde{K}(X) \otimes \mathbf{Q} \to \tilde{K}(X) \otimes \mathbf{Q}$ defined by the following relations where π_j is taken over $1 \leq j \leq p$, $j \neq i$,

$$\pi_i = \frac{\prod_j (\psi^k - k^{jm})}{\prod_j (k^{im} - k^{jm})}$$

Observe that $\pi_i \pi_i = \pi_i$, $\pi_i \pi_j = 0$ for $i \neq j$, and $(\psi^k - k^{im})\pi_i = 0$. Each $y \in \tilde{K}(X) \otimes \mathbf{Q}$ has the form $y = \sum_i \pi_i(y)$ with $\psi^l(y) = \sum_i l^{mi} \pi_i(y)$.

First, we prove $\psi^p(y) \in p\tilde{K}(X) + \mathrm{Tors}\,\tilde{K}(X)$ for any $y \in \tilde{K}(X)$ if m does not divide $p - 1$. Since $\psi^p(y) = \sum_{1 \leq i \leq p} p^{im} \pi_i(y)$, it suffices to show that

$$\mathrm{ord}_p(p^{im}) = im > \mathrm{ord}_p\left(\prod_{j \neq i}(k^{im} - k^{jm})\right)$$

where k is chosen so that it generates $(\mathbf{Z}/2p^2)^*/(\pm 1)$ as above. Now we calculate for given i

$$\sum_{j \neq i} \mathrm{ord}_p(k^{im} - k^{jm}) = \sum_{j \neq i} \mathrm{ord}_p(k^{(i-j)m} - 1)$$

$$= \sum_{p-1 \,|\,(i-j)m} \mathrm{ord}_p(i-j)m + 1$$

Let $h = \text{g.c.d.}\,(p-1, m)$ and write $m = \alpha h$, $p - 1 = \beta h$ where $\alpha > 1$. Then for $1 \leq j \leq p$, $j \neq i$, as above

$$\mathrm{ord}_p\left(\prod_{j \neq i} k^{im} - k^{jm}\right) = \sum_{\beta\,|\,(i-j)} \mathrm{ord}_p((i-j)m)$$

$$= \sum_{\beta\,|\,(i-j)} \mathrm{ord}_p(m)$$

$$= h \cdot \mathrm{ord}_p(m) = h \cdot \mathrm{ord}_p(\alpha) < h \cdot \alpha = m$$

since $\alpha > 1$. This proves that $\psi^p(y) \in p\tilde{K}(X) + \text{Tors } \tilde{K}(X)$ for $y \in \tilde{K}(X)$ and $m \nmid p - 1$ since $\text{ord}_p(\text{denominator of } \pi_i) < im$.

Since $\psi^p(y) - y^p \in pK(X)$, it follows that $y^p \in pK(X) + \text{Tors } K(X)$ for all $y \in \tilde{K}(X)$ when m does not divide $p - 1$. Note that $H^{2mi}(X, \mathbf{Z}_{(p)}) \to H^{2mi}(X, \mathbf{F}_p)$ is an epimorphism. For $x \in H^{2m}(X, \mathbf{Z}_{(p)})$ there exists $y \in \tilde{K}(X) \otimes \mathbf{Z}_{(p)}$ with $ch(y) = x + x'$ where $x' \in \coprod_{2 \le i} H^{2mi}(X, \mathbf{Q})$. Now $y^p = p\bar{y}$ and $pch(\bar{y}) = ch(y^p) = (x + x')^p = x^p$. Since $ch(y) \in H^{2mp}(X, \mathbf{Q})$ and has no component in H^{2mj} for $j < p$, it follows that $ch(\bar{x}) \in H^{2mp}(X, \mathbf{Z}_{(p)})$. Finally mod p we have $x^p = 0$ and the pth power is zero in $\bar{H}^*(X, \mathbf{F}_p)$ when m does not divide $p - 1$.

(2) Let x_i be a generator of $H^{2mi}(X, \mathbf{Z}_{(p)})$ where $x = x_1$, $x' = x_p$ and $x_1^p = x_p$. Then $\tilde{K}(X) \otimes \mathbf{Z}_{(p)}$ has a basis $y_1, \ldots, y_p$ with $ch(y_1) = x_1 + cx_p$ for $c \in \mathbf{Q}$ and $ch(y_i) = x_i + (\text{higher terms})$. Now apply the Chern character ch to the relation $\psi^p(y_1) - y_1^p = p \sum_i c_i y_i$ where $c_i \in \mathbf{Z}_{(p)}$ to get

$$(p^m x_1 + p^{mp} c x_p) - (x_1 + cx_p)^p$$

$$= pc_1(x_1 + cx_p) + c_p p x_p + \sum_{2 \le i \le p-1} c_i p(x_i + \text{higher terms})$$

or

$$(p^m - pc_1)x_1 + (p^{mp}c - 1 - pc_1 c - c_p p)x_p$$

$$= \sum_{2 \le i \le p-1} c_i p(x_i + \text{higher terms}).$$

By the linear independence of $x_1, \ldots, x_p$ we see that $c_2 = 0$, $c_3 = 0$, $\ldots$, $c_{p-1} = 0$, $p^m = pc_1$, $p^{mp}c = 1 + p(c_1 c + c_p)$. Thus $p^m(p^{m(p-1)} - 1)c = 1 + pc_p \equiv 1 \pmod{p}$. Hence $-m \ge \text{ord}_p c$.

Next apply ch to the relation $\psi^k(y_1) - k^m y_1 = \sum_i b_i y_i$ to get

$$(k^m x_1 + k^{mp} c x_p) - (k^m x_1 + k^m c x_p) = b_p x_p + \sum_{i < p} b_i(x_i + \text{higher terms}).$$

Again by linear independence of $x_1, \ldots, x_p$ it follows that $b_1 = 0, \ldots, b_{p-1} = 0$ and $b_p = k^m(k^{m(p-1)} - 1)c \in \mathbf{Z}_{(p)}$. For k a generator of $(\mathbf{Z}/2p^2)^*/(\pm 1)$, it follows that $\text{ord}_p(k^{m(p-1)} - 1) \ge -\text{ord}_p(c) \ge m$. By (6.1) we have $m = 1$ for p odd and $m = 1, 2,$ or 4 for $p = 2$.

(3) For n odd and $n > 1$, $H^n(X, \mathbf{Z}_{(2)}) = \mathbf{Z}_{(2)} u$ and $H^{2n}(X, \mathbf{Z}_{(2)}) = \mathbf{Z}_{(2)} v$ but $u^2 = 0$. Hence the same relation holds over $\mathbf{F}_2$ which implies (3). This proves the theorem.

Finally there is one special case for the prime 2 which we can exclude using the methods of K-theory.

6.3 Theorem. *Let X be a finite complex with $H^n(X, \mathbf{F}_2) = \mathbf{F}_2 x$, $H^{2n}(X, \mathbf{F}_2) = \mathbf{F}_2 x^2$, $H^{3n}(X, \mathbf{F}_2) = \mathbf{F}_2 x^3$, $x^4 = 0$, and other cohomology groups equal to zero mod 2. Then $n = 1, 2,$ or 4.*

Proof. By taking a skeleton Y of X we have a finite cell complex with $H^n(Y, \mathbf{F}_2) = \mathbf{F}_2 x$, $H^{2n}(Y, \mathbf{F}_2) = \mathbf{F}_2 x^2$, and other cohomology groups equal to zero mod 2. By (6.2) it follows that $n = 2, 4$, or 8. We have only to show $n = 8$ is impossible.

Let x denote a generator of $H^8(X, \mathbf{Z}_{(2)})$ over $\mathbf{Z}_{(2)}$. Then x^2 and x^3 are generators of $H^{16}(X, \mathbf{Z}_{(2)})$ and $H^{24}(X, \mathbf{Z}_{(2)})$ over $\mathbf{Z}_{(2)}$ respectively. Now $\tilde{K}(X) \otimes \mathbf{Z}_{(2)}$ has a basis y_1, y_2, y_3 over $\mathbf{Z}_{(2)}$ such that

$$ch(y_1) = x + ax^2 + bx^3, \qquad ch(y_2) = x^2 + cx^3, \qquad ch(y_3) = x^3.$$

Since $ch(y_1^2) = (x + ax^2 + bx^3)^2 = x^2 + 2ax^3$ and $ch(y_2) = x^2 + cx^3$, it follows that $ch(y_1^2 - y_2) = (2a - c)x^3$ and $2a - c \in \mathbf{Z}_{(2)}$ because $y_1^2 - y_2 \in \tilde{K}(X) \otimes \mathbf{Z}_{(2)}$.

Applying ch to the relation $\psi^2(y_1) - y_1^2 = 2(a_1 y_1 + a_2 y_2 + a_3 y_3)$, we derive the relation

$$(2^4 x + 2^8 a x^2 + 2^{12} b x^3) - (x^2 + 2ax^3)$$

$$= 2a_1(x + ax^2 + bx^3) + 2a_2(x^2 + cx^3) + 2a_3 x^3$$

Hence $2a_1 = 2^4$ and $2^8 a - 1 = 2^4 a + 2a_2$ or $2^4(2^4 - 1)a = 1 + 2a_2$. From this we deduce $\mathrm{ord}_2 a = -4$ and so $\mathrm{ord}_2 c = -3$ since $2a - c \in \mathbf{Z}_{(2)}$. From the coefficients of x^3 we have $2^{12}b - 2a = 2^4 b + 2ca_2 + 2a_3$ so that $\mathrm{ord}_2(b) + 4 = \mathrm{ord}_2(2^4(2^8 - 1)b) = \mathrm{ord}_2(2a) < \mathrm{ord}_2(2ca_2 + 2a_3)$. Thus $\mathrm{ord}_2(b) = -4 + 1 - 4 = -7$.

Now apply ch to the relation $\psi^k(y_1) - k^4 y_1 = b_1 y_1 + b_2 y_2 + b_3 y_3$ for $b_i \in \mathbf{Z}_{(2)}$ and we have

$$(k^4 x + k^8 a x^2 + k^{12} b x^3) - (k^4 x + k^4 a x^2 + k^4 b x^3)$$

$$= b_1(x + ax^2 + bx^3) + b_2(x^2 + cx^3) + b_3 x^3$$

First $b_1 = 0$ and $b_2 = k^4(k^4 - 1)a$. Thus $\mathrm{ord}_2 b_2 = (2 + 2) - 4 = 0$ by (6.1). Moreover $k^4(k^8 - 1)b = b_2 c + b_3$ and $-2 = 5 + (-7) = \mathrm{ord}_2(k^8 - 1) + \mathrm{ord}_2 b = \mathrm{ord}_2((b_2 c + b_3)) = \mathrm{ord}_2(c) = -3$ which is a contradiction. This proves the theorem.

Now we apply the previous theorems to questions of nonzero Hopf invariant and properties of H-space structures on odd spheres.

6.4 Theorem. *The* mod p *Hopf invariant* $H_{(p)}: \pi_{2mp}(S^{2m+1}) \to \mathbf{Z}/p$ *is trivial unless*

(1) $p = 2$ *and* $m = 1, 2$, *and* 4 (*theorem of Adams*),
(2) $p > 2$ *and* $m = 1$ (*Liulevicius, Shimada, Yamoshita*).

Proof. For $p = 2$ we use (2), (3)(6.2) and condition (3)(5.4). For $p > 2$ again we use (2)(6.2) and (3)(5.4), but we must observe that $ch: K(F_n(p)) \to H^{ev}(F_n(p), \mathbf{Q})$ factors through $H^{ev}(F_n(p), \mathbf{Z}_{(p)})$ and $ch: K(F_n(p)) \otimes \mathbf{Z}_{(p)} \to H^{ev}(F_n(p), \mathbf{Z}_{(p)})$ is an isomorphism, because $S^2 F_n(p)$ localized at p is a wedge of spheres. In (4.9) with (5.2) and (5.5) we showed that the mod p Hopf invariant was nontrivial in the cases described above. This proves the theorem.

In chap. 15, §4 a more elementary derivation of the above theorem is given.

6.5 Remark. Of the four H-spaces S^0, S^1, S^3, and S^7 we saw in (4.9) that the first three S^0, S^1, and S^3 have classifying spaces $\mathbf{P}_\infty(\mathbf{R})$, $\mathbf{P}_\infty(\mathbf{C})$, and $\mathbf{P}_\infty(\mathbf{H})$ respectively. By (6.3) the H-space S^7 has no classifying space, and in fact more is true, any H-space structure on S^7 is not homotopy associative. If an H-space G is homotopy associative, then $B_3(\mathbf{C})$, the third step in the classifying space construction, exists. For $G = S^7$ this has a cell decomposition $B_3(S^7) = S^8 \cup e^{16} \cup e^{24}$ with the third power $H^8(B_3(S^7)) \to H^{24}(B_3(S^7))$ nonzero. In (6.3) we showed that such a space cannot exist.

6.6 Remark. In (4.10) we remarked that each odd sphere $S^{2n-1}_{(p)}$ localized at p admits an H-space structure. By (1)(6.2) we see that if $S^{2n-1}_{(p)}$ has a classifying space, then n divides $p - 1$. D. Sullivan proved that if n divides $p - 1$, then the localized space $S^{2n-1}_{(p)}$ has a classifying space. He considered the space $K(\mathbf{Z}_p, 2)$ where $\mathbf{Z}_p$ is the p-adic integers and the subgroup Γ of $\mathbf{Z}_p^*$ (the p-adic units) of order n. Then Γ acts freely on a model of $K(\mathbf{Z}_p, 2)$ and $X = K(\mathbf{Z}_p, 2)/\Gamma$ has cohomology $H^*(X, \mathbf{F}_p) = S(x, 2n)$ and $\pi_1(X) = \Gamma$. After completing X at p to X_p, we have a space with $\pi_1(X_p) = 0$ and $H^*(X_p, \mathbf{F}_p) = S(x, 2n)$. The map $S^{2n-1} \to \Omega(X_p)$ defines a homotopy equivalence of spaces localized at p and X_p is a classifying space for $S^{2n-1}_{(p)}$.

7. Double Suspension Sequences

In this section we use the notation F_n for the space $F_n(p)$ the fibre of $f_n : \Omega S^{2n+1} \to \Omega S^{2np+1}$ introduced in and before (3.5) under the notation $f_{(p)}$ for f_n. Note that $S^{2n} \subset F_n$ as a subspace.

The main goal of this section is to analyze the filtration

$$S^{2n-1} \subset \Omega F_n \subset \Omega^2 S^{2n+1}$$

and calculate $\pi_*(\Omega^2 S^{2n+1}, S^{2n-1})_{(p)}$ in terms of $\pi_*(\Omega^2 S^{2n+1}, \Omega F_n)_{(p)}$ and $\pi_*(\Omega F_n, S^{2n-1})_{(p)}$. These last two relative homotopy groups will be shown to be isomorphic to certain homotopy groups of spheres as in (4.1) and (4.2) for the single suspension sequence. The term $\pi_*(\Omega^2 S^{2n+1}, S^{2n-1})_{(p)}$ is used to analyze the kernel and cokernel of the double suspension morphism $E^2 : \pi_*(S^{2n-1})_{(p)} \to \pi_*(\Omega^2 S^{2n+1})_{(p)}$.

The following proposition is immediate.

7.1 Proposition. *The map* $\Omega f_n : (\Omega^2 S^{2n+1}, \Omega F_n) \to (\Omega^2 S^{2np+1}, *)$ *induces an isomorphism*

$$(\Omega f_n)_* \colon \pi_*(\Omega^2 S^{2n+1}, \Omega F_n) \to \pi_*(\Omega^2 S^{2np+1}, *).$$

For the main task the study of the pair $(\Omega F_n, S^{2n-1})$, we consider the first nonzero relative group $\pi_i(\Omega^2 S^{2n+1}, S^{2n-1})_{(p)}$. The following sequence is exact for $n > 1$ by (2.4).

$$0 \to \pi_{2np-2}(\Omega^2 S^{2n+1}, S^{2n-1})_{(p)} \overset{\partial}{\to} \pi_{2pn-3}(S^{2n-1})_{(p)} \to \pi_{2pn-3}(\Omega^2 S^{2n+1})_{(p)} \to$$

$$\Bigg\downarrow \text{Hurwicz}$$

$$H_{2np-2}(\Omega^2 S^{2n+1}, S^{2n-1})_{(p)} = \mathbf{Z}/p$$

7.2 Notations. Let $\tau_n \in \pi_{2np-3}(S^{2n-1})_{(p)}$ denote the image under ∂ of a generator of $\pi_{2np-2}(\Omega^2 S^{2n+1}, S^{2n-1})_{(p)} = \mathbf{Z}/p$. This is nonzero exactly when $n > 1$, and since the image of τ_n in $\pi_{2np-3}(\Omega^2 S^{2n+1})$ is zero, it follows immediately that the double suspension $S^2 \tau_n$ is null homotopic. The space T_n is defined as the mapping cone of τ_n where

$$S^{2np-3} \overset{\tau_n}{\to} S^{2n-1} \to T_n = S^{2n-1} \cup_{\tau_n} e^{2np-2}$$

Observe that $S^2 T_n = S^{2n+1} \vee S^{2np-1}$.

7.3 Proposition. *Assume* $n > 1$. *The inclusion* $i_n \colon S^{2n-1} \to \Omega F_n$ *extends to a map* $j_n \colon T_n \to \Omega F_n$ *such that* $j_{n*} \colon H_i(T_n, \mathbf{F}_p) \to H_i(\Omega F_n, \mathbf{F}_p)$ *is a monomorphism which is an isomorphism in degrees* $i \leq 4pn - 3$.

Proof. Consider the following commutative diagram

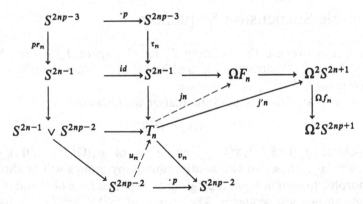

The map j'_n exists because $S^2 \tau_n$ is null homotopic, and since $(\Omega f_n) j'_n$ is null homotopic for reasons of connectivity, the map j_n exists making the diagram homotopy commutative. The map u_n exists since $pS(\tau_n)$ is null homotopic.

For the second statement we need only prove that j_{n*} is an isomorphism for $i = 2pn - 2$. For this consider the following commutative diagram.

$$
\begin{array}{ccccc}
H_{2np-2}(T_n) & \xrightarrow{\sim} & H_{2np-2}(T_n, S^{2n-1}) & \xrightarrow{\sim} & \pi_{2np-2}(T_n, S^{2n-1})_{(p)} \\
{\scriptstyle j_{n*}}\big\downarrow & & \big\downarrow & & {\scriptstyle w}\big\downarrow{\scriptstyle \wr} \quad \searrow{\scriptstyle \partial} \\
H_{2np-2}(\Omega F_n) & \xrightarrow{\sim} & H_{2np-2}(\Omega F_n, S^{2n-1}) & \xleftarrow{\sim} & \pi_{2np-2}(\Omega F_n, S^{2n-1})_{(p)} \searrow \\
& & & & \qquad\qquad \to \pi_{2np-3}(S^{2n-1})_{(p)} \\
{\scriptstyle \wr}\big\downarrow & & \big\downarrow{\scriptstyle \wr u} & & {\scriptstyle u}\big\downarrow{\scriptstyle \wr} \quad \nearrow \\
H_{2np-2}(\Omega^2 S^{2n+1}) & \xrightarrow{\sim} & H_{2np-2}(\Omega^2 S^{2n+1}, S^{2n-1}) & \xleftarrow{\sim} & \pi_{2np-2}(\Omega^2 S^{2n+1}, S^{2n-1})_{(p)}
\end{array}
$$

Since $\partial: \pi_{2np-2}(T_n, S^{2n-1})_{(p)} \to \pi_{2np-3}(S^{2n-1})_{(p)}$ is a monomorphism, w and v are monomorphisms of cyclic groups of order p and hence isomorphisms. Hence the diagram consists of isomorphisms and j_{n*} is an isomorphism. This proves the proposition.

Now we consider a decomposition of $S\Omega F_n$ localized at p which is similar to the decomposition of $S\Omega S(X)$ considered in (2.2).

7.4 Proposition. *Localized at p the space $S\Omega F_n$ has the homotopy type of a wedge of ST_n and a sequence of spheres S^i where $i > 2np - 2$ all localized at p.*

Proof. The map $j_n: T_n \to \Omega F_n$ defines by loop multiplication a map $g_m: (T_n)^m \to \Omega F_n$ and this map is a surjection in homology mod p in degrees $\leq (2pn - 2)m$. Suspend this map to get

$$
\begin{array}{ccc}
S((T_n)^m) & \xrightarrow{S(g_m)} & S\Omega F_n \\
\big\downarrow{\scriptstyle \sigma}\big\uparrow & \nearrow{\scriptstyle h_m} & \\
S(\bigwedge_m T_n) & &
\end{array}
$$

where $h_m = S(g_m)\sigma$ and s is the right inverse of the projection $S((T_n)^m) \to S(\bigwedge_m T_n)$ given in (2.1). Now collecting the various maps h_m we have a map

$$
\bar{h}: \bigvee_{1 \leq m} S(\bigwedge_m T_n) \to S\Omega F_n.
$$

which is a surjection on homology mod p.

Next observe that $S(T_n \wedge T_n) = ST_n \wedge T_n$ is a wedge of spheres $S^{4n-1} \vee S^{2n(p+1)-2} \vee S^{2n(p+1)-2} \vee S^{4np-3}$. Since $S^2\tau_n$ is null homotopic, the four cell complex $T_n \wedge T_n$ has the form

$$
T_n \wedge T_n = (S^{4n-2} \vee S^{2n(p+1)-3} \vee S^{2n(p+1)-3}) \cup_f e^{4np-4}
$$

where $f: S^{4np-5} \to S^{4n-2}$ is the attaching map for the top cell. Since $S^2 T_n \wedge T_n$ is a wedge of four spheres, it follows that $S^2 f$ is null homotopic. The homotopy class of Sf is in $\pi_{4np-4}(S^{2n-1})$, and the suspension morphism from odd to even spheres is injective on the odd prime components

$$\pi_{4np-4}(S^{2n-1})_{(p)} \to \pi_{4np-3}(S^{2n})_{(p)}.$$

Hence the class of $S(f)$ is zero, and the space $S(T_n \wedge T_n)$ has the homotopy type of a wedge of four spheres. When the map $\bar{h}$ restricts to a $\mod p$ homology isomorphism

$$h: ST_n \vee X \to S\Omega F_n$$

where X is a subwedge of spheres of $\bigvee_{2 \leq m} S(\bigwedge_m T_n)$. After localizing at p, we see that the map h is the desired homotopy equivalence which proves the proposition.

Note that the homology of $S\Omega F_n$ determines which dimensions the spheres appear in the wedge decomposition of $S\Omega F_n$, namely of the form $(2np - 2)i + 1$ for $i \geq 2$ and $(2np - 2)i + (2n - 1) + 1$ for $i \geq 1$.

Let $g'_n: S\Omega F_n \to S^{2np-1}$ be the composite of the projection $S\Omega F_n \to ST_n$ given by the previous splitting and $Sv_n: ST_n \to SS^{2np-2} = S^{2np-1}$ where v_n is the quotient $T_n \wedge T_n/S^{2n-1} = S^{2np-2}$. Then $\mod p$ the morphisms $H_{2np-1}(Sv_n)$ and $H_{2np-1}(g'_n)$ are isomorphisms. Let $g_n: \Omega F_n \to \Omega S^{2np-1}$ denote a fibre map adjoint to g'_n with $g_n(S^{2n-1}) = *$.

7.5 Proposition. *With the above notations, the inclusion $S^{2n-1} \to g_n^{-1}(*)$ localized at p is a homotopy equivalence where $g_n^{-1}(*)$ is the fibre of g_n.*

Proof. The following is a monomorphism onto its second factor where the cohomology is over any $\mathbf{Z}_{(p)}$ − algebra R,

$$S'(y, 2np - 2) = H^*(\Omega S^{2np-1}) \overset{g_n^*}{\to} H^*(\Omega F_n)$$

$$= E(2, 2n - 1) \otimes S'(x, 2np - 2)$$

Since $g_n^*(y) = x$, this holds over the integers localized at p, and since there is no p-torsion, it holds over any $\mathbf{Z}_{(p)}$ − algebra R. Thus $S^{2n-1} \to g_n^{-1}(*)$ localized at p is a homology isomorphism which implies the proposition.

Now we put together into one theorem the implications of having the fibrations: $f_n: \Omega S^{2n+1} \to \Omega S^{2np+1}$ and $g_n: \Omega F_n \to \Omega S^{2np-1}$ with fibres F_n and S^{2n-1} respectively given above. This is the double suspension theorem of Toda.

7.6 Theorem. *(Toda) For an odd sphere S^{2n-1} we have the following exact triangles and isomorphisms g_{n*} and f_{n*}.*

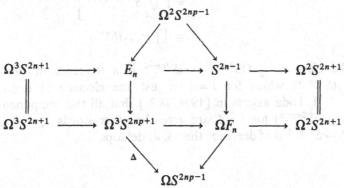

$$\begin{array}{ccc}
\pi_*(S^{2n-1}) & \longrightarrow & \pi_*(\Omega^2 S^{2n+1}) \\
\end{array}$$

$$\text{deg}=-1$$

$$\pi_*(\Omega^2 S^{2n+1}, S^{2n-1})$$

$$\pi_*(\Omega F_n, S^{2n-1}) \xleftarrow{\ \text{deg}=-1\ } \pi_*(\Omega^2 S^{2n+1}, \Omega F_n)$$

$$g_{n*} \downarrow \wr \qquad\qquad \wr \downarrow f_{n*}$$

$$\pi_*(\Omega S^{2np-1}) \qquad\qquad \pi_*(\Omega^2 S^{2np+1})$$

8. Study of the Boundary Map $\Delta\colon \Omega^3 S^{2np+1} \to \Omega S^{2np-1}$

By (7.1) we have the following fibre sequence

$$\Omega^3 S^{2n+1} \to \Omega^3 S^{2np+1} \to \Omega F_n \to \Omega^2 S^{2n+1} \to \Omega^2 S^{2np+1}$$

and by (7.5) the fibre sequence

$$\Omega^2 S^{2np-1} \to S^{2n-1} \to \Omega F_n \to S^{2np-1}$$

Let E_n denote the fibre of $S^{2n-1} \to \Omega^2 S^{2n+1}$ yielding the fibre sequence

$$\Omega^3 S^{2n+1} \to E_n \to S^{2n-1} \to \Omega^2 S^{2n+1}$$

From the factorization $S^{2n-1} \to \Omega F_n \to \Omega^2 S^{2n+1}$ these diagrams fit together into the following diagram of fibre sequences

$$\Omega^2 S^{2np-1}$$

$$\begin{array}{ccccccc}
\Omega^3 S^{2n+1} & \longrightarrow & E_n & \longrightarrow & S^{2n-1} & \longrightarrow & \Omega^2 S^{2n+1} \\
\| & & \downarrow & & \downarrow & & \| \\
\Omega^3 S^{2n+1} & \longrightarrow & \Omega^3 S^{2np+1} & \longrightarrow & \Omega F_n & \longrightarrow & \Omega^2 S^{2n+1}
\end{array}$$

$$\Delta \qquad\qquad \Omega S^{2np-1}$$

The map $\Omega^2 S^{2np-1} \to E_n$ exists because the composite $\Omega^2 S^{2np-1} \to S^{2n-1} \to \Omega^2 S^{2n+1}$ is null homotopic.

Composing the map $\Omega^3 S^{2np+1} \to \Omega F_n$ in the second row with the double suspension map $E^2\colon \Omega S^{2np-1} \to \Omega^3 S^{2np+1}$ yields a map $\Omega S^{2pn-1} \to \Omega F_n$ such that the following diagram is commutative from the lower part of the above diagram.

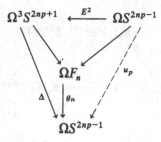

8.1 Proposition. *The map $u_p: \Omega S^{2np-1} \to \Omega S^{2np-1}$ localized at p is of degree p and $\Delta \circ E^2 = u_p$.*

Proof. The second statement is immediate and for the first recall that $H_{2np-2}(\Omega^2 S^{2n+1}, \mathbf{Z}_{(p)}) = \mathbf{Z}/p$ by (2.4). Since the following sequence of homology groups over $\mathbf{Z}_{(p)}$ is exact,

$$\mathbf{Z}_{(p)} = H_{2np-2}(\Omega^3 S^{2np+1}) \to \mathbf{Z}_{(p)} = H_{2np-2}(\Omega F_n) \to H_{2np-2}(\Omega^2 S^{2n+1}) = \mathbf{Z}/p$$

we see that $H_{2np-2}(\Omega^3 S^{2np+1}) = \mathbf{Z}_{(p)} \to \mathbf{Z}_{(p)} = H_{2np-2}(\Omega F_n)$ is multiplication by p up to a unit. Since E^2 and g_n induce isomorphisms on H_{2np-2} over $\mathbf{Z}_{(p)}$, it follows that $H_{2np-2}(u_p)$ is multiplication by p.

8.2 Remark. Using the decomposition of $S\Omega S^{n+1}$ given in (3.3), we have the following description of $[\Omega S^{n+1}, \Omega S^{n+1}]$.

$$[\Omega S^{n+1}, \Omega S^{n+1}] = [S\Omega S^{n+1}, S^{n+1}]$$

$$= \left[\bigvee_{1 \leq i} S^{ni+1}, S^{n+1} \right]$$

$$= \prod_{1 \leq i} \pi_{ni+1}(S^{n+1})$$

Hence the class of $u_p: \Omega S^{2np-1} \to \Omega S^{2np-1}$ is a sequence of elements in $\pi_{2(np-1)i+1}(S^{2np-1})$ which for $i = 1$ is just the element of degree p in $\pi_{2np-1}(S^{2np-1})$. Toda asserts in [1956, (8.3)] that all the components of u_p in $\pi_{2(np-1)i+1}(S^{2np-1})$ for $i > 1$ are zero. In other words $u_p = \Omega v_p$ where $v_p: S^{2np-1} \to S^{2np-1}$ is of degree p, that is, u_p deloops.

Bibliography

Adams, J. F.:
1. On the non-existence of elements of Hopf invariant one, *Bull. Am. Math. Soc.*, **64:** 279–282 (1958).
2. On the structure and applications of the Steenrod algebra, *Comment, Math. Helv.*, **32:** 180–214 (1958).
3. On the non-existence of elements of Hopf invariant one, *Ann. Math.*, **72:** 20–104 (1960).
4. On Chern characters and the structure of the unitary group, *Proc. Cambridge Phil. Soc.*, **57:** 189–199 (1961).
5. Vector-fields on spheres, *Bull. Am. Math. Soc.*, **68:** 39–41 (1962).
6. Vector fields on spheres, *Ann. Math.*, **75:** 603–632 (1962).
7. On the groups $J(X)$ I, II, III, and IV *Topology*, **2:** 181–195 (1963), **3:** 137–171, 193–222 (1965), and **5:** 21–71, (1966).

Adams, J. F., and G. Walker:
1. On complex Stiefel manifolds, *Proc. Cambridge Phil. Soc.*, **61:** 81–103 (1965).

Adem, J.:
1. The iteration of the Steenrod squares in algebraic topology, *Proc. Natl. Acad. Sci. U.S.A.*, **38:** 720–726 (1952).
2. Relations on iterated reduced powers, *Proc. Natl. Acad. Sci. U.S.A.*, **39:** 636–638 (1953).
3. The relations on Steenrod powers of cohomology classes, in "Algebraic Geometry and Topology, A Symposium in Honor of S. Lefschetz," pp. 191–238, Princeton University Press, Princeton, N.J., 1957.

Atiyah, M. F.:
1. Complex analytic connections in fibre bundles, *Trans. Am. Math. Soc.*, **85:** 181–207 (1957).
2. Thom complexes, *Proc. London Math. Soc.*, (3), **11:** 291–310 (1961).
3. Immersions and embeddings of manifolds, *Topology*, **1:** 125–132 (1962).
4. Power operations in K-theory, Quart. J. Math. (2), **17:** 165–93 (1966).
5. K-theory and reality, Quart. J. Math. (2), **17:** 367–86 (1966).

6. *K*-theory, Benjamin, Inc. 1967.
7. Geometry of Yang-Mills fields, Lezioni Fermiane, Accademia Nazionale dei Lincei & Scuola Normale Superiore, Pisa (1979).

Atiyah, M. F., and R. Bott:
1. A Lefschetz fixed-point formula for elliptic complexes I, II: *Ann. of Math.*, **86**: 374–407 (1966) and **88**: 451–491 (1968).
2. The Yang-Mills equations over Riemann surfaces., *Phil. Trans. R. Soc. Lond. A* **308**: 523–615 (1982).
3. The moment map and equivariant cohomology, *Topology*, **23**: 1–28 (1984).

Atiyah, M. F., R. Bott, and A. Shapiro:
1. Clifford modules, *Topology*, **3** (Supplement 1): 3–38 (1964).

Atiyah, M. F., and F. Hirzebruch:
1. Riemann-Roch theorems for differential manifolds, *Bull. Am. Math. Soc.*, **65**: 276–281 (1959).
2. Vector bundles and homogeneous spaces, in Differential Geometry, *Am. Math. Soc. Proc. Symp. Pure Math.*, **3**: 7–38 (1961).
3. Cohomologie-Operationen und charakteristische Klassen, *Math. Z.*, **77**: 149–187 (1961).
4. Analytic cycles on complex manifolds, *Topology*, **1**: 25–45 (1962).
5. Quelques théorème de non-plongement pour les variétés différentiables, *Bull. Soc. Math. France*, **87**: 383–396 (1959).
6. The Riemann-Roch theorem for analytic embeddings, *Topology*, **1**: 151–166 (1962).
7. Characteristische Klassen und Anwendungen, Monographies de L'Enseignement mathématique, Genève, Suisse, n°11, pp. 71–96, 1962.

Atiyah, M. F., N. J. Hitchen, and I. M. Singer:
1. Self-duality in four-dimensional Riemannian geometry, *Proc. R. Soc. Lond. A* **362**: 425–461 (1978).

Atiyah, M. F., and J. D. S. Jones:
1. Topological aspects of Yang-Mills theory, *Commun. Math. Phys.* **61**: 97–118 (1978).

Atiyah, M. F., and E. Rees:
1. Vector bundles on projective 3-space. *Inventiones Math.* **35**: 131–53 (1976).

Atiyah, M. F., and G. B. Segal:
1. Exponential isomorphisms for λ-rings, Quart. J. Math. Oxford (2), **22**: 371–8 (1971).

Atiyah, M. F., and D. O. Tall:
1. Group representations, λ-rings and the *J*-homomorphism, *Topology*, **8**: 253–297 (1969).

Atiyah, M. F., and J. A. Todd:
1. On complex Stiefel manifolds, *Proc. Cambridge Phil. Soc.*, **56**: 342–353 (1960).

Becker, J., and D. H. Gohlieb:
1. The Transfer Map and Fibre Bundles, (To appear).

Borel, A.:
1. Le plan projectif des octaves et les sphères comme espaces homogènes, *Compt. Rend.*, **230**: 1378–1380 (1950).
2. Impossibilité de fibrer une sphère par un produit de sphères, *Compt. Rend.*, **231**: 943–945 (1950).
3. Sur la cohomologie des variétés de Stiefel et de certains groups de Lie, *Compt. Rend.*, **232**: 1628–1630 (1951).

4. La transgression dans les espaces fibres principaux, *Compt. Rend.*, **232**: 2392–2394 (1951).
5. Sur la cohomologie des espaces fibres principaux et des espaces homogènes de groups de Lie compacts, *Ann. Math.*, **57**: 115–207 (1953).
6. Topology of Lie groups and characteristic classes, *Bull. Am. Math. Soc.*, **61**: 397–432 (1955).
7. Topics in the Homology Theory of Fibre Bundles, Springer-Verlag Lecture Notes, **36** (1967).

Borel, A., and F. Hirzebruch:
1. Characteristic classes and homogeneous spaces: I, II, III, *Am. J. Math*, **80**: 458–538 (1958), **81**: 315–382 (1959), and **82**: 491–504 (1960).

Bott, R.:
1. An application of the Morse theory to the topology of Lie-groups, *Bull. Soc. Math. France*, **84**: 251–281 (1956).
2. Homogeneous vector bundles, *Ann. Math.*, (2) **66**: 203–248 (1957).
3. The stable homotopy of the classical groups, *Ann. Math.*, **70**: 313–337 (1959).
4. Quelques remarques sur les théorèmes de periodicité de topologie, Lille, 1959, *Bull. Soc. Math. France*, **87**: 293–310 (1959).
5. Vector fields on spheres and allied problems, Monographies de L'Enseignement mathématique, Genève, Suisse, n°11, pp. 25–38.
6. Lectures on $K(X)$, mimeographed notes, Harvard University, Cambridge, Mass., 1962.
7. A note on the KO-theory of sphere bundles, *Bull. Am. Math. Soc.*, **68**: 395–400 (1962).

Bourbaki, N.:
1. Eléments de mathématique, Livre III, "Topologie générale," Hermann & Cie, Paris, 1953.

Brown, E., Jr.,
1. Cohomology theories, Annals of Math. **75**: 467–484 (1962).

Cartan, H.:
1. Sur les groupes d'Eilenberg-MacLane $K(\pi, n)$: I, Méthode des constructions, *Proc. Natl. Acad. Sci. U.S.A.*, **40**: 467–471 (1954).
2. Sur les groupes d'Eilenberg-MacLane: II, *Proc. Natl. Acad. Sci. U.S.A.*, **40**: 704–707 (1954).
3. Sur l'itération des opérations de Steenrod, *Comment. Math. Helv.*, **29**: 40–58 (1955).

Cartan, H., and S. Eilenberg:
1. "Homological Algebra," Princeton University Press, Princeton, N.J., 1956.

Chern, S. S.:
1. Characteristic classes of Hermitian manifolds, *Ann. Math.*, **47**: 85–121 (1946).
2. On the characteristic ring of a differentiable manifold, *Acad. Sinica Sci. Record*, **2**: 1–5 (1947).
3. On the characteristic classes of Riemannian manifolds, *Proc. Natl. Acad. Sci. U.S.A.*, **33**: 78–82 (1947).
4. On the multiplication in the characteristic ring of a sphere bundle, *Ann. Math.*, (2) **49**: 362–372 (1948).
5. On curvature and characteristic classes of a Riemann manifold, *Abh. Math. Sem. Univ. Hamburg*, **20**: 117–126 (1955).

Chern, S. S., and Sze-Tsen Hu:
1. Parallelizability of principal fibre bundles, *Trans. Am. Math. Soc.*, **67**: 304–309 (1949).

Chern, S. S., and E. Spanier:
1. The homology structure of sphere bundles, *Proc. Natl. Acad. Sci. U.S.A.*, **36:** 248–255 (1950).

Chern, S. S., and Yi-Fone Sun:
1. The embedding theorem for fibre bundles, *Trans. Am. Math. Soc.*, **67:** 286–303 (1949).

Chern, S. S., F. Hirzebruch, and J. P. Serre:
1. On the index of a fibred manifold, *Proc. Am. Math. Soc.*, **8:** 587–596 (1957).

Chevalley, C.:
1. "Theory of Lie Groups," Princeton University Press, Princeton, N.J., 1946.
2. "The Algebraic Theory of Spinors," Columbia University Press, New York, 1954.

Dold, A.:
1. Uber fasernweise Homotopieaquivalenz von Faserraumen, *Math. Z.*, **62:** 111–126 (1955).
2. Vollstandigkeit der Wuschen relation zwischen den Stiefel-Whitneyschen Zahlen diffenzierbarer Mannigfaltigkeiten, *Math. Z.*, **65:** 200–206 (1956).
3. Erzeugende der Thomschen Algebra N, *Math. Z.*, **65:** 25–35 (1956).
4. Partitions of unity in the theory of fibrations, *Ann. Math.*, **78:** 223–255 (1963).
5. Halbexakte Humotopiefunktoren, Springer-Verlag Lecture Notes, **12** (1966).

Dold, A., and R. Lashof:
1. Principal quasifibrations and fibre homotopy equivalence of bundles, *Illinois J. Math.*, **3:** 285–305 (1959).

Eckmann, B.:
1. Beweis des Satzes von Hurwitz-Radon, *Comment. Math. Helv.* **15:** 358–366 (1942).

Eilenberg, S., and N. Steenrod:
1. "Foundations of Algebraic Topology," Princeton University Press, Princeton, N.J., 1952.

Fox, R. H.:
1. On fibre spaces: I, *Bull. Am. Math. Soc.*, **49:** 555–557 (1943).
2. On fibre spaces: II, *Bull. Am. Math. Soc.*, **49:** 733–735 (1943).
3. On topologies for function spaces, *Bull. Am. Math. Soc.*, **51:** 429–432 (1945).

Freudenthal, H.:
1. Uber die Klassen der Spharenabbildungen, *Comp. Math.*, **5:** 299–314 (1937).

Friedlander, E.:
1. Fibrations in etale homotopy theory, Publications Mathematiques I.H.E.S., Vol. 42 (1972).

Godement, R.:
1. Théorie des faisceaux," Hermann & Cie, Paris, 1958.

Griffiths, P., and J. Harris.
1. Princeples of algebraic geometry, *New York: Wiley*, 1978.

Grothendieck, A.:
1. Sur quelques points d'algèbre homologique, *Tohoku Math. J.*, (2)**9:** 119–221 (1957).

Harder, G., and M. S. Narasimhan:
1. On the cohomology groups of moduli spaces of vector bundles over curves. *Math. Annalen*, **212:** 215–148 (1975).

Hardy, G. H., and E. M. Wright:
1. "An Introduction to the Theory of Numbers, Oxford University Press, Fair Lawn, N.J., 1954.

Hilton, P. J.:
1. "An Introduction to Homotopy Theory," Cambridge Tracts in Mathematics and Mathematical Physics, no. 43, Cambridge University Press, New York, 1953.

Hirsch, M. W.:
1. Immersions of manifolds, *Trans. Am. Math. Soc.*, **93**: 242–276 (1959).

Hirzebruch, F.:
1. "Topological Methods in Algebraic Geometry," 3d ed., Springer-Verlag, Berlin, 1966.
2. A Riemann-Roch theorem for differentiable manifolds, Séminaire Bourbaki, n°177, 1959.

Hopf, H.:
1. Über die Abbildungen der 3-Sphäre auf die Kugelflache, *Math. Ann.*, **104**: 637–665 (1931).
2. Über die Abbildungen von Sphären auf Sphären niedrigerer Dimension, *Fundam. Math.*, **25**: 427–440 (1935).
3. Über die Topologie der Gruppen-Mannigfaltigkeiten und ihre Verallgemeinerungen, *Ann. Math.*, (2)**42**: 22–52 (1941).
4. Introduction à la théorie des espaces fibres. Collogue de topologie (espaces fibres), pp. 9–14, Brussels, Paris, 1950.
5. Sur une formule de la théorie des espaces fibres. Collogue de topologie (espaces fibres), pp. 117–121, Brussels, Paris, 1950.

Hu, Sze-Tsen:
1. "Homotopy Theory," Academic Press Inc., New York, 1959.

Huebsch, W.:
1. On the covering homotopy theorem, *Ann. Math.*, (2)**61**: 555–563 (1955).
2. Covering homotopy, *Duke Math. J.*, **23**: 281–291 (1956).

Hunt, G.:
1. A theorem of Elie Cartan, *Proc. Am. Math. Soc.*, **7**: 307–308 (1956).

Hurewicz, W.:
1. On the concept of fibre space, *Proc. Natl. Acad. Sci. U.S.A.*, **41**: 956–961 (1955).
2. Beitrage zur Topologie der Deformationen I–IV, *Proc. Koninkl. Akad. Wetenschap Amsterdam*, **38**: 112–119, 521–528 (1935), and **39**: 117–126, 215–224 (1936).

Hurewicz, W., and N. Steenrod:
1. Homotopy relations in fibre spaces, *Proc. Natl. Acad. Sci. U.S.A.*, **27**: 60–64 (1941).

James, I. M.:
1. Multiplication on spheres: I, *Proc. Am. Math. Soc.*, **8**: 192–196 (1957).
2. Multiplication on spheres: II, *Trans. Am. Math. Soc.*, **84**: 545–558 (1957).
3. The intrinsic join: a study of the homotopy groups of Stiefel manifolds, *Proc. London Math. Soc.*, (3)**8**: 507–535 (1958).
4. Cross-sections of Stiefel manifolds, *Proc. London Math. Soc.*, (3)**8**: 536–547 (1958).
5. Spaces associated to Stiefel manifolds, *Proc. London Math. Soc.*, **3**: 115–140 (1959).

James, I. M., and J. H. C. Whitehead:
1. Note on fibre spaces, *Proc. London Math. Soc.*, **4**: 129–137 (1954).
2. The homotopy theory of sphere bundles over spheres: I, *Proc. London Math. Soc.*, (3)**4**: 196–218 (1954).
3. The homotopy theory of sphere bundles over spheres: II, *Proc. London Math. Soc.*, (3)**5**: 148–166 (1955).

Kervaire, M. A.:
1. Relative characteristic classes, *Am. J. Math.*, **79:** 517–558 (1957).
2. Pontrjagin classes, *Am. J. Math.*, **80:** 632–638 (1958).
3. Obstructions and characteristic classes, *Am. J. Math.*, **81:** 773–784 (1959).

MacLane, S.:
1. Categorical algebra, *Bull. Am. Math. Soc.*, **71:** 40–106 (1965).
2. Categories for the working mathematician, Graduate texts in mathematics **5**, Springer-Verlag, 1971.

Miller, C. E.:
1. The topology of rotation groups, *Ann. Math.*, **57:** 90–114 (1953).

Milnor, J.:
1. Construction of universal bundles: I, *Ann. Math.*, (2)**63:** 272–284 (1956).
2. Construction of universal bundles: II, *Ann. Math.*, (2)**63:** 430–436 (1956).
3. On manifolds homeomorphic to the 7-sphere, *Ann. Math.*, (2)**64:** 399–405 (1956).
4. The theory of characteristic classes, mimeographed, Princeton University, Princeton, N.J., 1957.
5. The Steenrod algebra and its dual, *Ann. Math.*, (2)**67:** 150–171 (1958).
6. On spaces having the homotopy type of CW-complex, *Trans. Am. Math. Soc.*, **90:** 272–280 (1959).
7. Some consequences of a theorem of Bott, *Ann. Math.*, **68:** 444–449 (1958).

Milnor, J., and M. A. Kervaire:
1. Bernoulli numbers, homotopy groups, and a theorem of Rohlin, *Proc. Intern. Congr. Math.*, 1958.

Milnor, J., and E. Spanier:
1. Two remarks on fibre homotopy type, *Pacific J. Math.*, **10:** 585–590 (1960).

Miyazaki, H.:
1. The paracompactness of CW-complexes, *Tohoku Math. J.*, (2), **4:** 309–313 (1952).

Moore, J. C.:
1. Séminaire Henri Cartan, 1954/55, exposé 22.
2. The double suspension and p-primary components of the homotopy groups of spheres, Boletin de la Sociedad Matematica Mexicana, 1956, vol. 1, pp. 28–37.

Morin, B.:
1. Champs de vecteurs sur les sphères d'après J. F. Adams, Séminaire Bourbaki, no. 233, 1961–1962.

Narasimhan, M. S., and S. Ramanan:
1. Existence of universal connections, *American Journal of Mathematics*, **83:** 563–572 (1961).
2. Existence of universal connections II, *American Journal of Mathematics*, **85:** 223–231 (1963).

Narasimhan, M. S., and C. S. Seshadri:
1. Stable and unitary vector bundles on a compact Riemann surface, *Ann. of Math.*, **82:** 540–567 (1965).

Pontrjagin, L.:
1. "Topological Groups," Princeton University Press, Princeton, N.J., 1939.
2. Characteristic cycles on differentiable manifolds, *Mat. Sb. N.S.*, (63), **21:** 233–284 (1947).

Puppe, D.:
1. Homotopiemengen und ihre induzierten Abbildungen. I. *Math. Z.*, **69**: 299–344 (1958).
2. Faserraume, University of Saarland, Saarbrucken, Germany, 1964.

Quillen, D.:
1. Some remarks on etale homotopy theory and a conjecture of Adams, *Topology*, **7** (1968), 111–116.
2. The Adams conjecture, *Topology*, **10** (1971), 67–80.
3. Cohomology of groups, Acts, *Congrês Intern. Math.* 1970, t.2: 47–51.
4. Higher algebraic K-theory I, *Springer Lecture Notes in Math.* **341**: 85–147 (1973).
5. Superconnection character forms and the Cayley transform, *Topology*, **27**: 211–238 (1988).
6. Algebra cochains and cyclic cohomology, *Pub. Math. de IHES*, **68** 139–174 (1990).

Sanderson, B. J.:
1. Immersions and embeddings of projective spaces, *Proc. London Math. Soc.*, (3)**14**: 137–153 (1964).
2. A non-immersion theorem for real projective space, *Topology*, **2**: 209–211 (1963).

Segal, G.:
1. Configuration spaces and iterated loop spaces, *Invent. Math.* **21**: 213–221 (1973).
2. Categories and cohomology theories, *Topology*, **13**: 293–312 (1974).

Seifert, H.:
1. Topologie 3-dimensionaler gefaserter Räume, *Acta Math.*, **60**: 147–238 (1932).

Serre, J. P.:
1. Homologie singulière des espaces fibres, *Ann. Math.*, **54**: 425–505 (1951).
2. Groupes d'homotopie et classes de groupes abéliens, *Ann. Math.*, **58**: 258–294 (1953).
3. Cohomologie, modulo 2 des complexes d'Eilenberg-MacLane, *Comment. Math. Helv.*, **27**: 198–232 (1953).

Shih, W.:
1. Appendix II, Seminar on the Atiyah-Singer index theorem by R. S. Palais, Annals of Mathematics Studies, Number 57.

Spanier, E.:
1. A formula of Atiyah and Hirzebruch, *Math. Z.*, **80**: 154–162 (1962).
2. Function spaces and duality, *Ann. Math.*, **70**: 338–378 (1959).
3. "Algebraic Topology," McGraw-Hill Book Company, New York, 1966.

Spanier, E., and J. H. C. Whitehead:
1. Duality in homotopy theory, *Mathematika*, **2**: 56–80 (1955).

Stasheff, J.:
1. A classification theorem for fibre spaces, *Topology*, **2**: 239–246 (1963).

Steenrod, N. E.:
1. Classification of sphere bundles, *Ann. Math.*, **45**: 294–311 (1944).
2. "Topology of Fibre Bundles," Princeton Mathematical Series, Princeton University Press, Princeton, N.J., 1951.
3. Cohomology operations, *Ann. Math. Studies* 50, 1962.

Steenrod, N. E., and J. H. C. Whitehead:
1. Vector fields on the n-sphere, *Proc. Natl. Acad. Sci. U.S.A.*, **37**: 58–63 (1951).

Stiefel, E.:
1. Richtungsfelder und Fernparakkekusnus in n-dimensionalen Mannigfaltigkeiten, *Comment. Math. Helv.*, **8**: 3–51 (1936).

Thom, R.:
1. Espaces fibres en sphères et carrés de Steenrod, *Ann. Ecole Norm. Sup.*, **69**: 109–182 (1952).
2. Quelques propriétés globales des variétés différentiables, *Comment. Math. Helv.*, **28**: 17–86 (1954).

Thomason, R.:
1. Algebraic K-theory and étale cohomology, *Ann. Scient. Ec. Norm. Sup.* **13**: 437–552 (1980).

Toda, H.:
1. Generalized Whitehead products and homotopy groups of spheres, *J. Inst. Polytech., Osaka City Univ.*, **3**: 43–82 (1952).
2. On the double suspension E^2, J-Inst. Polytech., Osaka City Univ., Ser. A, **7**: 103–145 (1956).
3. Reduced join and Whitehead product, *J. Inst. Polytech., Osaka City Univ., Ser. A*, **8**: 15–30 (1957).

Waldhausen, F.:
1. Algebraic K-theory of spaces, localization and the chromatic filtration of stable homotopy, *SLN* **1051**: 173–195 (1984).

Whitehead, G. W.:
1. On the homotopy groups of spheres and rotation groups, *Ann. Math.*, **43**: 634–640 (1942).
2. Homotopy properties of the real orthogonal groups, *Ann. Math.*, (2)**43**: 132–146 (1942).
3. On families of continuous vector fields over spheres, *Ann. Math.*, (2)**47**: 779–785 (1946).
4. A generalization of the Hopf invariant, *Ann. Math.*, **51**: 192–238 (1950).
5. Generalized homology theories, *Trans. Am. Math. Soc.*, **102**: 227–283 (1962).

Whitehead, J. H. C.:
1. On the groups $\pi_r(V_{n,m})$ and sphere bundles, *Proc. London Math. Soc.*, (2)**48**: 243–291 (1944).
2. Combinatorial homotopy: I, *Bull. Am. Math. Soc.*, **55**: 213–245 (1949).
3. Combinatorial homotopy: II, *Bull. Am. Math. Soc.*, **55**: 453–496 (1949).

Whitney, H.:
1. Sphere spaces, *Proc. Natl. Acad. Sci. U.S.A.*, **21**: 462–468 (1935).
2. Topological properties of differentiable manifolds, *Bull. Am. Math. Soc.*, **43**: 785–805 (1937).
3. On the theory of sphere bundles, *Proc. Natl. Acad. Sci. U.S.A.*, **26**: 148–153 (1940).
4. On the topology of differentiable manifolds, "Lectures in Topology," The University of Michigan Press, Ann Arbor, Mich., 1941.

Wu, Wen-Tsun:
1. On the product of sphere bundles and the duality theorem modulo two, *Ann. Math.*, **49**: 641–653 (1948).
2. Sur les classes caractéristiques d'un espace fibrées en spheres, *Compt. Rend.*, **227**: 582–584 (1948).
3. Les i-carrés dans une variété grassmannienne, *Compt. Rend.*, **230**: 918–920 (1950).
4. Classes caractéristiques et i-carrés d'une variété, *Compt. Rend.*, **230**: 508–511 (1950).

5. Sur les classes caractéristiques des structures fibriés sphériques, Actualitiés Sci. Inc., no. 1183—Publ. Inst. Math. Univ. Strasbourg 11, pp. 5–98, 155–156, Hermann & Cie, Paris, 1952.

Yokota, I.:
1. On the cellular decompositions of unitary groups, *J. Inst. Polytech., Osaka City Univ.*, 7: 30–49 (1956).

Index

Graduate Texts in Mathematics

continued from page ii

Printed in the United States
By Bookmasters

Printed in the United States
By Bookmasters